# SOLIDWORKS® 2019
# and Engineering Graphics

**Randy H. Shih**
Oregon Institute of Technology

**SDC**
**Publications**

**SDC Publications**
P.O. Box 1334
Mission, KS 66222
913-262-2664
www.SDCpublications.com
Publisher: Stephen Schroff

**Examination Copies**
Books received as examination copies are for review purposes only and may not be made available for student use. Resale of examination copies is prohibited.

**Electronic Files**
Any electronic files associated with this book are licensed to the original user only. These files may not be transferred to any other party.

**Trademarks**
SOLIDWORKS is a registered trademark of Dassault Systèmes Corporation.
Windows are either registered trademarks or trademarks of Microsoft Corporation. All other trademarks are trademarks of their respective holders.

The author and publisher of this book have used their best efforts in preparing this book. These efforts include the development, research and testing of the material presented. The author and publisher shall not be liable in any event for incidental or consequential damages with, or arising out of, the furnishing, performance, or use of the material.

ISBN-13: 978-1-63057-240-2
ISBN-10: 1-63057-240-3

Printed and bound in the United States of America.

# Preface

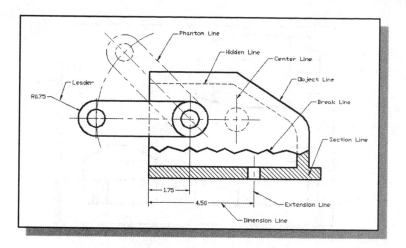

The primary goal of **SOLIDWORKS 2019 and Engineering Graphics: An Integrated Approach** is to introduce Engineering Graphics with the use of a modern Computer Aided Design package – SOLIDWORKS 2019. This text is intended to be used as a training guide for students and professionals. The chapters in this text proceed in a pedagogical fashion to guide you from constructing basic shapes to making complete sets of engineering drawings. This text takes a hands-on, exercise-intensive approach to all the important concepts of Engineering Graphics, as well as in-depth discussions of parametric feature-based CAD techniques. This textbook contains a series of fifteen chapters with detailed step-by-step tutorial style lessons, designed to introduce beginning CAD users to the graphic language used in all branches of technical industry. This book does not attempt to cover all of SOLIDWORKS 2019's features, only to provide an introduction to the software. It is intended to help you establish a good basis for exploring and growing in the exciting field of Computer Aided Engineering.

# Acknowledgments

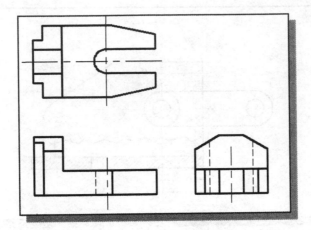

This book would not have been possible without a great deal of support. First, special thanks to two great teachers, Prof. George R. Schade of University of Nebraska-Lincoln and Mr. Denwu Lee, who taught me the fundamentals, the intrigue and the sheer fun of Computer Aided Engineering.

The effort and support of the editorial and production staff of SDC Publications is gratefully acknowledged. I would especially like to thank Stephen Schroff for the support and helpful suggestions during this project.

I am grateful that the Mechanical and Manufacturing Engineering Technology Department of Oregon Institute of Technology has provided me with an excellent environment in which to pursue my interests in teaching and research. I would especially like to thank Emeritus Professor Charles Hermach and Professor Brian Moravec for the helpful comments and encouragement.

Finally, truly unbounded thanks are due to my wife Hsiu-Ling and our daughter Casandra for their understanding and encouragement throughout this project.

Randy H. Shih
Klamath Falls, Oregon
Winter, 2019

# Table of Contents

Preface                                                                                         i
Acknowledgments                                                                       ii
Table of Contents                                                                      iii
Certified SOLIDWORKS Associate Examination Reference Guide        xv
Tips about Taking the SOLIDWORKS Certified Associate Examination   xxi

### Chapter 1
## Introduction

Introduction                                                                         1-2
Drawing in CAD Systems                                                       1-4
Development of Computer Geometric Modeling                      1-4
Feature-Based Parametric Modeling                                       1-8
Getting Started with SOLIDWORKS                                      1-9
Starting SOLIDWORKS                                                         1-9
SOLIDWORKS Screen Layout                                             1-12
Menu Bar                                                                             1-12
Menu Bar Pull-down Menus                                                 1-13
Heads-up View Toolbar                                                        1-13
Features Toolbar                                                                  1-13
Sketch Toolbar                                                                     1-13
Feature Manager-Design Tree/Property Manager                 1-14
Configuration Manager/DimXpert Manager/Display Manager   1-14
Graphics Area                                                                       1-15
Reference Triad                                                                    1-15
Origin                                                                                  1-15
Confirmation Corner                                                            1-15
Graphics Cursor or Crosshairs                                            1-15
Message and Status Bar                                                      1-15
Using the SOLIDWORKS Command Manager                      1-16
Mouse Buttons                                                                     1-17
[Esc] - Canceling Commands                                               1-17
SOLIDWORKS Help System                                               1-18
Leaving SOLIDWORKS                                                       1-18
Creating a CAD File Folder                                                 1-19

### Chapter 2
## Parametric Modeling Fundamentals

Introduction                                                                         2-3
The Adjuster Design                                                             2-4
Starting SOLIDWORKS                                                       2-4
SOLIDWORKS Part Modeling Window Layout                     2-5
Units Setup                                                                          2-6
Step 1: Determine/Set up the Base Feature                          2-7

Sketching Plane – It is an XY CRT, but an XYZ world          2-7
Creating Rough Sketches          2-9
Step 2: Creating a Rough Sketch          2-10
Graphics Cursors          2-10
Geometric Relation Symbols          2-12
Step 3: Apply/Modify Relations and Dimensions          2-13
Viewing Functions – Zoom and Pan          2-15
Delete an Existing Geometry of the Sketch          2-16
Modifying the Dimensions of the Sketch          2-17
Step 4: Completing the Base Solid Feature          2-18
Isometric View          2-19
Rotation of the 3-D Model – Rotate View          2-19
Rotation and Panning – Arrow keys          2-21
Dynamic Viewing – Quick Keys          2-22
3D Rotation          2-24
Viewing Tools – Heads-up View Toolbar          2-24
View Orientation          2-25
Display Style          2-26
Orthographic vs. Perspective          2-26
Customizing the Heads-up View Toolbar          2-26
Step 5-1: Adding an Extruded Boss Feature          2-27
Step 5-2: Adding an Extruded Cut Feature          2-31
Step 5-3: Adding another Cut Feature          2-33
Save the Model          2-35
Questions          2-36
Exercises          2-37

## Chapter 3
# Constructive Solid Geometry Concepts

Introduction          3-3
Binary Tree          3-4
The Locator Design          3-5
Modeling Strategy – CSG Binary Tree          3-6
Starting SOLIDWORKS and Activating the Command Manager          3-7
GRID and SNAP Intervals Setup          3-9
Base Feature          3-10
Repositioning Dimensions          3-12
Completing the Base Solid Feature          3-13
Creating the Next Solid Feature          3-14
Creating an Extruded Cut Feature          3-17
Creating a Hole with the Hole Wizard          3-20
Creating a Rectangular Extruded Cut Feature          3-23
Using the View Selector          3-25
Questions          3-27
Exercises          3-28

**Chapter 4**
# Geometric Constructions

Geometric Constructions                                                    4-3
Geometric Constructions - Classical Methods                                4-4
- Bisection of a Line or Arc                                               4-4
- Bisection of an Angle                                                    4-5
- Transfer of an Angle                                                     4-6
- Dividing a Given Line into a Number of Equal Parts                       4-7
- Circle through Three Points                                              4-8
- A Line Tangent to a Circle                                               4-9
- Line Tangent to a circle from a Given Point                              4-10
- Circle of a Given Radius Tangent to Two Given Lines                      4-11
- Circle of a Given Radius Tangent to an Arc and a Line                    4-12
- Circle of a Given Radius Tangent to Two Arcs                             4-13
Starting SOLIDWORKS                                                        4-14
Geometric Construction – CAD Method                                        4-16
- Bisection of a Line or Arc                                               4-16
Dimensions and Relations                                                   4-18
Geometric Symbols available in parametric sketching                        4-19
- Bisection of an Angle                                                    4-20
- Dividing a Given Line into a Number of Equal Parts                       4-22
- Circle through Three Points                                              4-29
- Line Tangent to a circle from a Given Point                              4-30
- Circle of a Given Radius Tangent to Two Given Lines                      4-31
Adding Geometric Relations and Fully Defined Geometry                      4-34
Starting SOLIDWORKS                                                        4-34
Over-Defining and Driven Dimensions                                        4-41
Deleting Existing Relations                                                4-42
Using the Fully Define Sketch Tool                                         4-43
Adding Additional Geometry                                                 4-44
Adding New Equations                                                       4-49
Questions                                                                  4-51
Exercises                                                                  4-52

**Chapter 5**
# Feature Design Tree

Introduction                                                               5-3
Starting SOLIDWORKS                                                        5-4
Creating a User-Defined Part Template                                      5-5
The *Saddle Bracket* Design                                                5-9
Modeling Strategy                                                          5-10
The SOLIDWORKS *Feature Manager Design Tree*                               5-11
Creating the Base Feature                                                  5-11
Adding the Second Solid Feature                                            5-14

Creating a 2D Sketch                                            5-15
Renaming the Part Features                                      5-17
Adjusting the Width of the Base Feature                         5-18
Adding a Cut Feature                                            5-19
Creating a Rectangular Extruded Cut Feature                     5-22
History-Based Part Modifications                                5-23
A Design Change                                                 5-24
Feature Manager Design Tree Views                               5-26
Questions                                                       5-28
Exercises                                                       5-29

## Chapter 6
# Geometric Construction Tools

Introduction                                                    6-3
The Gasket Design                                               6-3
Modeling Strategy                                               6-4
Starting SOLIDWORKS                                             6-5
Creating a 2D Sketch                                            6-6
Editing the Sketch by Dragging the Entities                     6-8
Adding Additional Relations                                     6-10
Using the *Trim* and *Extend* Commands                          6-11
Adding Dimensions with the Fully Define Sketch Tool             6-14
Fully Defined Geometry                                          6-16
Creating Fillets and Completing the Sketch                      6-17
Profile Sketch                                                  6-18
Redefining the Sketch and Profile using Contour Selection       6-19
Create an OFFSET Extruded Cut Feature                           6-23
Questions                                                       6-27
Exercises                                                       6-28

## Chapter 7
# Orthographic Projections and Multiview Constructions

Introduction                                                    7-3
Basic Principles of Projection                                  7-4
Orthographic Projection                                         7-4
Multiview Orthographic Projection                               7-5
First-Angle Projection                                          7-6
Rotation of the Horizontal and Profile Planes                   7-7
The 3D Adjuster Model and 1$^{st}$ angle projection             7-8
General Procedure: 1$^{St}$ Angle Orthographic Projection       7-9
Example 2: 1$^{St}$ Angle Orthographic Projection               7-10
Chapter 7 - 1$^{St}$ Angle Orthographic Sketching Exercise 1    7-11
Chapter 7 - 1$^{St}$ Angle Orthographic Sketching Exercise 2    7-13
Chapter 7 - 1$^{St}$ Angle Orthographic Sketching Exercise 3    7-15
Chapter 7 - 1$^{St}$ Angle Orthographic Sketching Exercise 4    7-17

Chapter 7 - 1$^{St}$ Angle Orthographic Sketching Exercise 5          7-19
Chapter 7 - 1$^{St}$ Angle Orthographic Sketching Exercise 6          7-21
Third-Angle Projection          7-23
Rotation of the Horizontal and Profile Planes          7-24
The 3D Adjuster Model and 3$^{rd}$ angle projection          7-25
The Glass Box and the Six Principal Views          7-26
General Procedure: 3$^{rd}$ Angle Orthographic Projection          7-28
Example 2: 3$^{rd}$ Angle Orthographic Projection          7-29
Example 3: 3$^{rd}$ Angle Orthographic Projection          7-30
Chapter 7 - 3$^{rd}$ Angle Orthographic Sketching Exercise 1          7-31
Chapter 7 - 3$^{rd}$ Angle Orthographic Sketching Exercise 2          7-33
Chapter 7 - 3$^{rd}$ Angle Orthographic Sketching Exercise 3          7-35
Chapter 7 - 3$^{rd}$ Angle Orthographic Sketching Exercise 4          7-37
Chapter 7 - 3$^{rd}$ Angle Orthographic Sketching Exercise 5          7-39
Chapter 7 - 3$^{rd}$ Angle Orthographic Sketching Exercise 6          7-41
Alphabet of Lines          7-43
Precedence of Lines          7-45
The U-Bracket Design          7-46
Starting SOLIDWORKS          7-46
Applying the BORN Technique          7-47
Creating the 2D Sketch of the Base Feature          7-48
Creating the First Extrude Feature          7-55
The Implied Parent/Child Relationships          7-55
Creating the Second Solid Feature          7-56
Creating the First Extruded Cut Feature          7-59
Creating the Second Extruded Cut Feature          7-60
Examining the Parent/Child Relationships          7-62
Modify a Parent Dimension          7-63
A Design Change          7-64
Feature Suppression          7-65
A Different Approach to the CENTER_DRILL Feature          7-66
Suppress the Rect_Cut Feature          7-67
Creating a Circular Extruded Cut Feature          7-68
A Flexible Design Approach          7-70
Drawings from Parts and Associative Functionality          7-71
Drawing Mode          7-72
Setting Document Properties          7-74
Setting Sheet Properties using the Pre-Defined Sheet Formats          7-75
Creating Three Standard Views          7-76
Repositioning Views          7-76
Adding a New Sheet          7-78
Adding a Base View          7-79
Adding an Isometric View using the View Palette          7-82
Adjusting the View Scale          7-83
Questions          7-85
Exercises          7-86

**Chapter 8**
# Dimensioning and Notes

Introduction                                                              8-3
Dimensioning Standards and Basic Terminology                              8-4
Selection and Placement of Dimensions and Notes                           8-5
Machined Holes                                                            8-12
Baseline and Chain Dimensioning                                           8-15
Dimensioning and Tolerance Accumulation                                   8-16
(1) Tolerance Accumulation - Baseline Dimensioning                        8-17
(2) Tolerance Accumulation - Chain Dimensioning                           8-18
(3) Avoid Tolerance Accumulation Feature Functionality Manufacturing      8-19
Dimensioning Tools in SOLIDWORKS                                          8-20
The U-Bracket Design                                                      8-20
Starting SOLIDWORKS                                                       8-21
Displaying Feature Dimensions                                             8-22
Repositioning, Appearance, and Hiding of Feature Dimensions               8-23
Adding Additional Dimensions – Reference Dimensions                       8-25
Tangent Edge Display                                                      8-27
Adding Center Marks, Center Lines, and Sketch Objects                     8-27
Edit Sheet vs. Edit Sheet Format                                          8-32
Modify the Title Block                                                    8-32
Property Links                                                            8-35
Associative Functionality – Modifying Feature Dimensions                  8-39
Saving the Drawing File                                                   8-42
Saving a Drawing Template                                                 8-42
Questions                                                                 8-45
Exercises                                                                 8-46

**Chapter 9**
# Tolerancing and Fits

Precision and Tolerance                                                   9-2
Methods of Specifying Tolerances – English System                         9-3
Nomenclature                                                              9-4
Example 9.1                                                               9-5
Fits between Mating parts                                                  9-6
Selective Assembly                                                        9-7
Basic Hole and Basic Shaft Systems                                        9-7
American National Standard Limits and Fits – Inches                       9-8
Example 9.2 Basic Hole System                                             9-13
Example 9.3 Basic Hole System                                             9-14
Example 9.4 Basic Shaft System                                            9-15
Example 9.5 Basic Shaft System                                            9-16
Tolerancing – Metric System                                               9-17
Metric Tolerances and Fits Designation                                    9-18
Preferred ISO Metric Fits                                                 9-19

Example 9.6 Metric Hole Basis System     9-20
Example 9.7 Shaft Basis System     9-21
Updating the U-Bracket Drawing     9-22
Determining the Tolerances Required     9-23
Questions     9-25
Exercises     9-26

**Chapter 10**

# Pictorials and Sketching

Engineering Drawings, Pictorials and Sketching     10-2
Isometric Sketching     10-7
Chapter 10 - Isometric Sketching Exercise 1     10-9
Chapter 10 - Isometric Sketching Exercise 2     10-11
Chapter 10 - Isometric Sketching Exercise 3     10-13
Chapter 10 - Isometric Sketching Exercise 4     10-15
Chapter 10 - Isometric Sketching Exercise 5     10-17
Chapter 10 - Isometric Sketching Exercise 6     10-19
Chapter 10 - Isometric Sketching Exercise 7     10-21
Chapter 10 - Isometric Sketching Exercise 8     10-23
Chapter 10 - Isometric Sketching Exercise 9     10-25
Chapter 10 - Isometric Sketching Exercise 10     10-27
Oblique Sketching     10-29
Chapter 10 - Oblique Sketching Exercise 1     10-31
Chapter 10 - Oblique Sketching Exercise 2     10-33
Chapter 10 - Oblique Sketching Exercise 3     10-35
Chapter 10 - Oblique Sketching Exercise 4     10-37
Chapter 10 - Oblique Sketching Exercise 5     10-39
Chapter 10 - Oblique Sketching Exercise 6     10-41
Perspective Sketching     10-43
One-Point Perspective     10-44
Two-Point Perspective     10-45
Chapter 10 - Perspective Sketching Exercise 1     10-47
Chapter 10 - Perspective Sketching Exercise 2     10-49
Chapter 10 - Perspective Sketching Exercise 3     10-51
Chapter 10 - Perspective Sketching Exercise 4     10-53
Chapter 10 - Perspective Sketching Exercise 5     10-55
Chapter 10 - Perspective Sketching Exercise 6     10-57
Questions     10-59
Exercises     10-60

**Chapter 11**

# Section Views & Symmetrical Features in Designs

Introduction     11-3
General Rules of Section Views     11-5

Section Drawing Types    11-6
- Full Section    11-6
- Half Section    11-6
- Offset Section    11-7
- Broken-Out Section    11-7
- Aligned Section    11-8
- Half Views    11-8
- Thin Sections    11-8
- Revolved Section    11-9
- Removed Section    11-9
- Conventional Breaks    11-10
- Ribs and Webs in Sections    11-10
- Parts Not Sectioned    11-10
Section Views in SOLIDWORKS    11-11
A Revolved Design: PULLEY    11-11
Modeling Strategy - A Revolved Design    11-12
Starting SOLIDWORKS    11-13
Creating the Base Feature    11-13
Creating the Revolved Feature    11-17
Mirroring Features    11-17
Creating an Extruded Cut Feature using Construction Geometry    11-19
Circular Pattern    11-24
Drawing Mode – Defining a New Border and Title Block    11-26
Creating a New Drawing Template    11-30
Creating Views    11-31
Retrieve Dimensions – Model Items Command    11-34
Save the Drawing File    11-35
Associative Functionality – A Design Change    11-36
Adding Centerlines to the Pattern Feature    11-38
Completing the Drawing    11-40
Questions    11-43
Exercises    11-44

## Chapter 12
# Auxiliary Views and Reference Geometry

Introduction    12-3
Normal View of an Inclined Surface    12-4
Construction Method I – Folding Line Method    12-6
Construction Method II – Reference Plane Method    12-8
Partial Views    12-10
Reference Geometry in SOLIDWORKS    12-11
Auxiliary Views in 2D Drawings    12-11
The Rod-Guide Design    12-11
Modeling Strategy    12-12
Starting SOLIDWORKS    12-13

Applying the BORN Technique ............................................................. 12-13
Creating the Base Feature .................................................................. 12-14
Creating an Angled Reference Plane .................................................. 12-16
Creating an Extruded Feature on the Reference Plane ....................... 12-19
Using the Convert Entities Option ..................................................... 12-19
Completing the Solid Feature ............................................................ 12-25
Creating an Offset Reference Plane .................................................. 12-26
Creating another Extruded Cut Feature using the Reference Plane .... 12-27
Starting a New 2D Drawing and Adding a Base View ....................... 12-29
Creating an Auxiliary View ............................................................... 12-30
Displaying Feature Dimensions ........................................................ 12-32
Adjusting the View Scale ................................................................. 12-34
Repositioning, Appearance, and Hiding of Feature Dimensions ....... 12-34
Tangent Edge Display ...................................................................... 12-37
Adding Center Marks and Center Lines ............................................ 12-37
Controlling the View and Sheet Scales ............................................. 12-40
Completing the Drawing Sheet ......................................................... 12-41
Editing the Isometric view ............................................................... 12-42
Adding a General Note ..................................................................... 12-42
Questions ......................................................................................... 12-44
Exercises ......................................................................................... 12-45

## Chapter 13
# Introduction to 3D Printing

What is 3D Printing ........................................................................... 13-2
Development of 3D Printing Technologies ......................................... 13-3
Primary types of 3D Printing processes ............................................ 13-6
Stereolithography ......................................................................... 13-6
Fused Deposition Modeling & Fused Filament Fabrication ........... 13-7
Laser Sintering / Laser Melting .................................................... 13-8
Primary 3D Printing Materials for FDM and FFF ............................. 13-9
From 3D model to 3D printed Part ................................................... 13-11
Starting SOLIDWORKS ................................................................... 13-12
Export the Design as an STL file ...................................................... 13-13
Using the 3D Printing software to create the 3D Print ....................... 13-18
Questions ........................................................................................ 13-26

## Chapter 14
# Threads and Fasteners

Introduction ..................................................................................... 14-2
Screw-Thread Terminology ............................................................... 14-3
Thread Forms ................................................................................... 14-5
Thread Representations ..................................................................... 14-6
• Detailed Representation .................................................................. 14-7

- Schematic Representation                                     14-8
- Simplified Representation                                    14-9
Thread Specification – English Units                          14-10
Unified Thread Series                                         14-11
Thread Fits                                                   14-12
Thread Specification – Metric                                 14-12
Thread Notes Examples                                         14-13
Specifying Fasteners                                          14-14
Commonly Used Fasteners                                       14-15
Drawing Standard Bolts                                        14-17
Bolt and Screw Clearances                                     14-17
Fasteners using SOLIDWORKS' Design Library                    14-18
ANSI Inch - Machine Screw                                     14-18
ANSI Metric - Machine Screw                                   14-20
Questions                                                     14-22

## Chapter 15
# Working Drawings

General Engineering Design Process                            15-3
Working Drawings                                              15-4
Detail Drawings                                               15-4
Assembly Drawings                                             15-5
Bill of Materials (BOM) and Parts List                        15-5
Drawing Sizes                                                 15-6
Drawing Sheet Borders and Revisions Block                     15-6
Title Blocks                                                  15-7
Working Drawings with SOLIDWORKS                              15-8
Assembly Modeling Methodology                                 15-8
The Shaft Support Assembly                                    15-9
Parts                                                         15-9
Creating the Collar with the Chamfer Command                  15-9
Creating the Bearing and Base-Plate                           15-11
Creating the Cap-Screw                                        15-12
Starting SOLIDWORKS                                           15-13
Document Properties                                           15-13
Inserting the First Component                                 15-14
Inserting the Second Component                                15-15
Degrees of Freedom                                            15-16
Assembly Mates                                                15-16
Apply the First Assembly Mate                                 15-18
Apply a Second Mate                                           15-19
Constrained Move                                              15-21
Apply a Third Mate                                            15-22
Inserting the Third Component                                 15-24
Applying Concentric and Coincident Mates                      15-24

Assemble the Cap-Screws using SmartMates                          15-26
Exploded View of the Assembly                                     15-30
Save the Assembly Model                                           15-32
Editing the Components                                            15-32
Set up a Drawing of the Assembly Model                            15-34
Creating a Bill of Materials                                      15-35
Editing the Bill of Materials                                     15-37
Completing the Assembly Drawing                                   15-41
Exporting the Bill of Materials                                   15-43
Questions                                                         15-44
Exercises                                                         15-45

**Chapter 16**

# CSWA Exam Preparation

Tips about Taking the Certified SOLIDWORKS Associate Examination   16-3
Introduction                                                      16-4
The Part Problem                                                  16-5
Strategy for Aligning the Part to the Default Axis System         16-6
Creating the Base Feature                                         16-6
Creating a New View Orientation                                   16-9
Create Reference Planes and a Reference Axis                      16-10
Selecting the Material and Viewing the Mass Properties            16-16
The Assembly Problem                                              16-19
Creating the Parts                                                16-20
Creating the Assembly                                             16-21
Creating a Reference Coordinate System                            16-27
View the Mass Properties                                          16-29
Questions                                                         16-31
Exercises                                                         16-32

## Appendix

A.  Running and Sliding Fits – American National Standard
B.  Preferred Metric Fits –ISO standard
C.  UNIFIED NATIONAL THREAD FORM
D.  METRIC THREAD FORM
E.  FASTENERS (INCH SERIES)
F.  METRIC FASTENERS
G.  BOLT AND SCREW CLEARANCE HOLES
H.  REFERENCES

## Index

## Certified SOLIDWORKS Associate Exam Overview

The Certified SOLIDWORKS Associate (CSWA) Exam is a performance-based exam. The examination is comprised of 10 – 20 questions to be completed in three hours. The test items will require you to use the SOLIDWORKS software to perform specific tasks and then answer questions about the tasks.

Performance-based testing is defined as *Testing by Doing*. This means you actually perform the given task then answer the questions regarding the task. Performance-based testing is widely accepted as a better way of ensuring the user has the skills needed, rather than just recalling information.

The CSWA examination is designed to test specific performance tasks in the following areas:

### Sketch Entities – Lines, Rectangles, Circles, Arcs, Ellipses, Centerlines
Objectives: Creating Sketch Entities.

| Certification Examination<br>Performance Task | Covered in this book on<br>Chapter – Page |
|---|---|
| Sketch Command | 2-10 |
| Line Command | 2-10 |
| Exit Sketch | 2-17 |
| Circle Command, Center Point Circle | 2-31 |
| Rectangle Command | 3-11 |
| Circle Command | 3-14 |
| Edit Sketch | 5-23 |
| Sketch Fillet | 5-25 |
| Centerline | 7-49 |
| Tangent Arc | 7-51 |
| Centerpoint Arc | 7-56 |
| Construction Geometry | 11-19 |
| Construction Lines | 11-21 |

### Sketch Tools – Offset, Convert, Trim
Objectives: Using Sketch Tools.

| Certification Examination<br>Performance Task | Covered in this book on<br>Chapter – Page |
|---|---|
| Trim and Extend Commands | 6-11 |
| Trim to Closest | 6-12 |
| Convert Entities | 6-23 |
| Offset Entities | 6-23 |

Dynamic Mirror ........................................................7-48
Mirror Entities...........................................................7-60

**Certified Associate Reference Guide**

## Sketch Relations

Objectives: Using Geometric Relations.

| Certification Examination Performance Task | Covered in this book on Chapter – Page |
| --- | --- |
| Horizontal Relation ............................................... | 2-11 |
| Geometric Relation Symbols ................................... | 2-12 |
| Preventing Relations with [Ctrl] Key ...................... | 2-12 |
| View Sketch Relations............................................ | 2-19 |
| Add Relation Command .......................................... | 4-18 |
| Geometric Sketch Relations Summary ..................... | 4-19 |
| Applying a Fix Relation.......................................... | 4-22 |
| Applying a Tangent Relation ................................... | 4-32 |
| Fully Defined Geometry ......................................... | 4-34 |
| Applying a Vertical Relation .................................. | 4-39 |
| Deleting Relations................................................. | 4-42 |
| Applying a Relation by Pre-Selecting Entities .......... | 4-47 |
| Applying a Coincident Relation............................... | 4-47 |
| Relations Settings................................................. | 4-48 |
| Applying a Collinear Relation ................................. | 12-22 |

## Boss and Cut Features – Extrudes, Revolves, Sweeps, Lofts

Objectives: Creating Basic Swept Shapes.

| Certification Examination Performance Task | Covered in this book on Chapter – Page |
| --- | --- |
| Extruded Boss/Base................................................ | 2-27 |
| Merge Result Option............................................... | 2-30 |
| Extruded Cut........................................................ | 2-31 |
| Extruded Cut/Through All ....................................... | 2-32 |
| Base Feature........................................................ | 3-10 |
| Reverse Direction Option ....................................... | 3-16 |
| Hole Wizard......................................................... | 3-20 |
| Rename Feature .................................................... | 5-17 |
| Edit Feature........................................................ | 5-23 |
| Selected Contours Option ...................................... | 6-19 |
| Suppress Features................................................. | 7-65 |
| Unsuppress Features ............................................. | 7-65 |
| Edit Sketch Plane ................................................. | 7-66 |
| Revolved Boss/Base.............................................. | 11-13 |

## Fillets and Chamfers
Objectives: Creating Fillets and Chamfers.

| Certification Examination<br>Performance Task | Covered in this book on<br>Chapter – Page |
| --- | --- |
| Chamfer Feature | 15-10 |

## Linear, Circular, and Fill Patterns
Objectives: Creating Patterned Features.

| Certification Examination<br>Performance Task | Covered in this book on<br>Chapter – Page |
| --- | --- |
| Mirror Feature | 11-17 |
| Circular Pattern | 11-24 |

## Dimensions
Objectives: Applying and Editing Smart Dimensions.

| Certification Examination<br>Performance Task | Covered in this book on<br>Chapter – Page |
| --- | --- |
| Smart Dimension | 2-13 |
| Dimension Standard | 2-14 |
| Modify Smart Dimension | 2-14 |
| Reposition Smart Dimension | 3-12 |
| Equations | 4-28 |
| Smart Dimension – Angle | 4-40 |
| Dimensional Values and Dimensional Variables | 4-41 |
| Driven Dimensions | 4-41 |
| Fully Define Sketch Tool | 4-43 |
| View Equations | 4-49 |
| Global Variables | 4-49 |
| Show Feature Dimensions | 5-8 |

Certified Associate Reference Guide

## Feature Conditions – Start and End

Objectives: Controlling Feature Start and End Conditions.

| Certification Examination Performance Task | Covered in this book on Chapter – Page |
|---|---|
| Extruded Boss/Base, Blind | 2-27 |
| Extruded Cut, Through All | 2-32 |
| Extruded Cut, Up to Next | 3-23 |
| Extruded Boss/Base, Mid-Plane | 5-13 |
| Extruded Boss/Base, Up to Surface | 5-16 |

## Mass Properties

Objectives: Obtaining Mass Properties for Parts and Assemblies.

| Certification Examination Performance Task | Covered in this book on Chapter – Page |
|---|---|
| Mass Properties Tool | 16-16 |
| View Mass Properties of a Part | 16-17 |
| Relative to Default Coordinate System | 16-17 |
| View Mass Properties of an Assembly | 16-29 |
| Relative to Reference Coordinate System | 16-29 |

## Materials

Objectives: Applying Material Selection to Parts.

| Certification Examination Performance Task | Covered in this book on Chapter – Page |
|---|---|
| Edit Material Command | 15-37 |

## Inserting Components

Objectives: Inserting Components into an Assembly.

| Certification Examination Performance Task | Covered in this book on Chapter – Page |
|---|---|
| Creating an Assembly File | 15-13 |
| Inserting a Base Component | 15-14 |
| Inserting Additional Components | 15-16 |
| Editing Parts in an Assembly | 15-37 |

## Standard Mates – Coincident, Parallel, Perpendicular, Tangent, Concentric, Distance, Angle

Objectives: Applying Standard Mates to Constrain Assemblies.

| Certification Examination<br>Performance Task | Covered in this book on<br>Chapter – Page |
| --- | --- |
| Assembly Mates | 15-16 |
| Coincident Mate, using Faces | 15-19 |
| Aligned Option | 15-19 |
| Anti-Aligned Option | 15-19 |
| Coincident Mate, using Temporary Axes | 15-19 |
| Coincident Mate, using Planes | 15-22 |
| Concentric Mate | 15-24 |
| SmartMates | 15-26 |
| Distance Mate | 16-23 |
| Angle Mate, using Faces | 16-25 |

## Reference Geometry – Planes, Axis, Mate References

Objectives: Creating Reference Planes, Axes, and Mate References.

| Certification Examination<br>Performance Task | Covered in this book on<br>Chapter – Page |
| --- | --- |
| Reference Axis | 12-17 |
| Reference Plane, At Angle Option | 12-16 |
| Reference Plane, Offset Distance Option | 12-26 |
| Reference Coordinate System | 16-28 |

## Drawing Sheets and Views

Objectives: Creating and Setting Properties for Drawing Sheets; Inserting and Editing Standard Views.

| Certification Examination<br>Performance Task | Covered in this book on<br>Chapter – Page |
| --- | --- |
| Section Views | 11-31 |
| Cutting Plane Line | 11-31 |
| Projected View Command | 11-32 |
| Auxiliary Views | 12-30 |
| Controlling View and Sheet Scales | 12-40 |
| Creating a Drawing from an Assembly | 15-34 |

## Annotations
Objectives: Creating Annotations.

| Certification Examination<br>Performance Task | Covered in this book on<br>Chapter – Page |
|---|---|
| Annotation Toolbar | 8-22 |
| Displaying Feature Dimensions | 8-22 |
| Model Items Command | 8-22 |
| Repositioning Dimensions | 8-23 |
| Reference Dimensions | 8-25 |
| Hide Dimensions | 8-25 |
| Center Mark | 8-27 |
| Centerline | 8-30 |
| Note Command | 8-34 |
| Property Links | 8-34 |
| Center Mark, Circular Option | 11-37 |
| AutoBalloon Command | 15-41 |
| Bill of Materials | 15-35 |

Certified Associate Reference Guide

**Notes:**

## Tips about Taking the Certified SOLIDWORKS Associate (CSWA) Examination

1. **Study**: The first step to maximize your potential on an exam is to sufficiently prepare for it. You need to be familiar with the SOLIDWORKS package, and this can only be achieved by doing designs and exploring the different commands available. The Certified SOLIDWORKS Associate (CSWA) exam is designed to measure your familiarity with the SOLIDWORKS software. You must be able to perform the given task and answer the exam questions correctly and quickly.

2. **Make Notes**: Take notes of what you learn either while attending classroom sessions or going through study material. Use these notes as a review guide before taking the actual test.

3. **Time Management**: The examination has a time limit. Manage the time you spend on each question. Always remember you do not need to score 100% to pass the exam. Also, keep in mind that some questions are weighed more heavily and may take more time to answer. You can flip back and forth to view different problems during the test time by using the arrow buttons. If you encounter a question you cannot answer in a reasonable amount of time, use the *Save As* feature in SOLIDWORKS to save a copy of the file, and move on to the next question. You can return to any question and enter or change the answer as long as you do not hit the [**End Examination**] button.

4. **Use the *SOLIDWORKS Help System***: If you get confused and can't think of the answer, remember the *SOLIDWORKS Help System* is a great tool to confirm your considerations. In preparing for the exam, familiarize yourself with the help utility organization (e.g., Contents, Index, Search options).

5. **Use Internet Search**: Use of an internet search utility is allowed during the test. If a test question requires general knowledge, for example definitions of engineering or drafting concepts (stress, yield strength, auxiliary view, etc.), remember the internet is available as a tool to assist in your considerations.

6. **Use Common Sense**: If you are unable to get the correct answer and unable to eliminate all distracters, then you need to select the best answer from the remaining selections. This may be a task of selecting the best answer from amongst several correct answers, or it may be selecting the least incorrect answer from amongst several poor answers.

7. **Be Cautious and Don't Act in Haste**: Devote some time to ponder and think of the correct answer. Ensure that you interpret all the options correctly before selecting from available choices. Don't go into panic mode while taking a test. Use the *Arrow Buttons* to review each question. When you are confident that you have answered all questions, end the examination using the [**End Examination**] button to submit your answers for scoring. You will receive a score report once you have submitted your answers.

8. **Relax before Exam**: In order to avoid last minute stress, make sure that you arrive 10 to 15 minutes early and relax before taking the exam.

# Chapter 1
# Introduction

## Learning Objectives

- ♦ **Introduction to Engineering Graphics**
- ♦ **Development of Computer Geometric Modeling**
- ♦ **Feature-Based Parametric Modeling**
- ♦ **Startup Options**
- ♦ **SOLIDWORKS Screen Layout**
- ♦ **User Interface & Mouse Buttons**
- ♦ **SOLIDWORKS Online Help**

## Introduction

**Engineering Graphics**, also known as **Technical Drawing**, is the technique of creating accurate representations of designs, an *engineering drawing*, for construction and manufacturing. An **engineering drawing** is a type of drawing that is technical in nature, used to fully and clearly define requirements for engineered items, and is usually created in accordance with standardized conventions for layout, nomenclature, interpretation, appearance, size, etc. A skilled practitioner of the art of engineering drawing is known as a *draftsman* or *draftsperson*.

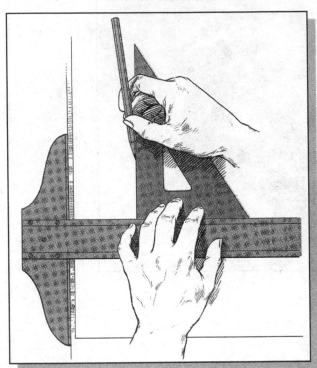

The basic mechanics of drafting is to use a pencil and draw on a piece of paper. For engineering drawings, papers are generally placed on a drafting table and additional tools are used. A T-square is one of the standard tools commonly used with a drafting table.

A T-square is also known as a *sliding straightedge*; parallel lines can be drawn simply by moving the T-square and running a pencil along the T-square's straightedge. The T-square is more typically used as a tool to hold other tools such as triangles. For example, one or more triangles can be placed on the T-square and lines can be drawn at chosen angles on the paper.

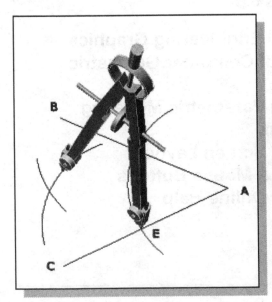

In addition to the triangles, other tools are used to draw curves and circles. The most common among these are the compass, used for drawing simple arcs and circles, and the French curve, typically a piece of plastic with a complex curve on it. A spline is a rubber coated articulated metal that can be manually bent to almost any curve.

This basic drafting system requires an accurate table and constant attention to the positioning of the tools. A common error draftspersons faced was allowing the triangles to push the top of the T-square down slightly, thereby throwing off all angles. Drafting in general was already a time-consuming process.

A solution to these problems was the introduction of the "**drafting machine**," a device that allowed the draftsperson to have an accurate right angle at any point on the page quite quickly. These machines often included the ability to change the angle, thereby removing the need for the triangles as well.

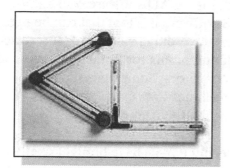

In addition to the mechanics of drawing the lines onto a piece of paper, drafting requires an understanding of geometry and the professional skills of the specific designer. At one time, drafting was a sought-after job considered one of the more demanding and highly-skilled positions of the trade. Today, the mechanics of the drafting task have been largely automated and greatly accelerated through the use of **computer aided design** (CAD) systems. Proficiency in using CAD systems has also become one of the more important requirements for engineers and designers.

## Drawing in CAD Systems

**Computer Aided Design** (CAD) is the process of creating designs with the aid of computers. This includes the generation of computer models, analysis of design data and the creation of necessary drawings. SOLIDWORKS is a CAD software package developed by *Dassault Systèmes*. The SOLIDWORKS software is a tool that can be used for design and drafting activities. Two-dimensional and three-dimensional models created in SOLIDWORKS can be transferred to other computer programs for further analysis and testing. The computer models can also be used in manufacturing equipment such as machining centers, lathes, mills or rapid prototyping machines to manufacture the product.

Rapid changes in the field of **computer aided engineering** (CAE) have brought exciting advances to the industry. Recent advances have made the long-sought goal of reducing design time, producing prototypes faster, and achieving higher product quality closer to a reality.

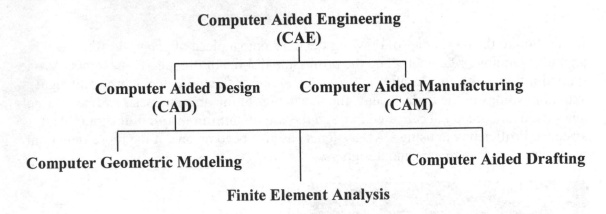

## Development of Computer Geometric Modeling

**Computer geometric modeling** is a relatively new technology, and its rapid expansion in the last fifty years is truly amazing. Computer-modeling technology has advanced along with the development of computer hardware. The first-generation CAD programs, developed in the 1950s, were mostly non-interactive; CAD users were required to create program-codes to generate the desired two-dimensional (2D) geometric shapes. Initially, the development of CAD technology occurred mostly in academic research facilities. The Massachusetts Institute of Technology, Carnegie-Mellon University, and Cambridge University were the leading pioneers at that time. The interest in CAD technology spread quickly and several major industry companies, such as General Motors, Lockheed, McDonnell, IBM, and Ford Motor Co., participated in the development of interactive CAD programs in the 1960s. Usage of CAD systems was primarily in the automotive industry, aerospace industry, and government agencies that developed their own programs for their specific needs. The 1960s also marked the beginning of the development of finite element analysis methods for computer stress analysis and computer aided manufacturing for generating machine tool paths.

The 1970s are generally viewed as the years of the most significant progress in the development of computer hardware, namely with the invention and development of **microprocessors**. With the improvement in computing power, new types of 3D CAD programs that were user-friendly and interactive became reality. CAD technology quickly expanded from very simple **computer aided drafting** to very complex **computer aided design**. The use of 2D and 3D wireframe modelers was accepted as the leading edge technology that could increase productivity in industry. The development of surface modeling and solid modeling technologies was taking shape by the late 1970s, but the high cost of computer hardware and programming slowed the development of such technology. During this period, the available CAD systems all required room-sized mainframe computers that were extremely expensive.

In the 1980s, improvements in computer hardware brought the power of mainframes to the desktop at less cost and with more accessibility to the general public. By the mid-1980s, CAD technology had become the main focus of a variety of manufacturing industries and was very competitive with traditional design/drafting methods. It was during this period of time that 3D solid modeling technology had major advancements, which boosted the usage of CAE technology in industry.

The introduction of the *feature-based parametric solid modeling* approach, at the end of the 1980s, elevated CAD/CAM/CAE technology to a new level. In the 1990s, CAD programs evolved into powerful design/manufacturing/management tools. CAD technology has come a long way, and during these years of development, modeling schemes progressed from two-dimensional (2D) wireframe to three-dimensional (3D) wireframe, to surface modeling, to solid modeling and, finally, to feature-based parametric solid modeling.

The first-generation CAD packages were simply 2D **computer aided drafting** programs, basically the electronic equivalents of the drafting board. For typical models, the use of this type of program would require that several views of the objects be created individually as they would be on the drafting board. The 3D designs remained in the designer's mind, not in the computer database. Mental translations of 3D objects to 2D views were required throughout the use of these packages. Although such systems have some advantages over traditional board drafting, they are still tedious and labor intensive. The need for the development of 3D modelers came quite naturally, given the limitations of the 2D drafting packages.

The development of three-dimensional modeling schemes started with three-dimensional (3D) wireframes. Wireframe models are models consisting of points and edges, which are straight lines connecting between appropriate points. The edges of wireframe models are used, similar to lines in 2D drawings, to represent transitions of surfaces and features. The use of lines and points is also a very economical way to represent 3D designs.

The development of the 3D wireframe modeler was a major leap in the area of computer geometric modeling. The computer database in the 3D wireframe modeler contains the locations of all the points in space coordinates, and it is typically sufficient to create just one model rather than multiple views of the same model. This single 3D model can then be viewed from any direction as needed. Most 3D wireframe modelers allow the user to create projected lines/edges of 3D wireframe models. In comparison to other types of 3D modelers, the 3D wireframe modelers require very little computing power and generally can be used to achieve reasonably good representations of 3D models. However, because surface definition is not part of a wireframe model, all wireframe images have the inherent problem of ambiguity. Two examples of such ambiguity are illustrated.

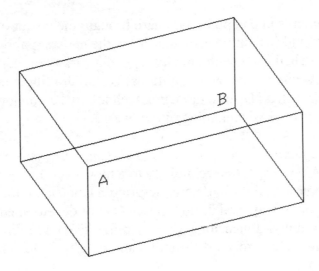

**Wireframe Ambiguity**: Which corner is in front, A or B?

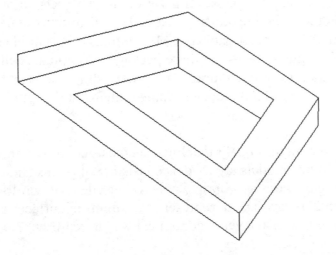

**A non-realizable object**: Wireframe models contain no surface definitions.

Surface modeling is the logical development in computer geometry modeling to follow the 3D wireframe modeling scheme by organizing and grouping edges that define polygonal surfaces. Surface modeling describes the part's surfaces but not its interiors. Designers are still required to interactively examine surface models to ensure that the various surfaces on a model are contiguous throughout. Many of the concepts used in 3D wireframe and surface modelers are incorporated in the solid modeling scheme, but it is solid modeling that offers the most advantages as a design tool.

In the solid modeling presentation scheme, the solid definition includes nodes, edges, and surfaces, and it is a complete and unambiguous mathematical representation of a precisely enclosed and filled volume. Unlike the surface modeling method, solid modelers start with a solid or use topology rules to guarantee that all of the surfaces are stitched together properly. Two predominant methods for representing solid models are **constructive solid geometry** (CSG) representation and **boundary representation** (B-rep).

The CSG representation method can be defined as the combination of 3D solid primitives. What constitutes a "primitive" varies somewhat with the software but typically includes a rectangular prism, a cylinder, a cone, a wedge, and a sphere. Most solid modelers also allow the user to define additional primitives, which are shapes typically formed by the basic shapes. The underlying concept of the CSG representation method is very straightforward; we simply **add** or **subtract** one primitive from another. The CSG approach is also known as the machinist's approach, as it can be used to simulate the manufacturing procedures for creating the 3D object.

In the B-rep representation method, objects are represented in terms of their spatial boundaries. This method defines the points, edges, and surfaces of a volume, and/or issues commands that sweep or rotate a defined face into a third dimension to form a solid. The object is then made up of the unions of these surfaces that completely and precisely enclose a volume.

By the 1980s, a new paradigm called *concurrent engineering* had emerged. With concurrent engineering, designers, design engineers, analysts, manufacturing engineers, and management engineers all work together closely right from the initial stages of the design. In this way, all aspects of the design can be evaluated and any potential problems can be identified right from the start and throughout the design process. Using the principles of concurrent engineering, a new type of computer modeling technique appeared. The technique is known as the *feature-based parametric modeling technique*. The key advantage of the *feature-based parametric modeling technique* is its capability to produce very flexible designs. Changes can be made easily, and design alternatives can be evaluated with minimal effort. Various software packages offer different approaches to feature-based parametric modeling, yet the end result is a flexible design defined by its design variables and parametric features.

## Feature-Based Parametric Modeling

One of the key elements in the SOLIDWORKS solid modeling software is its use of the **feature-based parametric modeling technique**. The feature-based parametric modeling approach has elevated solid modeling technology to the level of a very powerful design tool. Parametric modeling automates the design and revision procedures by the use of parametric features. Parametric features control the model geometry by the use of design variables. The word *parametric* means that the geometric definitions of the design, such as dimensions, can be varied at any time during the design process. Features are predefined parts or construction tools for which users define the key parameters. A part is described as a sequence of engineering features, which can be modified and/or changed at any time. The concept of parametric features makes modeling more closely match the actual design-manufacturing process than the mathematics of a solid modeling program. In parametric modeling, models and drawings are updated automatically when the design is refined.

Parametric modeling offers many benefits:

- **We begin with simple, conceptual models with minimal detail; this approach conforms to the design philosophy of "shape before size."**

- **Geometric constraints, dimensional constraints, and relational parametric equations can be used to capture design intent.**

- **The ability to update an entire system, including parts, assemblies and drawings after changing one parameter of complex designs.**

- **We can quickly explore and evaluate different design variations and alternatives to determine the best design.**

- **Existing design data can be reused to create new designs.**

- **Quick design turn-around.**

With parametric modeling, designers and engineers can maximize the productivity of the design and engineering resources to create products better, faster, and more cost-effectively. The tools available in parametric modeling are designed to cover the entire design process, from design and validation to technical communications and data management. The intuitive design interface and integrated software work together and provide designers and engineers the freedom to focus on innovation.

## Getting Started with SOLIDWORKS

- SOLIDWORKS is composed of several application software modules (these modules are called *applications*) all sharing a common database. In this text, the main concentration is placed on the solid modeling modules used for part design. The general procedures required in creating solid models, engineering drawings, and assemblies are illustrated.

## Starting SOLIDWORKS

How to start SOLIDWORKS depends on the type of workstation and the particular software configuration you are using. With most *Windows* systems, you can select **SOLIDWORKS** on the *Start* menu or select the **SOLIDWORKS** icon on the desktop. Consult your instructor or technical support personnel if you have difficulty starting the software.

❖ The program takes a while to load, so be patient. The tutorials in this text are based on the assumption that you are using the default settings in SOLIDWORKS. If your system has been customized for other uses, some of the settings may appear differently and not work with the step-by-step instructions in the tutorials. Contact your instructor and/or technical support personnel to restore the default software configuration.

Once the program is loaded into memory, the *SOLIDWORKS* program window appears. This screen contains the *Welcome to SOLIDWORKS* dialog box, the *Menu Bar* and the *Task Pane*. The *Menu Bar* contains a subset of commonly used tools from the *Menu Bar* toolbar (New, Open, Save, etc.), the *SOLIDWORKS* menus, the *SOLIDWORKS Search* oval, and a flyout menu of *Help* options.

The *task pane* appears to the right of the main screen; several options are available through the *task pane*, such as the *Design Library* and the *File Explorer*. The icons for these options appear below the **SOLIDWORKS Resources** icon. To collapse the *task pane*, click anywhere in the main area of the *SOLIDWORKS* window.

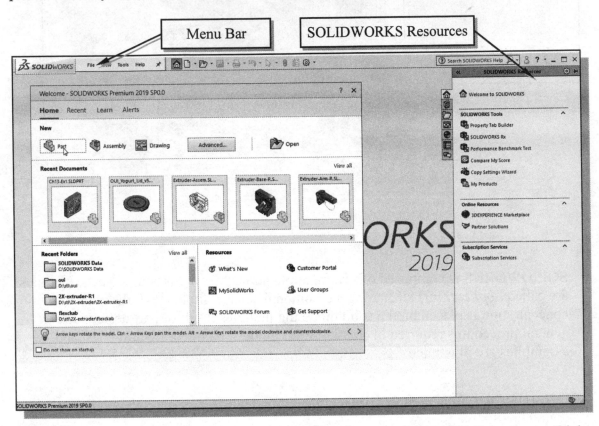

To the left side of the main window is the *Welcome to SOLIDWORKS* dialog box, which contains four tabs, *Home*, *Recent*, *Learn* and *Alert*. The Home tab can be used to quickly start a new document or open recently modified files. Note that this dialog box can be toggled on/off by hitting [Ctrl + F2].

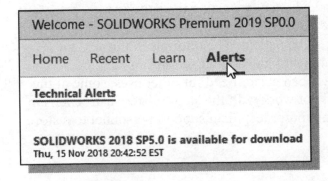

Click on the ***Alerts*** tab to display technical alerts, such as new updates and known issues, from SOLIDWORKS.

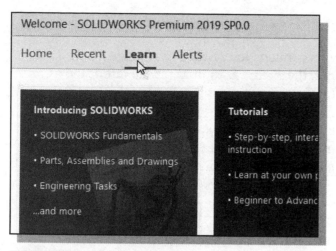

Click on the *Learn* tab to display the built-in SOLIDWORKS learning tools.

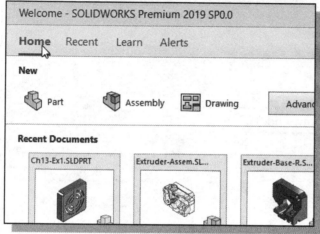

Switch back to the Home tab and notice the top section in the *Home* tab is the *New* file option, which allows us to start a new *SOLIDWORKS* PART file, a new ASSEMBLY file, or a new DRAWING file.

A part is a single three-dimensional (3D) solid model. Parts arc the basic building blocks in modeling with *SOLIDWORKS*. An assembly is a 3D arrangement of parts (components) and/or other assemblies (subassemblies). A drawing is a 2D representation of a part or an assembly.

➢ Select the **Part** icon as shown. Click **OK** in the *New SOLIDWORKS Document* dialog box to open a new part file.

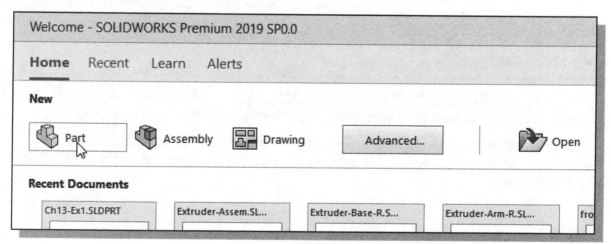

## SOLIDWORKS Screen Layout

The default *SOLIDWORKS* drawing screen contains the *Menu Bar*, the *Heads-up View* toolbar, the *Feature Manager Design Tree*, the *Features* toolbar (at the left of the window by default), the *Sketch* toolbar (at the right of the window by default), the graphics area, the *task pane* (collapsed to the right of the graphics area in the figure below), and the *Status Bar*. A line of quick text appears next to the icon as you move the *mouse cursor* over different icons. You may resize the *SOLIDWORKS* drawing window by clicking and dragging the edge of the window, or relocate the window by clicking and dragging the *window title* area.

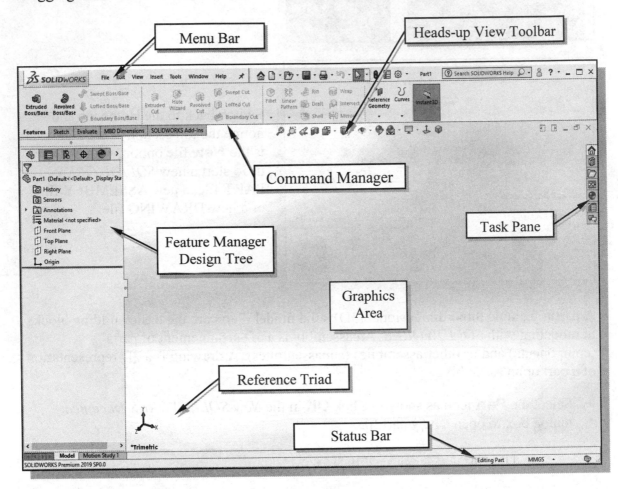

- **Menu Bar**

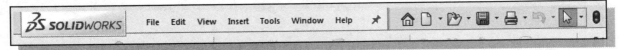

In the default view of the *Menu Bar*, only the toolbar options are visible. The default *Menu Bar* consists of a subset of frequently used commands from the *Menu Bar* toolbar as shown above.

- **Menu Bar Pull-down Menus**

To display the *pull-down* menus, move the cursor over or click the *SOLIDWORKS* logo. The pull-down menus contain operations that you can use for all modes of the system.

- **Heads-up View Toolbar**

The *Heads-up View* toolbar allows us quick access to frequently used view-related commands, such as Zoom, Pan and Rotate. Note: You cannot hide or customize the *Heads-up View* toolbar.

- **Features Toolbar**

By default, the *Features* toolbar is displayed at the top of the *SOLIDWORKS* window. The *Features* toolbar allows us quick access to frequently used feature-related commands, such as Extruded Boss/Base, Extruded Cut, and Revolved Boss/Base.

- **Sketch Toolbar**

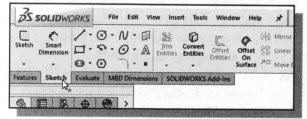

By default, the *Sketch* toolbar can be accessed by clicking on the **Sketch** tab in the *Ribbon* toolbar area. The *Sketch* toolbar provides tools for creating the basic geometry that can be used to create features and parts.

- **Feature Manager-Design Tree/Property Manager/Configuration Manager/ DimXpert Manager/Display Manager**

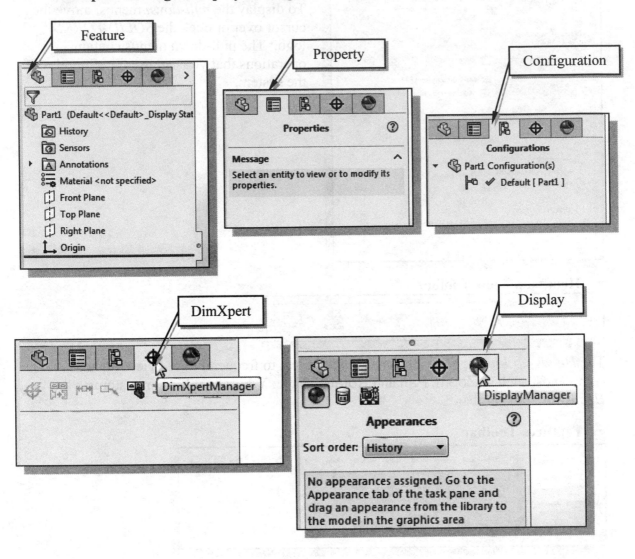

The left panel of the *SOLIDWORKS* window is used to display the *Feature Manager-Design Tree*, the *Property Manager*, the *Configuration Manager*, and the *DimXpert Manager*. These options can be chosen by selecting the appropriate tab at the top of the panel. The *Feature Manager Design Tree* provides an overview of the active part, drawing, or assembly in outline form. It can be used to show and hide selected features, filter contents, and manage access to features and editing. The *Property Manager* opens automatically when commands are executed or entities are selected in the graphics window, and is used to make selections, enter values, and accept commands. The *Configuration Manager* is used to create, select and view multiple configurations of parts and assemblies. The *DimXpert Manager* lists the tolerance features defined using the *SOLIDWORKS* 'DimXpert for parts' tools. The *Display Manager* lists the appearance, decals, lights, scene, and cameras applied to the current model. From the *Display Manager*, we can view applied content, and add, edit, or delete items. The *Display Manager* also provides access to *Photo View* options if the module is available.

- **Graphics Area**

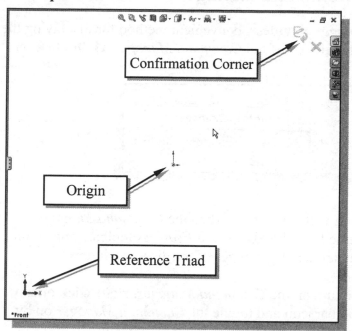

The graphics area is the area where models and drawings are displayed.

- **Reference Triad**

The *Reference Triad* appears in the graphics area of part and assembly documents. The triad is shown to help orient the user when viewing models and is for reference only.

- **Origin**

The *Origin* represents the (0,0,0) coordinate in a model or sketch. A model origin appears blue; a sketch origin appears red.

- **Confirmation Corner**

The *Confirmation Corner* offers an alternate way to accept features.

- **Graphics Cursor or Crosshairs**

The *graphics cursor* shows the location of the pointing device in the graphics window. During geometric construction, the coordinate of the cursor is displayed in the *Status Bar* area, located at the bottom of the screen. The cursor's appearance depends on the selected command or option.

- **Message and Status Bar**

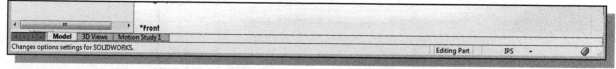

The *Message and Status Bar* area displays a single-line description of a command when the cursor is on top of a command icon. This area also displays information pertinent to the active operation. In the figure above, the cursor coordinates are displayed while in the *Sketch* mode.

# Using the *SOLIDWORKS* Command Manager

The *SOLIDWORKS Command Manager* provides a convenient method for displaying the most commonly used toolbars. To toggle on/off the *Command Manager*, **right click** on any toolbar icon, and select in the list.

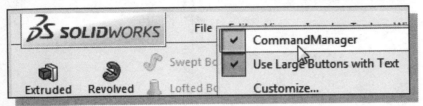

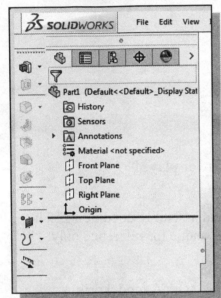

You will notice that, when the *Command Manager* is turned off, the *Sketch* and *Features* toolbars appear on the left and right edges of the display window.

To turn on the *Command Manager*, **right click** on any toolbar icon and toggle the *Command Manager* on by selecting it at the top of the pop-up menu.

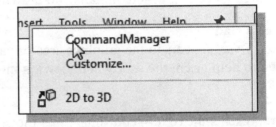

The *Command Manager* is a context-sensitive toolbar that dynamically updates based on the user's selection. When you click a tab below the *Command Manager*, it updates to display the corresponding toolbar. For example, if you click the **Sketches** tab, the *Sketch* toolbar appears. By default, the *Command Manager* has toolbars embedded in it based on the document type.

The default display of the *Command Manager* is shown below. You will notice that, when the *Command Manager* is used, the *Sketch* and *Features* toolbars no longer appear on the left and right edges of the display window.

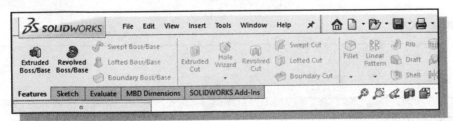

**Important Note:** The illustrations in this text use the *Command Manager*. If a user prefers to use the standard display of toolbars, the only change is that it may be necessary to activate the appropriate toolbars prior to selecting a command.

## Mouse Buttons

SOLIDWORKS utilizes the mouse buttons extensively. When learning the interactive environment of SOLIDWORKS, it is important to understand the basic functions of the mouse buttons.

- **Left mouse button**
  The **left mouse button** is used for most operations, such as selecting menus and icons, or picking graphic entities. One click of the button is used to select icons, menus and form entries, and to pick graphic items.

- **Right mouse button**
  The **right mouse button** is used to bring up additional options that are available in a context-sensitive pop-up menu. These menus provide shortcuts to frequently used commands.

- **Middle mouse button/wheel**
  The middle mouse button/wheel can be used to Rotate (hold down the wheel button and drag the mouse), Pan (hold down the wheel button and drag the mouse while holding down the [**Ctrl**] key), or Zoom (hold down the wheel button and drag the mouse while holding down the [**Shift**] key) real time. Spinning the wheel allows zooming to the position of the cursor.

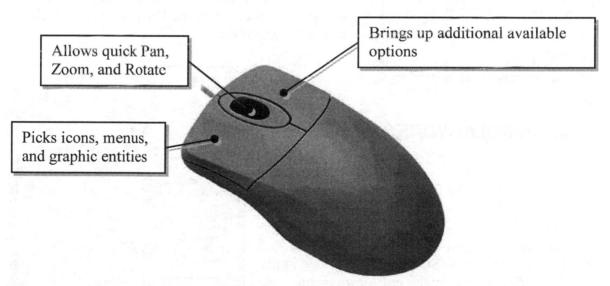

Allows quick Pan, Zoom, and Rotate

Brings up additional available options

Picks icons, menus, and graphic entities

## [Esc] – Canceling Commands

The [**Esc**] key is used to cancel a command in SOLIDWORKS. The [**Esc**] key is located near the top left corner of the keyboard. Sometimes, it may be necessary to press the [**Esc**] key twice to cancel a command; it depends on where we are in the command sequence. For some commands, the [**Esc**] key is used to exit the command.

## SOLIDWORKS Help System

❖ Several types of online help are available at any time during a SOLIDWORKS session. SOLIDWORKS provides many online help functions, such as:

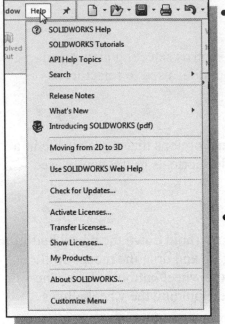

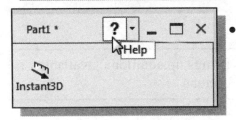

- The **Help** menu: Click on the **Help** option in the *Menu Bar* to access the SOLIDWORKS *Help* menu system. (Note: Move the cursor over the SOLIDWORKS logo in the *Menu Bar* to display the pull-down menu options.) The **SOLIDWORKS Help** option provides general help information, such as command options and command references. The **SOLIDWORKS Tutorials** option provides a collection of tutorials illustrating different SOLIDWORKS operations.

- The SOLIDWORKS Tutorials can also be accessed from the SOLIDWORKS *Resources task pane* on the right side of the screen by selecting **Tutorials**.

- The **SOLIDWORKS Help** option can also be accessed by clicking on the **Help** icon at the right end of the *Menu Bar*.

## Leaving SOLIDWORKS

➤ To leave SOLIDWORKS, use the left-mouse-button and click on **File** at the top of the SOLIDWORKS screen window, then choose **Exit** from the pull-down menu. (Note: Move the cursor over the SOLIDWORKS logo in the *Menu Bar* to display the pull-down menu options.)

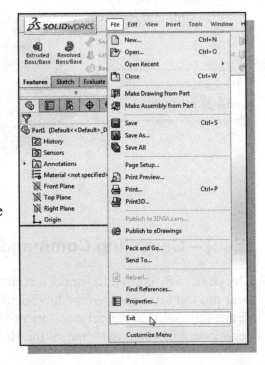

## Creating a CAD File Folder

It is a good practice to create a separate folder to store your CAD files. You should not save your CAD files in the same folder where the SOLIDWORKS application is located. It is much easier to organize and back up your project files if they are in a separate folder. Making folders within this folder for different types of projects will help you organize your CAD files even further. When creating CAD files in SOLIDWORKS, it is strongly recommended that you *save* your CAD files on the hard drive.

➢ To create a new folder in the *Windows* environment:

1. On the *desktop* or under the *My Documents* folder in which you want to create a new folder...

2. *Right-click* once to bring up the option menu, then select **New→ Folder**.

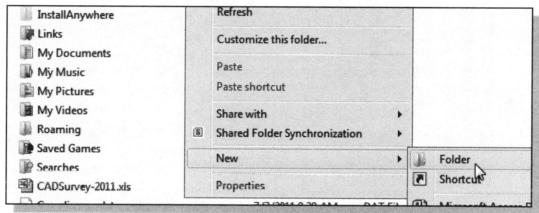

3. Type a name, such as ***Engr-Graphics***, for the new folder, and then press **[ENTER]**.

**Notes:**

# Chapter 2
# Parametric Modeling Fundamentals

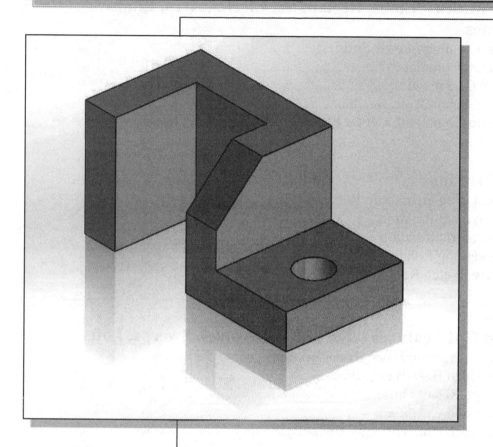

## Learning Objectives

♦ **Create Simple Extruded Solid Models**
♦ **Understand the Basic Parametric Modeling Procedure**
♦ **Units Setup**
♦ **Create 2D Sketches**
♦ **Understand the "Shape before Size" Design Approach**
♦ **Use the Dynamic Viewing Commands**
♦ **Create and Edit Parametric Dimensions**

## Certified SOLIDWORKS Associate Exam Objectives Coverage

### Sketch Entities – Lines, Rectangles, Circles, Arcs, Ellipses, Centerlines

Objectives: Creating Sketch Entities.

Sketch Command ..........................................................2-10
Line Command..............................................................2-10
Exit Sketch ..................................................................2-17
Circle Command, Center Point Circle .......................2-31

### Sketch Relations

Objectives: Using Geometric Relations.

Horizontal Relation .....................................................2-11
Geometric Relation Symbols .....................................2-12
Preventing Relations with [Ctrl] Key .........................2-12
View Sketch Relations................................................2-19

### Boss and Cut Features – Extrudes, Revolves, Sweeps, Lofts

Objectives: Creating Basic Swept Features.

Extruded Boss/Base, Blind ........................................2-27
Merge Result Option...................................................2-30
Extruded Cut................................................................2-31
Extruded Cut, Through All .........................................2-32

### Dimensions

Objectives: Applying and Editing Smart Dimensions.

Dimension, Smart Dimension .....................................2-13
Dimension Standard ....................................................2-14
Modify Smart Dimension ...........................................2-14

### Feature Conditions – Start and End

Objectives: Controlling Feature Start and End Conditions.

Extruded Boss/Base, Blind ........................................2-27
Extruded Cut, Through All .........................................2-31

# Introduction

The **feature-based parametric modeling** technique enables the designer to incorporate the original **design intent** into the construction of the model. The word *parametric* means the geometric definitions of a design, such as dimensions, can be varied at any time in the design process. Parametric modeling is accomplished by identifying and creating the key features of the design with the aid of computer software. The design variables, described in the sketches and described as parametric relations, can then be used to quickly modify/update the design.

In SOLIDWORKS, the parametric part modeling process involves the following steps:

1.  **Determine the type of the base feature, the first solid feature, of the design. Note that *Extrude*, *Revolve*, or *Sweep* operations are the most common types of base features.**

2.  **Create a rough two-dimensional sketch of the basic shape of the base feature of the design.**

3.  **Apply/modify constraints and dimensions to the two-dimensional sketch.**

4.  **Transform the two-dimensional parametric sketch into a 3D feature.**

5.  **Add additional parametric features by identifying feature relations and complete the design.**

6.  **Perform analyses/simulations, such as finite element analysis (FEA) or cutter path generation (CNC), on the computer model and refine the design as needed.**

7.  **Document the design by creating the desired 2D/3D drawings.**

The approach of creating two-dimensional sketches of the three-dimensional features is an effective way to construct solid models. Many designs are in fact the same shape in one direction. Computer input and output devices we use today are largely two-dimensional in nature, which makes this modeling technique quite practical. This method also conforms to the design process that helps the designer with conceptual design along with the capability to capture the ***design intent***. Most engineers and designers can relate to the experience of making rough sketches on restaurant napkins to convey conceptual design ideas. SOLIDWORKS provides many powerful modeling and design-tools, and there are many different approaches to accomplishing modeling tasks. The basic principle of **feature-based modeling** is to build models by adding simple features one at a time. In this chapter, the general parametric part modeling procedure is illustrated; a very simple solid model with extruded features is used to introduce the SOLIDWORKS user interface. The display viewing functions and the basic two-dimensional sketching tools are also demonstrated.

## The *Adjuster* Design

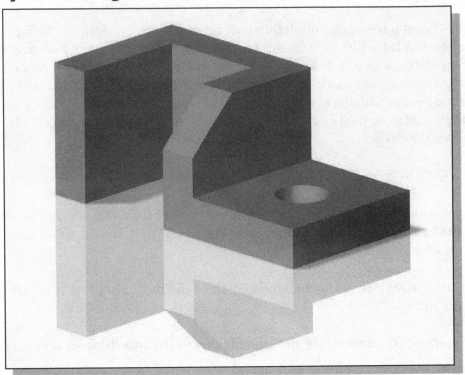

## Starting SOLIDWORKS

1. Select the **SOLIDWORKS** option on the *Start* menu or select the **SOLIDWORKS** icon on the desktop to start *SOLIDWORKS*. The *SOLIDWORKS* main window will appear on the screen.

2. Select **Part** by clicking on the first icon in the *New SOLIDWORKS Document* dialog box as shown.

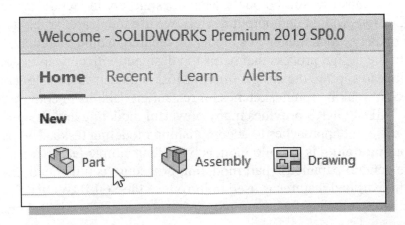

## SOLIDWORKS Part Modeling Window Layout

The default *part modeling window* contains the *Menu Bar*, the *Heads-up View* toolbar, the *Feature Manager Design Tree*, the *Command Manager* toolbar, the graphics area, the *task pane* (collapsed to the right of the graphics area in the figure below), and the *Status Bar*. A line of quick text appears next to the icon as you move the *mouse cursor* over different icons. You may resize the SOLIDWORKS drawing window by clicking and dragging on the edge of the window, or relocate the window by clicking and dragging on the *window title* area.

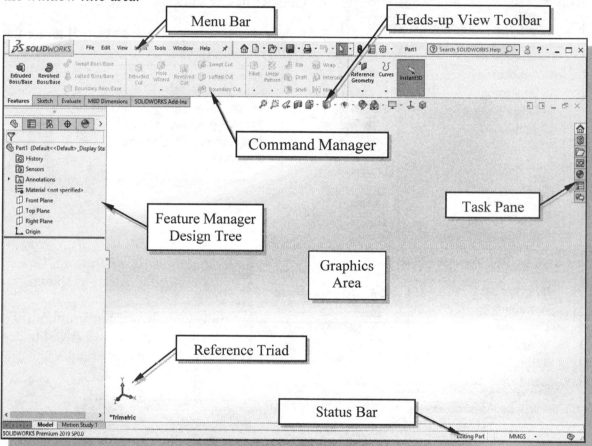

3. In the *Standard* toolbar area, **right-click** on any icon and activate **Command Manager** in the option list if necessary.

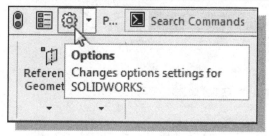

4. Select the **Options** icon from the *Menu* toolbar to open the *Options* dialog box.

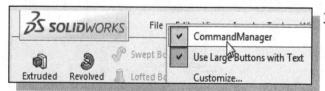

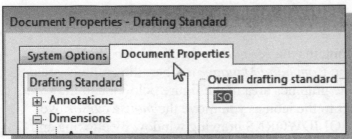

5. Select the **Document Properties** tab as shown.

6. Set the *Overall drafting standard* to **ISO** to reset the default setting.

7. Click **Units** as shown in the figure.

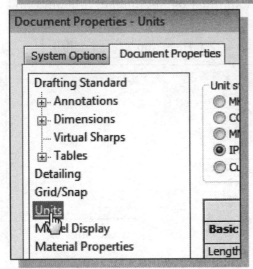

8. Select **IPS (inch, pound, second)** under the *Unit system* options.

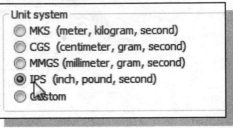

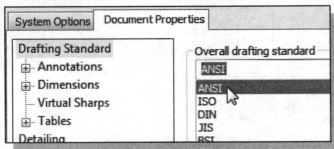

9. Set the *Overall drafting standard* to **ANSI** as shown.

10. On your own, confirm/modify all standards to use **ANSI** under the *Dimensions* group.

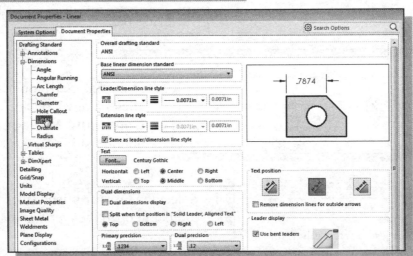

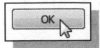

11. Click **OK** in the *Options* dialog box to accept the selected settings.

## Step 1: Determine/Set up the Base Solid Feature

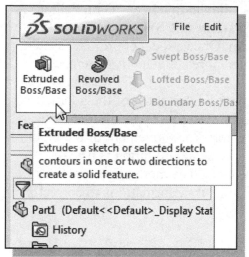

1. In the *Features* toolbar select the **Extruded Boss/Base** command by clicking once with the left-mouse-button on the icon.

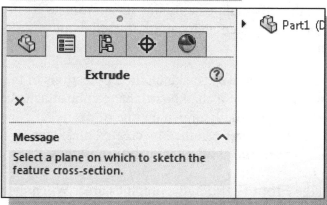

- Note the *Extrude Property Manager* is displayed in the left panel. We have activated the extrude feature; SOLIDWORKS requires the use of an existing 2D sketch or the creation of a new 2D sketch.

## Sketching Plane – It is an XY CRT, but an XYZ World

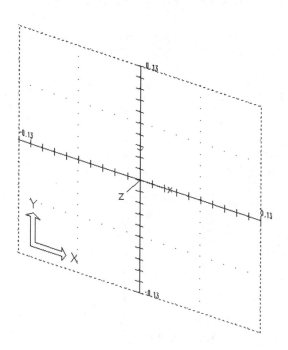

Design modeling software is becoming more powerful and user friendly, yet the system still does only what the user tells it to do. When using a geometric modeler, we therefore need to have a good understanding of what its inherent limitations are.

In most 3D geometric modelers, 3D objects are located and defined in what is usually called **world space** or **global space**. Although a number of different coordinate systems can be used to create and manipulate objects in a 3D modeling system, the objects are typically defined and stored using the world space. The world space is usually a **3D Cartesian coordinate system** that the user cannot change or manipulate.

In engineering designs, models can be very complex, and it would be tedious and confusing if only the world coordinate system were available. Practical 3D modeling systems allow the user to define **Local Coordinate Systems (LCS)** or **User Coordinate Systems (UCS)** relative to the world coordinate system. Once a local coordinate system is defined, we can then create geometry in terms of this more convenient system.

Although objects are created and stored in 3D space coordinates, most of the geometric entities can be referenced using 2D Cartesian coordinate systems. Typical input devices such as a mouse or digitizer are two-dimensional by nature; the movement of the input device is interpreted by the system in a planar sense. The same limitation is true of common output devices, such as CRT displays and plotters. The modeling software performs a series of three-dimensional to two-dimensional transformations to correctly project 3D objects onto the 2D display plane.

The SOLIDWORKS *sketching plane* is a special construction approach that enables the planar nature of the 2D input devices to be directly mapped onto the 3D coordinate system. The *sketching plane* is a local coordinate system that can be aligned to an existing face of a part or a reference plane.

Think of the sketching plane as the surface on which we can sketch the 2D sections of the parts. It is similar to a piece of paper, a white board, or a chalkboard that can be attached to any planar surface. The first sketch we create is usually drawn on one of the established datum planes. Subsequent sketches/features can then be created on sketching planes that are aligned to existing **planar faces of the solid part** or **datum planes**.

1. Move the cursor over the edge of the Front Plane in the graphics area. When the Front Plane is highlighted, click once with the **left-mouse-button** to select the Front Plane as the sketch plane for the new sketch. The *sketching plane* is a reference location where two-dimensional sketches are created. The *sketching plane* can be any planar part surface or datum plane.

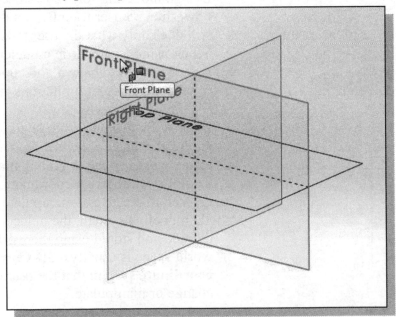

## Creating Rough Sketches

Quite often during the early design stage, the shape of a design may not have any precise dimensions. Most conventional CAD systems require the user to input the precise lengths and locations of all geometric entities defining the design, which are not available during the early design stage. With *parametric modeling*, we can use the computer to elaborate and formulate the design idea further during the initial design stage. With SOLIDWORKS, we can use the computer as an electronic sketchpad to help us concentrate on the formulation of forms and shapes for the design. This approach is the main advantage of *parametric modeling* over conventional solid-modeling techniques.

As the name implies, a ***rough sketch*** is not precise at all. When sketching, we simply sketch the geometry so that it closely resembles the desired shape. Precise scale or lengths are not needed. SOLIDWORKS provides us with many tools to assist us in finalizing sketches. For example, geometric entities such as horizontal and vertical lines are set automatically. However, if the rough sketches are poor, it will require much more work to generate the desired parametric sketches. Here are some general guidelines for creating sketches in SOLIDWORKS:

- **Create a sketch that is proportional to the desired shape.** Concentrate on the shapes and forms of the design.

- **Keep the sketches simple.** Leave out small geometry features such as fillets, rounds and chamfers. They can easily be placed using the Fillet and Chamfer commands after the parametric sketches have been established.

- **Exaggerate the geometric features of the desired shape.** For example, if the desired angle is 85 degrees, create an angle that is 50 or 60 degrees. Otherwise, SOLIDWORKS might assume the intended angle to be a 90-degree angle.

- **Draw the geometry so that it does not overlap.** The geometry should eventually form a closed region. *Self-intersecting* geometry shapes are not allowed.

- **The sketched geometric entities should form a closed region.** To create a solid feature, such as an extruded solid, a closed region is required so that the extruded solid forms a 3D volume.

➢ **Note:** The concepts and principles involved in *parametric modeling* are very different, and sometimes they are totally opposite, to those of conventional computer aided drafting. In order to understand and fully utilize the functionality of SOLIDWORKS, it will be helpful to take a *Zen* approach to learning the topics presented in this text: **Have an open mind and temporarily forget your experiences using conventional Computer Aided Drafting systems.**

## Step 2: Creating a Rough Sketch

> ➢ The *Sketch* toolbar provides tools for creating the basic geometry that can be used to create features and parts.

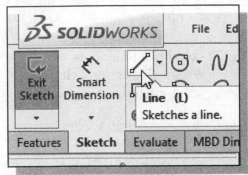

1. Move the graphics cursor to the **Line** icon in the *Sketch* toolbar. A *Help tip* box appears next to the cursor and a brief description of the command is displayed at the bottom of the drawing screen: "*Sketches a line.*" Select the icon by clicking once with the **left-mouse-button**.

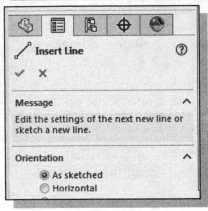

- Note the *Insert Line Feature Manager* is displayed in the left panel with different options related to the active command.

## Graphics Cursors

> ➢ Notice the cursor changes from an arrow to a pencil when graphical input is expected.

1. Left-click a starting point for the shape, roughly near the lower left side of the graphics window.

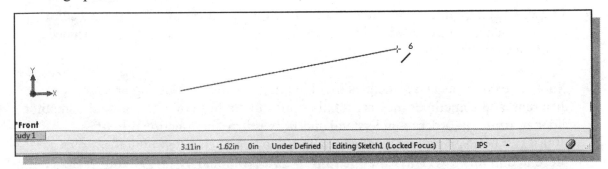

2. As you move the graphics cursor, you will see a digital readout next to the cursor. This readout gives you the line length. In the *Status Bar* area at the bottom of the window, the readout gives you the cursor location. Move the cursor around and you will notice different symbols appear at different locations.

3. Move the graphics cursor toward the right side of the graphics window to create a horizontal line as shown below. Notice the geometric relation symbol displayed. When the **Horizontal** relation symbol is displayed, left-click to select **Point 2**.

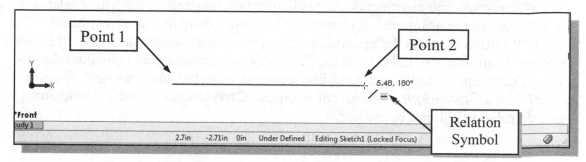

4. Complete the sketch as shown below, creating a closed region **ending at the starting point** (Point 1). Do not be overly concerned with the actual size of the sketch. Note that all line segments are sketched horizontally or vertically.

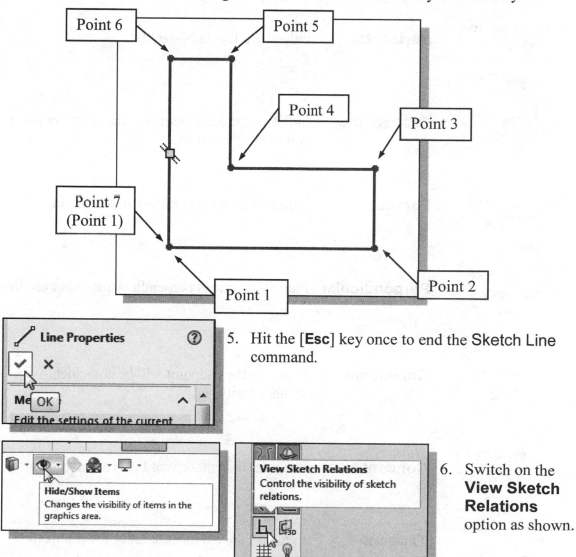

5. Hit the [**Esc**] key once to end the Sketch Line command.

6. Switch on the **View Sketch Relations** option as shown.

# Geometric Relation Symbols

While sketching, SOLIDWORKS automatically displays different visual clues, or symbols, to show you alignments, perpendicularities, tangencies, etc. These relations are used to capture the *design intent* by creating relations where they are recognized. SOLIDWORKS displays the governing geometric rules as models are built. To prevent relations from forming, hold down the [**Ctrl**] key while creating an individual sketch curve. For example, while sketching line segments with the Line command, endpoints are joined with a *Coincident relation*, but when the [**Ctrl**] key is pressed and held, the inferred relation will not be created.

| | | |
|---|---|---|
| 1.182, 90° | **Vertical** | indicates a line is vertical |
| 1.863, 180° | **Horizontal** | indicates a line is horizontal |
| 0.641 | **Dashed line** | indicates the alignment is to the center point or endpoint of an entity |
| 51.83 | **Parallel** | indicates a line is parallel to other entities |
| 0.338 | **Perpendicular** | indicates a line is perpendicular to other entities |
| 5.151 | **Coincident** | indicates the endpoint will be coincident with another entity |
| | **Concentric** | indicates the cursor is at the center of an entity |
| 0.689 | **Tangent** | indicates the cursor is at tangency points to curves |

## Step 3: Apply/Modify Relations and Dimensions

➢ As the sketch is made, SOLIDWORKS automatically applies some of the geometric constraints (such as horizontal, parallel, and perpendicular) to the sketched geometry. We can continue to modify the geometry, apply additional constraints, and/or define the size of the existing geometry. In this example, we will illustrate adding dimensions to describe the sketched entities.

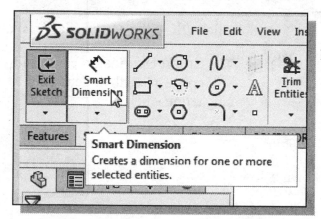

1. Move the cursor to the second icon of the *Sketch* toolbar; it is the **Smart Dimension** icon. Activate the command by left-clicking once on the icon.

2. Select the bottom horizontal line by left-clicking once on the line.

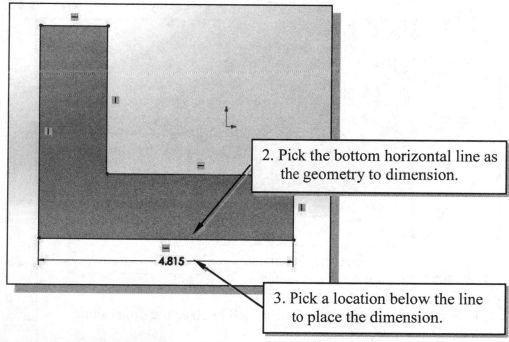

2. Pick the bottom horizontal line as the geometry to dimension.

3. Pick a location below the line to place the dimension.

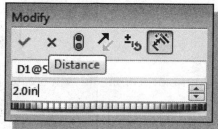

3. Enter **2.0** in the *Modify* dialog box.

4. Left-click **OK** (green check mark) in the *Modify* dialog box to save the current value and exit the dialog. Notice the associated geometry is adjusted.

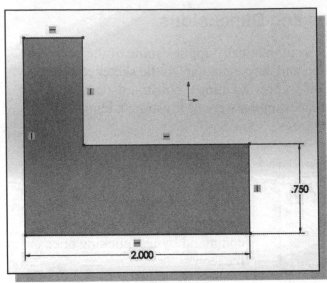

5. Select the lower right-vertical line.

6. Pick a location toward the right of the sketch to place the dimension.

7. Enter **0.75** in the *Modify* dialog box.

8. Click **OK** in the *Modify* dialog box.

❖ The Smart Dimension command will create a length dimension if a single line is selected.

9. Select the top-horizontal line as shown below.

10. Select the bottom-horizontal line as shown below.

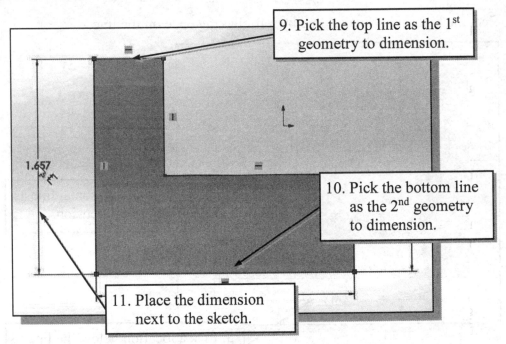

9. Pick the top line as the 1st geometry to dimension.

10. Pick the bottom line as the 2nd geometry to dimension.

11. Place the dimension next to the sketch.

11. Pick a location to the left of the sketch to place the dimension.

12. Enter **2.0** in the *Modify* dialog box.

13. Click **OK** in the *Modify* dialog box.

❖ When two parallel lines are selected, the Smart Dimension command will create a dimension measuring the distance between them.

14. On your own, repeat the above steps and create an additional dimension for the top line. Make the dimension **0.75**.

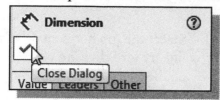

15. Click the **OK** icon in the *Property Manager* as shown, or hit the [**Esc**] key once, to end the Smart Dimension command.

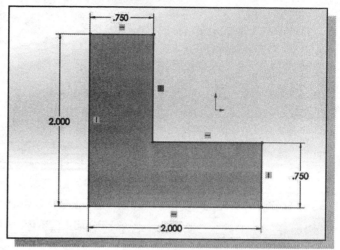

# Viewing Functions – Zoom and Pan

SOLIDWORKS provides a special user interface that enables convenient viewing of the entities in the graphics window. There are many ways to perform the **Zoom** and **Pan** operations.

1. Hold the [**Ctrl**] function key down. While holding the [**Ctrl**] function key down, press the mouse wheel down and drag the mouse to **Pan** the display. This allows you to reposition the display while maintaining the same scale factor of the display.

2. Hold the [**Shift**] function key down. While holding the [**Shift**] function key down, press the mouse wheel down and drag the mouse to **Zoom** the display. Moving downward will reduce the scale of the display, making the entities display smaller on the screen. Moving upward will magnify the scale of the display.

3. Turning the mouse wheel can also adjust the scale of the display. Turn the mouse wheel forward. Notice the scale of the display is reduced, making the entities display smaller on the screen.

4. Turn the mouse wheel backward. Notice the scale of the display is magnified. (**NOTE:** Turning the mouse wheel allows zooming to the position of the cursor.)

5. On your own, use the options above to change the scale and position of the display.

6. Press the **F** key on the keyboard to automatically fit the model to the screen.

## Delete an Existing Geometry of the Sketch

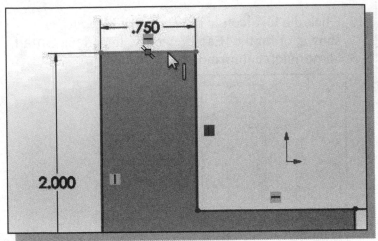

1. Select the top horizontal line by **left-clicking** once on the line.

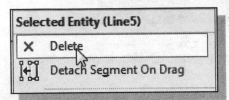

2. Click once with the **right-mouse-button** to bring up the *option menu*.

3. Select **Delete** from the option list as shown.

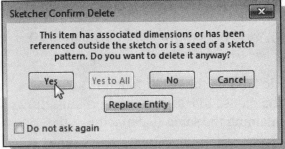

4. Click **Yes** to proceed with deleting the selected line.

   ❖ Note that the height dimensions and the horizontal constraint attached to the geometry are also deleted.

5. Activate the **Undo** command by left-clicking once on the icon in the standard toolbar to bring back the deleted line.

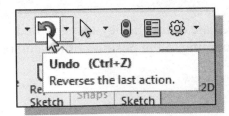

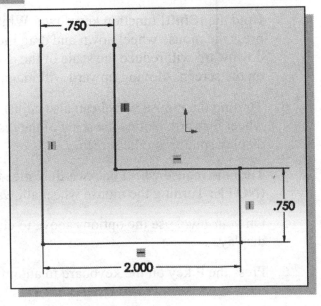

## Modifying the Dimensions of the Sketch

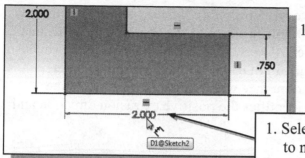

1. ·Select the dimension that is at the bottom of the sketch by **double-clicking** with the **left-mouse-button** on the dimension text.

1. Select this dimension to modify.

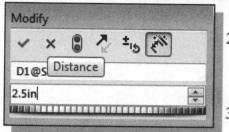

2. In the *Modify* window, the current length of the line is displayed. Enter **2.5** to reset the length of the line.

3. Click on the **OK** icon to accept the entered value.

➢ SOLIDWORKS will now update the profile with the new dimension value.

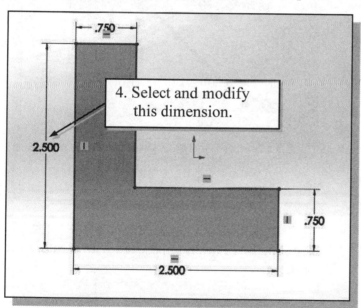

4. Select and modify this dimension.

4. On your own, repeat the above steps and adjust the left vertical dimension to **2.5** so that the sketch appears as shown.

5. Press the [**Esc**] key once to exit the Dimension command.

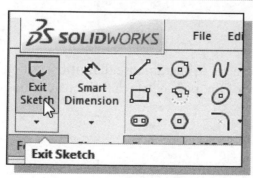

6. Click once with the **left-mouse-button** on the **Exit Sketch** icon on the *Sketch* toolbar to exit the sketch

## Step 4: Complete the Base Solid Feature

Now that the 2D sketch is completed, we will proceed to the next step: creating a 3D part from the 2D profile. Extruding a 2D profile is one of the common methods that can be used to create 3D parts. We can extrude planar faces along a path. We can also specify a height value and a tapered angle. In SOLIDWORKS, each face has a positive side and a negative side; the current face we're working on is set as the default positive side. This positive side identifies the positive extrusion direction and it is referred to as the face's *normal*.

1.  In the *Extrude Property Manager*, enter **2.5** as the extrusion distance. Notice that the completed sketch region is automatically selected as the extrusion profile.

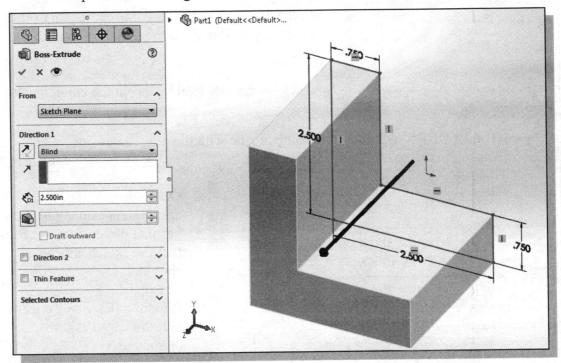

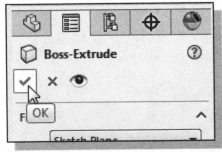

2.  Click on the **OK** button to proceed with creating the 3D part.

> Note that all dimensions disappeared from the screen. All parametric definitions are stored in the **SOLIDWORKS database** and any of the parametric definitions can be re-displayed and edited at any time.

## Isometric View

❖ SOLIDWORKS provides many ways to display views of the three-dimensional design. We will first orient the model to display in the *isometric view*, by using the *View Orientation* pull-down menu on the *Heads-up View* toolbar.

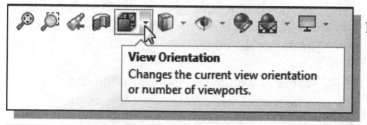

1. Select the **View Orientation** button on the *Heads-up View* toolbar by clicking once with the left-mouse-button.

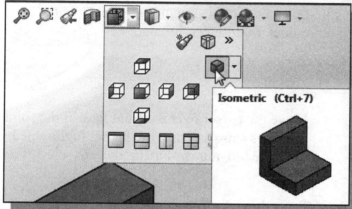

2. Select the **Isometric** icon in the *View Orientation* pull-down menu.

❖ Note that many other view-related commands are also available under the **View** *pull-down menu*.

## Rotation of the 3D Model – Rotate View

The Rotate View command allows us to rotate a part or assembly in the graphics window. Rotation can be around the center mark, free in all directions, or around a selected entity (vertex, edge, or face) on the model.

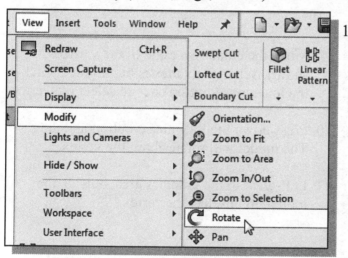

1. Move the cursor over the **SOLIDWORKS** logo to display the pull-down menus. Select **View → Modify → Rotate** from the pull-down menu as shown.

2. Move the cursor inside the graphics area. Press down the left-mouse-button and drag in an arbitrary direction; the Rotate View command allows us to freely rotate the solid model.

- The model will rotate about an axis normal to the direction of cursor movement. For example, drag the cursor horizontally across the screen and the model will rotate about a vertical axis.

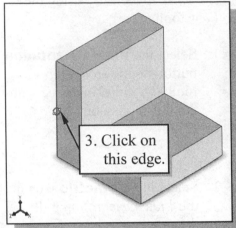

3. Move the cursor over one of the vertical edges of the solid model as shown. When the edge is highlighted, click the **left-mouse-button** once to select the edge.

4. Press down the left-mouse-button and drag. The model will rotate about this edge.

5. Left-click in the graphics area, outside the model, to unselect the edge.

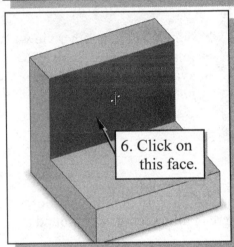

6. Move the cursor over the front face of the solid model as shown. When the face is highlighted, click the **left-mouse-button** once to select the face.

7. Press down the left-mouse-button and drag. The model will rotate about the direction normal to this face.

8. Left-click in the graphics area, outside the model, to unselect the face.

9. Move the cursor over one of the vertices as shown. When the vertex is highlighted, click the left-mouse-button once to select the vertex.

10. Press down the left-mouse-button and drag. The model will rotate about the vertex.

11. Left-click in the graphics area, outside the model, to unselect the vertex.

12. Press the [**Esc**] key once to exit the Rotate View command.

13. On your own, reset the display to the *isometric* view.

## Rotation and Panning – *Arrow Keys*

> ➢ SOLIDWORKS allows us to easily rotate a part or assembly in the graphics window using the **arrow** keys on the keyboard.

- Use the **arrow** keys to rotate the view horizontally or vertically. The **left-right** keys rotate the model about a vertical axis. The **up-down** keys rotate the model about a horizontal axis.

- Hold down the [**Alt**] key and use the **left-right arrow** keys to rotate the model about an axis normal to the screen, i.e., to rotate clockwise and counter-clockwise.

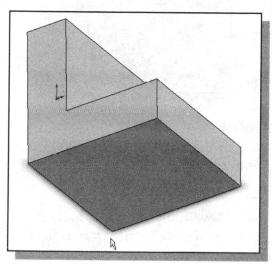

1. Hit the **left arrow** key. The model view rotates by a pre-determined increment. The default increment is 15°. (This increment can be set in the *Options* dialog box.) On your own, use the **left-right** and **up-down arrow** keys to rotate the view.

2. Hold down the [**Alt**] key and hit the **left arrow** key. The model view rotates in the clockwise direction. On your own use the **left-right** and **up-down arrow** keys, and the [**Alt**] key plus the **left-right arrow** keys, to rotate the view.

3. On your own, reset the display to the *isometric* view.

- Hold down the [**Shift**] key and use the **left-right** and **up-down arrow** keys to rotate the model in 90° increments.

4. Hold down the [**Shift**] key and hit the **right arrow** key. The view will rotate by 90°. On your own use the [**Shift**] key plus the **left-right arrow** keys to rotate the view.

5. Select the **Front** icon in the *View Orientation* pull-down menu as shown to display the **Front** view of the model.

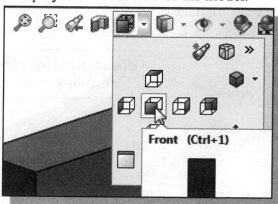

6.  Hold down the [**Shift**] key and hit the left arrow key. The view rotates to the Right side view.

7.  Hold down the [**Shift**] key and hit the down arrow key. The view rotates to the Top view.

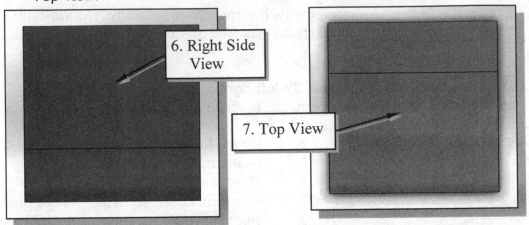

6. Right Side View

7. Top View

8.  On your own, reset the display to the **Isometric** view.

•   Hold down the [**Ctrl**] key and use the left-right and up-down arrow keys to Pan the model in increments.

9.  Hold down the [**Ctrl**] key and hit the left arrow key. The view Pans, moving the model toward the left side of the screen. On your own use [**Ctrl**] key plus the left-right and up-down arrow keys to Pan the view.

## Dynamic Viewing – Quick Keys

We can also use the function keys on the keyboard and the mouse to access the *Dynamic Viewing* functions.

❖ **Panning**

**(1) Hold the Ctrl key; press and drag the mouse wheel**

Hold the [**Ctrl**] function key down, and press and drag with the mouse wheel to **Pan** the display. This allows you to reposition the display while maintaining the same scale factor of the display.

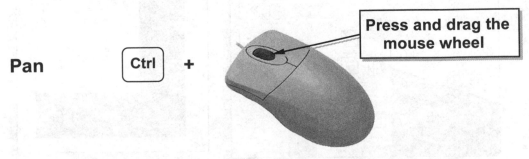

**Pan**        Ctrl   +

Press and drag the mouse wheel

**(2) Hold the Ctrl key; use arrow keys**

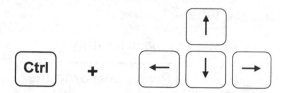

❖ **Zooming**

**(1) Hold the Shift key; press and drag the mouse wheel**

Hold the [**Shift**] function key down, and press and drag with the mouse wheel to **Zoom** the display. Moving downward will reduce the scale of the display, making the entities display smaller on the screen. Moving upward will magnify the scale of the display.

**Zoom**          Shift

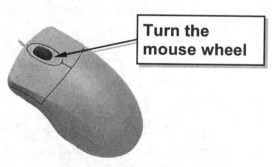

Press and drag the mouse wheel

**(2) Turning the mouse wheel**

Turning the mouse wheel can also adjust the scale of the display. Turning forward will reduce the scale of the display, making the entities display smaller on the screen. Turning backward will magnify the scale of the display.

- Turning the mouse wheel allows zooming to the position of the cursor.

- The cursor position, inside the graphics area, is used to determine the center of the scale of the adjustment.

Turn the mouse wheel

**(3) Z key or Shift + Z key**

Pressing the [**Z**] key on the keyboard will zoom out. Holding the [**Shift**] function key and pressing the [**Z**] key will zoom in.

### 3D Rotation

#### (1) Press and drag the mouse wheel

Press and drag with the mouse wheel to rotate the display.

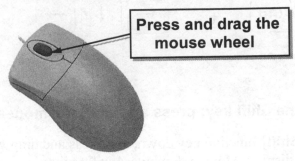

**Press and drag the mouse wheel**

#### (2) Use arrow keys

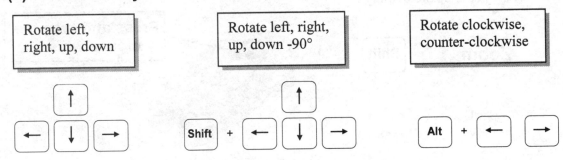

| Rotate left, right, up, down | Rotate left, right, up, down -90° | Rotate clockwise, counter-clockwise |

## Viewing Tools – Heads-up View Toolbar

The *Heads-up View* toolbar is a transparent toolbar that appears in each viewport and provides easy access to commonly used tools for manipulating the view. The default toolbar is described below.

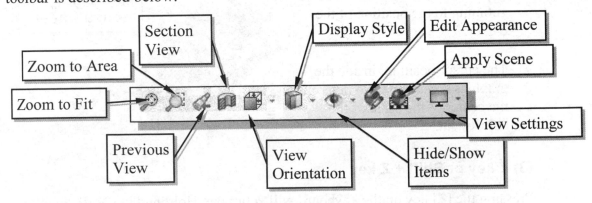

**Zoom to Fit** – Adjusts the view so that all items on the screen fit inside the graphics window.

**Zoom to Area** – Use the cursor to define a region for the view; the defined region is zoomed to fill the graphics window.

**Previous View** – Returns to the previous view.

**Section View** – Displays a cutaway of a part or assembly using one or more section planes.

**View Orientation** – This allows you to change the current view orientation or number of viewports.

**Display Style** – This can be used to change the display style (shaded, wireframe, etc.) for the active view.

**Hide/Show Items** – This pull-down menu is used to control the visibility of items (axes, sketches, relations, etc.) in the graphics area.

**Edit Appearance** – Modifies the appearance of entities in the model.

**Apply Scene** – Cycles through or applies a specific scene.

**View Settings** – Allows you to toggle various view settings (e.g., shadows, perspective).

## View Orientation

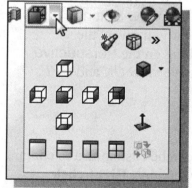

1.  Click on the **View Orientation** icon on the *Heads-up View* toolbar to reveal the view orientation and number of viewports options.

❖  Standard view orientation options: **Front**, **Back**, **Left**, **Right**, **Top**, **Bottom**, **Isometric**, **Trimetric** or **Dimetric** icons can be selected to display the corresponding standard view.

  **Normal to** – In a part or assembly, zooms and rotates the model to display the selected plane or face. You can select the element either before or after clicking the **Normal to** icon.

❖  The icons across the bottom of the pull-down menu allow you to display a single viewport (the default) or multiple viewports.

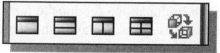

## Display Style

1. Click on the **Display Style** icon on the *Heads-up View* toolbar to reveal the display style options.

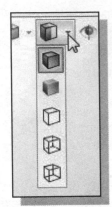

**Shaded with Edges** – Allows the display of a shaded view of a 3D model with its edges.

**Shaded** – Allows the display of a shaded view of a 3D model.

**Hidden Lines Removed** – Allows the display of the 3D objects using the basic wireframe representation scheme. Only those edges which are visible in the current view are displayed.

**Hidden Lines Visible** – Allows the display of the 3D objects using the basic wireframe representation scheme in which all the edges of the model are displayed, but edges that are hidden in the current view are displayed as dashed lines (or in a different color).

**Wireframe** – Allows the display of 3D objects using the basic wireframe representation scheme in which all the edges of the model are displayed.

## Orthographic vs. Perspective

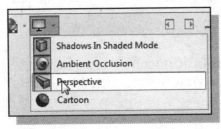

1. Besides the basic display modes, we can also choose an orthographic view or perspective view of the display. Clicking on the View Settings icon on the *Heads-up View* toolbar will reveal the Perspective icon. Clicking on the Perspective icon toggles the perspective view *ON* and *OFF*.

## Customizing the Heads-up View Toolbar

1. The *Heads-up View* toolbar can be customized to contain icons for user-preferred view options. **Right-click** anywhere on the *Heads-up View* toolbar to reveal the display menu option list. Click on the **View (Heads-up)** icon to turn *OFF* the display of the toolbar. Notice the **Customize** option is also available to add/remove different icons.

➢ On your own, use the different options described in the above sections to familiarize yourself with the 3D viewing/display commands. Reset the display to the standard **Isometric view** before continuing to the next section.

## Step 5-1: Adding an Extruded Feature

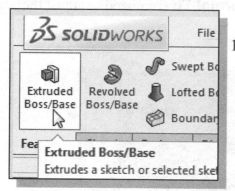

1.  In the *Features* toolbar select the **Extruded Boss/ Base** command by clicking once with the left-mouse-button on the icon.

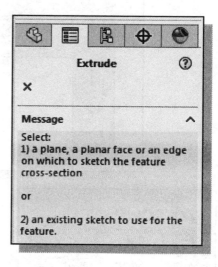

*   In the *Extrude Manager* area, SOLIDWORKS indicates the two options to create the new extrusion feature. We will select the back surface of the base feature to align the sketching plane.

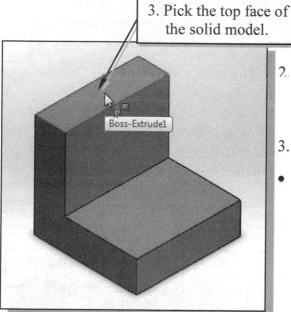

3. Pick the top face of the solid model.

2.  On your own, use one of the rotation quick keys/mouse-button and view the back face of the model as shown.

3.  Pick the top face of the 3D solid object.

*   Note that SOLIDWORKS automatically establishes a *User Coordinate System* (UCS), and records its location with respect to the part on which it was created.

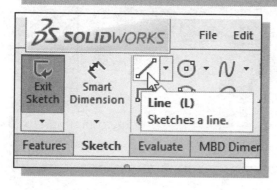

4.  Select the **Line** command by clicking once with the **left-mouse-button** on the icon in the *Sketch* toolbar.

➤   To illustrate the usage of dimensions in parametric sketches, we will intentionally create a sketch away from the desired location.

5. Create a sketch with segments perpendicular/parallel to the existing edges of the solid model as shown below. (Hint: Use the **Normal To** command to adjust the display normal to the sketching plane direction.)

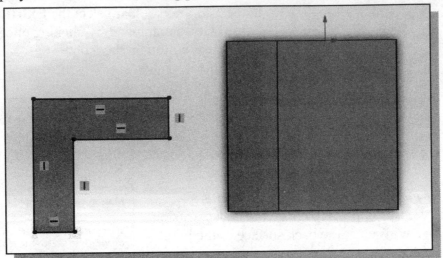

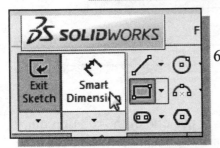

6. Select the **Smart Dimension** command in the *Sketch* toolbar. The Smart Dimension command allows us to quickly create and modify dimensions.

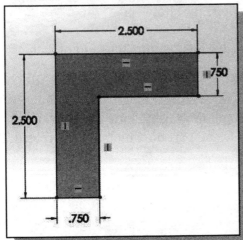

7. On your own, create and modify the size dimensions to describe the size of the sketch as shown in the figure.

8. Create the two location dimensions, accepting the default values, to describe the position of the sketch relative to the top corner of the solid model as shown.

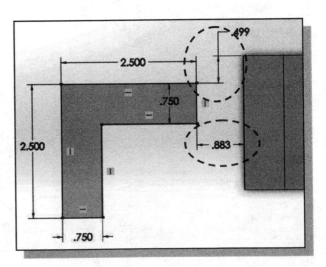

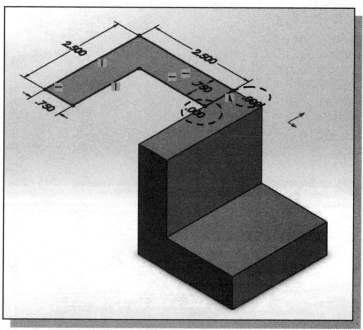

9. On your own, modify the two location dimensions to **0.0** and **0.0** as shown in the figure.

➢ In parametric modeling, the dimensions can be used to quickly control the size and location of the defined geometry.

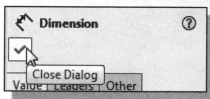

10. Select **Close Dialog** in the *Dimension Property Manager* to end the **Smart Dimension** command.

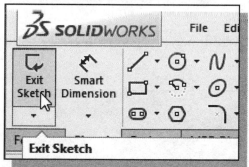

11. Click **Exit Sketch** in the *Sketch* toolbar to end the **Sketch** mode.

12. In the *Boss-Extrude Property Manager*, enter **2.5** as the extrude *Distance* as shown.

13. Click the **Reverse Direction** button in the *Property Manager* as shown. The extrude preview should appear as shown.

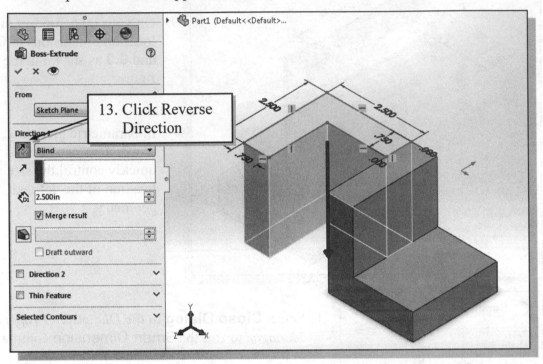

14. Confirm the **Merge result** option is activated as shown.

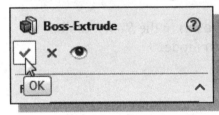

15. Click on the **OK** button to proceed with creating the extruded feature.

## Step 5-2: Adding a Cut Feature

- Next, we will create a **cut** feature that will be added to the existing solid object.

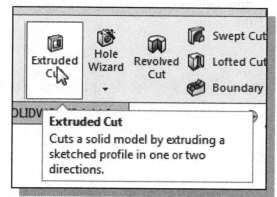

1. In the *Features* toolbar select the **Extruded Cut** command by clicking once with the left-mouse-button on the icon.

2. Pick the vertical face of the last feature we created, as shown.

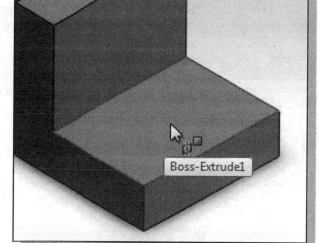

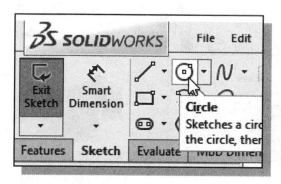

3. Select the **Circle** command by clicking once with the **left-mouse-button** on the icon in the *Sketch* toolbar.

4. Create a circle roughly at the center of the surface as shown.

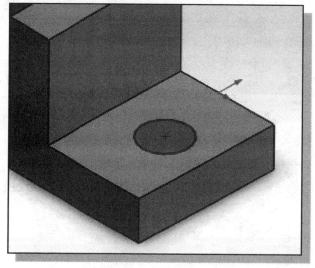

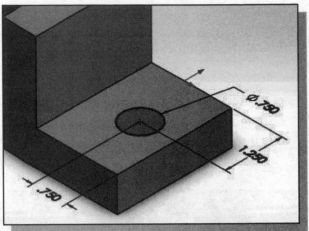

5.  On your own, create and modify the dimensions of the sketch as shown in the figure.

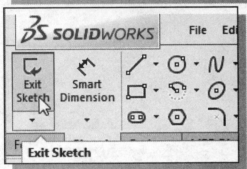

6.  Click **Exit Sketch** in the *Sketch* toolbar to end the **Sketch** mode.

7.  Adjust the display by using the **Isometric** option in the *View Orientation* panel as shown. (Quick Key: **Ctrl + 7**)

-   Next, we will create and profile another sketch, a rectangle, which will be used to create another extrusion feature that will be added to the existing solid object.

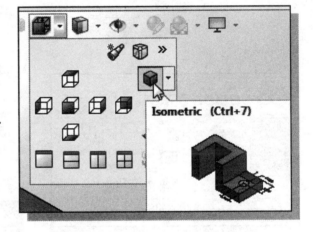

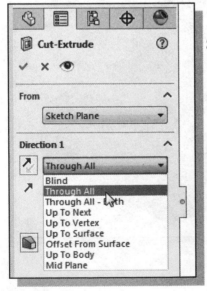

8.  Set the *Extents* option to **Through All** as shown. The *Through All* option instructs the software to calculate the extrusion distance and assures the created feature will always cut through the full length of the model.

-   Note the different options available to create the extruded cut feature.

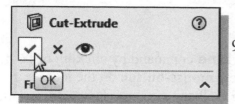

9. Click on the **OK** button to proceed with creating the extruded feature.

## Step 5-3: Adding another Cut Feature

- Next, we will create and profile a triangle, which will be used to create a **cut** feature that will be added to the existing solid object.

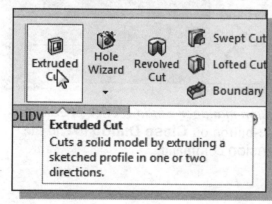

1. In the *Features* toolbar select the **Extruded Cut** command by clicking once with the left-mouse-button on the icon.

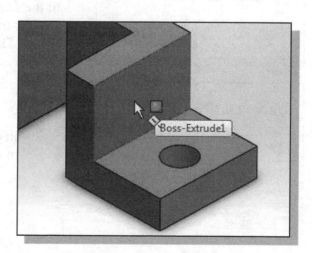

2. Pick the vertical face of the first feature we created, as shown.

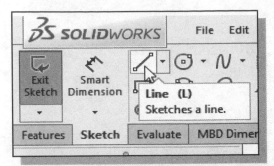

3. Select the **Line** command by clicking once with the **left-mouse-button** on the icon in the *Sketch* toolbar.

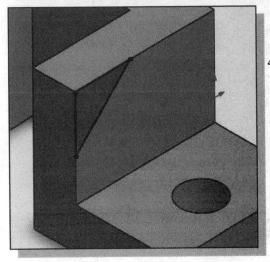

4. Start at the upper left corner and create three line segments to form a small triangle as shown.

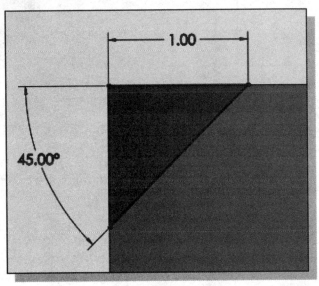

5. On your own, create and modify the two dimensions of the sketch as shown in the figure. (Hint: create the angle dimension by selecting the two adjacent lines as shown and place the angular dimension inside the desired quadrant.)

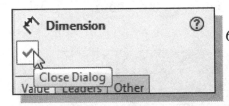

6. In the *Dimension manager window*, click once with the left-mouse-button on **Close Dialog** to end the Smart Dimension command.

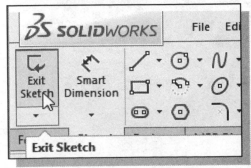

7. In the *Features* toolbar, select **Exit Sketch** in the pop-up menu to end the Sketch mode.

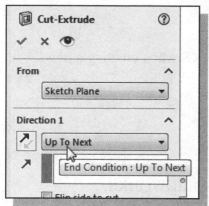

8. Set the *Extents* option to **Up To Next face/body** as shown. The *Up To Next* option instructs the software to calculate the extrusion distance and assures the created feature will always cut through the proper length of the model.

9. Click on the **OK** button to proceed with creating the extruded feature.

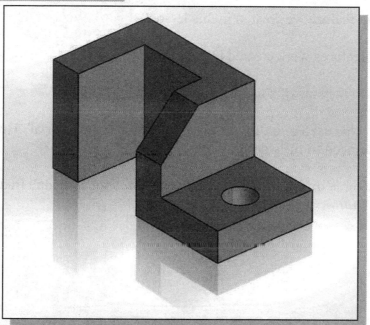

## Save the Model

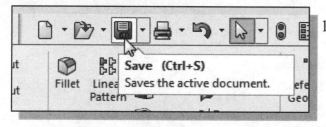

1. Select **Save** in the *Quick Access* toolbar, or you can also use the "**Ctrl-S**" combination (hold down the "Ctrl" key and hit the "S" key once) to save the part.

2. In the pop-up window, select the directory to store the model in and enter ***Adjuster*** as the name of the file.

3. Click on the **Save** button to save the file.

❖ You should form a habit of saving your work periodically, just in case something might go wrong while you are working on it. In general, one should save one's work at an interval of every 15 to 20 minutes. One should also save before making any major modifications to the model.

## Questions:

1.  What is the first thing we should set up in SOLIDWORKS when creating a new model?

2.  Describe the general *parametric modeling* procedure.

3.  Describe the general guidelines in creating *rough sketches*.

4.  List two of the geometric constraint symbols used by SOLIDWORKS.

5.  What was the first feature we created in this lesson?

6.  How many solid features were created in the tutorial?

7.  How do we control the size of a feature in parametric modeling?

8.  Which command was used to create the last cut feature in the tutorial? How many dimensions do we need to fully describe the cut feature?

9.  List and describe three differences between *parametric modeling* and traditional 2D *computer aided drafting* techniques.

# Exercises:

1.  **Inclined Support** (Thickness: **.5**)

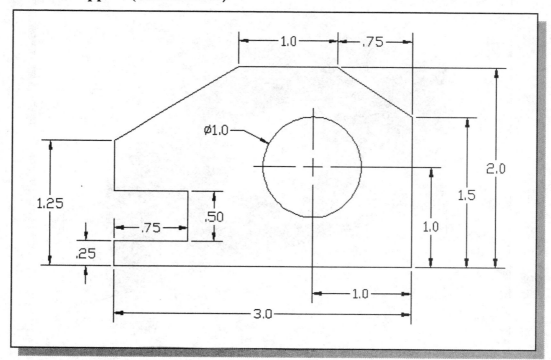

2.  **Spacer Plate** (Thickness: **.125**)

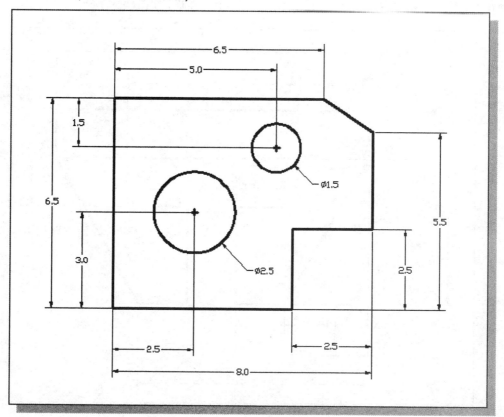

3. **Positioning Stop**

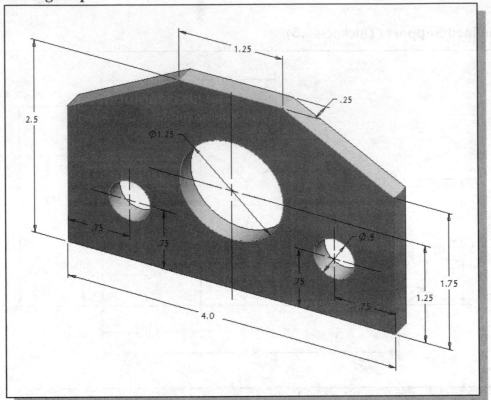

4. **Latch Clip** (Dimensions are in inches. Thickness: **0.25** inches.)

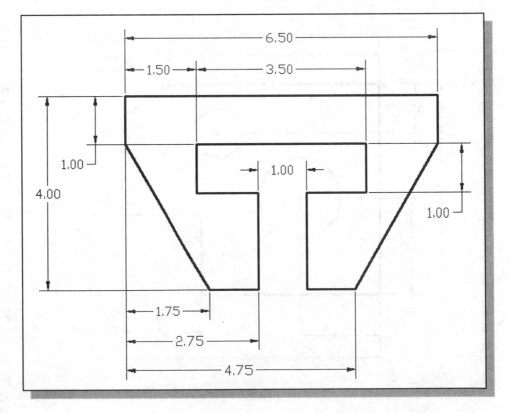

5. **Slider Block**

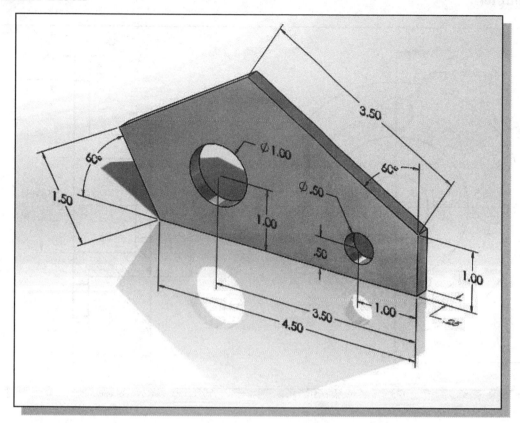

6. **Angle Lock**

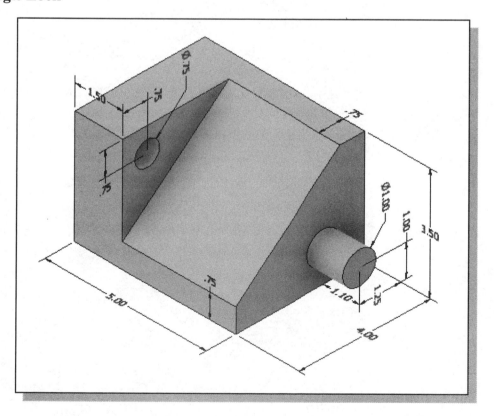

## 7.  Coupler

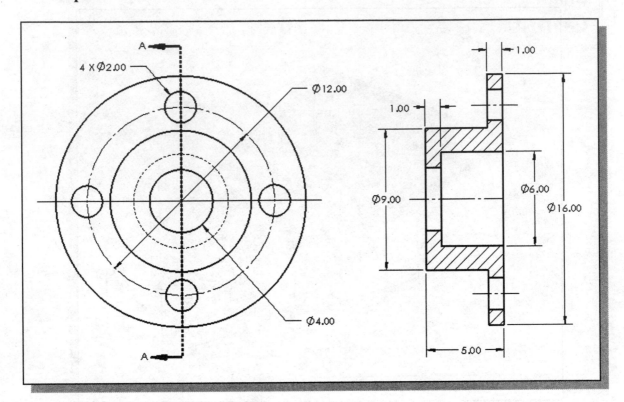

# Chapter 3
# Constructive Solid Geometry Concepts

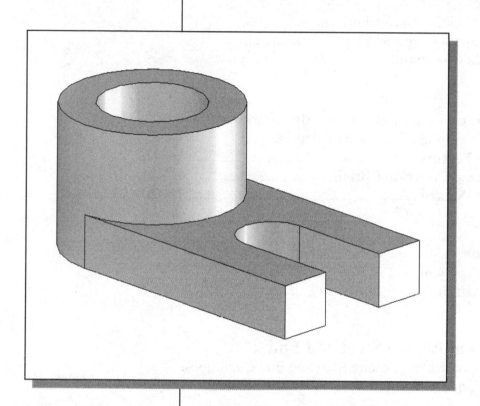

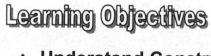

## Learning Objectives

- ◆ **Understand Constructive Solid Geometry Concepts**
- ◆ **Create a Binary Tree**
- ◆ **Understand the Basic Boolean Operations**
- ◆ **Use the SOLIDWORKS CommandManager User Interface**
- ◆ **Set up GRID and SNAP Intervals**
- ◆ **Understand the Importance of Order of Features**
- ◆ **Use the Different Extrusion Options**

| **Certified SOLIDWORKS Associate Exam Objectives Coverage** |

Certified Associate Reference Guide

### Sketch Entities – Lines, Rectangles, Circles, Arcs, Ellipses, Centerlines
Objectives:  Creating Sketch Entities.
**Rectangle Command**.................................................3-11
**Circle Command** .......................................................3-14

### Boss and Cut Features – Extrudes, Revolves, Sweeps, Lofts
Objectives:  Creating Basic Swept Features.
**Base Feature** ...............................................................3-10
**Reverse Direction Option** .......................................3-16
**Hole Wizard**................................................................3-20

### Dimensions
Objectives: Applying and Editing Smart Dimensions.
**Reposition Smart Dimension** ..................................3-12

### Feature Conditions – Start and End
Objectives: Controlling Feature Start and End Conditions.
**Extruded Cut, Up to Next** ........................................3-23

## Introduction

In the 1980s, one of the main advancements in **solid modeling** was the development of the **Constructive Solid Geometry** (CSG) method. CSG describes the solid model as combinations of basic three-dimensional shapes (**primitive solids**). The basic primitive solid set typically includes Rectangular-prism (Block), Cylinder, Cone, Sphere, and Torus (Tube). Two solid objects can be combined into one object in various ways using operations known as **Boolean operations**. There are three basic Boolean operations: **JOIN (Union)**, **CUT (Difference)**, and **INTERSECT**. The *JOIN* operation combines the two volumes included in the different solids into a single solid. The *CUT* operation subtracts the volume of one solid object from the other solid object. The *INTERSECT* operation keeps only the volume common to both solid objects. The CSG method is also known as the **Machinist's Approach**, as the method is parallel to machine shop practices.

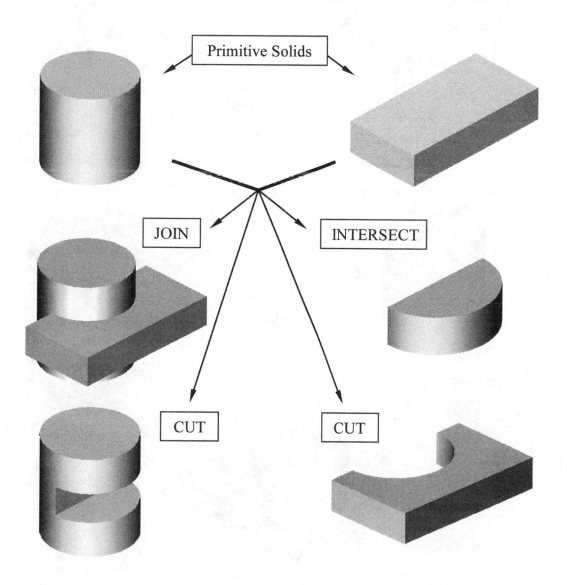

## Binary Tree

The CSG is also referred to as the method used to store a solid model in the database. The resulting solid can be easily represented by what is called a **binary tree**. In a binary tree, the terminal branches (leaves) are the various primitives that are linked together to make the final solid object (the root). The binary tree is an effective way to keep track of the *history* of the resulting solid. By keeping track of the history, the solid model can be re-built by re-linking through the binary tree. This provides a convenient way to modify the model. We can make modifications at the appropriate links in the binary tree and re-link the rest of the history tree without building a new model.

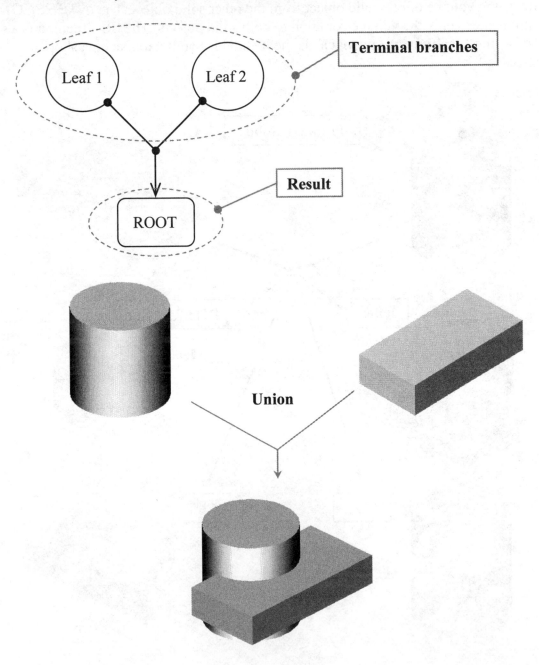

## The *Locator* Design

The CSG concept is one of the important building blocks for feature-based modeling. In SOLIDWORKS, the CSG concept can be used as a planning tool to determine the number of features that are needed to construct the model. It is also a good practice to create features that are parallel to the manufacturing process required for the design. With parametric modeling, we are no longer limited to using only the predefined basic solid shapes. In fact, any solid features we create in SOLIDWORKS are used as primitive solids; parametric modeling allows us to maintain full control of the design variables that are used to describe the features. In this lesson, a more in-depth look at the parametric modeling procedure is presented. The equivalent CSG operation for each feature is also illustrated.

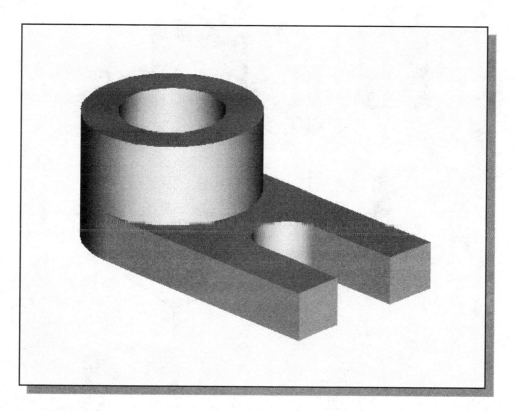

> ➤ Before going through the tutorial on your own, make a sketch of a CSG binary tree of the **Locator** design using only two basic types of primitive solids: cylinder and rectangular prism. In your sketch, how many *Boolean operations* will be required to create the model? What is your choice of the first primitive solid to use, and why? Take a few minutes to consider these questions and do the preliminary planning by sketching on a piece of paper. Compare the sketch you make to the CSG binary tree steps shown on page 3-6. Note that there are many different possibilities in combining the basic primitive solids to form the solid model. Even for the simplest design, it is possible to take several different approaches to creating the same solid model.

## Modeling Strategy – CSG Binary Tree

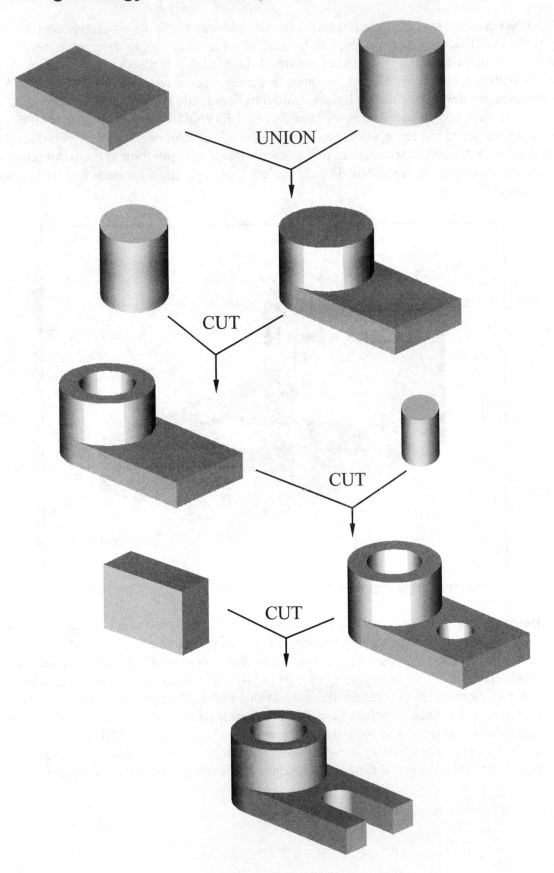

## Starting SOLIDWORKS and Activating the Command Manager

1.  Select the **SOLIDWORKS** option on the *Start* menu or select the **SOLIDWORKS** icon on the desktop to start *SOLIDWORKS*. The *SOLIDWORKS* main window will appear on the screen.

2.  Select **Part** to start a new part, by clicking on the first icon in the *Welcome - SOLIDWORKS Document* dialog box as shown.

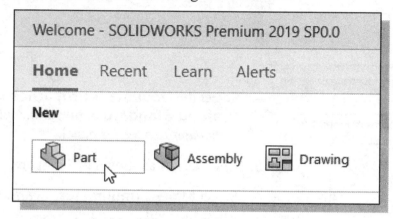

❖ Notice the *Part Modeling* window appears on the screen with the *Ribbon style* display of the *Command Manager*.

- Every object we construct in a CAD system is measured in units. We should determine the value of the units within the CAD system before creating the first geometric entities. For example, in one model, a unit might equal one millimeter of the real-world object; in another model, a unit might equal an inch. In most CAD systems, setting the model units does not always set units for dimensions. We generally set model units and dimension units to the same type and precision.

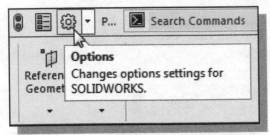

3.  Select the **Options** icon from the *Menu Bar* toolbar to open the *Options* dialog box.

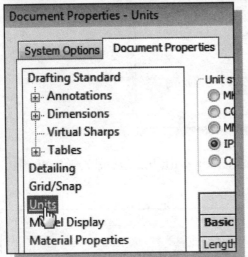

4.  Select the **Document Properties** tab and the **Drafting Standard** option at the left of the *Document Properties* panel.

5.  Click **Units** as shown in the figure.

6.  Select **MMGS (millimeter, gram, second)** under the *Unit system* options.

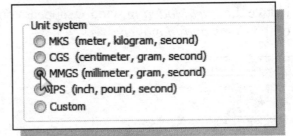

7.  Set the *Overall drafting standard* to **ISO** to reset the default setting.

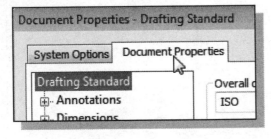

8.  Select **None** in the *Decimals* spin box for the *Length units* as shown to set the display to no decimal places. (Note the *Overall Drafting Standard* is automatically adjusted with our setting as shown.)

| Type | Unit | Decimals | Fractions | More |
|------|------|----------|-----------|------|
| **Basic Units** | | | | |
| Length | millimeters | None | | ... |
| Dual Dimension Length | inches | .12 | | ... |
| Angle | degrees | .12 | | |
| **Mass/Section Properties** | | | | |

## *GRID* and *SNAP* Intervals Setup

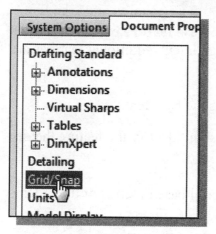

1. Click **Grid/Snap** as shown below.

2. Check the Display grid checkbox under the *Grid* options.

3. Uncheck the Dash checkbox under the *Grid* options.

4. Set the Major grid spacing to **50 mm** under the *Grid* options.

5. Set the Minor-lines per major to **5** under the *Grid* options.

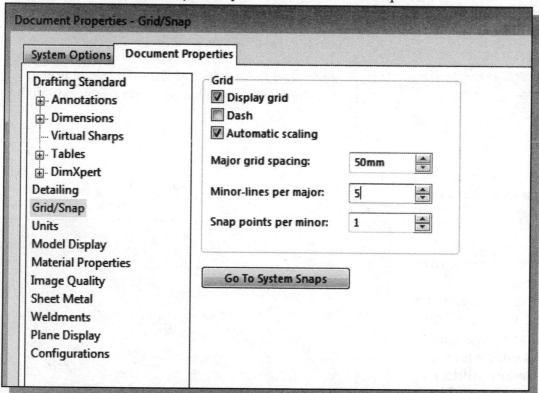

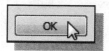

6. Click **OK** in the *Options* dialog box to accept the selected settings.

➢ Note that the above settings set the grid displaying in SOLIDWORKS. Although the **Snap to grid** option is available in SOLIDWORKS, its usage in parametric modeling is not recommended.

## Base Feature

In *parametric modeling*, the first solid feature is called the **base feature,** which usually is the primary shape of the model. Depending upon the design intent, additional features are added to the base feature.

These are some of the considerations involved in selecting the base feature:

• **Design intent** – Determine the functionality of the design; identify the feature that is central to the design.

• **Order of features** – Choose the feature that is the logical base in terms of the order of features in the design.

• **Ease of making modifications** – Select a base feature that is more stable and is less likely to be changed.

➢ A rectangular block will be created first as the base feature of the *Locator* design.

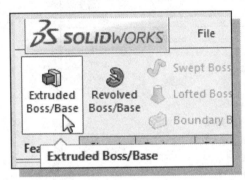

1. Select the **Extruded Boss/Base** button on the *Features* toolbar to create a new extruded feature.

2. Move the cursor over the edge of the Top Plane in the graphics area. When the Top Plane is highlighted, click once with the **left-mouse-button** to select the Top Plane (XZ Plane) as the sketch plane for the new sketch.

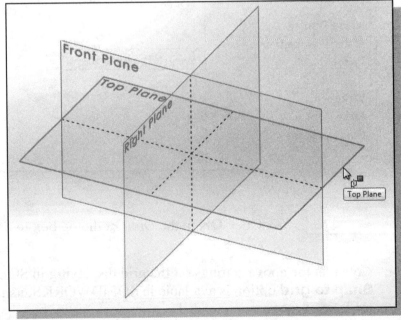

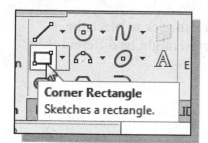

3. Select the **Corner Rectangle** command by clicking once with the **left-mouse-button** on the icon in the *Sketch* toolbar.

4. Create a rectangle of arbitrary size by selecting two locations on the screen as shown below.

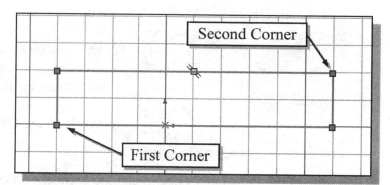

5. Inside the graphics window, click once with the **right-mouse-button** to bring up the option menu.

6. Choose **Select** to end the Rectangle command.

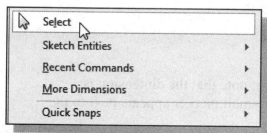

7. Select the **Smart Dimension** command by clicking once with the **left-mouse-button** on the icon in the *Sketch* toolbar.

8. The message "*Select one or two edges/vertices and then a text location*" is displayed in the *Status Bar* area at the bottom of the SOLIDWORKS window. Select the bottom horizontal line by left-clicking once on the line.

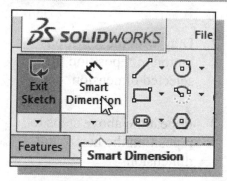

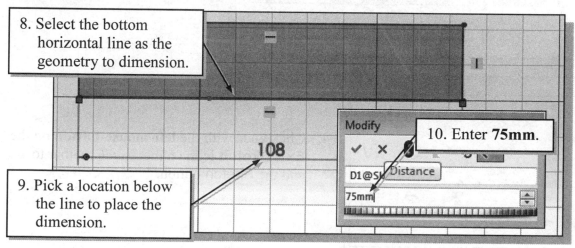

8. Select the bottom horizontal line as the geometry to dimension.

9. Pick a location below the line to place the dimension.

10. Enter **75mm**.

9. Move the graphics cursor below the selected line and left-click to place the dimension. (Note that the value displayed on your screen might be different than what is shown in the figure above.)

10. Enter **75** in the *Modify* dialog box.

11. Click **OK** in the *Modify* dialog box.

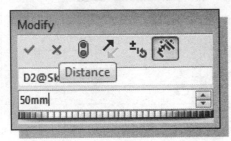

12. On your own, create the vertical size dimension of the sketched rectangle as shown. Enter **50** to set the length of the line.

13. Click on the **OK** icon to accept the entered value.

14. Hit the [**Esc**] key once to end the **Dimension** command.

## Repositioning Dimensions

1. Move the cursor near the vertical dimension; note that the dimension is highlighted. Move the cursor slowly until a small marker appears next to the cursor, as shown in the figure.

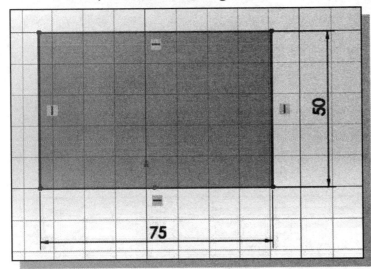

2. Drag with the left-mouse-button to reposition the selected dimension.

3. Repeat the above steps to reposition the horizontal dimension.

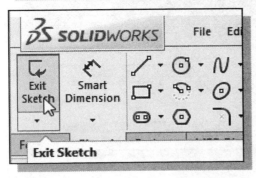

4. Click once with the **left-mouse-button** on the **Exit Sketch** icon on the *Sketch* toolbar to end the **Sketch** option.

## Completing the Base Solid Feature

1.  In the *Extrude Property Manager* panel, enter **15** as the extrusion distance. Notice that the sketch region is automatically selected as the extrusion profile.

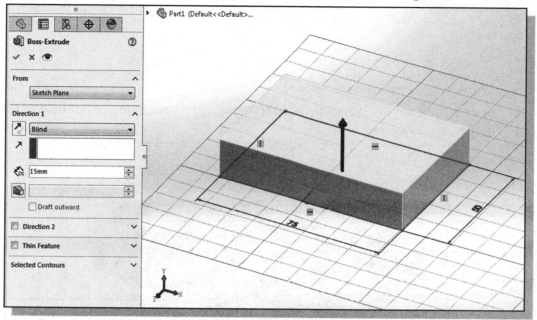

2.  Click on the **OK** button to proceed with creating the 3D part.

3.  Use the *Viewing* options to view the created part. On the *Heads-up View* toolbar, select **View Orientation** (to open the *View Orientation* pull-down menu) and select the **Isometric** icon to reset the display to the Isometric view before going to the next section.

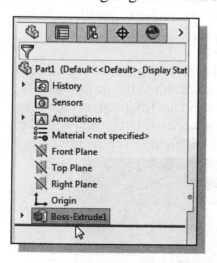

4.  Note the Extruded feature you created is listed in the *Feature Manager Design Tree* as shown. Note the Design Tree can be used to identify all features used to create the design.

## Creating the Next Solid Feature

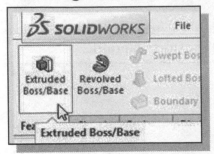

1. Select the **Extruded Boss/Base** button on the *Features* toolbar to create a new extruded feature.

2. Rotate the view by using the **up arrow** key to display the bottom face of the solid model as shown below.

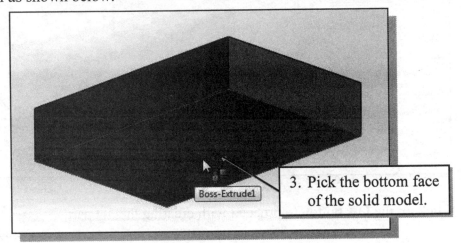

3. Pick the bottom face of the solid model.

3. Pick the bottom face of the 3D model as the sketching plane.

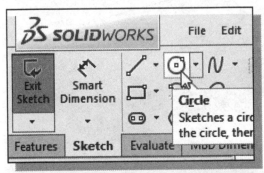

➢ Note that the sketching plane is aligned to the selected face.

4. Select the **Circle** command by clicking once with the left-mouse-button on the icon in the *Sketch* toolbar.

➢ We will align the center of the circle to the midpoint of the base feature.

5. Move the cursor along the shorter edge of the base feature; when the **midpoint** is highlighted and the Midpoint sketch relation icon appears, click once with the left-mouse-button to select the midpoint.

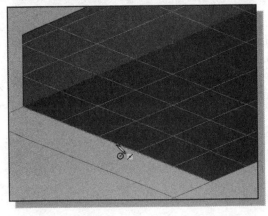

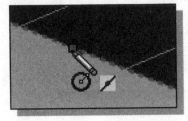

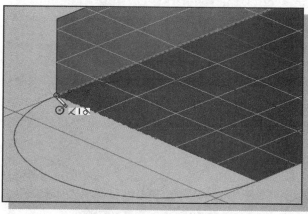

6. Move the cursor over the corner of the base feature; when the corner is highlighted, click once with the left-mouse-button to create a circle, as shown.

7. Press the [**Esc**] key once to end the Circle command.

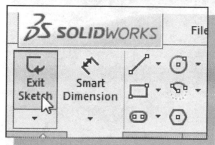

8. Click once with the **left-mouse-button** on the **Exit Sketch** icon on the *Sketch* toolbar to exit the Sketch option.

➢ Notice that the sketch region is automatically selected as the extrusion profile.

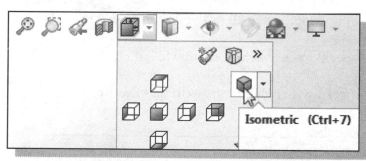

9. Select the **Isometric** icon in the *View Orientation* pull-down menu.

❖ Note that many other view-related commands are also available under the **View** *pull-down menu*.

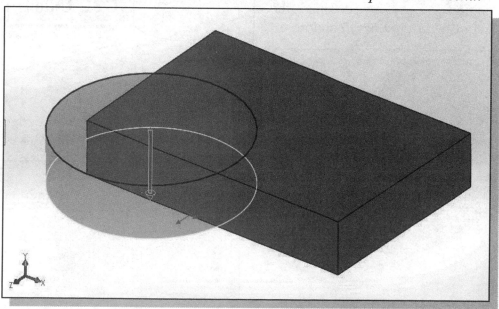

10. In the *Extrude Property Manager* panel, click the **Reverse Direction** icon to set the extrusion direction as shown below.

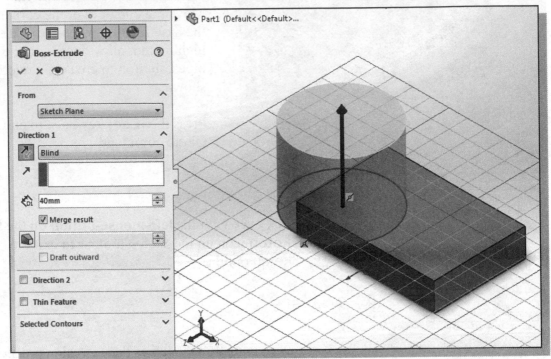

11. In the *Extrude PropertyManager* panel, enter **40** as the extrusion distance.

12. Confirm the **Merge result** checkbox is checked as shown.

13. Click on the **OK** button to proceed with creating the extruded feature.

- The two features are joined together into one solid part; the *CSG-Union* operation was performed. This is the effect of the *Merge result* option.

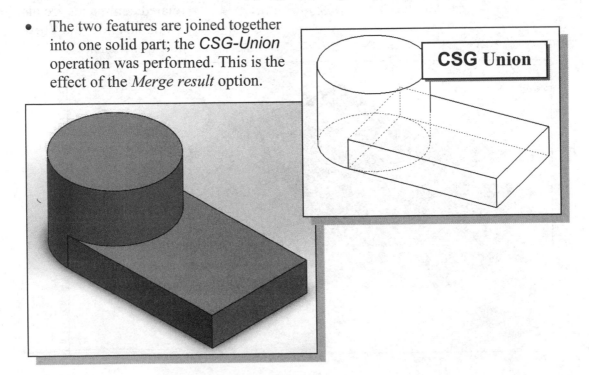

CSG Union

## Creating an Extruded Cut Feature

- We will create a circular cut as the next solid feature of the design. We will align the sketch plane to the top of the last cylinder feature.

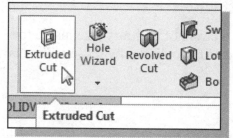

1. In the *Features* toolbar select the **Extruded Cut** command by clicking once with the left-mouse-button on the icon.

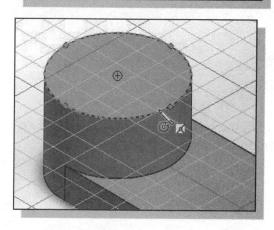

2. Pick the top face of the cylinder as shown.

    2. Pick the top face to align the sketch.

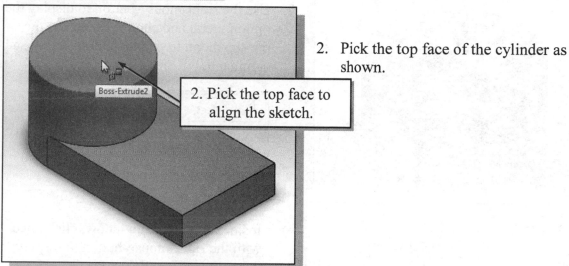

3. Select the **Circle** command by clicking once with the left-mouse-button on the icon in the *Sketch* toolbar.

4. Move the cursor over the circular edge of the top face as shown. (Do not click.) Notice the center and quadrant marks appear on the circle.

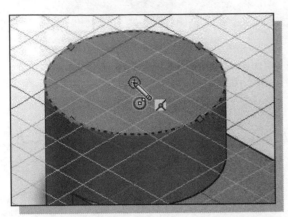

5.  Select the **Center** point of the top face of the 3D model by left-clicking once on the icon as shown.

6.  Sketch a circle of arbitrary size inside the top face of the cylinder by left-clicking as shown.

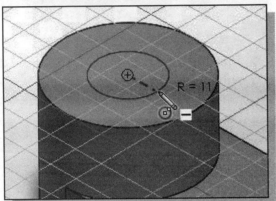

7.  Use the right-mouse-button to display the option menu and choose **Select** in the pop-up menu to end the Circle command.

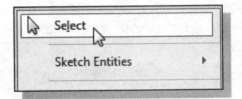

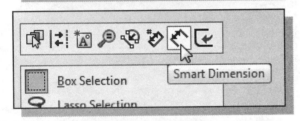

8.  Inside the graphics window, click once with the right-mouse-button to display the option menu. Select the **Smart Dimension** option in the pop-up menu.

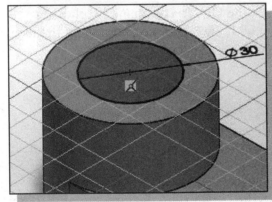

9.  Create a dimension to describe the size of the circle and set it to **30mm**.

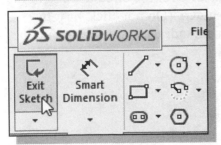

10. Click once with the **left-mouse-button** on the **Exit Sketch** icon on the *Sketch* toolbar to exit the Sketch mode.

➢ The *Cut-Extrude PropertyManager* is displayed in the left panel. Notice that the sketch region (the circle) is automatically selected as the extrusion profile.

11. In the *Cut-Extrude Property Manager* panel, click the arrow to reveal the pull-down options for the *End Condition* (the default end condition is 'Blind') and select **Through All** as shown.

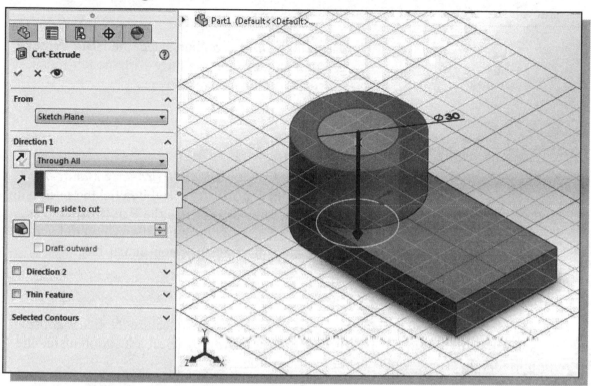

12. Click the **OK** button (green check mark) in the *Cut-Extrude PropertyManager* panel.

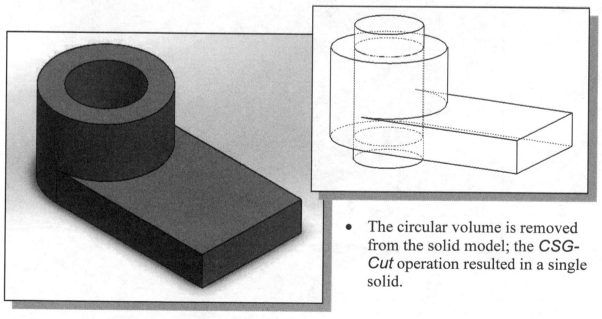

- The circular volume is removed from the solid model; the *CSG-Cut* operation resulted in a single solid.

## Creating a Hole with the *Hole Wizard*

- The last cut feature we created is a *sketched feature*, where we created a rough sketch and performed an extrusion operation. We can also create a hole using the SOLIDWORKS Hole Wizard. With the Hole Wizard, the hole feature does not need a sketch and can be created automatically. Holes, fillets, chamfers, and shells are all examples of features that do not require a sketch.

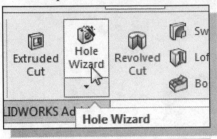

1. In the *Features* toolbar, select the **Hole Wizard** command by clicking once with the left-mouse-button on the icon as shown.

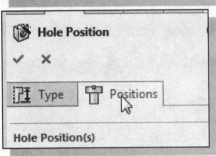

2. In the *Hole Specification Property Manager*, select the **Positions** panel by clicking once with the left-mouse-button on the **Positions** tab as shown. The Positions tab allows you to locate the hole on a planar or non-planar face.

3. Move the cursor over the horizontal surface of the base feature. Notice that the surface is highlighted. Click the **left mouse button** to select a location inside the horizontal surface as the position for the hole.

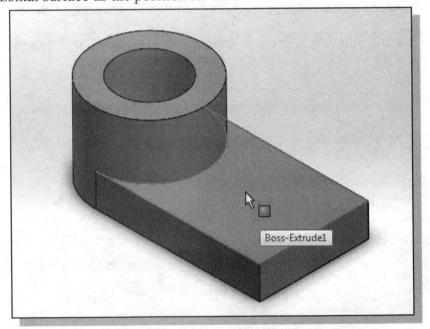

➤ Notice the *Sketch* toolbar is active and the **Point** button is now selected. The Point command has been automatically executed to allow the insert of a point to serve as the center for the hole. We will insert the point and use dimensions to locate it.

4. Move the cursor to a location on the horizontal surface of the base feature to move the center location and click the **left mouse button** to insert the point near the center of the surface.

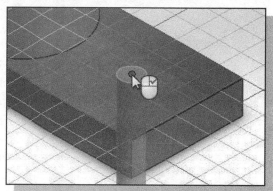

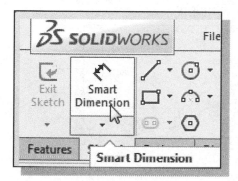

5. Select the **Smart Dimension** command by clicking once with the **left-mouse-button** on the icon in the *Sketch* toolbar.

6. Pick the center point by clicking once with the left-mouse-button as shown.

7. Pick the right-edge of the top face of the base feature by clicking once with the left-mouse-button as shown.

8. Select a location for the dimension by clicking once with the left-mouse-button as shown.

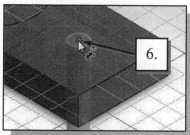
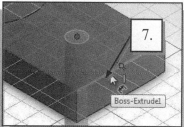
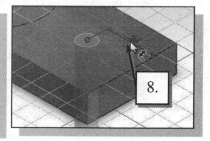

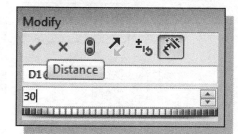

9. Enter **30** in the *Modify* dialog box, and select **OK**.

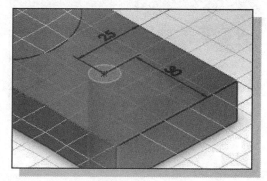

10. On your own, enter the additional dimension as shown (the dimension is **25** mm).

11. Press the [Esc] key once to end the Smart Dimension command.

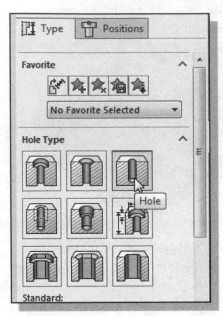

12. In *Hole Property Manager*, select the **Type** panel by clicking once with the left-mouse-button on the **Type** tab as shown.

13. Select the **Hole** icon under the *Hole Specification* option. (This is the default setting and is probably already selected.)

14. Select **ANSI Metric** and **Drill sizes** in the *Standard* option window.

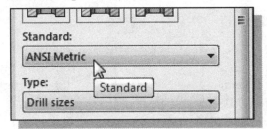

15. Set the *Size* option to a diameter of **20** mm.

16. Set the *End Condition* option to **Through All**.

17. Click the **OK** button (green check mark) in the *Hole Property Manager* to proceed with the *Hole* feature.

• The circular volume is removed from the solid model; the *CSG-Cut* operation resulted in a single solid.

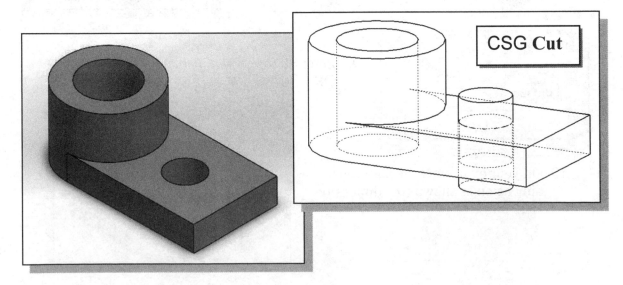

## Creating a Rectangular Extruded Cut Feature

- Next create a rectangular cut as the last solid feature of the *Locator*.

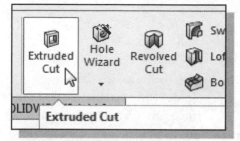

1. In the *Features* toolbar select the **Extruded Cut** command by clicking once with the left-mouse-button on the icon.

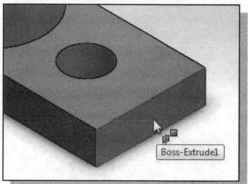

2. Pick the right face of the base feature as shown.

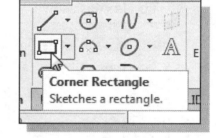

3. Select the **Corner Rectangle** command by clicking once with the left-mouse-button on the icon in the *Sketch* toolbar.

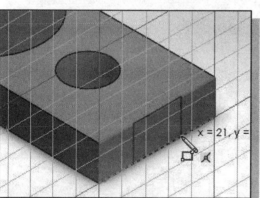

4. Create a rectangle that is aligned to the top and bottom edges of the base feature as shown.

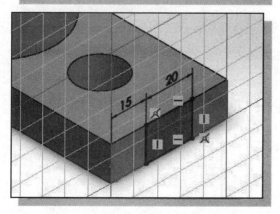

5. On your own, create and modify the two dimensions as shown. The dimensions are **15** mm and **20** mm.

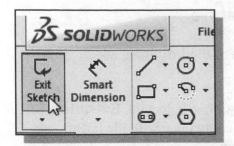

6.  Click once with the **left-mouse-button** on the **Exit Sketch** icon on the *Sketch* toolbar to exit Sketch option.

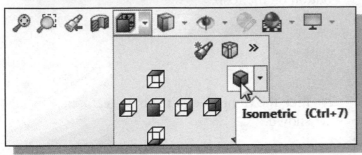

7.  Select the **Isometric** icon in the *View Orientation* pull-down menu.

❖  Note that many other view-related commands are also available under the **View** *pull-down menu.*

8.  In the *Extrude Property Manager* panel, click the arrow to reveal the pull-down options for the *End Condition* (the default end condition is Blind) and select **Up To Next** as shown.

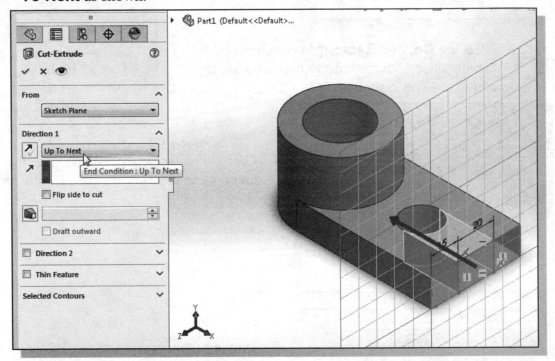

9.  Click the **OK** button (green check mark) in the *Cut-Extrude Property Manager* panel.

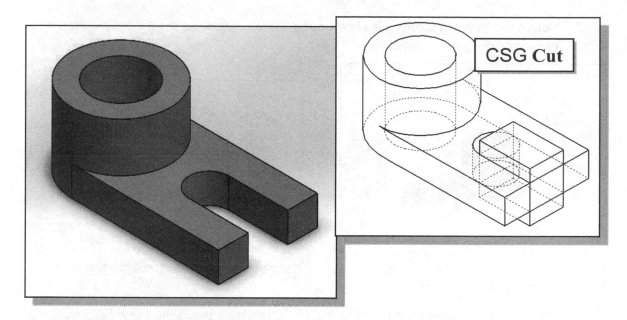

## Using the View Selector

The **View Selector** provides an in-context method to select standard and non-standard views.

1. Click on the **View Orientation** icon on the *Heads-Up View Toolbar* to reveal the view orientation options.

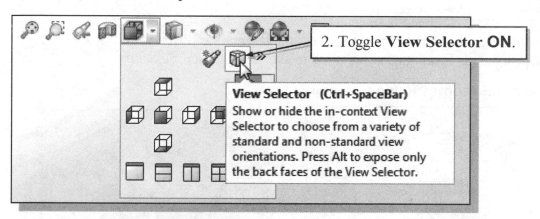

2. Toggle **View Selector ON**.

**View Selector  (Ctrl+SpaceBar)**
Show or hide the in-context View Selector to choose from a variety of standard and non-standard view orientations. Press Alt to expose only the back faces of the View Selector.

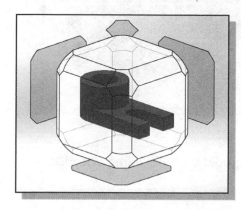

2. **Toggle** the View Selector **ON** by left-clicking on the **View Selector** icon in the *Orientation dialog box*.

➤ The View Selector appears in the graphics area. The View Selector provides an in-context method to select right, left, front, back, top, and isometric views of your model, as well as additional standard and isometric views.

3. Select the **bottom isometric view** on the *View Selector* as shown below. Note the corresponding view appears.

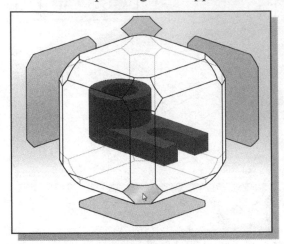 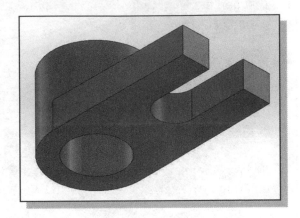

4. On your own, experiment with selecting other views using the *View Selector*. Notice that with the *View Selector* toggled **ON**, it automatically appears when the *View Orientation* option is selected on the *Heads-Up View* toolbar.

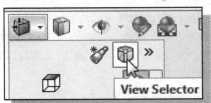

5. **Toggle** the View Selector **OFF** by left-clicking on the **View Selector** icon in the *Orientation dialog box*.

6. Hold down the **[Ctrl]** button and press the **[Spacebar]**. Notice the *View Selector* appears. This is an alternate method to activate the *View Selector*.

7. Select the **isometric view** as shown.

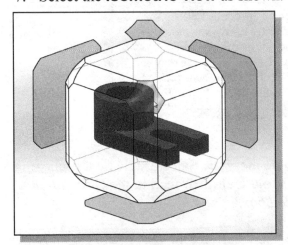 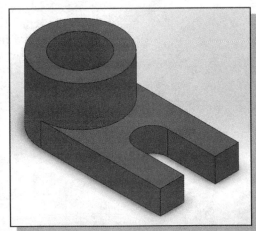

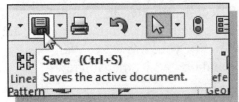

8. Save the model with the name *Locator*.

## Questions:

1. List and describe three basic *Boolean operations* commonly used in computer geometric modeling software.

2. What is a *primitive solid*?

3. What does *CSG* stand for?

4. Which *Boolean operation* keeps only the volume common to the two solid objects?

5. What is the main difference between creating an *EXTRUDED CUT feature* and creating a *HOLE feature* in SOLIDWORKS?

6. Using the CSG concepts, create *Binary Tree* sketches showing the steps you plan to use to create the two models shown on the next page:

Ex.1)

Ex.2)

## Exercises:

1. **L-Bracket** (Dimensions are in inches.)

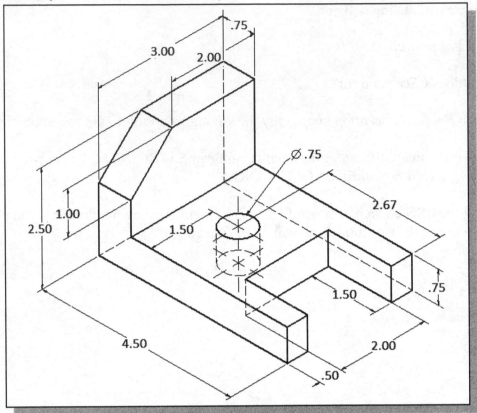

2. **Guide Plate** (Dimensions are in inches. Thickness: **0.25** inches. Boss height **0.125** inches.)

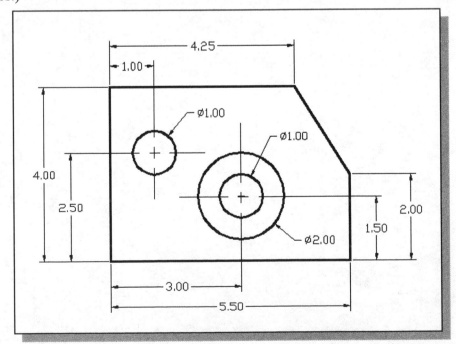

3. **Coupling Base** (Dimensions are in inches.)

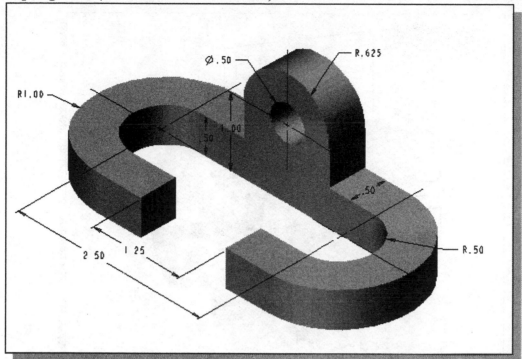

4. **Angle Slider** (Dimensions are in Millimeters.)

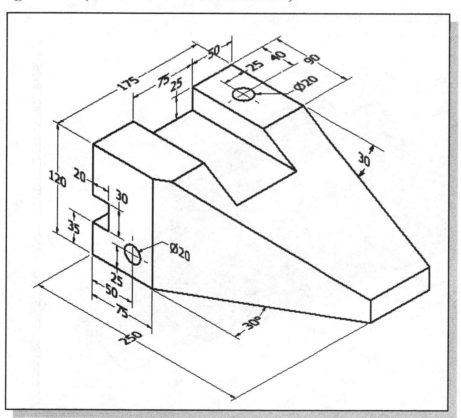

5. **Indexing Guide** (Dimensions are in inches.)

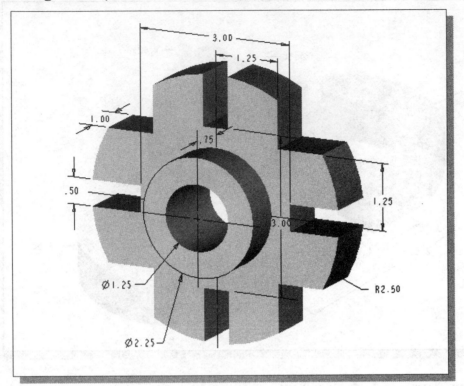

6. **Guide Block** (Dimensions are in inches.)

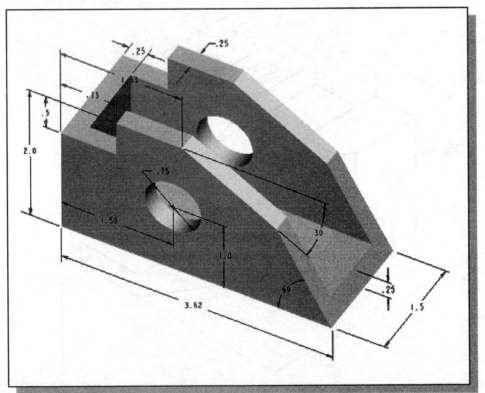

7. **Slide Base** (Dimensions are in inches.)

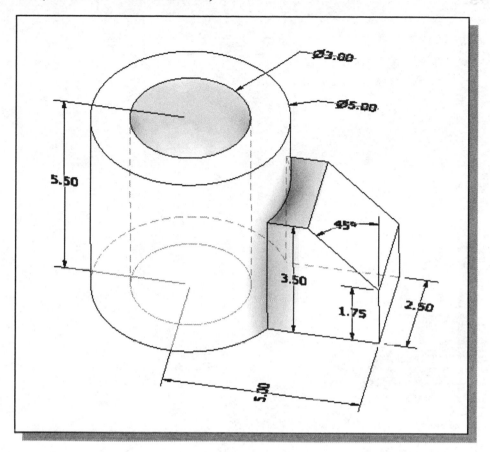

**Notes:**

# Chapter 4
# Geometric Constructions

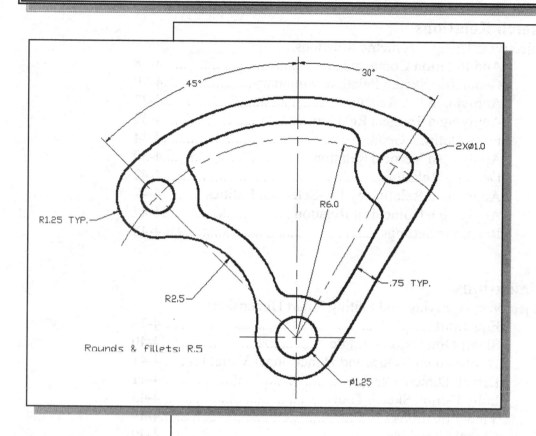

45°    30°

2XØ1.0

R6.0

R1.25 TYP.

.75 TYP.

R2.5

Rounds & fillets: R.5

Ø1.25

## Learning Objectives

♦ **Understand the Classic Geometric Construction Tools and Methods**
♦ **Create Geometric Relations**
♦ **Use Dimensional Variables**
♦ **Display, Add, and Delete Geometric Relations**
♦ **Understand and Apply Different Geometric Relations**
♦ **Display and Modify Parametric Relations**
♦ **Create Fully Defined Sketches**

| Certified **SOLIDWORKS** Associate Examination Objectives Coverage |

## Sketch Relations

Objectives: Using Geometric Relations.

Add Relation Command ...............................................4-18

Geometric Sketch Relations Summary .....................4-19

Applying a Fix Relation.............................................4-22

Applying a Tangent Relation .....................................4-32

Fully Defined Geometry ............................................4-34

Applying a Vertical Relation .....................................4-39

Deleting Relations......................................................4-42

Applying a Relation by Pre-Selecting Entities ..........4-47

Applying a Coincident Relation.................................4-47

Relations Settings......................................................4-48

## Dimensions

Objectives: Applying and Editing Smart Dimensions.

Equations..................................................................4-28

Smart Dimension – Angle..........................................4-40

Dimensional Values and Dimensional Variables ......4-41

Driven Dimensions ....................................................4-41

Fully Define Sketch Tool...........................................4-43

View Equations..........................................................4-49

Global Variables .......................................................4-49

# Geometric Constructions

The creation of designs usually involves the manipulation of geometric shapes. Traditionally, manual graphical construction uses simple hand tools like the T-square, straightedge, scales, triangles, compass, dividers, pencils, and paper. The manual drafting tools are designed specifically to assist in the construction of geometric shapes. For example, a T-square and drafting machine can be used to construct parallel and perpendicular lines very easily and quickly. Today, modern CAD systems provide designers much better control and accuracy in the construction of geometric shapes.

In technical drawings, many of the geometric shapes are constructed with specific geometric properties, such as perpendicularity, parallelism and tangency. For example, in the below drawing, quite a few **implied** geometric properties are present.

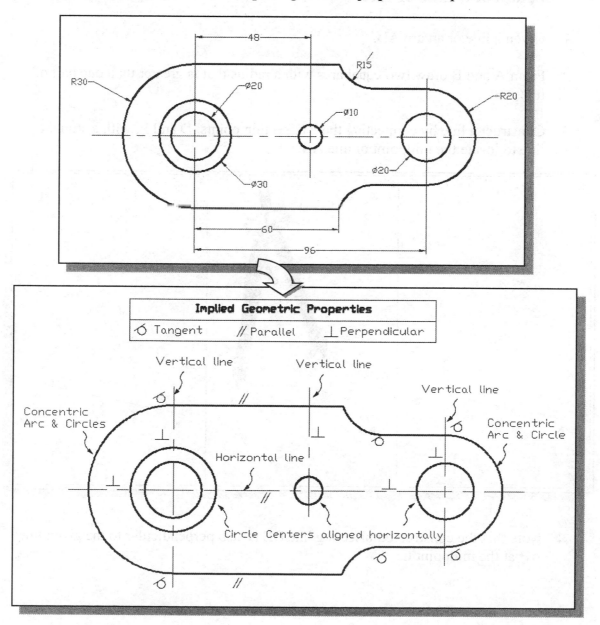

# Geometric Constructions – Classical Methods

Geometric constructions are done by applying geometric rules to graphics entities. Knowledge of the principles of geometric construction and its applications are essential to Designers, Engineers and CAD users.

For 2D drawings, it is crucial to be able to construct geometric entities at specified angles to each other, various plane figures, and other graphic representations. In this section, we will examine both the traditional graphical methods and the CAD methods of the basic geometric constructions commonly used in engineering graphics. This chapter provides information that will aid you in drawing different types of geometric constructions.

- ## Bisection of a Line or Arc

    1.  Given a line or an arc AB.

    2.  From A and B draw two equal arcs with a radius that is greater than one half of line AB.

    3.  Construct a line by connecting the intersection points, D and E, with a straight line to locate the midpoint of line AB.

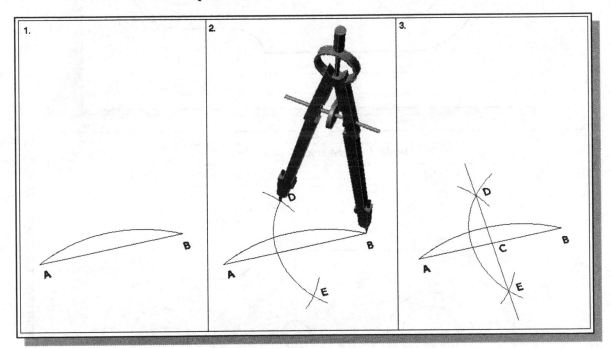

❖   Note that the constructed bisecting line DE is also perpendicular to the given line AB at the midpoint C.

- ## Bisection of an Angle

  1. Given an angle ABC.

  2. From A draw an arc with an arbitrary radius.

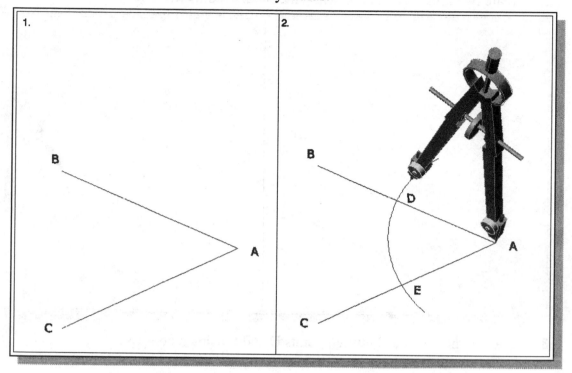

  3. Construct two equal radius arcs at D and E.

  4. Construct a straight line by connecting point A to the intersection of the two arcs.

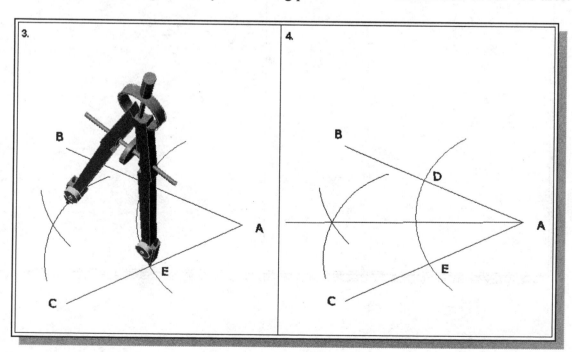

- **Transfer of an Angle**

    1. Given an angle ABC, transfer the angle to line XY.

    2. Create two arcs, at A and X, with an arbitrary radius R.

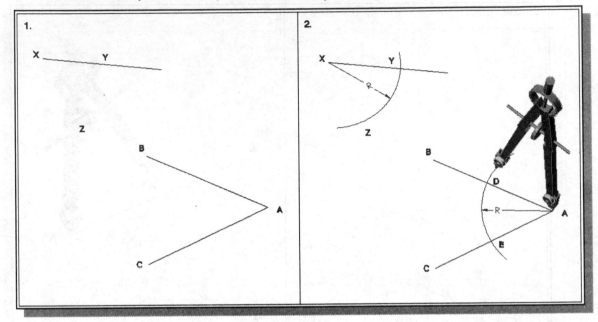

    3. Measure the distance between points D and E using a compass.

    4. Construct an arc at Y, using the distance measured in the previous step.

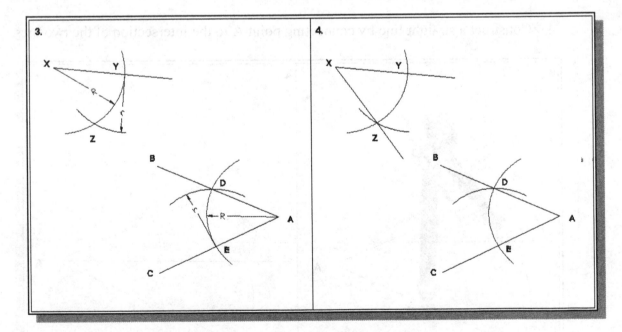

- **Dividing a Given Line into a Number of Equal Parts**

    1. Given a line AB; the line is to be divided into five equal parts.

    2. Construct another line at an arbitrary angle. Measure and mark five units along the line.

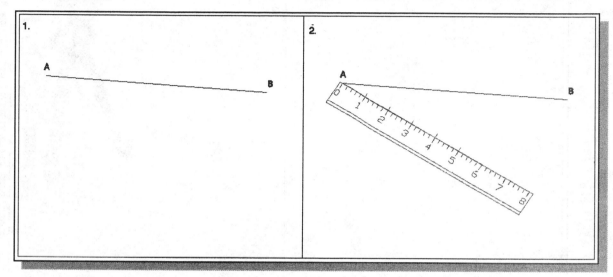

    3. Construct a line connecting the fifth mark to point B.

    4. Create four lines parallel to the constructed line through the marks.

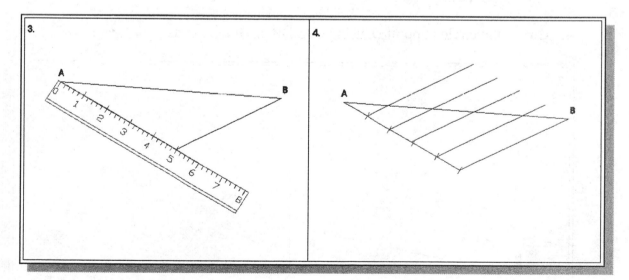

- **Circle through Three Points**

1.  Given three points A, B and C.

2.  Construct a bisecting line through line AB.

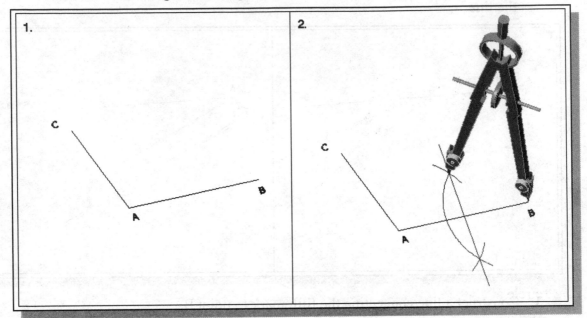

3.  Construct a second bisecting line through line AC. The two bisecting lines intersect at point D.

4.  Create the circle at point D, using DA, DB or DC as radius.

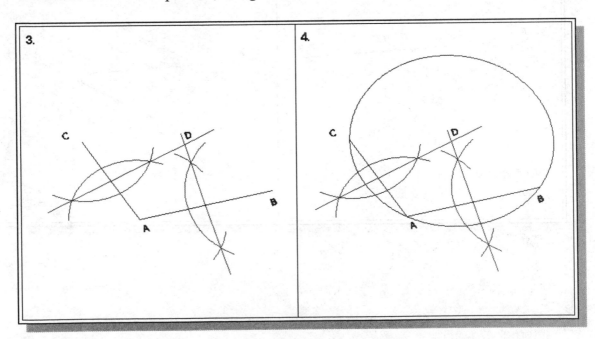

- **A Line Tangent to a Circle**

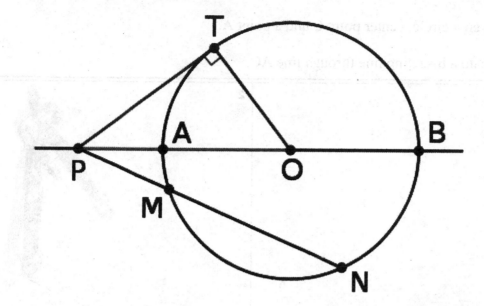

A line **tangent** to a circle intersects the circle at a single point. For comparison, **secant lines** intersect a circle at two points, whereas another line may not intersect a circle at all. In machine design, a smooth transition from surface to surface is typically desired, both for esthetic and functionality considerations. Tangency is therefore a common implied geometric property in Mechanical and Manufacturing Engineering practices.

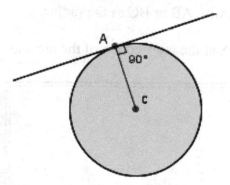

One unique property exists for the tangency between a line and a circle: *The radius of a circle is perpendicular to the tangent line through its endpoint on the circle's circumference.* Conversely, perpendicular to a radius through the same endpoint is a tangent line.

Note that tangency can also be established in between curves.

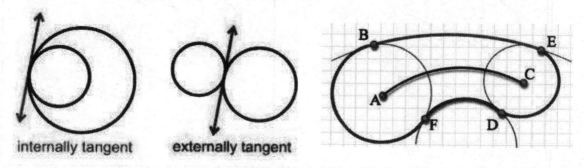

internally tangent        externally tangent

### • Line Tangent to a Circle from a Given Point

1. Given a circle, center point C and a point A.

2. Create a bisecting line through line AC.

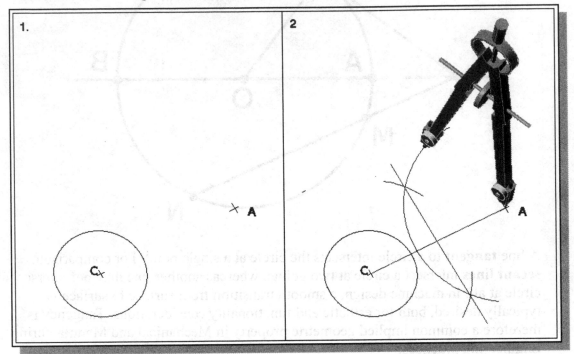

3. Create an arc at point B (midpoint on line AC), with AB or BC as the radius.

4. Construct the tangent line by connecting point A at the intersection of the arc and the circle.

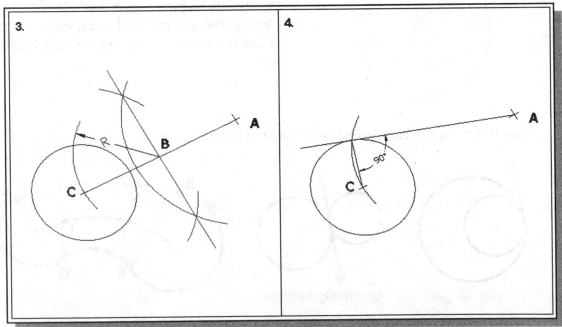

- **Circle of a Given Radius Tangent to Two Given Lines**

    1.  Given two lines and a given radius R.

    2.  Create a parallel line by creating two arcs of radius R, and draw a line tangent to the two arcs.

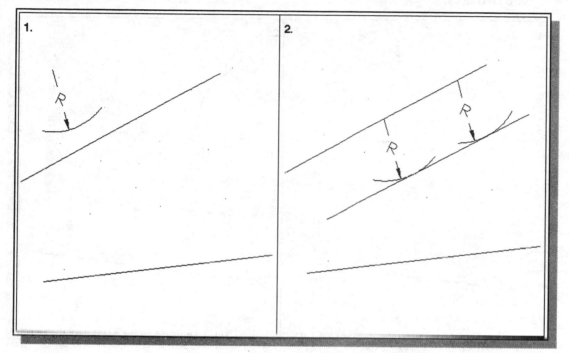

    3.  Create another line parallel to the bottom edge by first drawing two arcs.

    4.  Construct the required circle at the intersection of the two lines using the given radius R.

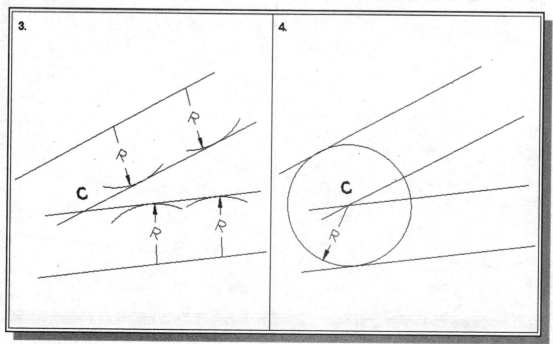

- ## **Circle of a Given Radius Tangent to an Arc and a Line**

   1.  Given a radius R, a line and an arc.

   2.  Create a parallel line by creating two arcs of radius R, and draw a line tangent to the two arcs.

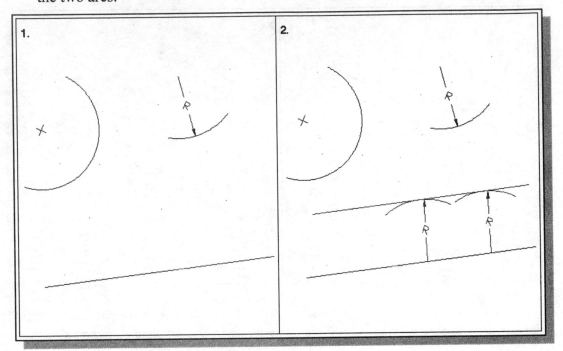

   3.  Create a concentric arc at the center of the given arc, using a radius that is r+R.

   4.  Create the desired circle at the intersection using the given radius R.

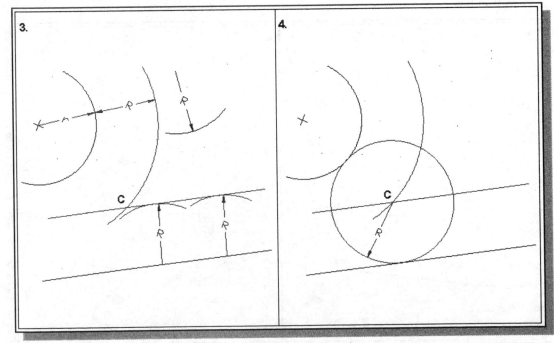

- ## Circle of a Given Radius Tangent to Two Arcs

    1. Given a radius R and two arcs.

    2. Create a concentric arc at the center of the small arc, using a radius that is R distance more than the original radius.

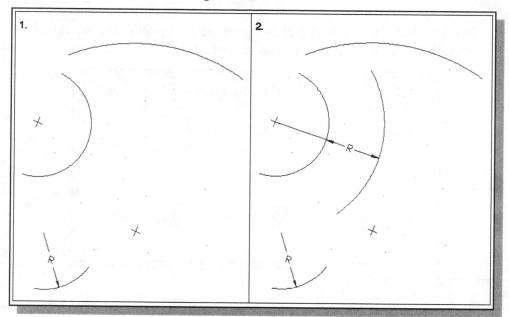

    3. Create another concentric arc at the center of the large arc, using a radius that is R distance smaller than the original radius.

    4. Create the desired circle at the intersection using the given radius R.

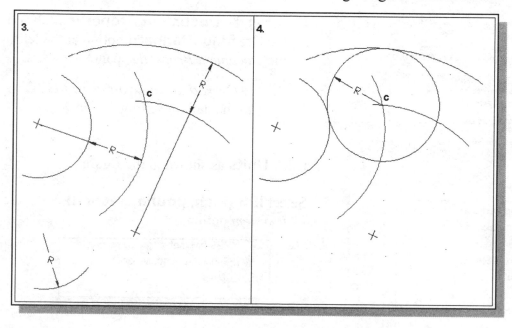

## Starting SOLIDWORKS

1.  Select the **SOLIDWORKS** option on the *Start* menu or select the **SOLIDWORKS** icon on the desktop to start SOLIDWORKS. The SOLIDWORKS main window will appear.

2.  Select **Part** to start a new part, by clicking on the first icon in the *Welcome - SOLIDWORKS Document* dialog box as shown.

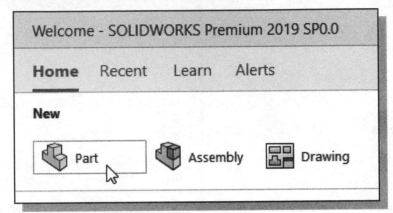

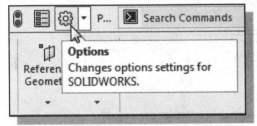

3.  Select the **Options** icon from the *Menu Bar* toolbar.

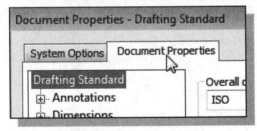

4.  Select the **Document Properties** tab and the **Drafting Standard** option at the left of the *Document Properties* panel.

5.  Set the *Overall drafting standard* to **ISO** to reset to the default settings.

6.  Click **Units** as shown in the figure.

7.  Select **IPS (inch, pound, second)** under the *Unit system* options.

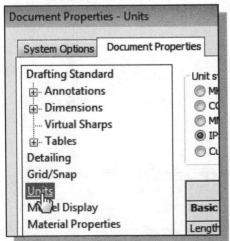

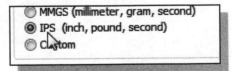

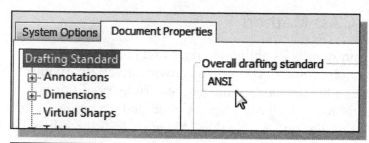

8. Select **ANSI** in the pull-down selection window under the *Overall drafting standard* panel as shown.

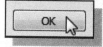

9. Click OK to accept the settings and exit the options dialog box.

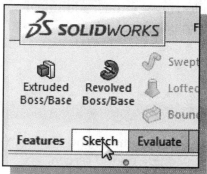

10. Select the **Sketch** tab on the *CommandManager* to display the *Sketch* toolbar.

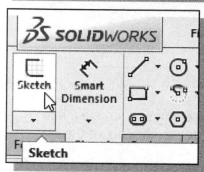

11. Select the **Sketch** button on the *Sketch* toolbar to create a new sketch.

12. Move the cursor over the edge of the **Front Plane** in the graphics area. When the Front Plane is highlighted, click once with the **left-mouse-button** to select the plane as the sketch plane for the new sketch.

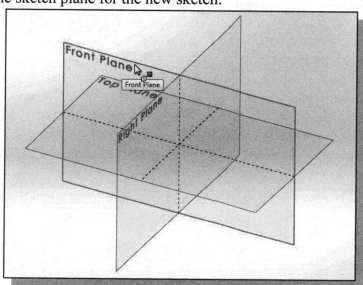

## Geometric Construction – CAD Method

The main characteristic of any CAD system is its ability to create and modify 2D/3D geometric entities quickly and accurately. Most CAD systems provide a variety of object construction and editing tools to relieve the designer of the tedious drudgery of this task, so that the designer can concentrate more on design content. A good understanding of the computer geometric construction techniques will enable the CAD users to fully utilize the capability of the CAD systems.

➢ Note that with CAD systems, besides following the classic geometric construction methods, quite a few options are also feasible. In the following sections, additional sketching/editing tools will be used to illustrate some of the classical geometric construction methods.

### • Bisection of a Line or Arc

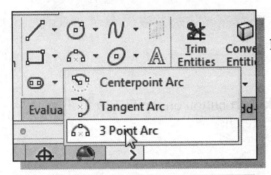

1. Use the **3 Point Arc** command and create an arbitrary arc at any angle on the screen.

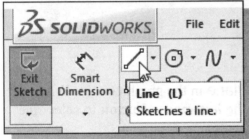

2. Using the **Line** command, create a line connecting the two endpoints of the arc.

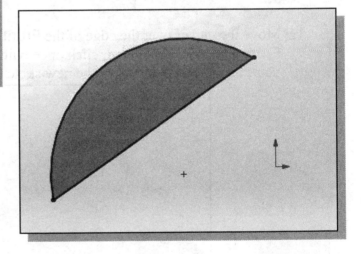

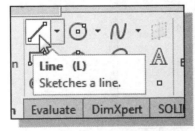

3. Activate the **Line** command again by clicking on the icon as shown.

4. Move the cursor along the arc and notice a red maker appears on the arc indicating the midpoint. Click once with the **left mouse button** to attach the first endpoint of the line at this location.

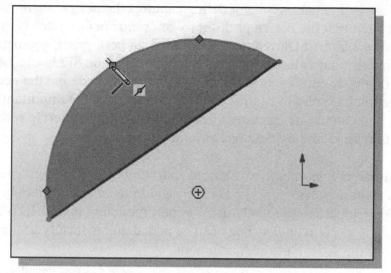

5. Move the cursor along the line and notice a red maker appears on the arc indicating the midpoint. Click once with the **left mouse button** to attach the first endpoint of the line at this location.

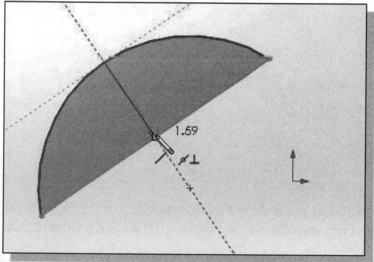

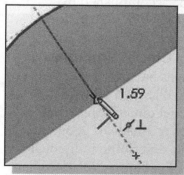

➤ Note the two small symbols next to the cursor show the geometric properties at the current location. In this case, the new line will be **perpendicular** to the existing line and the endpoint is also aligned to the **midpoint** of the existing line.

➤ The constructed bisecting line is perpendicular to the line and passes through the midpoints of both the line and arc.

## Dimensions and Relations

A primary and essential difference between parametric modeling and previous generation computer modeling is that parametric modeling captures the *design intent*. In the previous lessons, we have seen that the design philosophy of *"shape before size"* is implemented through the use of the **Smart Dimension** commands. In performing geometric constructions, dimensional values are necessary to describe the **SIZE** and **LOCATION** of constructed geometric entities. Besides using dimensions to define the geometry, we can also apply geometric rules to control geometric entities. More importantly, SOLIDWORKS can capture design intent through the use of **geometric relations**, **dimensional constraints** and **parametric relations**.

**Geometric relations** are geometric restrictions that can be applied to geometric entities; for example, *horizontal, parallel, perpendicular,* and *tangent* are commonly used *geometric relations* in parametric modeling. For part modeling in SOLIDWORKS, relations are applied to *2D sketches*. They can be added automatically as the sketch is created or by using the **Add Relation** command.

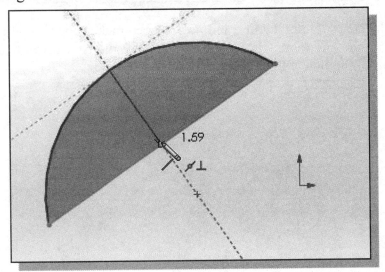

**Dimensional constraints** are used to describe the SIZE and LOCATION of individual geometric shapes. They are added using the SOLIDWORKS **Smart Dimension** command. One should also realize that depending upon the way the geometric relations and dimensional constraints are applied, the same results can be accomplished by applying different constraints to the geometric entities.

In SOLIDWORKS, **parametric relations** can be applied using **Global Variables** and **Equations**. Global variables are used when multiple dimensions have the same value. The dimension value is applied through the use of a named variable. SOLIDWORKS Equations are user-defined mathematical relations between model dimensions, using dimension names as variables. In parametric modeling, features are made of geometric entities with dimensional, geometric, and parametric constraints describing individual design intent. In this lesson, we will discuss the fundamentals of geometric relations and equations.

# Geometric Symbols available in parametric sketching

- Geometric relations in SOLIDWORKS for 2D sketches are summarized below.

| Icon | Relation | Entities Selected | Effect |
|------|----------|-------------------|--------|
| ⊥ | Perpendicular | Two lines | Causes selected lines to lie at right angles to one another. |
| ∖∖ | Parallel | Two or more lines | Causes selected lines to lie parallel to one another. |
| ♂ | Tangent | An arc, spline, or ellipse and a line or arc | Constrains two curves to be tangent to one another. |
| ⦦ | Coincident | A point and a line, arc, or ellipse | Constrains a point to a curve. |
| ╱ | Midpoint | Two lines or a point and a line | Causes a point to remain at the midpoint of a line or an arc. |
| ◎ | Concentric | Two or more arcs, or a point and an arc | Constrains selected items to the same center point. |
| ⫽ | Colinear | Two or more lines | Causes selected lines to lie along the same line. |
| — | Horizontal | One or more lines or two or more points | Causes selected items to lie parallel to the X-axis of the sketch coordinate system. |
| │ | Vertical | One or more lines or two or more points | Causes selected items to lie parallel to the Y-axis of the sketch coordinate system. |
| = | Equal | Two or more lines or two or more arcs | Constrains selected arcs/circles to the same radius or selected lines to the same length. |
| ⫿ | Fix | Any entity | Constrains selected entities to a fixed location relative to the sketch coordinate system. However, endpoints of a fixed line, arc, or elliptical segment are free to move along the underlying fixed curve. |
| ▣ | Symmetric | A centerline and two points, lines, arcs or ellipses | Causes items to remain equidistant from the centerline, on a line perpendicular to the centerline. |
| ◖ | Coradial | Two or more arcs | Causes the selected arcs to share the same centerpoint and radius. |
| ⊠ | Intersection | Two lines and one point | Causes the point to remain at the intersection of the two lines. |
| ◁ | Merge Points | Two points (sketchpoints or endpoints) | Causes the two points to be merged into a single point. |

- **Bisection of an Angle**

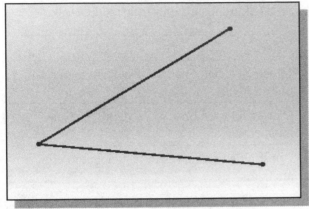

1. Using the **Line** command, create an arbitrary angle as shown in the figure.

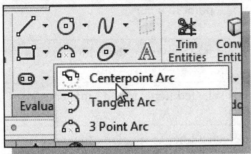

2. Activate the **Centerpoint Arc** command by clicking on the icon as shown.

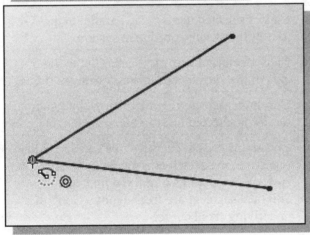

3. Create the arc by first selecting the vertex point of the angle as the center of the new arc.

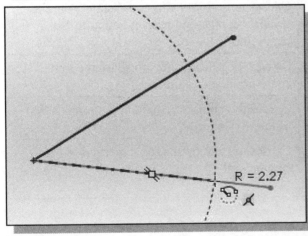

4. Move the cursor along the bottom leg of the angle and place the first endpoint on the line as shown. Confirm the other end of the arc intersects with the other leg as shown.

   ➤ What is the geometric property identified by SOLIDWORKS?

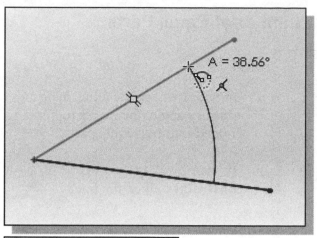

5.  Place the other endpoints of the arc on the other leg as shown.

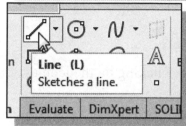

6.  Activate the **Line** command by clicking on the icon as shown.

7.  Create a bisecting line of the angle as shown.

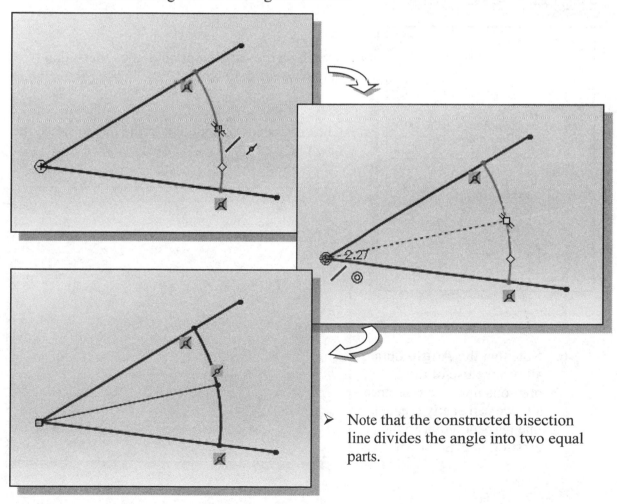

➢  Note that the constructed bisection line divides the angle into two equal parts.

- ## Dividing a Given Line into a Number of Equal Parts

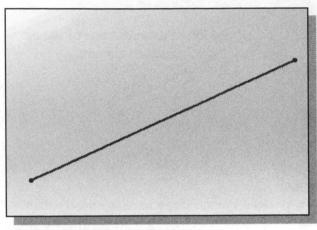

1. On your own, create a line at an arbitrary angle; the line is to be divided into five equal parts.

2. On your own, apply a **Fix constraint** to lock the line.

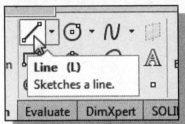

3. Activate the **Line** command by clicking on the icon as shown.

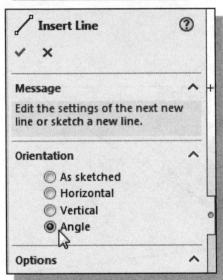

4. Switch on the **Angle** option in the Orientation options area as shown.

5. Click on the lower endpoint of the line to attach the new line.

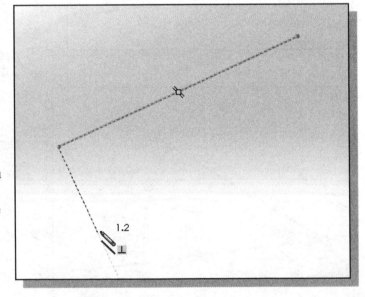

6. Note that the **Angle** option allows the use of the previous line as a reference and we can easily create a line perpendicular to the existing line as shown.

7. On your own, create four additional short line segments at arbitrary angles toward the right side of the previous line as shown. Note the last line segment on the right is a vertical line as the *vertical symbol* is displayed next to the line.

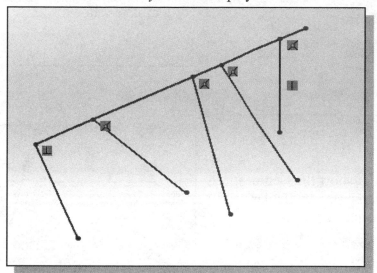

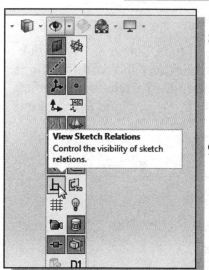

8. Select **View Sketch Relations** in the *View* toolbar as shown.

> Note the icon acts as a toggle switch; each click turns on or off the selected option.

9. On your own, experiment with the available visibility controls. Turn **On** the *View Sketch Relations* before proceeding to the next step.

10. **Right-mouse-click** once on the vertical icon to bring up the option list. Scroll down in the list and select **Delete** as shown.

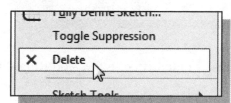

> Note that any of the applied geometric relations can be deleted.

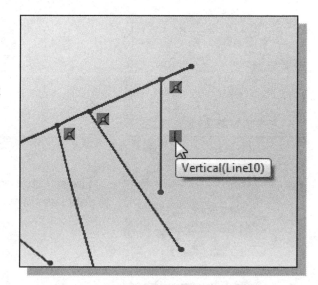

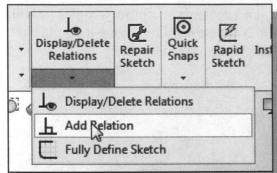

11. Select **Add Relation** in the *Sketch* toolbar as shown.

12. Select the first short line segment, the one with the perpendicular constraint, as shown.

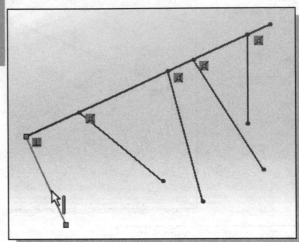

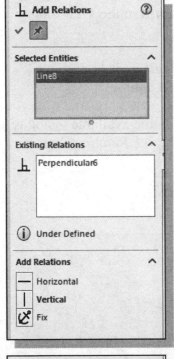

- In the *Add Relations Manager*, three panels are available: **Selected Entities**, **Existing Relations** and **Add Relations**. For the selected line, one perpendicular relation is applied, and we can also add *Horizontal*, *Vertical* and *Fix* relations to it.

13. Select the line next to the previously selected line.

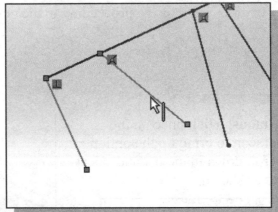

➢ The selected entities are identified in the Add Relations panel as shown.

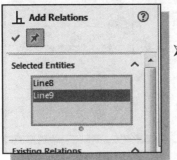

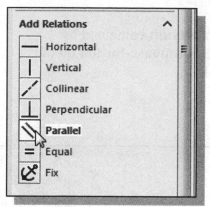

14. In the *Add Relations* panel, additional relations can be applied to the two selected lines. Click on the **Parallel** icon to add the relation to the selected lines as shown.

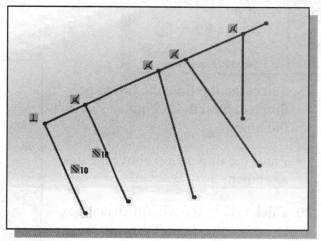

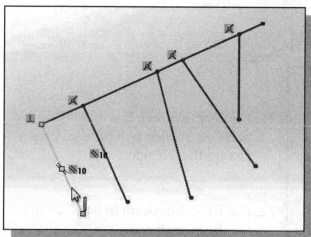

15. On your own, clear the items in the selected entities box and select the first short line segment and the other three non-parallel segments.

16. In the *Add Relations* panel, click on the **Parallel** icon to add the relation to the selected lines as shown.

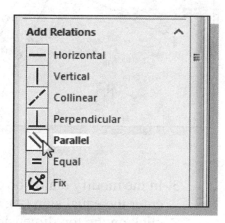

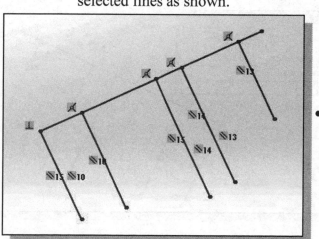

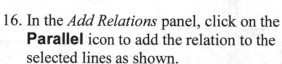

- In SOLIDWORKS, geometric relations can be applied to multiple selected entities.

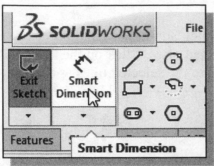

17. Select the **Smart Dimension** command by clicking once with the **left-mouse-button** on the icon in the *Sketch* toolbar.

18. Select the top line, the line to be divided, by left-clicking once on the line.

19. Place the dimension above the line as shown.

20. Click **OK** in the *Modify* dialog box to accept the default dimension. Note that the dimension on your screen may look different than what is shown in the figure.

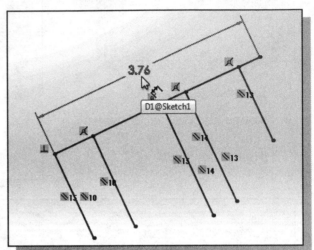

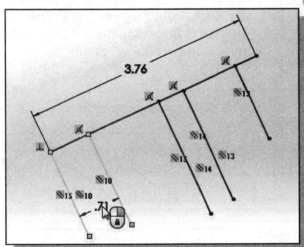

21. On your own, create a dimension between the first two parallel lines toward the left side.

22. Place the dimension in between the two lines as shown.

23. In the modify dialog box, enter the equal sign (**=**), and click on the top dimension as shown.

24. Enter **/5** (divided by 5) to create a parametric equation as shown.

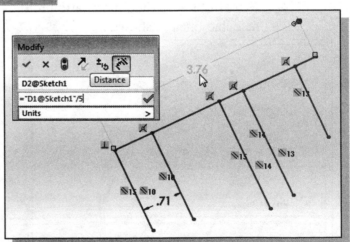

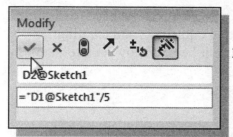

25. Click **OK** to accept the entered equation as shown.

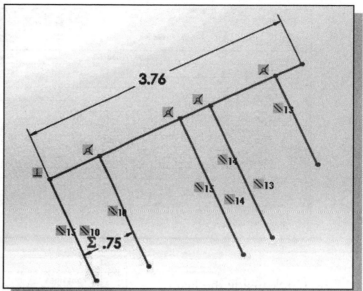

• Notice, the derived dimension value is displayed with **Σ** in front of the numbers.

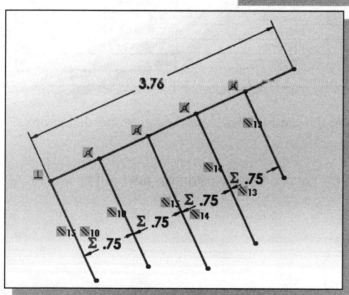

26. On your own, repeat the above steps and create the other three parametric relations as shown.

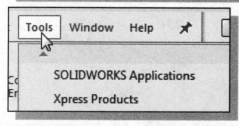

27. Move the cursor over the SOLIDWORKS icon on the *Menu Bar* to reveal the pull-down menus, click once with the left-mouse-button on the **Tools** icon and select **Equations** in the pull-down menu as shown.

• The *Equations, Global Variables, and Dimensions* dialog box is displayed. Notice the equation we created appears in the table under **'Equations'**.

- The *Equations, Global Variables, and Dimensions* dialog box can be used to perform all tasks related to equations, including editing existing equations, and creating additional design variables and equations. We will use it here to show an alternate method to add/edit an equation.

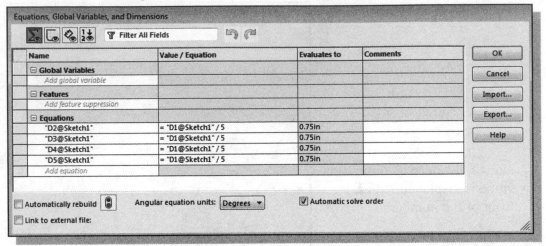

28. Click inside the last equation block and change the last equation to **/4** as shown.

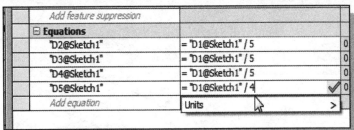

29. Click **OK** to accept the settings and close the *Equations* dialog box.

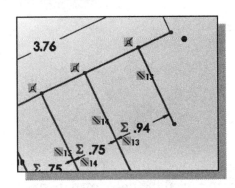

- Notice the dimension has not been updated yet.

30. On your own, double click on the modified dimension and change the equation back to **/5** and also adjust the top dimension to **3.5** to see the effect.

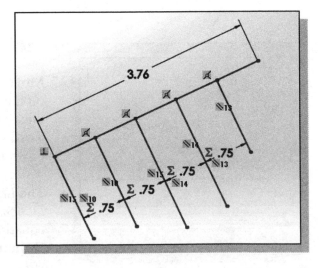

- **Circle through Three Points**

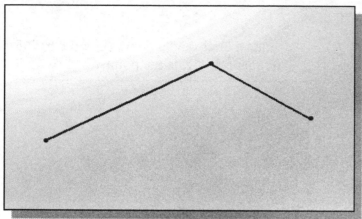

1. Create two arbitrary connected line segments as shown.

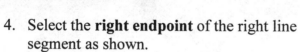

2. Select the **Perimeter Circle** command in the *Sketch* toolbar as shown. Note that the order of selecting the three points does not matter for this Circle command.

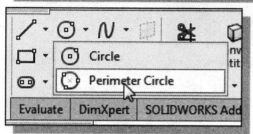

3. Select the **left endpoint** of the left line segment as shown.

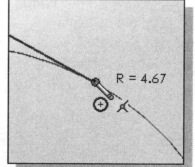

4. Select the **right endpoint** of the right line segment as shown.

5. Select the point on top to create the circle that passes through all three points.

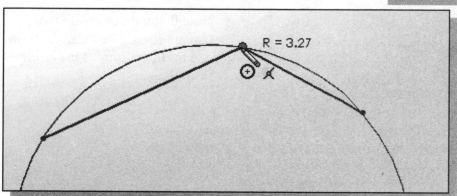

- ### Line Tangent to a Circle from a Given Point

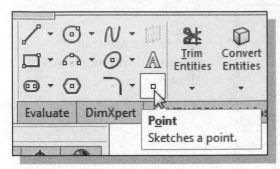

1. Create a circle and a point. (Use the Point command to create the point.)

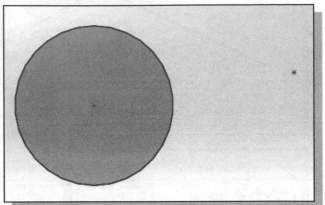

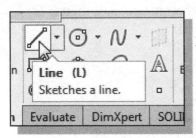

2. Activate the **Line** command in the *Sketch* toolbar as shown.

3. Select point **A** as the starting point of the new line.

4. Move the cursor on top of the circle and click on the circle when the Tangent symbol is displayed.

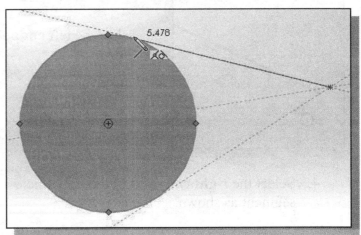

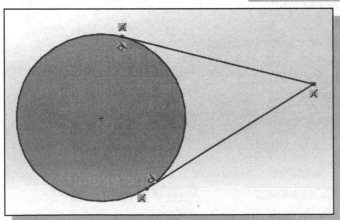

➢ Note that we can create two tangent lines, one to the top and one to the bottom of the circle, from the point.

- ## Circle of a Given Radius Tangent to two Given Lines

### Option I: Using the Add Relations option

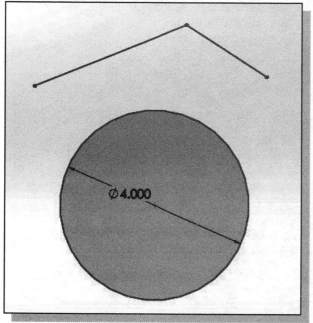

1. Create **two arbitrary line segments** as shown.

2. Create a **circle** below the two line segments as shown.

3. Use the **Smart Dimension** command and adjust the size of the circle to **4.00**.

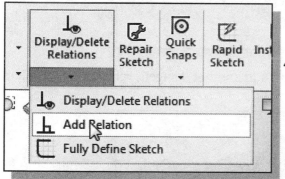

4. Select **Add Relation** in the *Sketch* toolbar as shown.

5. Select the two line segments and apply a **Fix** relation.

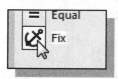

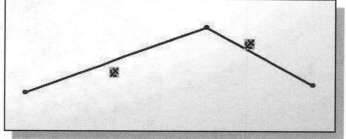

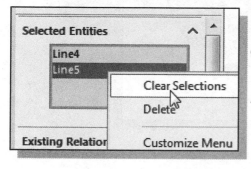

6. **Right-click** once inside the *Selected Entities* panel to bring up the option list.

7. Select **Clear Selections** to deselect all entities.

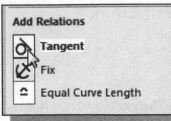

**Add Relations**

⊙ Tangent

✗ Fix

⌃ Equal Curve Length

8. Select the **line segment** on the left and the **circle**.

9. Click on the Tangent icon to apply a tangent relation between the two selected entities.

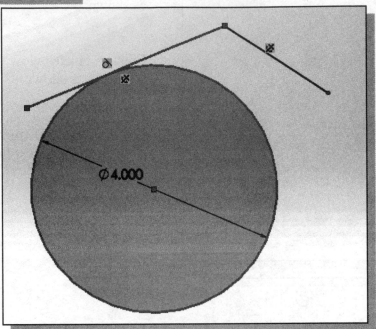

10. On your own, repeat the above steps and add another Tangent relation to complete the construction.

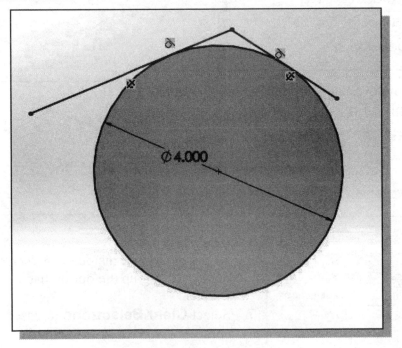

## Option II: Fillet command

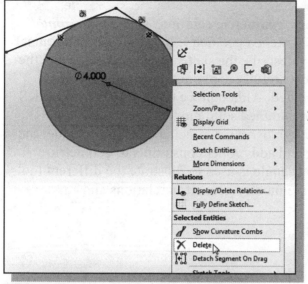

1. On your own, delete the circle by using the **Delete** option with the right-mouse-click option list as shown.

2. Click **Yes** to delete the circle.

3. Select **Sketch Fillet** in the *Sketch* toolbar as shown.

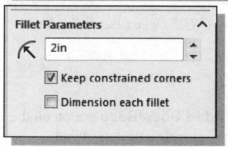

4. Enter **2** as the new radius of the **Fillet** command.

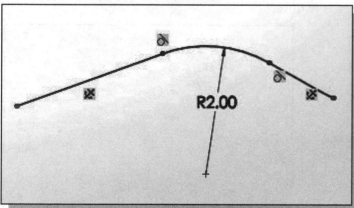

5. Select the two lines as the edges tangent to the fillet.

6. Click **OK** to create the Fillet.

➢ Note that the Fillet command automatically trims the edges as shown in the figure.

## Adding Geometric Relations and Fully Defined Geometry

In SOLIDWORKS, as we create 2D sketches, **geometric relations** such as *horizontal* and *parallel* are automatically added to the sketched geometry. In most cases, additional relations and dimensions are needed to fully describe the sketched geometry beyond the geometric relations added by the system. By carefully applying proper **geometric relations**, very intelligent models can be created. Although we can use SOLIDWORKS to build partially constrained or totally unconstrained solid models, the models may behave unpredictably as changes are made. In most cases, it is important to consider the design intent, develop a modeling strategy, and add proper constraints to geometric entities. In the following sections, a simple triangle is used to illustrate the different tools that are available in SOLIDWORKS to create/modify geometric relations and dimensional constraints.

## Starting SOLIDWORKS

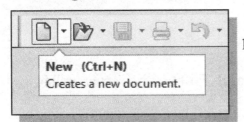

1. Select the **New** icon with a single click of the left-mouse-button on the *Menu Bar* toolbar.

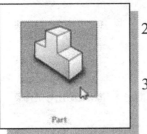

2. Select the **Part** icon with a single click of the left-mouse-button in the *New SOLIDWORKS Document* dialog box.

3. Select **OK** in the *New SOLIDWORKS Document* dialog box to open a new part document.

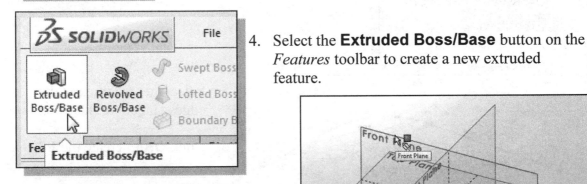

4. Select the **Extruded Boss/Base** button on the *Features* toolbar to create a new extruded feature.

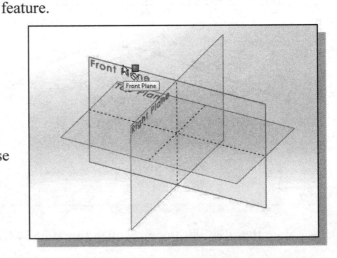

5. Click once with the **left-mouse-button** on the Front Plane to use as the sketch plane for the new sketch.

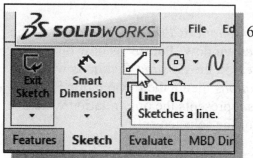

6.  Select the **Line** command by clicking once with the **left-mouse-button**.

7.  Create a triangle of arbitrary size positioned near the center of the screen as shown below. (Note that the base of the triangle is horizontal.)

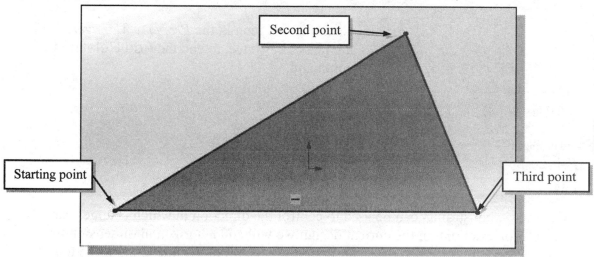

8.  Press the **[Esc]** key to exit the Line command.

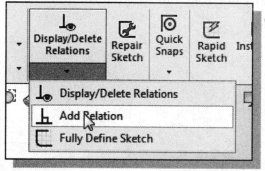

9.  **Left click** on the arrow below the **Display/Delete Relations** button on the *Sketch* toolbar to reveal additional sketch relation commands.

10. Select the **Add Relation** command from the pop-up menu.

11. Select the lower right corner of the triangle by clicking once with the **left-mouse-button**.

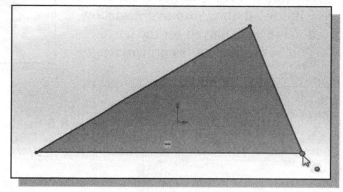

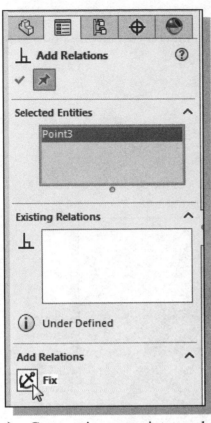

➢ Look at the *Add Relations Property Manager*. In the *Selected Entities* window '**Point3**' is now displayed. No entries appear in the *Existing Relations* window. In the *Add Relations* menu, the **Fix** relation is displayed. This represents the only relation which can be added for the selected entity.

12. Click once with the left-mouse-button on the **Fix** icon in the *Add Relations Property Manager* as shown. This activates the **Fix** relation.

13. Click the **OK** icon in the *Property Manager*, or hit the [**Esc**] key once, to end the Add Relations command.

➢ Geometric constraints can be used to control the direction in which changes can occur. For example, in the current design we will add a horizontal dimension to control the length of the horizontal line. If the length of the line is modified to a greater value, SOLIDWORKS will lengthen the line toward the left side. This is due to the fact that the Fix constraint will restrict any horizontal movement of the horizontal line toward the right side.

14. Select the horizontal line by clicking once with the **left-mouse-button**.

➢ Look at the *Line Properties Property Manager*. The **Horizontal** relation appears in the *Existing Relations* window. In the *Parameters* window, a value is displayed for the length. This is the current (as drawn) length of the line.

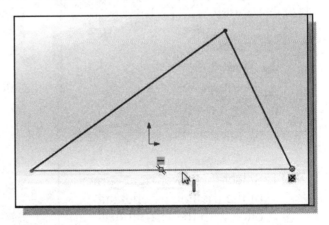

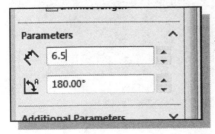

15. Enter **6.5** for the **Length** in the *Parameters* window as shown.

➤ Entering values in the *Parameters* panel of the *Property Manager* will set the length of the line to the entered value. However, it **will not define or constrain the length** to remain that value. We will demonstrate this next.

16. Move the cursor on top of the **lower left corner** of the triangle.

17. **Click and drag** the left corner of the triangle and note that the corner can be moved to a new location. Release the mouse button at a new location and notice the corner is adjusted only in the left or right direction.

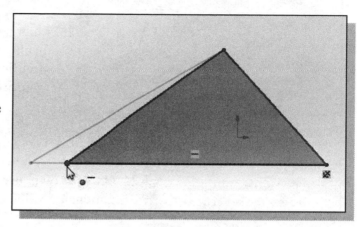

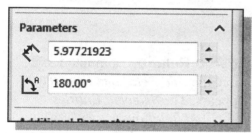

➤ Note that the two adjacent lines are automatically adjusted. On your own, confirm the Length of the horizontal line has changed to reflect the new location of the left end point.

➤ It is important to note the difference between entering a value in the *Parameters* window and the application of a constraining **Smart Dimension**. We will now add a horizontal Dimension to define the length of the bottom line.

18. Select the **Smart Dimension** command by clicking once with the **left-mouse-button** on the icon.

19. On your own, create and set the dimension to **6.5** as shown in the figure below.

20. Press the [**Esc**] key once to exit the Smart Dimension command.

21. On your own, try to drag the lower left corner of the triangle to a new location.

➤ Notice that you cannot drag the lower left corner to a new location. Its position is fully defined by the Fix relation, the Horizontal relation, and the length Dimension.

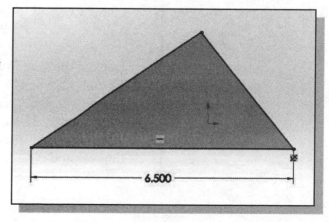

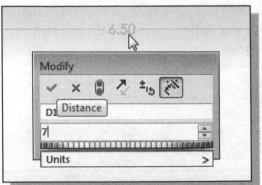

22. Double-click with the left-mouse-button on the dimension text in the graphics area to open the *Modify* dialog box.

23. Enter a value that is greater than the displayed value to observe the effects of the modification. Click the **OK** button in the *Modify* dialog box.

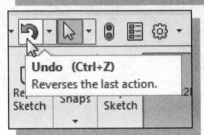

24. On your own, use the **Undo** command to reset the dimension value to the previous value.

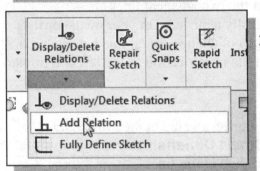

25. Activate the **Add Relation** command as shown. Notice the *Add Relations Property Manager* appears. The *Selected Entities* window in the *Add Relations PropertyManager* is blank because no entities are selected.

26. Select the inclined line on the right by clicking once with the **left-mouse-button** as shown in the figure below.

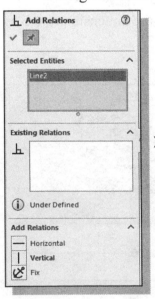

> Look at the *Add Relations PropertyManager*. In the *Selected Entities* text box '**Line2**' is now displayed. There are no relations in the *Existing Relations* text box. In the *Add Relations* panel, the **Horizontal**, **Vertical**, and **Fix** relations are displayed. These are the relations which can be added for the selected entity.

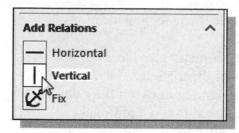

27. Click once with the left-mouse-button on the **Vertical** icon in the *Add Relations Property Manager* as shown. This activates the **Vertical** relation.

28. Click the **OK** icon in the *Property Manager*, or hit the [**Esc**] key once, to end the Add Relations command.

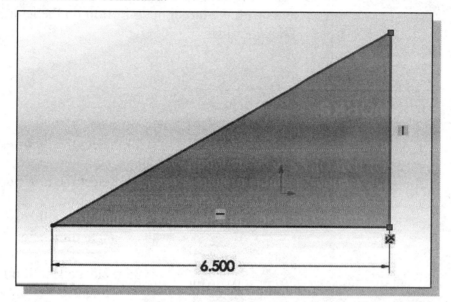

➢ One should think of the relations and dimensions as defining elements of the geometric entities. How many more relations or dimensions will be necessary to fully constrain the sketched geometry? Which relations or dimensions would you use to fully define the sketched geometry?

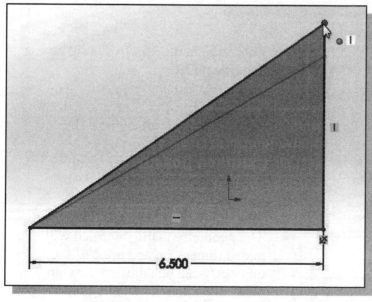

29. Move the cursor on top of the top corner of the triangle.

30. Click and drag the top corner of the triangle and note that the corner can be moved to a new location. Release the mouse button at a new location and notice the corner is adjusted only in an upward or downward direction. Note that the two adjacent lines are automatically adjusted to the new location.

31. On your own, experiment with dragging the other corners to new locations.

- The relations and dimensions that are applied to the geometry provide a full description for the location of the two lower corners of the triangle. The Vertical relation, along with the Fix relation at the lower right corner, does not fully describe the location of the top corner of the triangle. We will need to add additional information, such as the length of the vertical line or an angle dimension.

32. Select the **Smart Dimension** command by clicking once with the **left-mouse-button** on the icon in the *Sketch* toolbar.

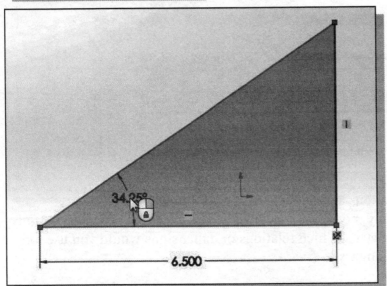

33. Select the *inclined line*.

34. Select the *horizontal line*. Selecting the two lines automatically executes an angle dimension.

35. Select a location for the dimension as shown.

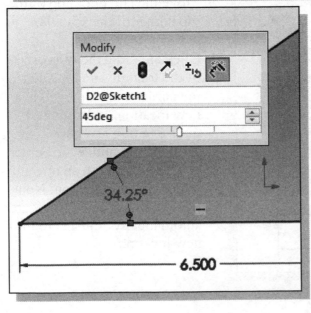

36. Enter **45°** in the *Modify* dialog box and select **OK**.

37. Press the [**Esc**] key once to exit the Smart Dimension command.

38. On your own, experiment with dragging the three corners of the triangle.

- The geometry is fully defined with the added dimension and none of the geometry can be adjusted through dragging with the cursor anymore.

## Over-Defining and Driven Dimensions

- We can use SOLIDWORKS to build partially defined solid models. In most cases, these types of models may behave unpredictably as changes are made. However, SOLIDWORKS will not let us over-define a sketch; additional dimensions can still be added to the sketch, but they are used as references only. These additional dimensions are called *driven dimensions*. *Driven dimensions* do not constrain the sketch; they only reflect the values of the dimensioned geometry. They are shaded differently (grey by default) to distinguish them from normal (parametric) dimensions. A *driven dimension* can be converted to a normal dimension only if another dimension or geometric relation is removed.

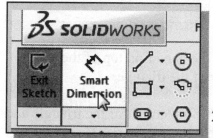

1. Select the **Smart Dimension** command in the *Sketch* toolbar.

2. Select the *vertical line*.

3. Pick a location that is to the right side of the triangle to place the dimension text.

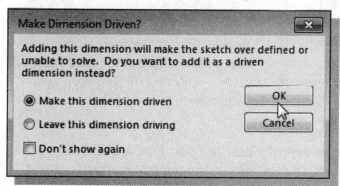

4. A *warning* dialog box appears on the screen stating that the dimension we are trying to create will over-define the sketch. Click on the **OK** button to proceed with the creation of a driven dimension.

5. Press the [**Esc**] key once to exit the Smart Dimension command.

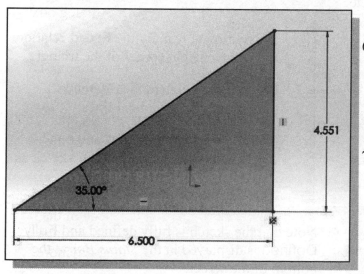

6. On your own, modify the angle dimension to **35°** and observe the changes of the 2D sketch and the driven dimension.

7. Press the [**Esc**] key to ensure that no objects are selected.

## Deleting Existing Relations

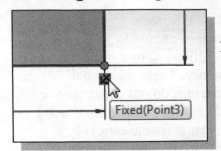

Fixed(Point3)

1. Move the cursor on top of the **Fixed** relation icon and **right-mouse-click** once to bring up the *option menu*.

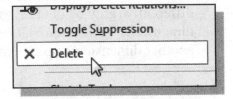

Toggle Suppression

X    Delete

2. Select **Delete** to remove the Fixed relation that is applied to the lower right corner of the triangle.

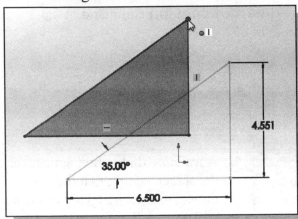

4.551

35.00°

6.500

3. Move the cursor on top of the top corner of the triangle.

4. Drag the top corner of the triangle and note that the entire triangle is free to move in all directions. Drag the corner toward the top right corner of the graphics window as shown in the figure. Release the mouse button to move the triangle to the new location.

5. On your own, experiment with dragging the other corners and/or the three line segments to new locations on the screen.

❖ **Dimensional constraints** are used to describe the SIZE and LOCATION of individual geometric shapes. **Geometric relations** are **geometric restrictions** that can be applied to geometric entities. The constraints applied to the triangle are sufficient to maintain its size and shape, but the geometry can be moved around; its location definition isn't complete.

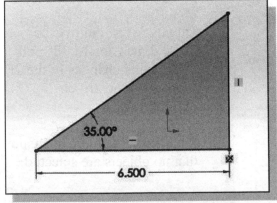

35.00°

6.500

6. On your own, reapply the **Fixed** relation to the lower right corner of the triangle.

7. On your own, **delete** the reference dimension on the vertical line.

8. Confirm the same relations and dimensions are applied on your sketch as shown.

Fully Defined    Editing Sketch1

❖ Note that the sketch is fully defined and **Fully Defined** is displayed in the *Status Bar* at the bottom of the screen.

## Using the Fully Define Sketch Tool

In SOLIDWORKS, the **Fully Define Sketch** tool can be used to calculate which dimensions and relations are required to fully define an under defined sketch. Fully defined sketches can be updated more predictably as design changes are implemented. One general procedure for applying dimensions to sketches is to use the **Smart Dimension** command to add the desired dimensions, and then use the **Fully Define Sketch** tool to fully constrain the sketch. It is also important to realize that different dimensions and geometric relations can be applied to the same sketch to accomplish a fully defined geometry.

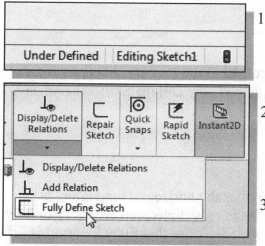

1. **Right click** once on the **6.5** linear dimension and select **Delete** in the option menu. Notice Under Defined is now displayed in the *Status Bar*.

2. **Left click** on the arrow below the **Display/Delete Relations** button on the *Sketch* toolbar to reveal additional sketch relation commands.

3. Activate the **Fully Define Sketch** command from the pop-up menu.

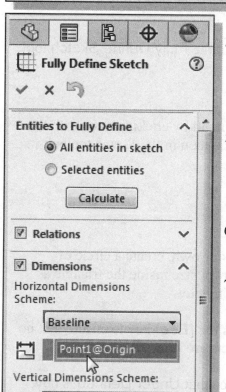

4. The *Fully Define Sketch Property Manager* appears. **If necessary**, click once with the **left-mouse-button** on the arrows to reveal the Dimensions option panel as shown.

5. Notice that under the *Horizontal Dimensions Scheme* option, Baseline is selected. The origin is selected as the default baseline datum and appears in the datum selection window as Point1@Origin.

6. Switch **On** the *Relations* and *Dimensions* options as shown.

7. Click once with the **left-mouse-button** in the *Datum* selection text box to select a different datum to serve as the baseline.

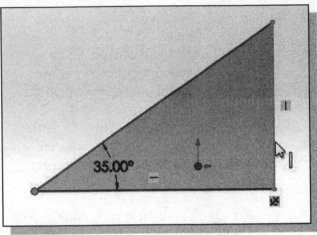

8. In the graphics window select the vertical line. Notice Line2 is now displayed in the *Datum* selection window as the *Horizontal Dimensions* baseline.

9. On your own, select the horizontal line as the appropriate baseline datum for the vertical dimensions.

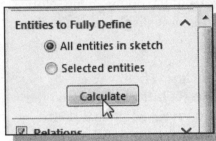

10. Select **Calculate** in the *Property Manager* to continue with the **Fully Define Sketch** command.

❖ Note that SOLIDWORKS automatically applies the horizontal dimension, using the vertical line as the baseline. Since this is the only dimension necessary to fully define the sketch, it is the only dimension added. Notice that **Fully Defined** is now displayed in the *Status Bar*.

11. Click the **OK** button to accept the results and exit the **Fully Define Sketch** tool.

## Adding Additional Geometry

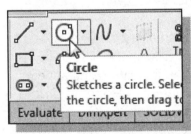

1. Select the **Circle** command by clicking once with the left-mouse-button on the icon in the *Sketch* toolbar.

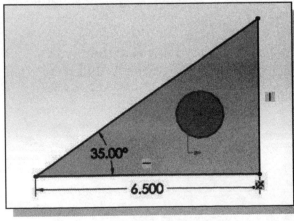

2. On your own, create a circle of arbitrary size inside the triangle as shown below.

3. Press the [**Esc**] key to ensure that no objects are selected.

❖ Notice that **Under Defined** is now again shown in the *Status Bar*.

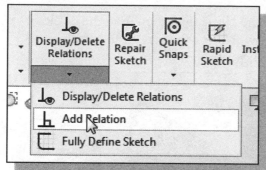

4. Select **Add Relation** from the *Sketch* toolbar.

5. Pick the circle by left-mouse-clicking once on the geometry.

6. Pick the inclined line.

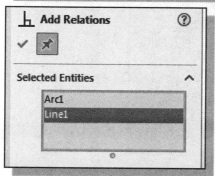

➢ Look at the *Add Relations Property Manager*. In the *Selected Entities* text box **Arc1** and **Line1** are now displayed. There are no relations in the *Existing Relations* window. In the *Add Relations* menu, the **Tangent** and **Fix** relations are displayed. These are the relations which can be added for the selected entities.

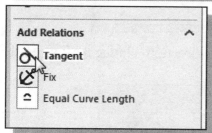

7. Click once with the left-mouse-button on the **Tangent** icon in the *Add Relations Property Manager* as shown. This activates the **Tangent** relation.

8. Click the **OK** icon in the *Property Manager*, or hit the [**Esc**] key once, to end the **Add Relations** command.

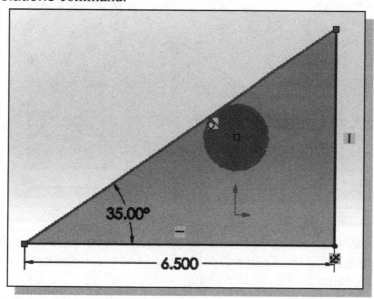

➢ How many more relations or dimensions do you think will be necessary to fully define the circle? Which relations or dimensions would you use to fully define the geometry?

9. Move the cursor on top of the right side of the circle, and then drag the circle toward the right edge of the graphics window. Notice the size of the circle is adjusted while the system maintains the Tangent relation.

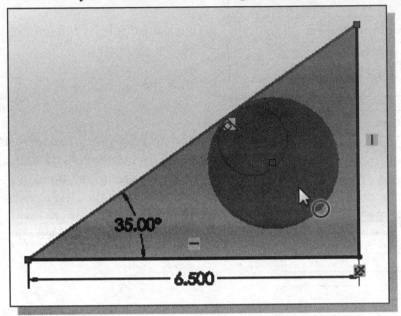

10. Drag the center of the circle toward the upper right direction. Notice the Tangent relation is always maintained by the system.

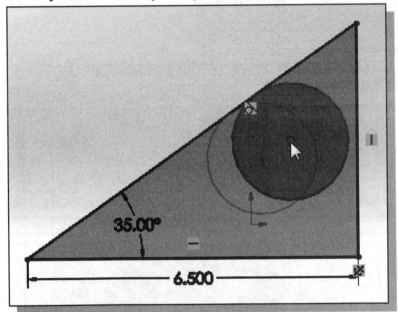

> On your own, experiment with adding additional relations and/or dimensions to fully define the sketched geometry. Use the **Undo** command to undo any changes before proceeding to the next section.

11. Press the [**Esc**] key to ensure that no objects are selected.

12. Inside the graphics window, click once with the **left-mouse-button** and select the **center of the circle**.

13. Hold down the **[Ctrl]** key and click once with the **left-mouse-button** on the **vertical line**. Holding the [Ctrl] key allows selecting the vertical line while maintaining selection of the center of the circle. This is a method to select multiple entities.

- Notice the *Properties Property Manager* for the selected entities is displayed. This provides an alternate way to control the relations for selected properties.

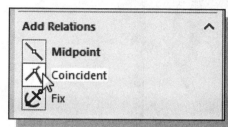

14. Click once with the left-mouse-button on the **Coincident** icon in the *Add Relations* panel in the *Property Manager* as shown. This activates the Coincident relation.

15. Click the **OK** icon in the *Property Manager*.

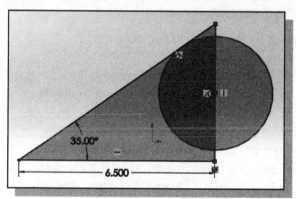

- Is the circle fully defined? (**HINT:** Drag the circle to see whether it is fully defined.)

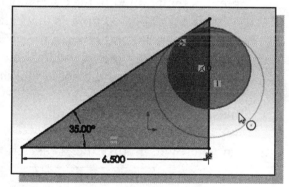

16. On your own, select the center of the circle <u>and</u> the vertical line in the graphics area. Notice the *Property Manager* appears.

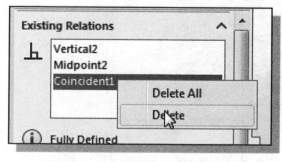

17. In the *Property Manager*, Coincident is listed under *Existing Relations*. Move the cursor over **Coincident** and click once with the **right-mouse-button**.

18. Select **Delete** in the pop-up menu by clicking once with the **left-mouse-button**.

19. Click the **OK** icon in the *Property Manager*.

20. On your own, add a **Coincident** relation between the center of the circle and the horizontal line.

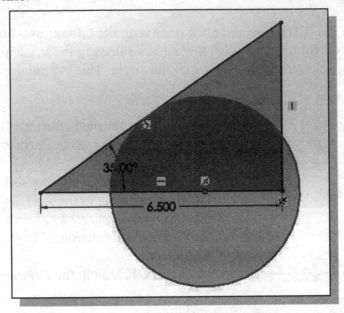

* Which relations or dimensions would you use to fully define the geometry?

  ❖ The application of different relations affects the geometry differently. The design intent is maintained in the CAD model's database and thus allows us to create very intelligent CAD models that can be modified/revised fairly easily. On your own, experiment and observe the results of applying different relations to the triangle. For example: (1) adding another **Fix** constraint to the top corner of the triangle; (2) deleting the horizontal dimension and adding another **Fix** relation to the left corner of the triangle; and (3) adding another **Tangent** relation and adding the size dimension to the circle.

21. On your own, modify the 2D sketch as shown below.

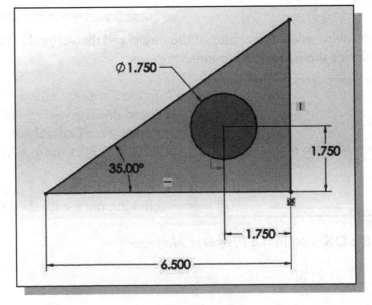

## Adding New Equations

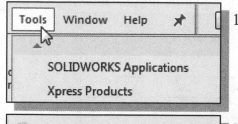

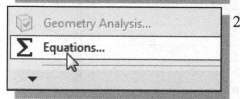

1. Move the cursor over the SOLIDWORKS icon on the *Menu Bar* to reveal the pull-down menus, click once with the left-mouse-button on the **Tools** icon and select **Equations** in the pull-down menu as shown.

2. The *Equations, Global Variables, and Dimensions* dialog box is displayed. Notice the equation we created appears in the table under **'Equations'**.

3. Switch to the **Equation View** and click the **Add equation** cell in the *Equations, Global Variables, and Dimensions* dialog box as shown. The cell becomes active, awaiting the start of a new equation.

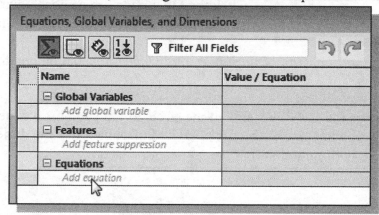

4. We will enter an equation to constrain the vertical location dimension for the circle equal to the size of the circle.

5. Move the cursor to the graphics area and click once with the **left-mouse-button** on the vertical (1.75″) location dimension. (**NOTE:** You may have to reposition the *Add Equation* window to uncover the dimension in the graphics area.) Notice the dimension name automatically appears in the new equation line of the *Equations, Global Variables, and Dimensions* dialog box.

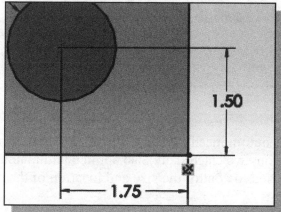

6. Click once with the **left-mouse-button** on the Size (1.75″) dimension of the circle. Notice the dimension name appears in the equation line as shown.

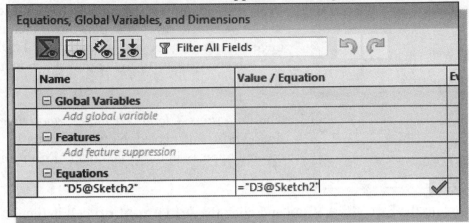

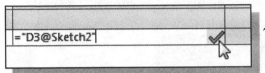

7. Click the **OK** button (green check mark) in the equation cell to accept the new equation.

8. Click **OK** in the *Equations, Global Variables, and Dimensions* dialog box to accept the equations.

9. On your own, change the size dimension of the circle to **1.5** and observe the changes to the location and size of the circle. (**HINT:** Double-click the dimension text to bring up the *Modify* window.)

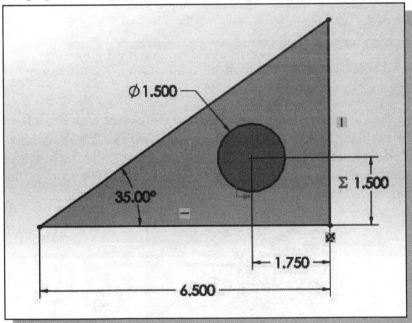

❖ SOLIDWORKS automatically adjusts the dimensions of the design, and the parametric relations we entered are also applied and maintained. The dimensional constraints are used to control the size and location of the hole.

## Questions:

1.  What is the difference between *dimensional* constraints and *geometric* relations?

2.  How can we confirm that a sketch is fully defined?

3.  How do we distinguish between derived dimensions and parametric dimensions on the screen?

4.  Describe the procedure to Display/Edit user-defined equations.

5.  List and describe three different geometric relations available in SOLIDWORKS.

6.  Does SOLIDWORKS allow us to build partially defined solid models? What are the advantages and disadvantages of building these types of models?

7.  Identify and describe the following constraints.

    (a)

    (b)

    (c)

    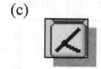

    (d)

    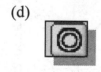

## Exercises:

(Create and establish three parametric relations for each of the following designs.)

1.  **C-Clip** (Dimensions are in inches. Plate thickness: **0.25 inches**.)

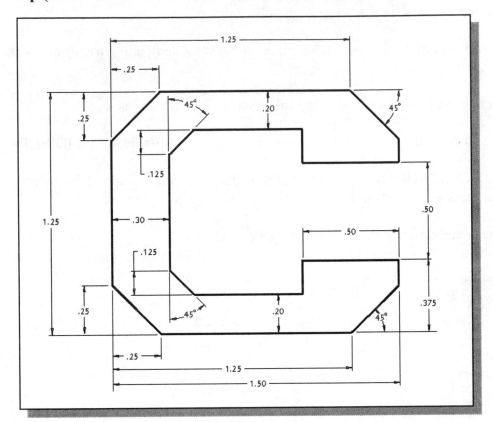

2.  **Swivel Base** (Dimensions are in millimeters. Base thickness: **10 mm**. Boss: **5 mm**.)

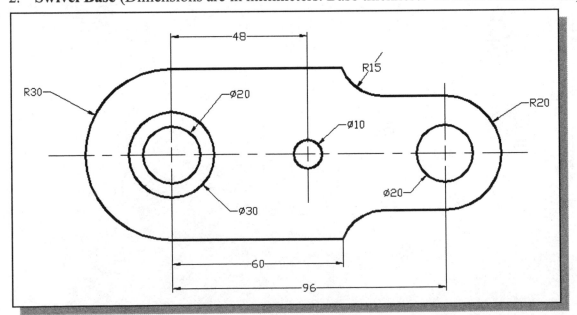

3. **Anchor Base** (Dimensions are in inches.)

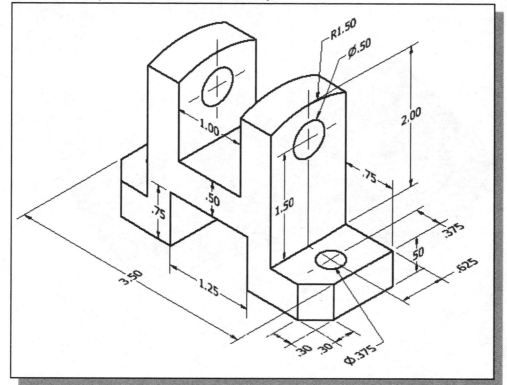

4. **Wedge Block** (Dimensions are in inches.)

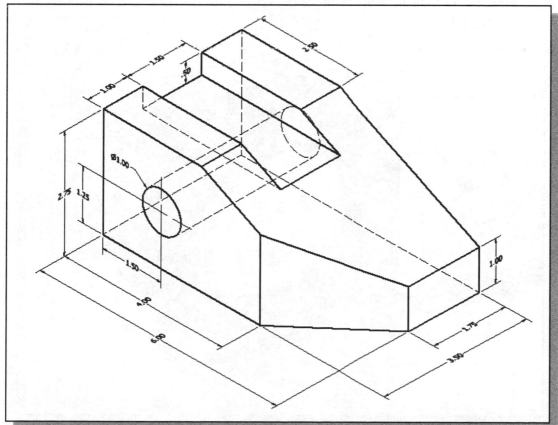

5.  **Hinge Guide** (Dimensions are in inches.)

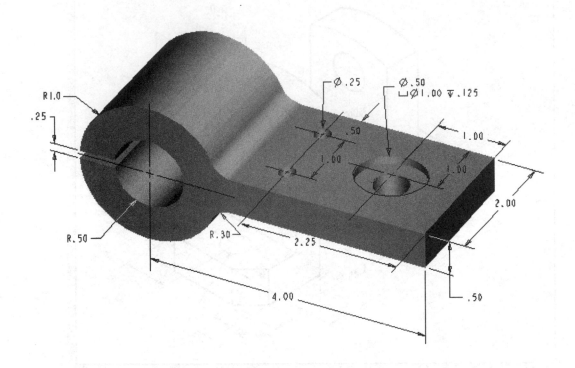

6.  **Pivot Holder** (Dimensions are in inches.)

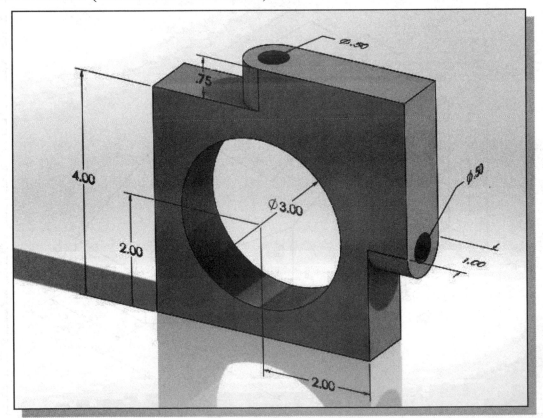

# Chapter 5
# Feature Design Tree

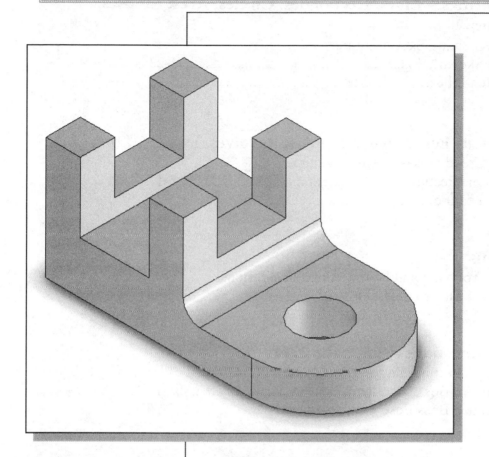

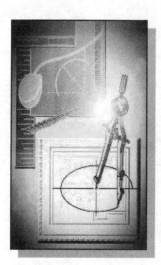

## Learning Objectives

- ◆ **Understand Feature Interactions**
- ◆ **Use the Feature Manager Design Tree**
- ◆ **Modify and Update Feature Dimensions**
- ◆ **Perform History-Based Part Modifications**
- ◆ **Change the Names of Created Features**
- ◆ **Implement Basic Design Changes**

## Certified SOLIDWORKS Associate Exam Objectives Coverage

### Sketch Entities – Lines, Rectangles, Circles, Arcs, Ellipses, Centerlines
Objectives: Creating Sketch Entities.
Edit Sketch ................................................................5-23
Sketch Fillet ..............................................................5-25


### Boss and Cut Features – Extrudes, Revolves, Sweeps, Lofts
Objectives: Creating Basic Swept Features.
Rename Feature ..........................................................5-17
Edit Feature ...............................................................5-23


### Dimensions
Objectives: Applying and Editing Smart Dimensions.
Show Feature Dimensions ..........................................5-8


### Feature Conditions – Start and End
Objectives: Controlling Feature Start and End Conditions.
Extruded Boss/Base, Mid-Plane ................................5-13
Extruded Boss/Base, Up to Surface ...........................5-16

Certified Associate Reference Guide

## Introduction

In SOLIDWORKS, the **design intents** are embedded into features in the **Feature Manager Design Tree**. The structure of the design tree resembles that of a **CSG binary tree**. A CSG binary tree contains only *Boolean relations*, while the SOLIDWORKS **design tree** contains all features, including *Boolean relations*. A design tree is a sequential record of the features used to create the part. This design tree contains the construction steps, plus the rules defining the design intent of each construction operation. In a design tree, each time a new modeling event is created previously defined features can be used to define information such as size, location, and orientation. It is therefore important to think about your modeling strategy before you start creating anything. It is important, but also difficult, to plan ahead for all possible design changes that might occur. This approach in modeling is a major difference of **FEATURE-BASED CAD SOFTWARE**, such as SOLIDWORKS, from previous generation CAD systems.

Sequential record of the construction steps

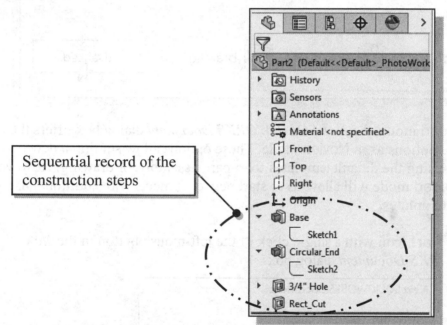

Feature-based parametric modeling is a cumulative process. Every time a new feature is added, a new result is created and the feature is also added to the design tree. The database also includes parameters of features that were used to define them. All of this happens automatically as features are created and manipulated. At this point, it is important to understand that all of this information is retained, and modifications are done based on the same input information.

In SOLIDWORKS, the design tree gives information about modeling order and other information about the feature. Part modifications can be accomplished by accessing the features in the design tree. It is therefore important to understand and utilize the feature design tree to modify designs. SOLIDWORKS remembers the history of a part, including all the rules that were used to create it, so that changes can be made to any operation that was performed to create the part. In SOLIDWORKS, to modify a feature, we access the feature by selecting the feature in the *Feature Manager Design Tree* window.

## Starting SOLIDWORKS

1. Select the **SOLIDWORKS** option on the *Start* menu or select the **SOLIDWORKS** icon on the desktop to start *SOLIDWORKS*. The *SOLIDWORKS* main window will appear on the screen.

2. In the *Welcome to SOLIDWORKS* dialog box, enter the **Advanced** mode; click once with the left-mouse-button on the **Advanced** icon.

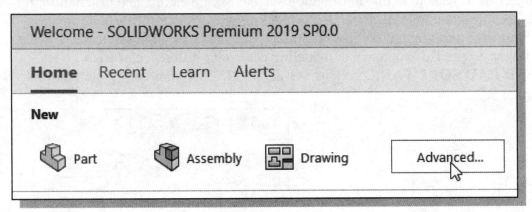

- In Advanced mode, the *New SOLIDWORKS Document* dialog box offers the same three options as in **Novice** mode. These options allow starting a new document using the default templates for a part, assembly, or drawing. However, the Advanced mode will allow us to start new documents with additional user-definable templates.

3. Select the **Part** icon with a single click of the left-mouse-button in the *New SOLIDWORKS Document* dialog box.

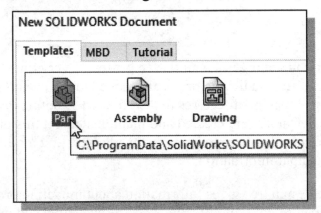

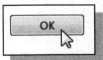

4. Select **OK** in the *New SOLIDWORKS Document* dialog box to open a new part document.

## Creating a User-Defined Part Template

❖ We will create a part template using ANSI standards for dimensions and English (inch, pound, second) units. In the future, using this template will eliminate the need to adjust these document settings each time a new part is started.

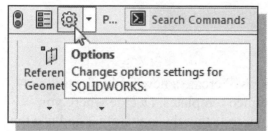

1. Select the **Options** icon from the *Menu* toolbar to open the *Options* dialog box.

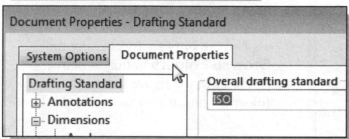

2. Select the **Document Properties** tab as shown in the figure.

3. Set the *Overall drafting standard* to **ISO** to reset the default setting.

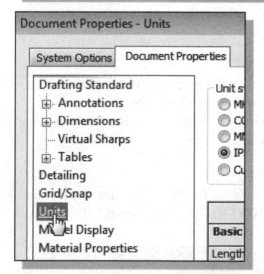

4. Click **Units** as shown in the figure.

5. Select **IPS (inch, pound, second)** under the *Unit system* options.

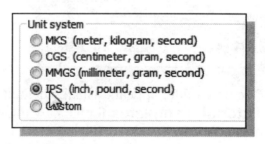

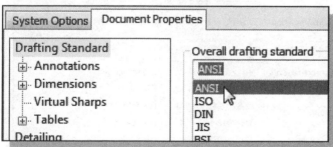

6. Set the *Overall drafting standard* to **ANSI** as shown. (This step will set all the drafting settings to the ANSI standard.)

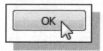

7. Click **OK** in the *Options* dialog box to accept the selected settings.

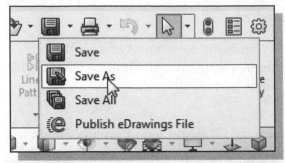

8. Click the arrow next to the **Save** icon in the *Menu Bar* to reveal the save options and select **Save As**.

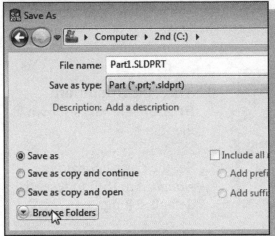

9. We will create a new folder for the user-defined templates. Decide where you want to locate this new folder and use the browser in the *Save As* dialog box to select the location. It may be necessary to expand the browser by clicking the **Browse Folders** button as shown. (**NOTE:** In the figure below, the **C:** folder is chosen.)

10. In the *Save As* dialog box, select the **New Folder** option by clicking once with the left-mouse-button on the icon as shown.

11. The new folder appears with the default name *New Folder*. Type the folder name **Tutorial_Templates** for the new folder. (**NOTE:** This folder could also be created using Windows Explorer, etc.)

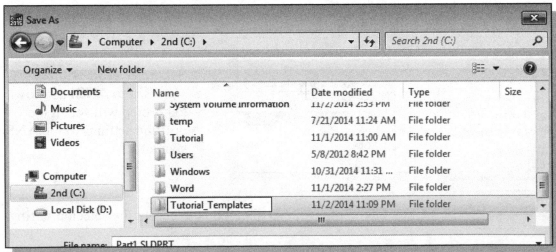

12. Left click on the **Hide Folders** button to hide the browser.

13. Under *Save as type*, select **Part Templates (\*.prtdot)**. Notice the browser automatically goes to the default *templates* folder.

14. Switch to the **Tutorial_Templates** folder you just created.

15. Enter the *File name* **Part_IPS_ANSI**.

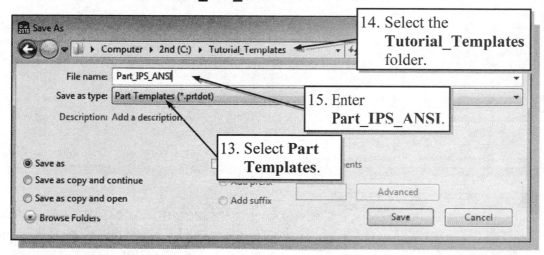

16. Click **Save** to save the new part template file.

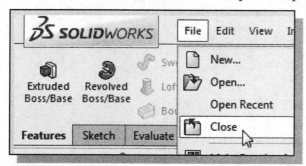

17. Select **Close** in the *File* pull-down menu, or use the key combination **Ctrl+W**, to close the document.

  ➤ We will now open a *new part document* using the template we just saved.

18. Select the **New** icon with a single click of the left-mouse-button on the *Menu Bar*.

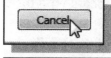

19. Notice that the new template does not appear as an option. Click **Cancel** in the *New SOLIDWORKS Document* dialog box.

20. Select the **Options** icon from the *Menu Bar* to open the *Options* dialog box.

21. Select **File Locations** under the *System Options* tab as shown.

22. Make sure **Document Templates** is selected as the *Show folders for:* option.

23. Click the **Add** button to add the directory with the user-defined templates to the list of folders containing document templates.

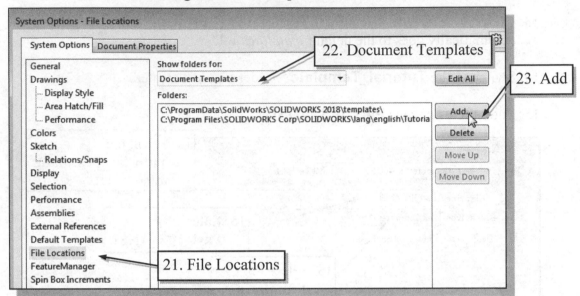

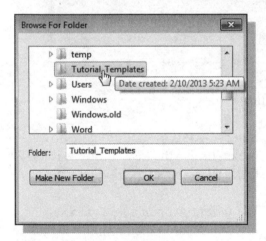

24. Locate and select the **Tutorial_Templates** folder using the browser, and click **OK** in the *Browse For Folder* dialog box.

25. Select **OK** in the *System Options* dialog box.

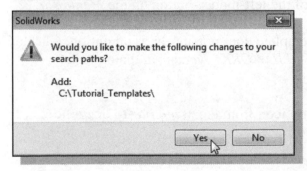

26. If a pop-up window appears with the question "*Would you like to make the following changes to your search paths?*" click **Yes**.

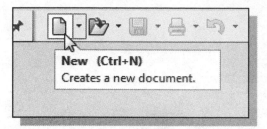

27. Select the **New** icon with a single click of the left-mouse-button on the *Menu Bar*. Notice the Tutorial_Templates folder now appears as a tab in the *New SOLIDWORKS Document* dialog box.

28. Select the **Tutorial_Templates** tab.

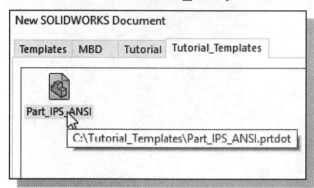

29. Notice the Part_IPS_ANSI template appears. Select the **Part_IPS_ANSI** template as shown.

30. Click on the **OK** button to open a new document.

## The *Saddle Bracket* Design

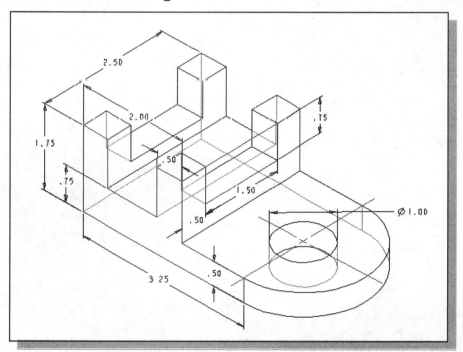

❖ Based on your knowledge of SOLIDWORKS so far, how many features would you use to create the design? Which feature would you choose as the **BASE FEATURE**, the first solid feature, of the model? What is your choice in arranging the order of the features? Would you organize the features differently if additional fillets were to be added in the design? Take a few minutes to consider these questions and do preliminary planning by sketching on a piece of paper. You are also encouraged to create the model on your own prior to following through the tutorial.

## Modeling Strategy

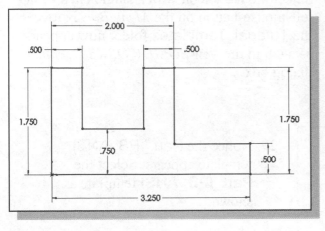

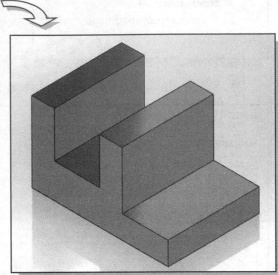

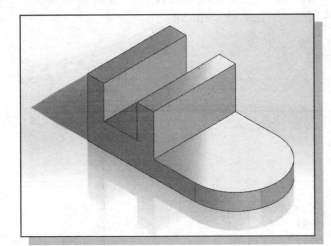

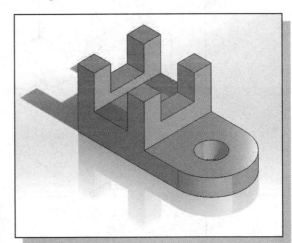

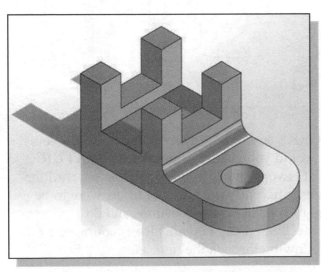

## The SOLIDWORKS Feature Manager Design Tree

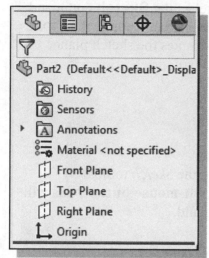

- In the SOLIDWORKS screen layout, the **Feature Manager Design Tree** is located to the left of the graphics window. SOLIDWORKS can be used for part modeling, assembly modeling, part drawings, and assembly presentation. The *Feature Manager Design Tree* window provides a visual structure of the features, relations, and attributes that are used to create the part, assembly, or scene. The *Feature Manager Design Tree* also provides right-click menu access for tasks associated specifically with the part or feature, and it is the primary focus for executing many of the SOLIDWORKS commands.

➢ The first item displayed in the *Feature Manager Design Tree* is the name of the part, which is also the filename. By default, the name "*PartX*" is used when we first start SOLIDWORKS. The parts are then numbered sequentially as new parts are created. If the part is saved with a new filename, this name will replace "*PartX*."

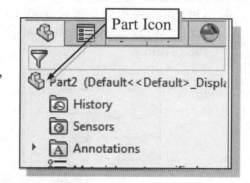

➢ The *Feature Manager Design Tree* can also be used to modify parts and assemblies by moving, deleting, or renaming items within the hierarchy. Any changes made in the *Feature Manager Design Tree* directly affect the part or assembly and the results of the modifications are displayed on screen instantly. The *Feature Manager Design Tree* also reports any problems and conflicts during the modification and updating procedure.

## Creating the Base Feature

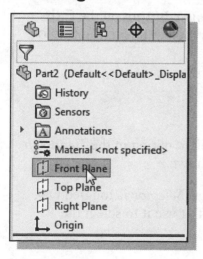

1. Move the graphics cursor to the **Front Plane** icon in the *Feature Manager Design Tree*. Notice the *Front Plane* icon is highlighted in the design tree and the *Front Plane* outline appears in the graphics area. Click once with the **left-mouse-button** to select the Front Plane.

- Note that with SOLIDWORKS, we can also create the sketch of a feature first then use the feature command to select the sketch and create the solid feature.

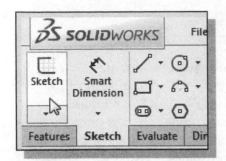

2. In the *Sketch* toolbar, select the **Sketch** command by left-clicking once on the icon. Notice the Front Plane automatically becomes the sketch plane because it was pre-selected.

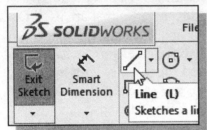

3. Select the **Line** icon on the *Sketch* toolbar by clicking once with the **left-mouse-button**; this will activate the Line command.

4. On your own, create and adjust the geometry by adding and modifying dimensions as shown below. (Note the lower left corner of the sketch is aligned to the origin.)

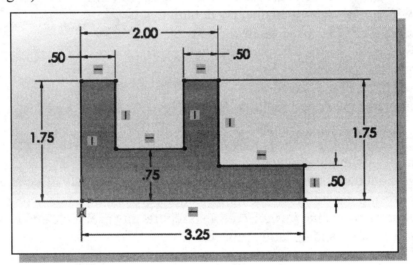

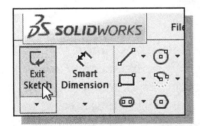

5. Click once with the **left-mouse-button** on the **Exit Sketch** icon on the *Sketch* toolbar to exit the Sketch option.

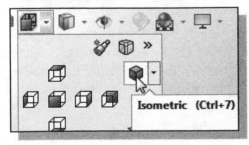

6. Select the **Isometric** option in the *Heads-up View* toolbar as shown.

➤ NOTE: If the *View Selector* is toggled ON, either toggle it off or use it to select the *Isometric* view.

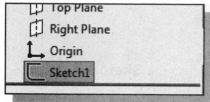

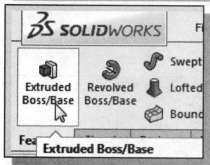

7. Make sure the sketch – Sketch1 – is selected in the *Feature Manager Design Tree*.

8. In the *Features* toolbar, select the **Extruded Boss/Base** command by clicking once with the left-mouse-button on the icon.

9. In the *Extrude Property Manager* panel, click the drop-down arrow to reveal the options for the *End Condition* (the default end condition is Blind) and select **Mid Plane** as shown.

10. In the *Extrude Property Manager* panel, enter **2.5** as the extrusion distance. Notice that the sketch region is automatically selected as the extrusion profile.

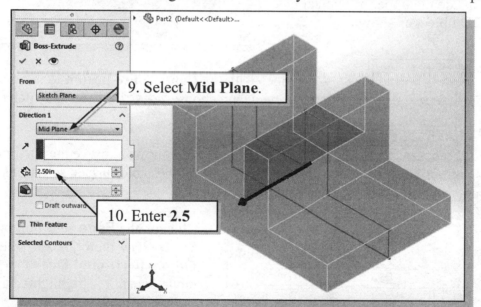

9. Select **Mid Plane**.

10. Enter **2.5**

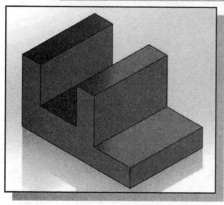

11. Click on the **OK** button to accept the settings and create the base feature.

➤ On your own, use the *Viewing* functions to view the 3D model. Also, notice the **Boss-Extrude** feature is added to the *Model Tree* in the *Feature Manager Design Tree* area.

## Adding the Second Solid Feature

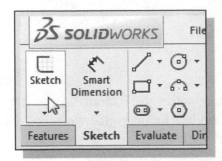

1. In the *Sketch* toolbar, select the **Sketch** command by left-clicking once on the icon.

2. Notice the left panel displays the *Edit Sketch Property Manager* with the instruction "*Select a plane on which to create a sketch for the entity.*" Move the graphics cursor on the 3D part and notice that SOLIDWORKS will automatically highlight feasible planes and surfaces as the cursor is on top of the different surfaces. Move the cursor inside the upper horizontal face of the 3D object as shown.

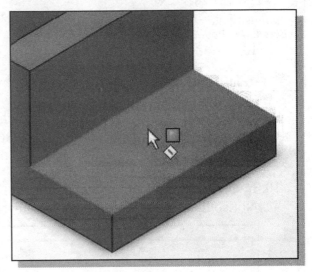

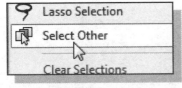

3. Click once with the **right-mouse-button** to bring up the option menu and choose **Select Other** to switch to the next feasible choice.

4. The *Select Other* pop-up dialog box appears. On your own, move the cursor over the options (e.g., Face) in the box to examine all possible surface selections.

Accept the bottom surface to align the sketching plane.

5. Click on the **Face** selection in the *Select Other* dialog box to select the **bottom horizontal face** of the solid model when it is highlighted as shown in the figure.

## Creating a 2D Sketch

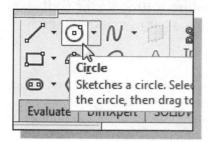

1. Select the **Circle** command by clicking once with the left-mouse-button on the icon in the *Sketch* toolbar.

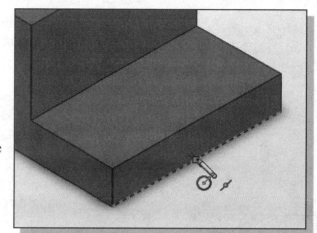

➢ We will align the center of the circle to the midpoint of the base feature.

2. Move the cursor along the shorter edge of the base feature; when the **Midpoint** is highlighted as shown in the figure, left-click to select this point.

3. Select the front corner of the base feature to create a circle as shown below.

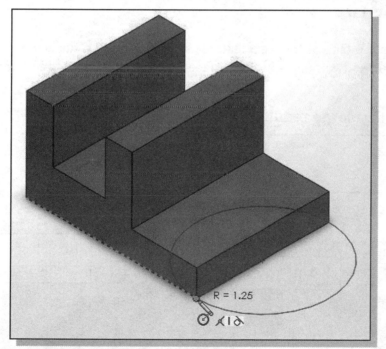

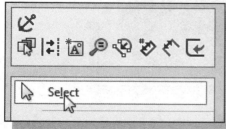

4. Inside the graphics window, click once with the right-mouse-button to display the option menu. Choose **Select** in the pop-up menu to end the Circle option.

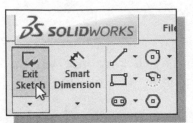

5. Click once with the **left-mouse-button** on the **Exit Sketch** icon on the *Sketch* toolbar to exit the Sketch option.

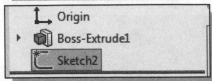

6. Make sure the sketch – Sketch2 – is selected in the *Feature Manager Design Tree*.

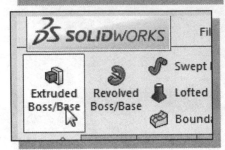

7. In the *Features* toolbar, select the **Extruded Boss/Base** command by clicking once with the left-mouse-button on the icon.

8. Click the **Reverse Direction** button in the *Property Manager* as shown. The extrude preview should appear as shown below.

9. In the *Extrude Property Manager* panel, click the drop-down arrow to reveal the pull options for the *End Condition* (the default end condition is Blind) and select **Up To Surface**.

10. Select the top face of the base feature as the termination surface for the extrusion. Notice Face<1> appears in the surface selection window in the *Extrude Property Manager*.

11. Confirm the **Merge result** checkbox is checked.

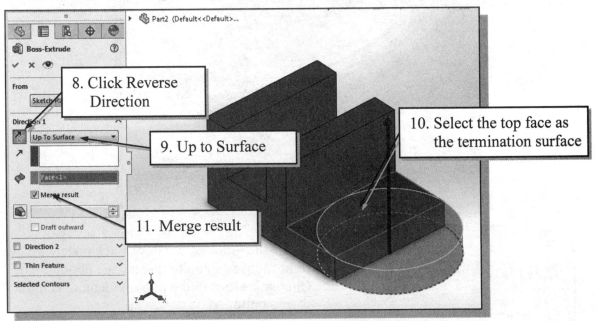

12. Click on the **OK** button to proceed with the Boss-Extrude operation.

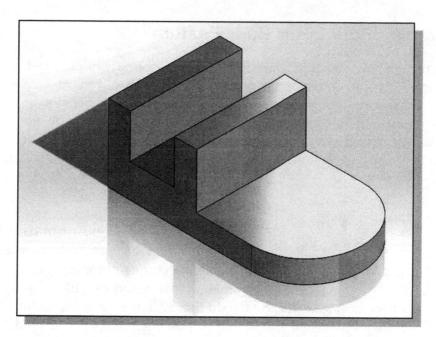

## Renaming the Part Features

Currently, our model contains two extruded features. The feature is highlighted in the display area when we select the feature in the *Feature Manager Design Tree* window. Each time a new feature is created, the feature is also displayed in the *Design Tree* window. By default, SOLIDWORKS will use generic names for part features. However, when we begin to deal with parts with a large number of features, it will be much easier to identify the features using more meaningful names. Two methods can be used to rename the features: 1. **clicking** twice on the name of the feature; and 2. using the **Properties** option. In this example, the use of the first method is illustrated.

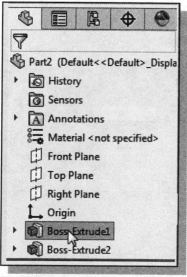

1. Select the first extruded feature in the *Model Browser* area by left-clicking once on the name of the feature, **Extrude1**. Notice the selected feature is highlighted in the graphics window.

2. Left-mouse-click on the feature name again to enter the **Edit** mode as shown.

3. Type **Base** as the new name for the first extruded feature and press the **[Enter]** key.

4. On your own, rename the second extruded feature to **Circular_End**.

## Adjusting the Width of the Base Feature

One of the main advantages of parametric modeling is the ease of performing part modifications at any time in the design process. Part modifications can be done through accessing the features in the design tree. SOLIDWORKS remembers the history of a part, including all the rules that were used to create it, so that changes can be made to any operation that was performed to create the part. For our *Saddle Bracket* design, we will reduce the size of the base feature from 3.25 inches to 3.0 inches, and the extrusion distance to 2.0 inches.

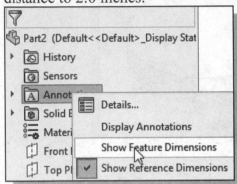

1. Inside the *Feature Manager Design Tree* area, **right-mouse-click** on **Annotations** to bring up the *option menu* and select the **Show Feature Dimensions** option in the pop-up menu. This option will allow us to view and edit dimensions by moving the cursor over the corresponding feature on the model.

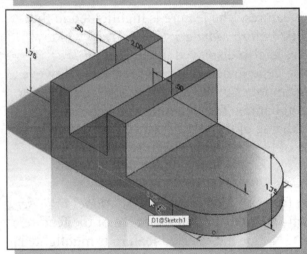

2. Move the cursor over the model to reveal the dimension of the **Base** feature.

3. Select the overall width of the **Base** feature, the **3.25** dimension value, by left-clicking on the dimension text, as shown.

4. Enter **3.0** in the *Modify* window and hit the **Enter** key once.

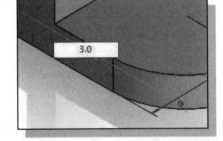

5. On your own, repeat the above steps and modify the extruded distance from **2.5** to **2.0**.

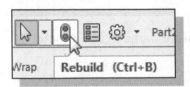

6. Click **Rebuild** in the *Menu Bar*.

   ➢ Note that SOLIDWORKS updates the model by re-linking all elements used to create the model. Any problems or conflicts that occur will also be displayed during the updating process.

7. **Right-mouse-click** on **Annotations** to bring up the option menu and de-activate the **Show Feature Dimensions** option in the pop-up menu.

## Adding a Cut Feature

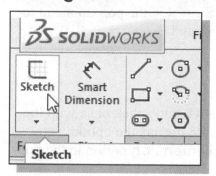

1. Select the **Sketch** button on the *Sketch* toolbar to create a new sketch.

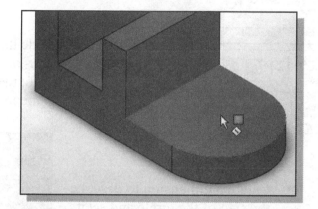

2. Pick the upper horizontal face as shown.

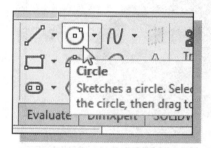

3. Select the **Circle** command by clicking once with the left-mouse-button on the icon in the *Sketch* toolbar.

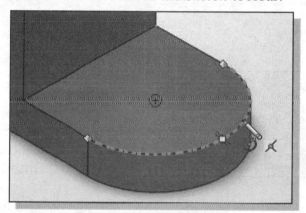

4. Move the cursor over the circular edge of the face as shown. (Do not click.) Notice the center and quadrant marks appear on the circle.

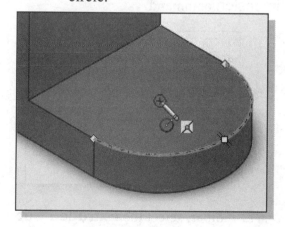

5. Select the **Center** point of the top face of the 3D model by left-clicking once on the icon as shown.

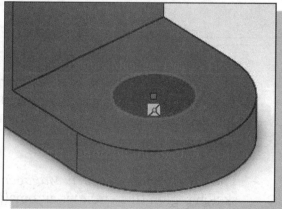

6. Sketch a circle of arbitrary size inside the top face of the cylinder as shown.

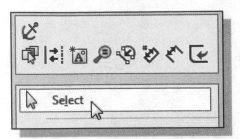

7.  Use the right-mouse-button to display the option menu and choose **Select** in the pop-up menu to end the Circle command.

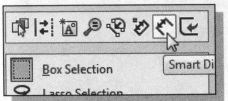

8.  Inside the graphics window, click once with the *right-mouse-button* to display the option menu. Select the **Smart Dimension** option in the pop-up menu.

9.  Create a dimension to describe the size of the circle and set it to **0.75 in**.

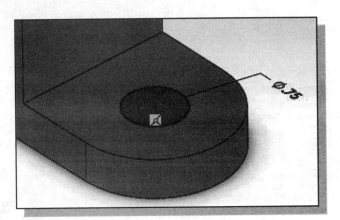

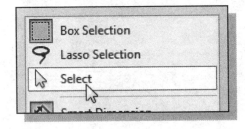

10. Inside the graphics window, click once with the right-mouse-button to display the option menu. Choose **Select** in the pop-up menu to end the Smart Dimension command.

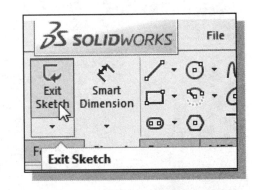

11. Click once with the **left-mouse-button** on the **Exit Sketch** icon on the *Sketch* toolbar to exit the Sketch option.

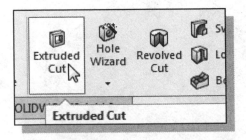

12. In the *Features* toolbar, select the **Extruded Cut** command by clicking once with the left-mouse-button on the icon.

➢ The *Cut-Extrude Property Manager* is displayed in the left panel. Notice that the sketch region (the circle) is automatically selected as the extrusion profile.

13. In the *Cut-Extrude Property Manager* panel, click the arrow to reveal the pull-down options for the *end condition* (the default end condition is Blind) and select **Through All** as shown.

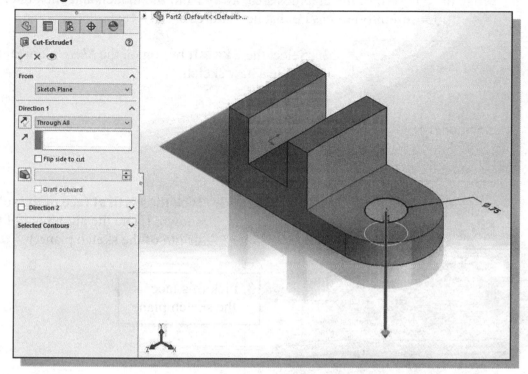

14. Click the **OK** button (green check mark) in the *Cut Extrude Property Manager* panel.

15. On your own, rename the feature as **3/4″ Hole** in the *Feature Manager Design Tree*.

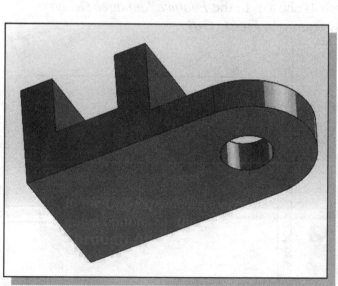

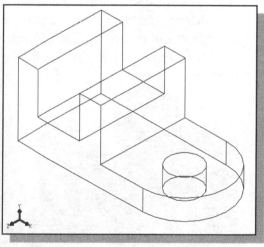

## Creating a Rectangular Extruded Cut Feature

1.  Move the cursor into the graphics area, away from the model, and click once with the left-mouse-button to ensure that no features are selected.

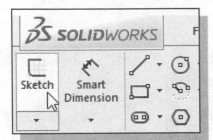

2.  Select the **Sketch** button on the *Sketch* toolbar to create a new sketch.

3.  Pick the **vertical face** of the solid as shown. (Note the alignment of the origin of the sketch plane.)

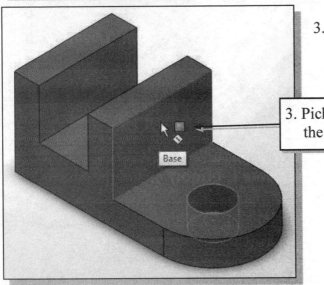

3. Pick this face as the sketch plane.

4.  On your own, sketch a rectangle (**1.0″ W x 0.75″ H** offset **0.5″** from the edge) as shown. Exit the sketch and create an **Extruded Cut** feature using the **Up To Next** option for the *End Condition* as shown. In the *Feature Manager Design Tree*, change the name of the new feature to **Rect_Cut**.

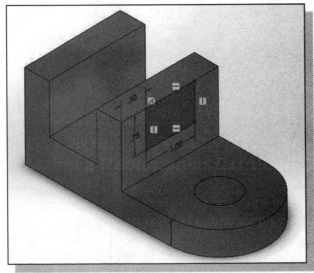

## History-Based Part Modifications

SOLIDWORKS uses the *history-based part modification* approach, which enables us to make modifications to the appropriate features and re-link the rest of the history tree without having to reconstruct the model from scratch. We can think of it as going back in time and modifying some aspects of the modeling steps used to create the part. We can modify any feature that we have created. As an example, we will adjust the depth of the rectangular cutout.

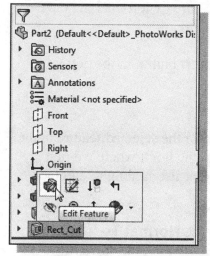

1. In the *Feature Manager Design Tree* window, select the last cut feature, **Rect_Cut**, by left-clicking once on the name of the feature.

2. In the *Feature Manager Design Tree* window, left-mouse-click once on the **Rect_Cut** feature.

3. Select the **Edit Feature** button in the pop-up menu. Notice the *Extrude Property Manager* appears on the screen.

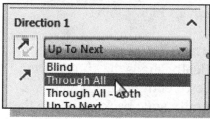

4. In the *Extrude Property Manager*, set the termination *End Condition* to the **Through All** option.

5. Click on the **OK** button to accept the settings.

- As can been seen, the history-based modification approach is very straightforward and it only took a few seconds to adjust the cut feature to the **Through All** option.

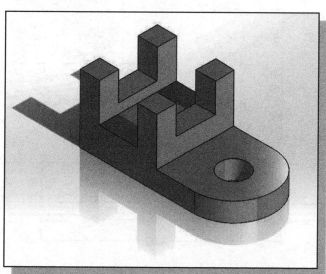

## A Design Change

❖ Engineering designs usually go through many revisions and changes. SOLIDWORKS provides an assortment of tools to handle design changes quickly and effectively. We will demonstrate some of the tools available by changing the **Base** feature of the design.

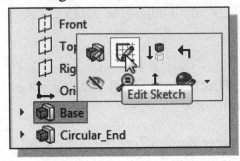

1. In the *Feature Manager Design Tree* window, select the **Base** feature by left-clicking once on the name of the feature.

2. Pick the **Edit Sketch** button in the pop-up menu.

❖ SOLIDWORKS will now display the original 2D sketch of the selected feature in the graphics window. We have literally gone back in time to the point where we first created the 2D sketch. Notice the feature being modified is also highlighted in the desktop *Feature Manager Design Tree*.

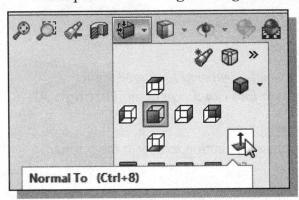

3. Click on the **Normal To** icon in the *View Orientation* pull-down menu on the *Heads-up View* toolbar.

• The **Normal To** command automatically aligns the *sketch plane* of a selected entity to the screen. We have literally gone back in time to the point where we first created the 2D sketch.

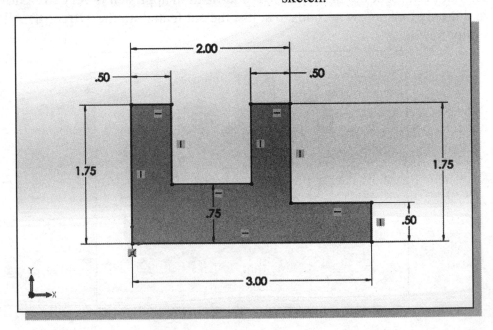

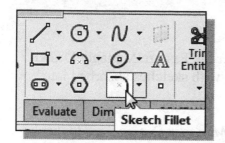

4.  Select the **Sketch Fillet** command in the *Sketch* toolbar.

5.  In the *Sketch Fillet Property Manager*, enter **0.25** as the new radius of the fillet. (**NOTE:** If the *Fillet Parameters* panel is minimized, click on the double arrows to expand the panel.)

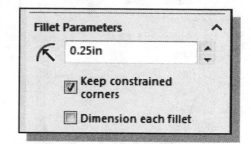

6.  Select the two edges as shown.

7.  Click the **OK** icon (green check mark) in the *Property Manager* to create the fillet.

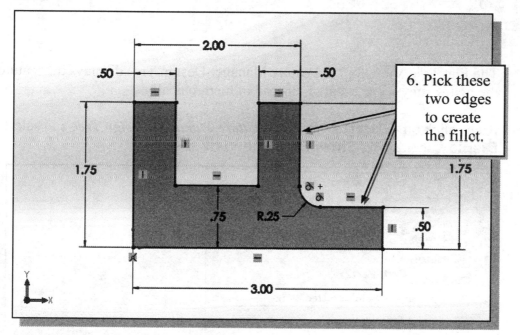

*   Note that the fillet is created automatically with the dimension attached. The attached dimension can also be modified through the history tree.

8.  Click the **OK** icon (green check mark) in the *Property Manager*, or hit the [**Esc**] key once, to end the **Sketch Fillet** command.

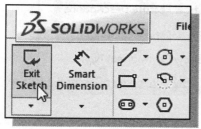

9.  Click on the **Exit Sketch** icon to exit the *Sketch* command.

*   Note that SOLIDWORKS will **Rebuild** the model by relinking all steps in the design tree.

## Feature Manager Design Tree Views

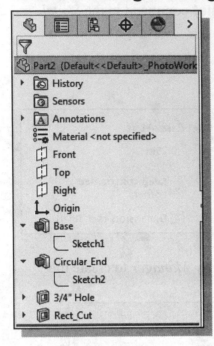

> ➢ The **Normal View** for the Feature Manager Design Tree is **hierarchical**, with sketches absorbed into features.

1. Expand the *Feature Manager Design Tree* to display the absorbed sketches as shown.

> ➢ The Flat Tree View for the Feature Manager Design Tree displays the features in the **order they were created**, instead of hierarchically.

2. Right-click on the **Part** icon in the *Feature Manager Design Tree* and select **Tree Display**, then select **Show Flat Tree View** as shown.

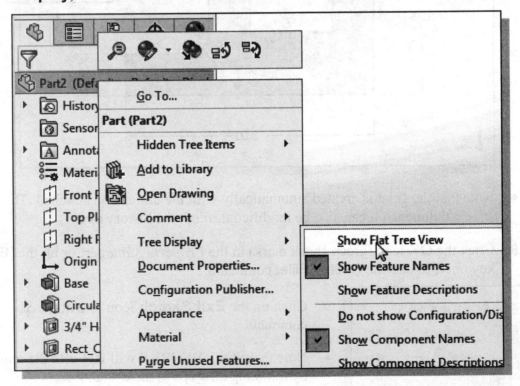

> Note the changes in the display of the *Feature Manager Design Tree*. Items are now shown in the order of creation.

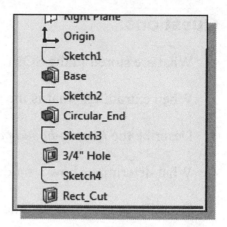

3.  Right-click on the **Part** icon in the *Feature Manager Design Tree* and select **Tree Display**, then **unselect** the **Show Flat Tree View** to return to the hierarchical display.

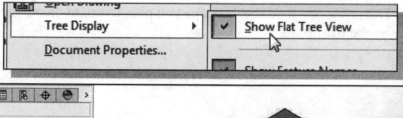

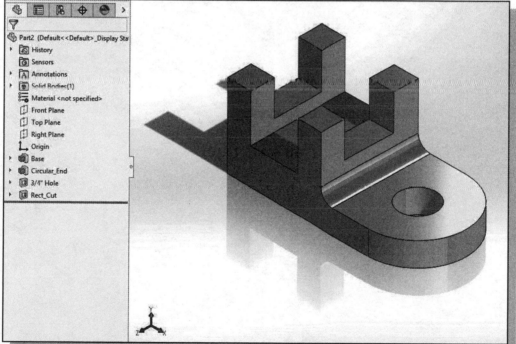

❖  In a typical design process, the initial design will undergo many analyses, testing, and reviews. The *history-based part modification* approach is an extremely powerful tool that enables us to quickly update the design. At the same time, it is quite clear that PLANNING AHEAD is also important in doing feature-based modeling.

4.  Save the part with the filename **Saddle_Bracket**. (**NOTE:** Do **not** save the part file in the Tutorial_Templates folder.)

## Questions:

1.  What are stored in the SOLIDWORKS *Feature Manager Design Tree*?

2.  When extruding, what is the difference between Blind and Through All?

3.  Describe the *history-based part modification* approach.

4.  What determines how a model reacts when other features in the model change?

5.  Describe the steps to rename existing features.

6.  Describe two methods available in SOLIDWORKS to *modify the dimension values* of parametric sketches.

7.  Create *Design Tree sketches* showing the steps you plan to use to create the two models shown on the next page:

Ex.1)

Ex.2)

# Exercises:

1.  **Tube Mount** (Dimensions are in inches.)

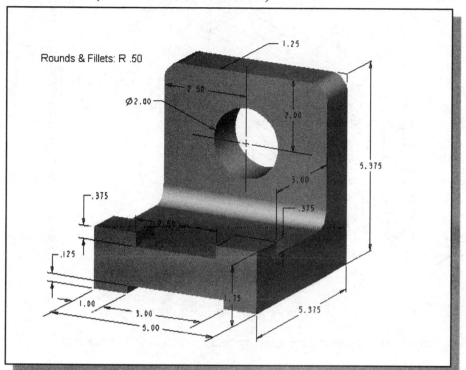

2.  **Hanger Jaw** (Dimensions are in inches.)

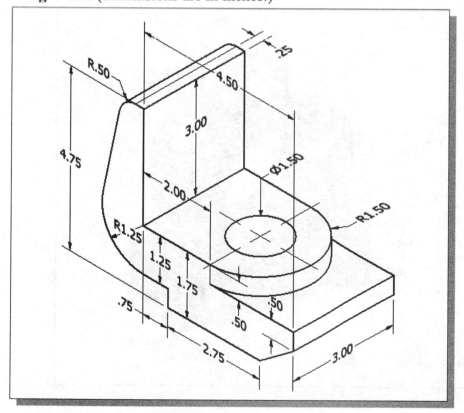

3. **Transfer Fork** (Dimensions are in inches.)

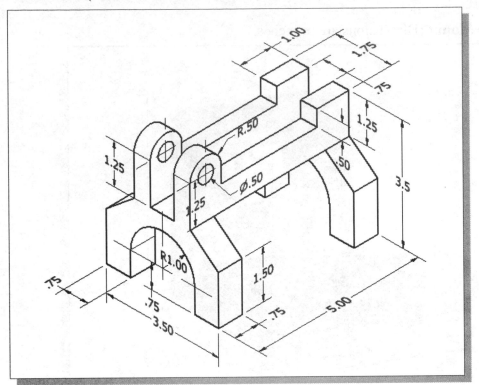

4. **Guide Slider** (Dimensions are in inches.)

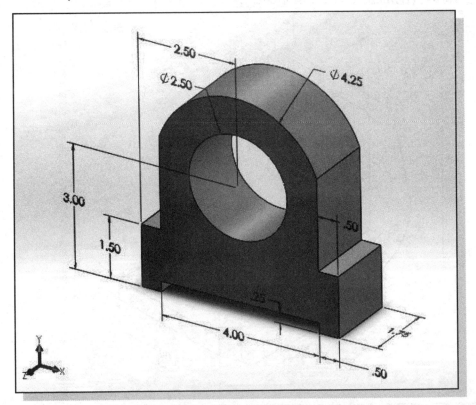

5.  **Shaft Guide** (Dimensions are in inches.)

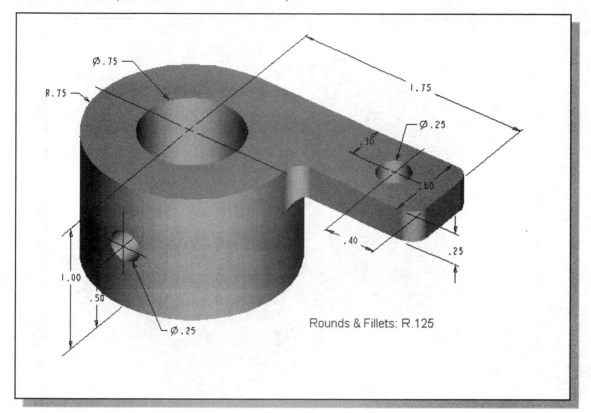

Rounds & Fillets: R.125

6.  **Support Fixture** (Dimensions are in inches.)

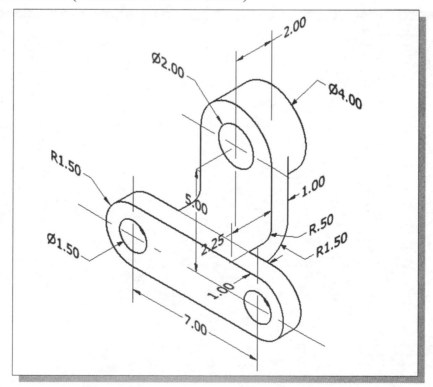

7.  **Idler Mount** (Dimensions are in inches. Center to center distance: 6.0)

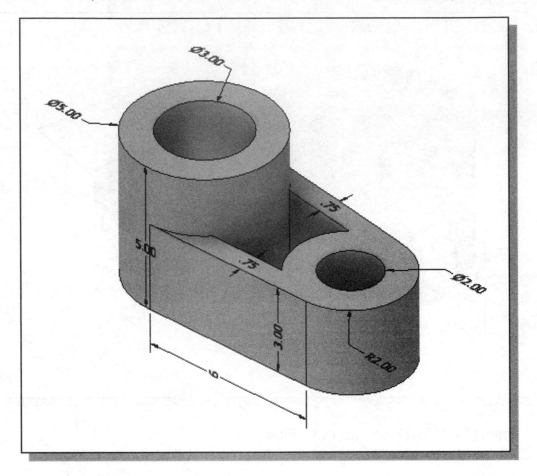

# Chapter 6
# Geometric Construction Tools

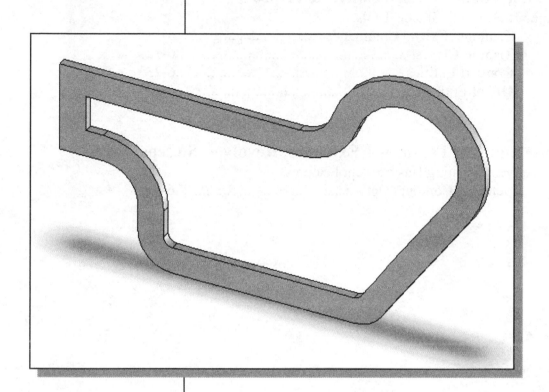

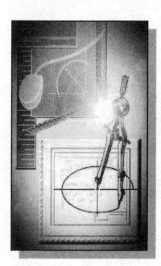

## Learning Objectives

- ◆ **Apply Geometry Relations**
- ◆ **Use the Trim/Extend Command**
- ◆ **Use the Offset Command**
- ◆ **Understand the Profile Sketch Approach**
- ◆ **Use the Convert Entities Command**
- ◆ **Understand and Use Reference Geometry**
- ◆ **Edit with Click and Drag**
- ◆ **Use the Fully Define Sketch Tool**
- ◆ **Use the Selected Contours Option**

## Certified SOLIDWORKS Associate Exam Objectives Coverage

### Sketch Tools – Offset, Convert, Trim

Objectives: Using Sketch Tools.

Trim and Extend Commands .......................................6-11
Trim to Closest .........................................................6-12
Convert Entities .........................................................6-23
Offset Entities  ..........................................................6-23


### Boss and Cut Features – Extrudes, Revolves, Sweeps, Lofts

Objectives: Creating Basic Swept Features.

Selected Contours Option ..........................................6-19

## Introduction

The main characteristics of solid modeling are the accuracy and completeness of the geometric database of the three-dimensional objects. However, working in three-dimensional space using input and output devices that are largely two-dimensional in nature is potentially tedious and confusing. SOLIDWORKS provides an assortment of two-dimensional construction tools to make the creation of wireframe geometry easier and more efficient. SOLIDWORKS includes sketch tools to create basic geometric entities such as lines, arcs, etc. These entities are grouped to define a boundary or profile. A *profile* is a closed region and can contain other closed regions. Profiles are commonly used to create extruded and revolved features. An *invalid profile* consists of self-intersecting curves or open regions. In this lesson, the basic geometric construction tools, such as Trim and Extend, are used to create profiles. Mastering the geometric construction tools along with the application of proper geometric and parametric relations is the true essence of *parametric modeling*.

## The *Gasket* Design

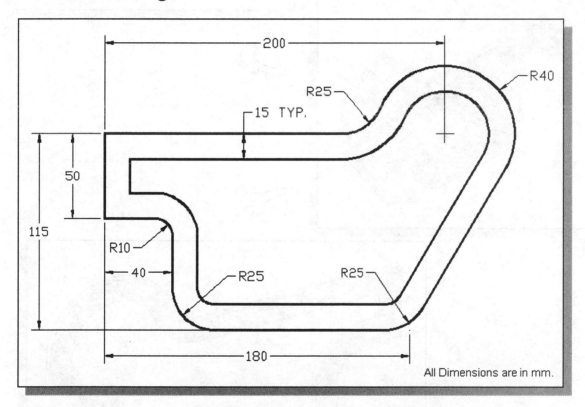

❖ Based on your knowledge of SOLIDWORKS so far, how would you create this design? What is the more difficult geometry involved in the design? Take a few minutes to consider a modeling strategy and do preliminary planning by sketching on a piece of paper. You are also encouraged to create the design on your own prior to following through the tutorial.

## Modeling Strategy

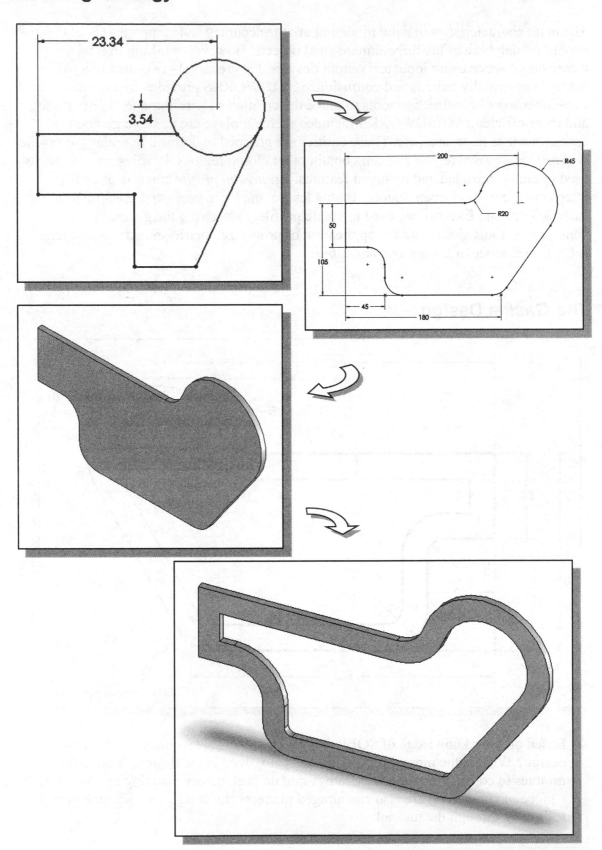

## Starting SOLIDWORKS

1. Select the **SOLIDWORKS** option on the *Start* menu or select the **SOLIDWORKS** icon on the desktop to start SOLIDWORKS. The SOLIDWORKS main window will appear on the screen.

2. Select **Part** to start a new part, by clicking on the first icon in the *Welcome - SOLIDWORKS Document* dialog box as shown.

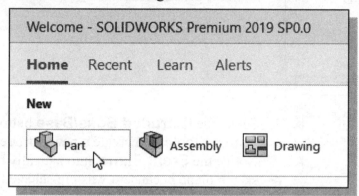

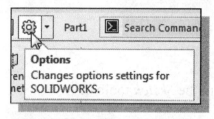

3. Select the **Options** icon from the *Menu Bar* to open the *Options* dialog box.·

4. Select the **Document Properties** tab.

5. Select **ISO** in the pull-down selection window under the *Overall drafting standard* panel to reset the settings.

6. Click **Units** as shown in the figure.

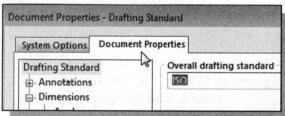

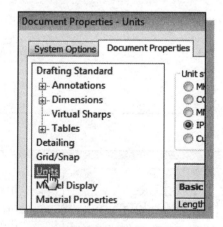

7. Select **MMGS (millimeter, gram, second)** under the *Unit system* options.

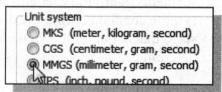

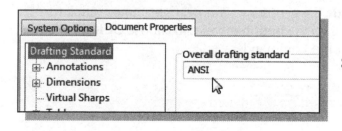

8. Select **ANSI** under the *Overall drafting standard* panel.

## Creating a 2D Sketch

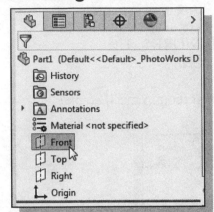

1. Click once with the **left-mouse-button** to select the **Front Plane** icon in the *Feature Manager Design Tree*. Notice the **Front Plane** icon is highlighted in the design tree and the *Front Plane* outline appears in the graphics area.

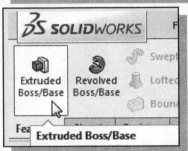

2. Select the **Extruded Boss/Base** button on the *Features* toolbar to create a new extruded feature. Notice the **Front Plane** automatically becomes the sketch plane because it was pre-selected.

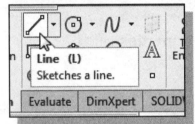

3. Click on the **Line** icon in the *Sketch* toolbar.

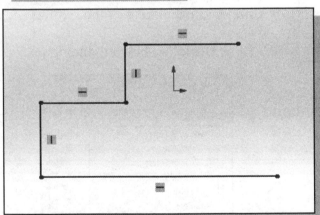

4. Create a sketch as shown in the figure. Start the sketch from the top right corner. The line segments are all parallel and/or perpendicular to each other. We will intentionally make the line segments of arbitrary length, as it is quite common during the design stage that not all of the values are determined.

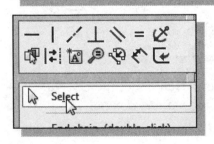

5. Inside the *graphics window*, right-mouse-click to bring up the option menu and choose **Select** to end the Line command.

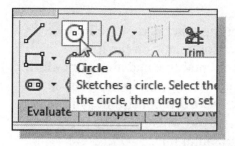

6. Select the **Circle** command by clicking once with the left-mouse-button on the icon in the *2D Sketch Panel*.

7. Pick a location that is above the bottom horizontal line as the center location of the circle.

8. Move the cursor toward the right and create a circle of arbitrary size, by clicking once with the left-mouse-button.

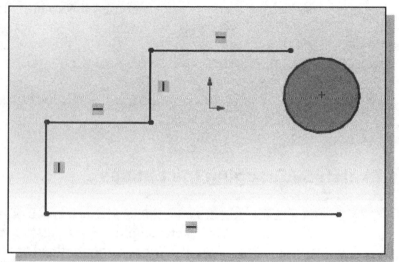

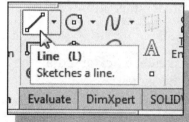

9. Click on the **Line** icon in the *Sketch* toolbar.

10. Move the cursor near the upper portion of the circle and, when the **Coincident** constraint symbol is displayed, click once with the **left-mouse-button** to select the starting point for the line.

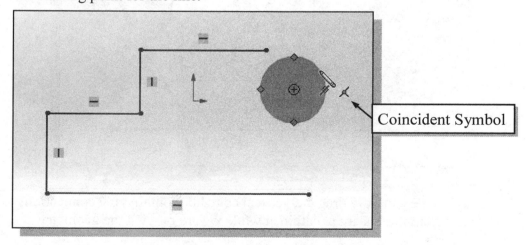

Coincident Symbol

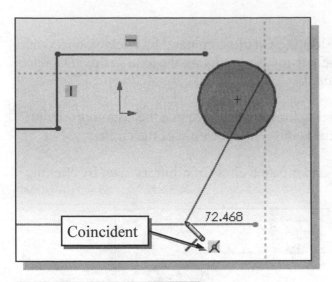

11. For the other end of the line, select a location that is on the lower horizontal line and about one-third from the right endpoint. Notice the **Coincident** constraint symbol is displayed when the cursor is on the horizontal line.

12. Inside the graphics window, right-mouse-click to bring up the option menu.

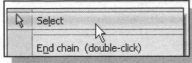

13. In the option menu, choose **Select** to end the Line command.

## Editing the Sketch by Dragging the Entities

❖ In SOLIDWORKS, we can click and drag any under-defined curve or point in the sketch to change the size or shape of the sketched profile. As illustrated in the previous chapter, this option can be used to identify under-defined entities. This *editing by dragging* method is also an effective visual approach that allows designers to quickly make changes.

1. Move the cursor on the lower left vertical edge of the sketch. Click and drag the edge to a new location that is toward the right side of the sketch.

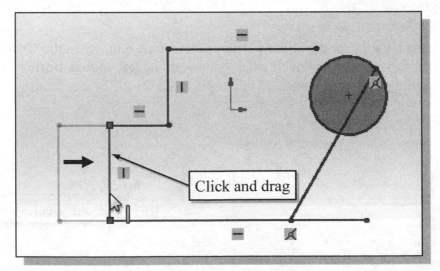

❖ Note that we can only drag the vertical edge horizontally; the connections to the two horizontal lines are maintained while we are moving the geometry.

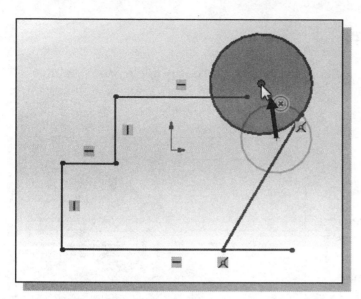

2. Click and drag the center point of the circle to a new location.

❖ Note that as we adjust the size and the location of the circle, the connection to the inclined line is maintained.

3. Click and drag the lower endpoint of the inclined line to a new location.

❖ Note that as we adjust the size and the location of the inclined line, the location of the bottom horizontal edge is also adjusted.

➢ Note that several changes occur as we adjust the size and the location of the inclined line. The location of the bottom horizontal line and the length of the vertical line are adjusted accordingly.

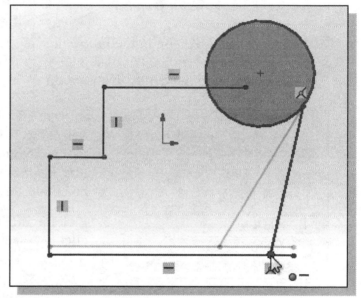

4. On your own, adjust the sketch so that the shape of the sketch appears roughly as shown.

❖ The *editing by dragging* method is an effective approach that allows designers to explore and experiment with different design concepts.

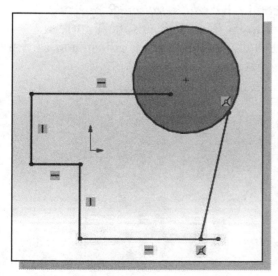

## Adding Additional Relations

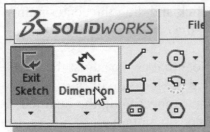

1. Choose **Smart Dimension** in the *Sketch* toolbar.

2. Add the horizontal location dimension from the top left vertical edge to the center of the circle as shown. (Do not be overly concerned with the dimensional value; we are still working on creating a *rough sketch*.)

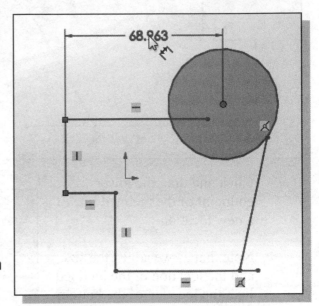

3. Click the **OK** icon in the *Property Manager*, or hit the [**Esc**] key once, to end the **Smart Dimension** command.

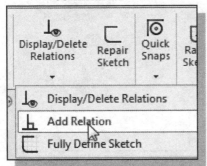

4. Select the **Add Relation** command in the *Sketch* toolbar. Notice the *Selected Entities* window in the *Add Relations Property Manager* is blank because no entities are selected.

5. Pick the **inclined line** and the **circle** by left-mouse-clicking once on the geometry.

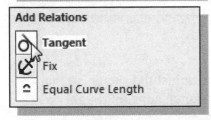

6. Click once with the left-mouse-button on the **Tangent** icon in the *Add Relations Property Manager* as shown. This activates the **Tangent** relation. The sketched geometry is adjusted as shown below.

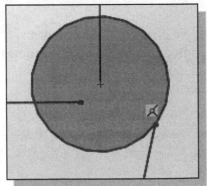  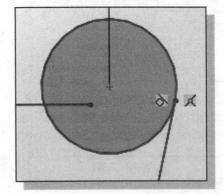

7. Click the **OK** icon in the *Property Manager*, or hit the **[Esc]** key once, to end the Add Relations command.

8. Click and drag the circle upward so that there is a gap between the circle and the top horizontal line as shown.

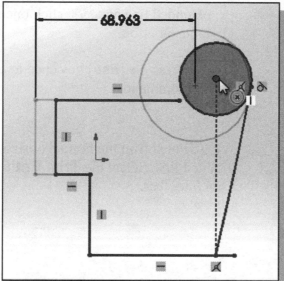

❖ Note that the Tangent relation to the inclined line is maintained during the drag & drop movement.

## Using the *Trim* and *Extend* Commands

➢ In the following sections, we will illustrate using the Trim and Extend commands to complete the desired 2D profile.

❖ The **Trim** and **Extend** commands can be used to shorten/lengthen an object so that it ends precisely at a boundary. As a general rule, SOLIDWORKS will try to clean up sketches by forming a closed region sketch.

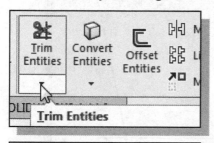

1. **Left-click** on the **arrow** below the **Trim Entities** button on the *Sketch* toolbar to reveal additional commands.

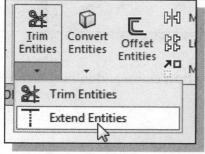

2. Select the **Extend Entities** command from the pop-up menu.

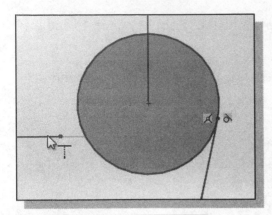

We will first extend the top horizontal line to the circle. Move the cursor near the right-hand endpoint of the top horizontal line. SOLIDWORKS will automatically display the possible result of the selection. When the extended line appears, click once with the left-mouse-button.

3.  Press the [**Esc**] key once to end the Extend command.

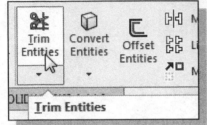

4.  We will next trim the bottom horizontal line to the inclined line. Select the **Trim Entities** icon on the *Sketch* toolbar.

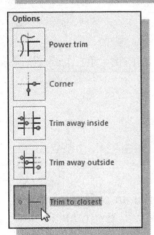

5.  Select the **Trim to closest** option in the *Trim Property Manager*.

6.  Move the cursor near the right-hand endpoint of the bottom horizontal line. The portion of the line that will be trimmed is highlighted. Click once with the **left-mouse-button** to trim the line.

7.  Click the **OK** icon in the *Property Manager*, or hit the [**Esc**] key once, to end the Trim command.

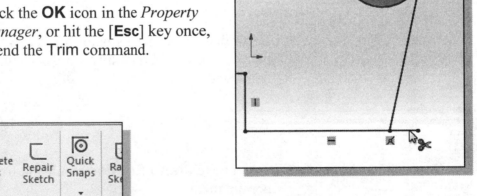

8.  Activate the **Add Relation** command in the *Sketch* toolbar.

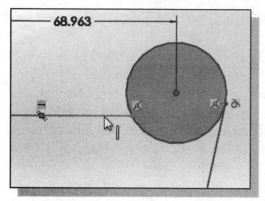

9. In the graphics area, select the **center point of the circle** and the **top horizontal line** as shown.

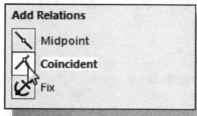

10. Click once with the left-mouse-button on the **Coincident** icon in the *Add Relations Property Manager* as shown. This activates the Coincident relation.

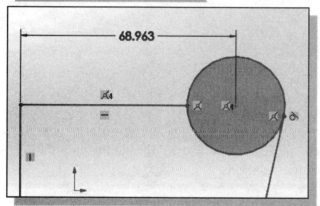

11. Click the **OK** icon in the *Property Manager*, or hit the [**Esc**] key once, to end the Add Relations command.

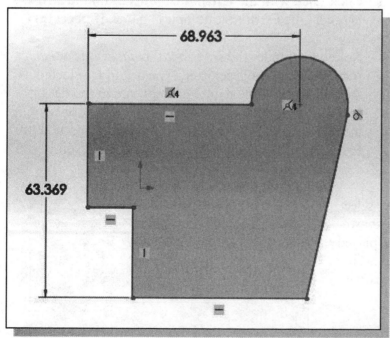

12. On your own, create the **height dimension** to the left of the sketch as shown.

13. On your own, **Trim** the circle as shown.

14. Hit the [**Esc**] key to end the Trim command.

## Adding Dimensions with the Fully Define Sketch Tool

In SOLIDWORKS, the **Fully Define Sketch** tool can be used to calculate which dimensions and relations are required to fully define an under defined sketch. Fully defined sketches can be updated more predictably as design changes are implemented. The general procedure for applying dimensions to sketches is to use the **Smart Dimension** command to add the desired key dimensions, and then use the **Fully Define Sketch** tool as a quick way to calculate all other sketch dimensions necessary. SOLIDWORKS remembers which dimensions were added by the user and which were calculated by the system, so that automatic dimensions do not replace the desired dimensions. In the following steps we will use the Fully Define Sketch tool to apply the dimensions necessary but will disable the automatic application of sketch relations.

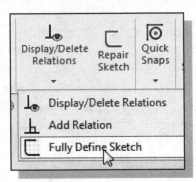

1. Select the **Fully Define Sketch** command in the *Sketch* toolbar.

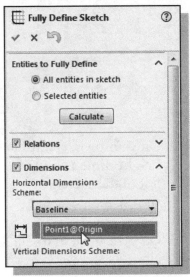

2. The *Fully Define Sketch Property Manager* appears. **Uncheck** the Relations box to disable the calculation of geometric relations.

3. Click once with the **left-mouse-button** on the arrows to reveal the Dimensions option panel, **if necessary**.

4. Notice that under the *Horizontal Dimensions Scheme* option, Baseline is selected. The origin is selected as the default baseline datum and appears in the datum selection window as Point1@Origin. Click once with the **left-mouse-button** in the *Datum* selection text box to select a different datum to serve as the baseline.

5. In the graphics window, select the left **vertical line** as shown. Notice Line2 is now displayed in the *Datum* selection text box as the *Horizontal Dimensions* baseline.

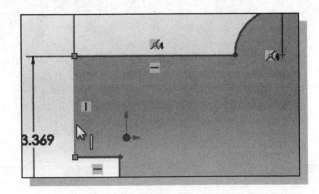

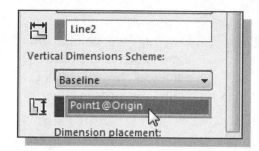

6.  In the *Property Manager*, under the *Vertical Dimensions Scheme* option panel, click once with the **left-mouse-button** in the *Datum* selection text box.

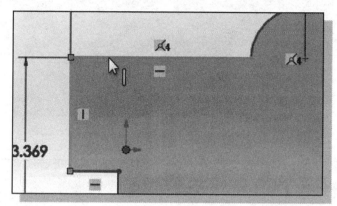

7.  In the graphics window, select the top horizontal line as shown. Notice Line1 is now displayed in the *Datum* selection text box as the *Vertical Dimensions* baseline.

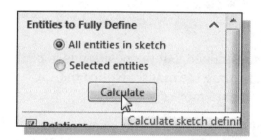

8.  Select **Calculate** in the *Property Manager* to continue with the Fully Define Sketch command.

9.  A dialog box appears with the statement "*Fully Define Sketch is complete but the sketch is still under defined.*" This is because the sketch is not fixed to a location in the coordinate system. Select **OK** to close the dialog box.

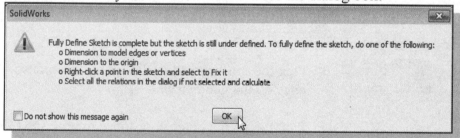

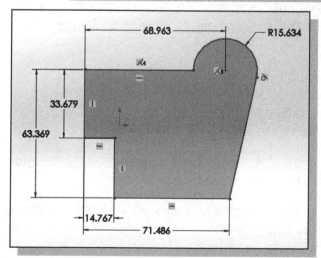

10. Click **OK** in the *Property Manager* to accept the results and exit the Fully Define Sketch command.

➢ The *Auto-dimension* results should appear as shown. (**NOTE:** Dimensional values on your screen will be different; this is only a rough sketch.)

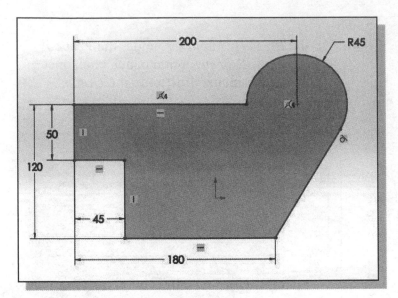

11. On your own, change the dimensions to the desired values as shown. (**HINT:** adjust the larger values first.)

## Fully Defined Geometry

The sketch is under defined. The shape is now fully constrained, but the position is not fixed. On your own, left-click on any sketch feature and drag the mouse. Notice the sketch shape does not change, but the entire sketch is free to move.

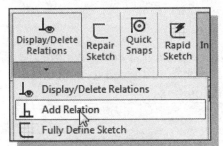

1. Press the [**Esc**] key to ensure no items are selected.

2. Select the **Add Relation** command from the pop-up menu on the *Sketch* toolbar.

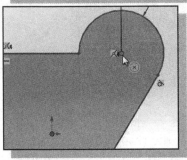

3. Pick the **centerpoint** of the arc by left-mouse-clicking once on the geometry.

4. Pick the sketch plane coordinate **origin**. (**NOTE:** If the origin symbol is not visible in the graphics area, turn *ON* the visibility by selecting Origins under the *View* toolbar.)

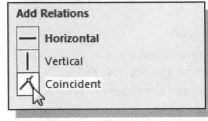

5. Click once with the left-mouse-button on the **Coincident** icon in the *Add Relations Property Manager* as shown.

6. Click the **OK** icon in the *Property Manager* to end the Add Relations command.

7. Press the [**F**] key to zoom the sketch to fit the graphics area. Note that the sketch is fully defined with the added relation. Fully Defined is displayed in the *Status Bar* and the color of the sketch features changes to **black**.

## Creating Fillets and Completing the Sketch

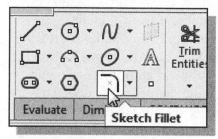

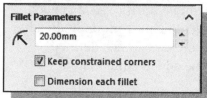

1. Click on the **Sketch Fillet** icon in the *Sketch* toolbar.

2. The *Sketch Fillet Property Manager* appears. Enter **20 mm** for the fillet radius as shown.

3. Notice "*Select a sketch vertex or entities to fillet*" is displayed in the *Message* panel of the *Property Manager*. Select the corner where the horizontal line meets the arc, as shown.

4. On your own, create the three additional fillets as shown in the figure below. Note that all the rounds and fillets are created with the same radius.

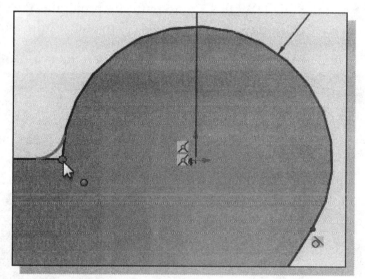

5. Click on the **OK** icon in the *Property Manager* to end the Sketch Fillet command.

6. Click once with the **left-mouse-button** on the **Exit Sketch** icon on the *Sketch* toolbar to exit the Sketch option.

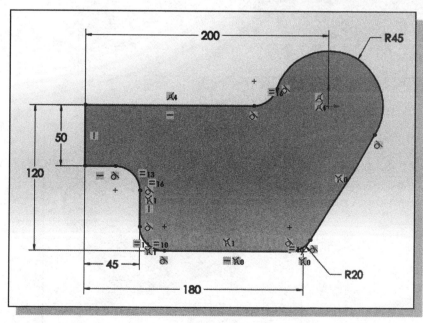

## Profile Sketch

❖ In SOLIDWORKS, *profiles* are closed regions that are defined from sketches. Profiles are used as cross sections to create solid features. For example, **Extrude**, **Revolve**, **Sweep**, **Loft**, and **Coil** operations all require the definition of at least a single profile. The sketches used to define a profile can contain additional geometric entities since the additional entities are consumed when the feature is created. To create a profile we can create single or multiple closed regions, or we can select existing solid edges to form closed regions. A profile cannot contain self-intersecting geometry; regions selected in a single operation form a single profile. As a general rule, we should dimension and constrain profiles to prevent them from unpredictable size and shape changes. SOLIDWORKS does allow us to create under-defined profiles; the dimensions and/or geometric relations can be added/edited later.

❖ SOLIDWORKS automatically highlights the closed region in the sketch and the defining geometry, which forms the profile required for the Extrude operation.

1. In the *Extrude Property Manager*, enter **5 mm** as the extrusion distance.

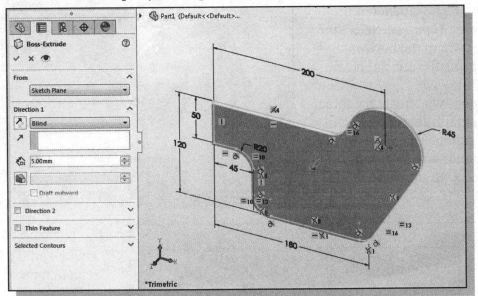

2. Click on the **OK** button to accept the settings and create the base feature.

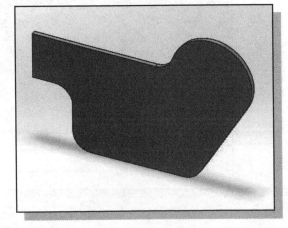

❖ Note that all the sketched geometric entities and dimensions are consumed and have disappeared from the screen when the feature is created.

## Redefining the Sketch and Profile using Contour Selection

- Engineering designs usually go through many revisions and changes. SOLIDWORKS provides an assortment of tools to handle design changes quickly and effectively. We will demonstrate some of the tools available by changing the base feature of the design. The profile used to create the extrusion is selected from the sketched geometry entities. In SOLIDWORKS, any profile can be edited and/or redefined at any time. It is this type of functionality in parametric modeling software that provides designers with greater flexibility and the ease to experiment with different design considerations.

- In the SOLIDWORKS *Property Manager* for features requiring definition of a profile, the **Selected Contours** selection window can be used to select sketch contours and model edges, and apply features to them. Contour Selection is a grouping mechanism that allows us to use a partial sketch to create features. It is a tool that helps maintain design intent by reducing the amount of trimming necessary to build the contour. In the previous section, the **Boss-Extrude** feature was created using a single continuous sketch contour to define the profile. In this section we will demonstrate the utility of the Contour Selection tool to generate a similar profile from a sketch containing self-intersecting geometry and multiple closed regions.

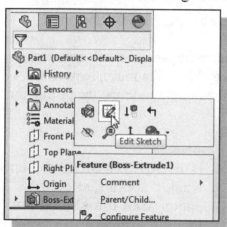

1. In the *Feature Manager Design Tree*, **right-mouse-click** once on the **Boss-Extrude1** feature to bring up the option menu; then pick the **Edit Sketch** icon in the pop-up menu as shown.

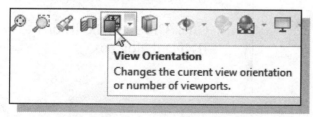

2. Select the **View Orientation** button on the *Heads-up View* toolbar to show the available pre-defined view orientations.

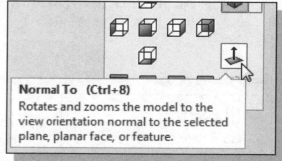

3. Select the **Normal to** icon to obtain a view normal to the sketch.

4. On your own, create the very rough sketch as shown in the figure below. We will intentionally under-define the new sketch to illustrate the flexibility of the system. The dimensions of the new sketch elements are not important. It is important that **tangent relations** are added where the circles meet tangent to lines or other circles. (Hint: Use the **Perimeter Circle** option for the *Circle* command.)

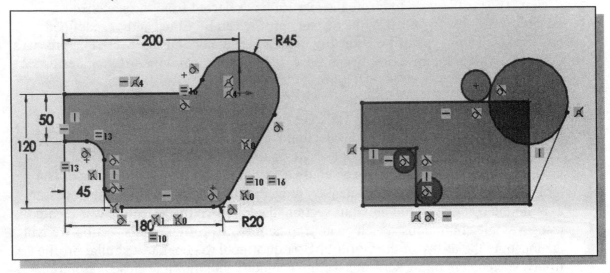

5. Select the **Exit Sketch** icon on the *Sketch* toolbar to exit the **Sketch** option.

➢ The SOLIDWORKS pop-up window may appear indicating an error will occur when SOLIDWORKS attempts to rebuild the model. This is due to the multiple possible contours for extrusion now present in the sketch. We will next redefine the profile for the base feature using the **Selected Contours** tool.

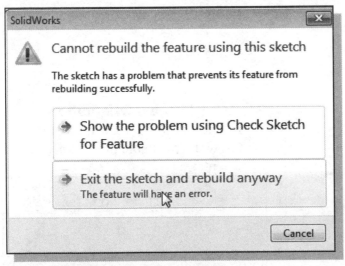

6. If the SOLIDWORKS pop-up window appears, select the option '**Exit the sketch and rebuild anyway**'.

7. Click **Close** in the *What's Wrong* dialog box.

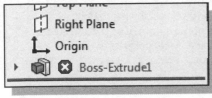

8. The red **x** next to the icon indicates the error in the feature.

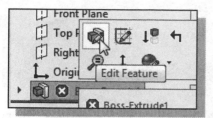

9. In the *Feature Manager Design Tree*, **left-mouse-click** once on the **Boss-Extrude1** feature to bring up the option menu, then pick the **Edit Feature** icon in the pop-up menu.

❖ SOLIDWORKS will now display the 2D sketch of the selected feature in the graphics window. We have literally gone back in time to the point where we define the extrusion feature. The original sketch and the new sketch we just created are wireframe entities recorded by the system as belonging to the same **SKETCH**; but only the **selected profile** entities are used to create the feature. We will now change the selected profile entities.

10. In the *Boss-Extrude1 Property Manager*, click on the arrow to reveal the *Selected Contours* option panel as shown (if necessary).

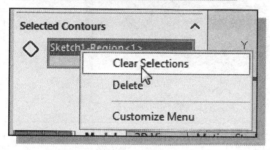

11. If the *Selected Contours* panel is not blank, move the cursor over the panel and click once with the **right mouse button**. In the pop-up option menu, select **Clear Selections**.

➤ **Sketch1** is displayed in the graphics area with no entities selected to define the profile for extrusion. The instruction "*Pick a sketch entity to define an open or closed contour. To define a region, pick inside an area bounded by sketch geometry*" appears in the *Status Bar*. The profile can be defined by either selecting sketch entities (lines, circles, etc.) to define a boundary, or by selecting bound areas for inclusion. We will use the latter option here.

12. Move the cursor inside the bound area shown in the figure. When the area is highlighted, click once with the **left-mouse-button** to select the area for inclusion in the profile. Notice **Sketch1-Region<1>** now appears in the *Selected Contours* panel on the *Property Manager*.

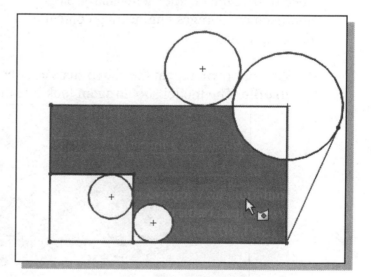

13. On your own, select additional areas for inclusion as shown. (If an incorrect area is selected, simply click on it again to unselect it. Be careful to click inside bound areas and not on the sketch entities. If difficulty is encountered, clear the selections and start over.)

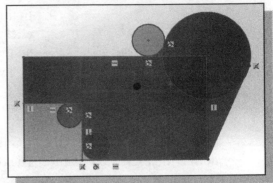

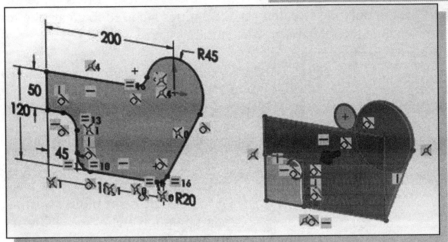

14. In the *Boss-Extrude1 Property Manager*, click on the **OK** button to accept the settings and update the solid feature. The feature is recreated using the newly sketched geometric entities.

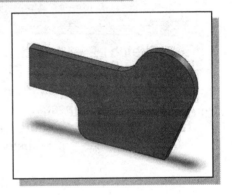

• The *Selected Contours* approach eliminates the need to trim the geometry manually. This approach encourages engineering content over drafting technique.

15. On your own, repeat the above steps and set the profile back to the **original profile**. The model should again look like the one shown on page 6-18.

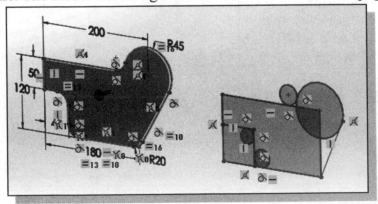

# Create an OFFSET Extruded Cut Feature

➢ To complete the design, we will create a cutout feature by using the **Offset** command. First we will set up the sketching plane to align with the front face of the 3D model.

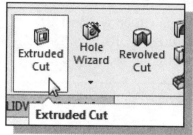

1. In the *Features* toolbar, select the **Extruded Cut** command by left-clicking once on the icon.

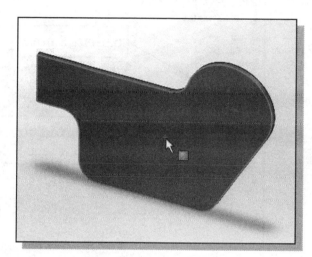

2. Notice the left panel displays the *Edit Sketch Property Manager* with the instruction "*Select: a plane, planar face ...*" **Select the front face** of the 3D model in the graphics window.

3. We will now convert the edge of the front face of the model into segments in the new sketch. **Select the front face again** as shown.

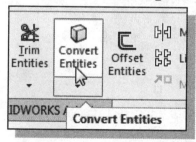

4. Select the **Convert Entities** icon in the *Sketch* toolbar.

5. Click **OK** to accept the selection.

➢ The **Convert Entities** command converts selected model features into sketch segments. Notice the edges of the selected surface have been converted to line and arc segments in the current sketch.

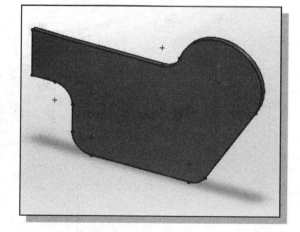

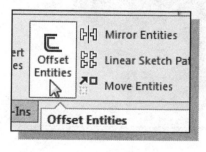

6. Select the **Offset Entities** icon in the *Sketch* toolbar.

7. In the *Offset Property Manager*, enter **5 mm** for the offset distance.

8. In the *Offset Property Manager*, select the options for *Add dimensions*, *Reverse*, and *Select chain* as shown.

9. In the graphics area, select any segment of the face outline which was just converted. Because the *Select chain* option is active, SOLIDWORKS will automatically select all of the connecting geometry to form a closed region.

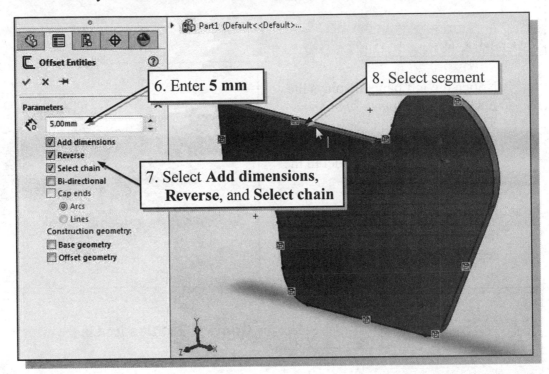

10. Click on the **OK** button to accept the settings and create the offset feature.

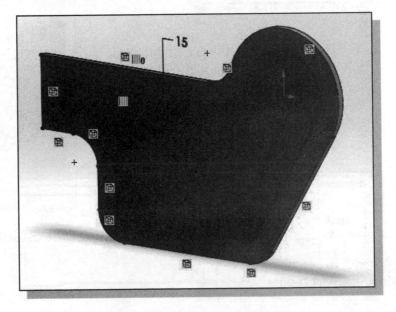

11. On your own, modify the offset dimension to **15 mm** as shown in the figure.

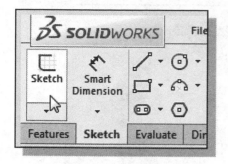

12. Click once with the **left-mouse-button** on the **Exit Sketch** icon on the *Sketch* toolbar to exit the Sketch option.

13. In the *Cut-Extrude Property Manager* panel, click the arrow to reveal the pull-down options for the *End Condition* (the default end condition is Blind) and select **Through All** as shown.

14. In the *Cut-Extrude Property Manager* panel, click on the arrow to reveal the *Selected Contours* option text box as shown. On your own, clear the selection in the box.

15. In the graphics area, move the cursor over the interior of the offset contour as shown in the figure below. Click once with the **left-mouse-button** to select the profile. Notice Sketch2-Region<1> now appears in the *Selected Contours* box on the *Property Manager*.

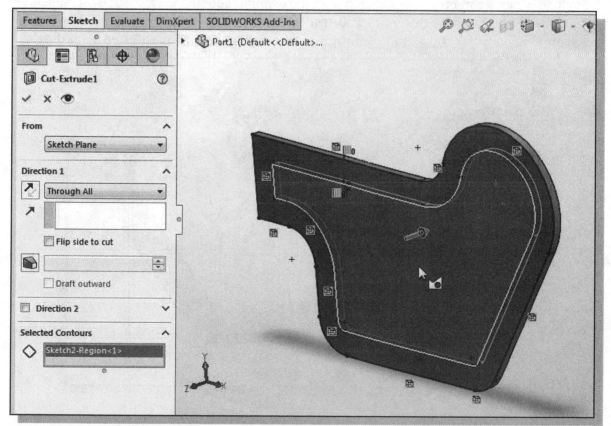

16. Click on the **OK** button to accept the settings and create the cut-extrude feature.

➢ The offset geometry is associated with the original geometry. On your own, adjust the overall height of the design to **100 millimeters** (i.e., change the 120 mm vertical dimension to 100 mm) and confirm that the offset geometry is adjusted accordingly.

17. Save the part with the filename *Gasket*.

## Questions:

1. Can we create a profile with extra 2D geometry entities?

2. How do we access the SOLIDWORKS **Edit Sketch** option?

3. How do we create a *profile* in SOLIDWORKS?

4. Can we build a profile that consists of self-intersecting curves?

5. Describe the procedure to create an offset copy of a sketched 2D geometry.

6. Describe the **Selected Contours** option in SOLIDWORKS.

7. Identify and briefly describe the following commands:

(a)

(b)

(c)

(d)

## Exercises:

1. **V-slide Plate** (Dimensions are in inches. Plate Thickness: **0.25**)

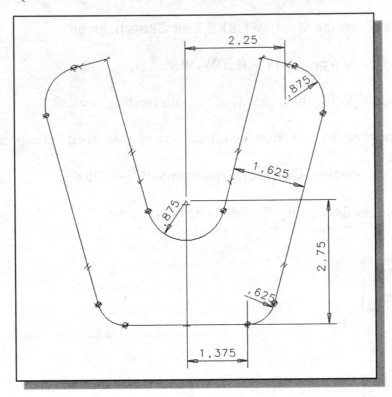

2. **Shaft Support** (Dimensions are in millimeters. Note the two R40 arcs at the base share the same center.)

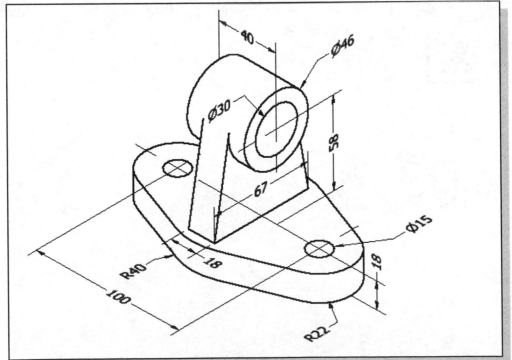

3. **Vent Cover** (Thickness: **0.125** inches. Hint: Use the Ellipse command.)

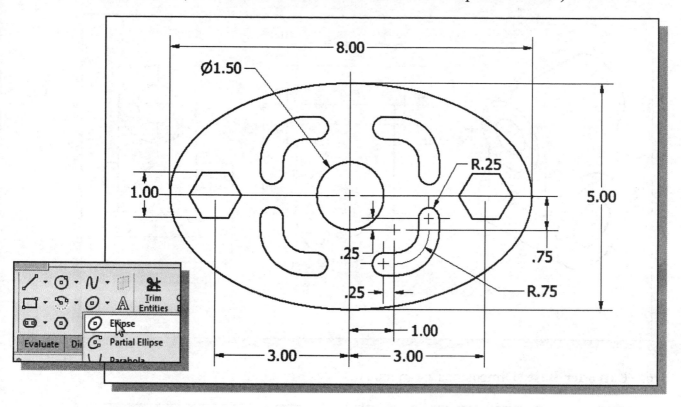

4   **Anchor Base** (Dimensions are in inches. Rounds R .5)

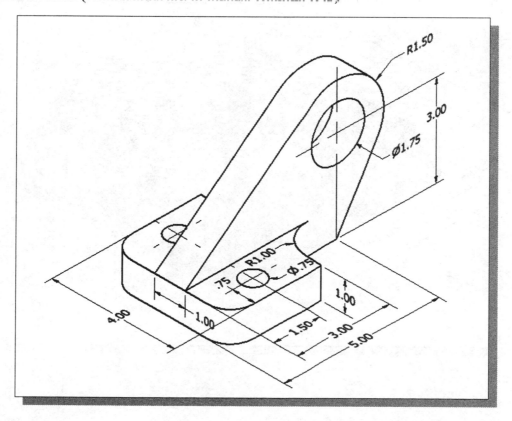

5.  **Tube Spacer** (Dimensions are in inches.)

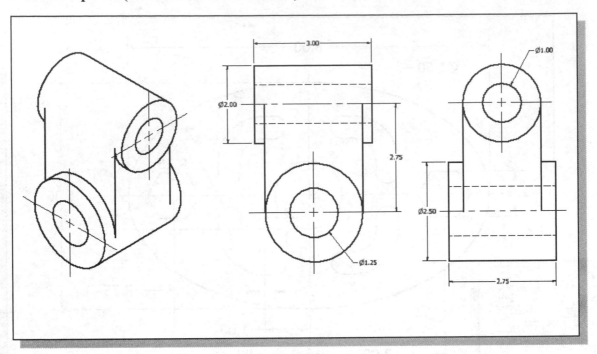

6.  **Extruder Base** (Dimensions are in mm.)

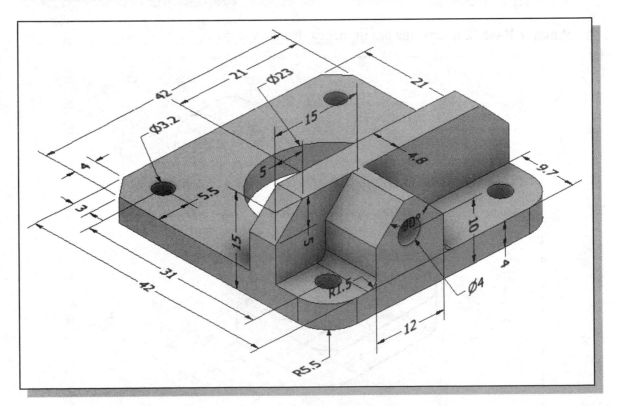

**Chapter 7**
# Orthographic Projection and Multiview Constructions

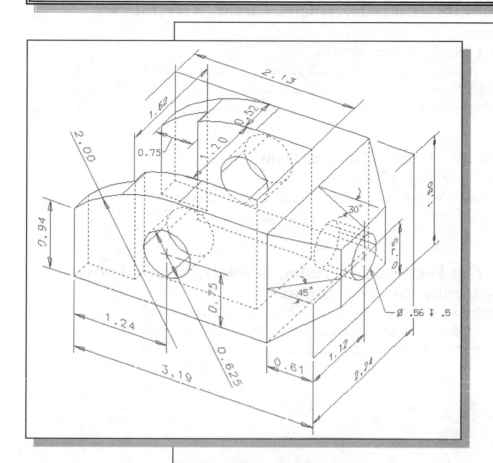

## Learning Objectives

- **Understand the Basic Orthographic Projection Principles**
- **Be able to Perform 1st and 3rd Angle Projections**
- **Understand the Concept and Usage of the BORN Technique**
- **Understand the Importance of Parent/Child Relations in Features**
- **Use the *Suppress Feature* Option**
- **Resolve Undesired Feature Interactions**

| Certified SOLIDWORKS Associate Examination Objectives Coverage |

## Sketch Entities – Lines, Rectangles, Circles, Arcs, Ellipses, Centerlines
Objectives: Creating Sketch Entities.
- Centerline ................................................................ 7-49
- Tangent Arc ............................................................. 7-51
- Centerpoint Arc ....................................................... 7-56

## Sketch Tools – Offset, Convert, Trim
Objectives: Using Sketch Tools.
- Dynamic Mirror ....................................................... 7-48
- Mirror Entities ........................................................ 7-60

## Boss and Cut Features – Extrudes, Revolves, Sweeps, Lofts
Objectives: Creating Basic Swept Features.
- Suppress Features .................................................... 7-65
- Unsuppress Features ................................................ 7-65
- Edit Sketch Plane .................................................... 7-66

**Certified Associate Reference Guide**

## Introduction

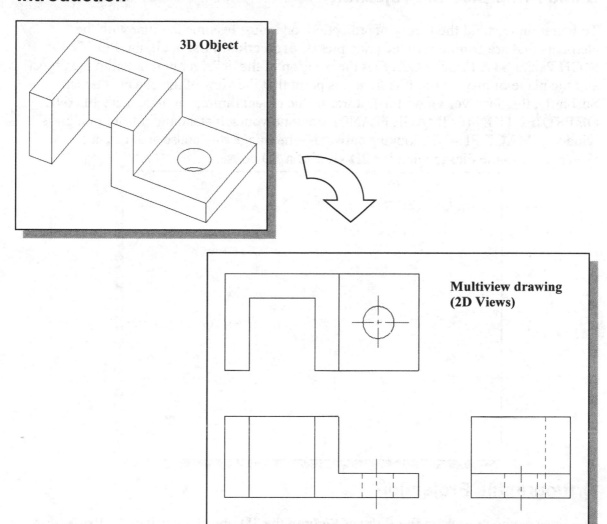

Most drawings produced and used in industry are *multiview drawings*. Multiview drawings are used to provide accurate three-dimensional object information on two-dimensional media, a means of communicating all of the information necessary to transform an idea or concept into reality. The standards and conventions of multiview drawings have been developed over many years, which equip us with a universally understood method of communication.

Multiview drawings usually require several orthographic projections to define the shape of a three-dimensional object. Each orthographic view is a two-dimensional drawing showing only two of the three dimensions of the three-dimensional object. Consequently, no individual view contains sufficient information to completely define the shape of the three-dimensional object. All orthographic views must be looked at together to comprehend the shape of the three-dimensional object. The arrangement and relationship between the views are therefore very important in multiview drawings. Before taking a more in-depth look into the multiview drawings, we will first look at the concepts and principles of projections.

## Basic Principles of Projection

To better understand the theory of projection, one must become familiar with the elements that are common to the principles of **projection**. First of all, the **POINT OF SIGHT** (aka **STATION POINT**) is the position of the observer in relation to the object and the plane of projection. It is from this point that the view of the object is taken. Secondly, the observer views the features of the object through an imaginary PLANE OF PROJECTION (or IMAGE PLANE). Imagine yourself standing in front of a glass window, IMAGE PLANE, looking outward—the image of a house at a distance is sketched on to the glass which is a 2D view of a 3D house.

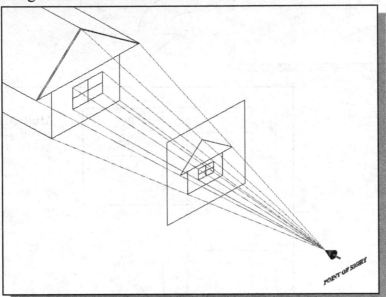

## Orthographic Projection

The lines connecting from the *Point of Sight* to the 3D object are called the **Projection Lines** or **Lines of Sight**. Note that in the above figure, the projection lines are connected at the point of sight, and the projected 2D image is smaller than the actual size of the 3D object.

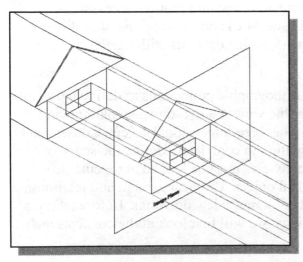

Now, if the *projection lines* are **parallel** to each other and the image plane is also **perpendicular** (*normal*) to the projection lines, the result is what is known as an **orthographic projection**. When the projection lines are parallel to each other, an accurate outline of the visible face of the object is obtained.

The term **orthographic** is derived from the word *orthos* meaning **perpendicular** or **90°**.

In *Engineering Graphics*, the projection of one face of an object usually will not provide an overall description of the object; other planes of projection must be used. To create the necessary 2D views, the *point of sight* is changed to project different views of the same object; hence, each view is from a different point of sight. If the point of sight is moved to the front of the object, this will result in the front view of the object. And then move the point of sight to the top of the object and looking down at the top, and then move to the right side of the object, as the case may be. Each additional view requires a new point of sight.

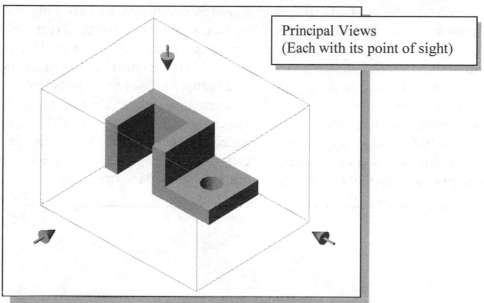

Principal Views
(Each with its point of sight)

## Multiview Orthographic Projection

In creating multiview orthographic projection, different systems of projection can be used to create the necessary views to fully describe the 3D object. In the figure below, two perpendicular planes are established to form the image planes for a multiview orthographic projection.

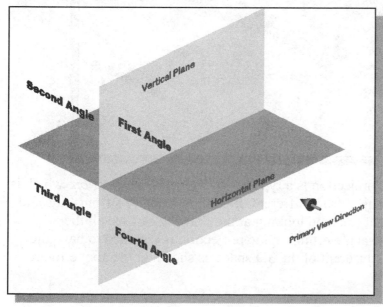

The angles formed between the horizontal and the vertical planes are called the **first**, **second**, **third** and **fourth angles**, as indicated in the figure. For engineering drawings, both **first angle projection** and **third angle projection** are commonly used.

## First-Angle Projection

❖ First-angle orthographic projection is commonly used in countries that use the metric (International System of Units: SI) system as the primary units of measurement. The metric system was first developed in 17th century France; most European countries adopted the metric system by the end of the 18th century. Today, most countries in the world have adopted and are using the metric SI system as the primary units of measurement. In 1875, the United States solidified its commitment to the development of the internationally recognized metric system by becoming one of the original seventeen signatory nations to the Metre Convention or the Treaty of the Metre. Over the last 50 years, there has been a quiet ongoing debate in Washington over metric conversion. Many people have said that the United States must adopt the Metric System or face irreparable economic harm. And, several efforts have been made to start conversion. Back in 1975, Congress passed the *Metric Conversion Act.* The act mandated a ten year voluntary conversion period starting in 1975. The full conversion to the metric system was expected to occur in 1985. Yet, today, in spite of years of arguments and Federal efforts, very little has happened. The USA is one of a handful of countries that still use the *Imperial Units* of Measurement.

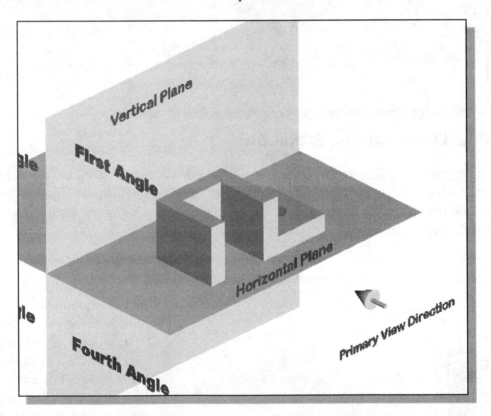

● First-angle orthographic projection is a type of *Orthographic Projection,* which is a way of drawing a 3D object from different directions. Usually a front, side and top view are drawn so that a person looking at the drawing can see all the important sides. The first-angle orthographic projection is assumed to have the object placed in the first quadrant of the 3D space as shown in the above figure.

❖ In first-angle projection, the object is placed in **front** of the image planes. And the views are formed by projecting to the image plane located at the back.

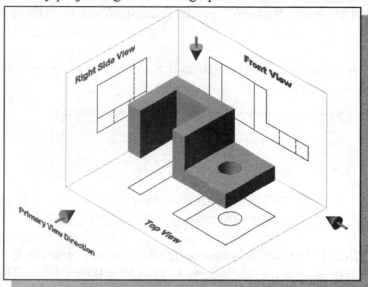

## Rotation of the Horizontal and Profile Planes

In order to draw all three views of the object on the same plane, the horizontal (*Top View*) and profile (*Right Side view*) are rotated into the same plane as the primary image plane (*Front View*). Dashed lines are used to indicate edges of the design that are not visible.

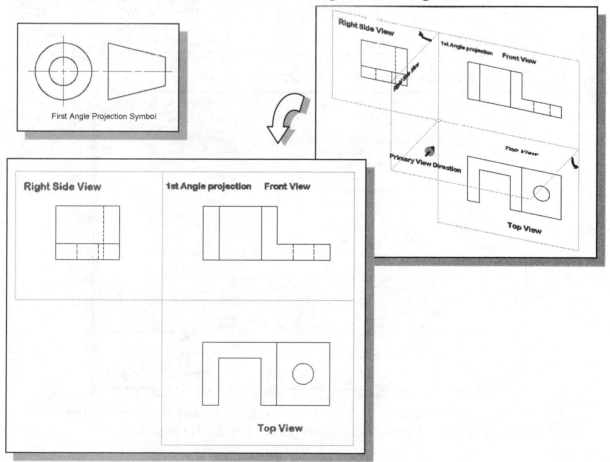

## The 3D Adjuster Model and 1st Angle Projection

❖ The **Adjuster1stAngle.SLDPRT** model is available on the SDC Publications website.

1. Launch your internet browser, such as the *MS Internet Explorer* or *Mozilla Firefox* web browsers.

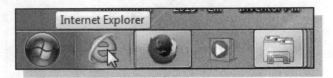

2. Download the SOLIDWORKS part file using the following *URL address*: www.SDCpublications.com/downloads/978-1-63057-240-2.

3. Select the **SOLIDWORKS** option on the *Start* menu or select the **SOLIDWORKS** icon on the desktop to start SOLIDWORKS. The SOLIDWORKS main window will appear.

4. Select the **Open** icon with a single click of the left-mouse-button on the *Menu Bar* toolbar.

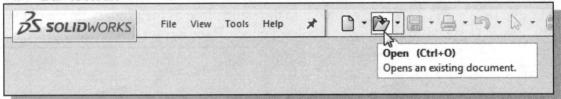

5. Select the downloaded file and click **Open** as shown.

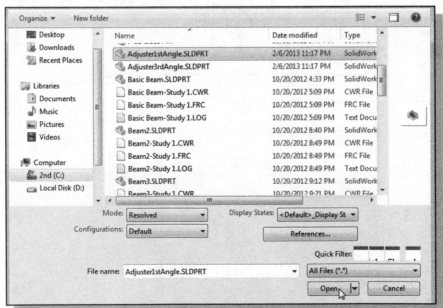

- On your own, use the real-time dynamic rotation feature and examine the relations of the 2D views, projection planes and the 3D object.

# General Procedure: 1ˢᵗ Angle Orthographic Projection

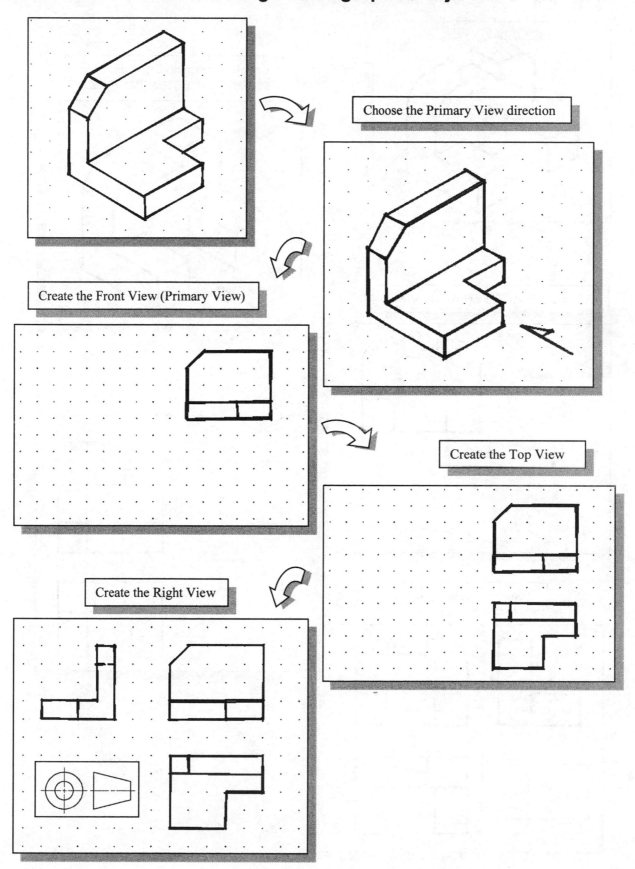

Choose the Primary View direction

Create the Front View (Primary View)

Create the Top View

Create the Right View

## Example 2: 1ˢᵗ Angle Orthographic Projection

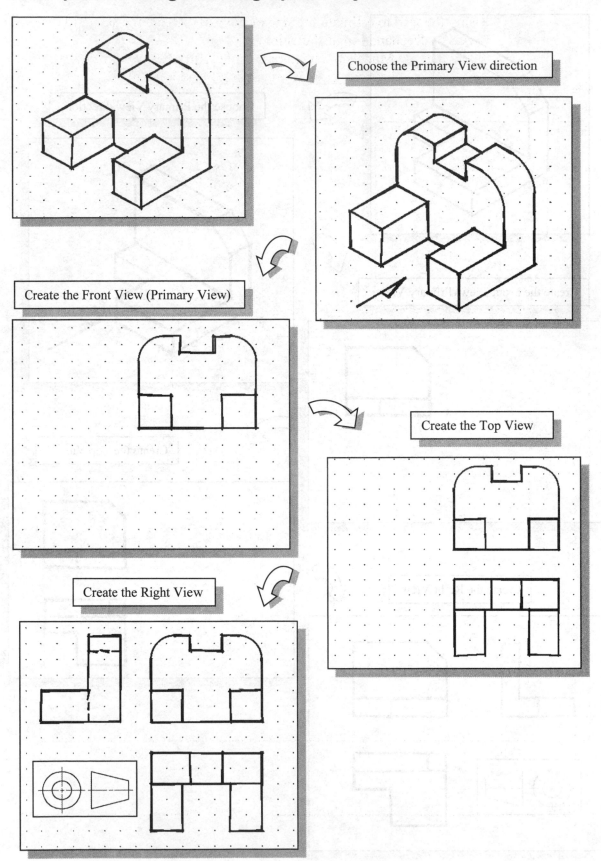

Choose the Primary View direction

Create the Front View (Primary View)

Create the Top View

Create the Right View

## Chapter 7 - 1<sup>st</sup> Angle Orthographic Sketching Exercise 1:

Using the grid to estimate the size of the parts, create the standard three views by freehand rough sketching.
(Note that 3D Models of chapter examples and exercises are available at www.SDCpublications.com/downloads/978-1-63057-240-2.)

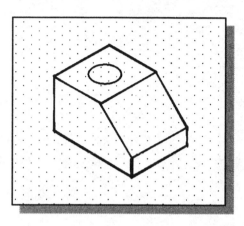

Name: _____     Date: _____

**Notes:**

## Chapter 7 - 1st Angle Orthographic Sketching Exercise 2:

Using the grid to estimate the size of the parts, create the standard three views by freehand rough sketching.
(Note that 3D Models of chapter examples and exercises are available at http://www.SDCpublications.com/downloads/978-1-63057-240-2.)

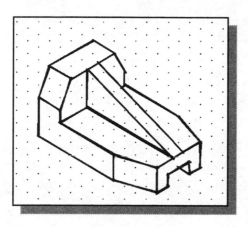

Name: _____    Date: _____

**Notes:**

## Chapter 7 - 1ˢᵗ Angle Orthographic Sketching Exercise 3:

Using the grid to estimate the size of the parts, create the standard three views by freehand rough sketching.
(Note that 3D Models of chapter examples and exercises are available at http://www.SDCpublications.com/downloads/978-1-63057-240-2.)

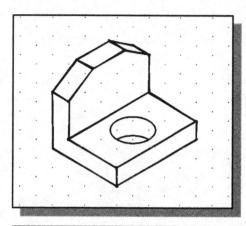

Name: _____        Date: _____

**Notes:**

## Chapter 7 - 1ˢᵗ Angle Orthographic Sketching Exercise 4:

Using the grid to estimate the size of the parts, create the standard three views by freehand rough sketching.
(Note that 3D Models of chapter examples and exercises are available at http://www.SDCpublications.com/downloads/978-1-63057-240-2.)

Name: _____        Date: _____

**Notes:**

## Chapter 7 - 1ˢᵗ Angle Orthographic Sketching Exercise 5:

Using the grid to estimate the size of the parts, create the standard three views by freehand rough sketching.
(Note that 3D Models of chapter examples and exercises are available at http://www.SDCpublications.com/downloads/978-1-63057-240-2.)

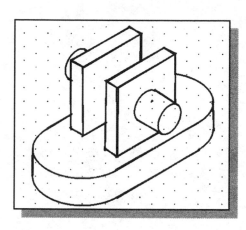

Name: _____     Date: _____

**Notes:**

## Chapter 7 - 1ˢᵗ Angle Orthographic Sketching Exercise 6:

Using the grid to estimate the size of the parts, create the standard three views by freehand rough sketching.
(Note that 3D Models of chapter examples and exercises are available at http://www.SDCpublications.com/downloads/978-1-63057-240-2.)

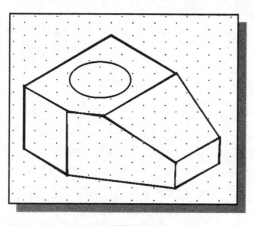

Name: _____        Date: _____

**Notes:**

## Third-Angle Projection

❖ In the United States, *third-angle projection* is commonly used in Engineering Drawings.

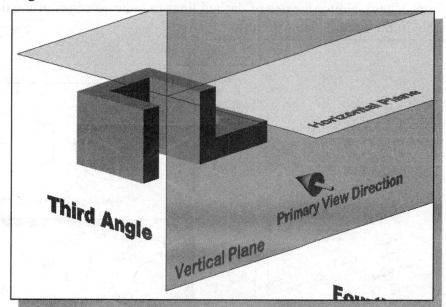

❖ In *third-angle projection*, the image planes are placed in between the object and the observer. And the views are formed by projecting to the image plane located in front of the object.

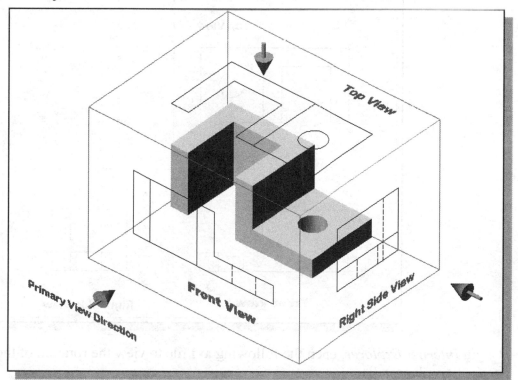

## Rotation of the Horizontal and Profile Planes

In order to draw all three views of the object on the same plane, the horizontal (Top View) and profile (Right Side view) are rotated into the same plane as the primary image plane (Front View). Notice the use of dashed lines to indicate edges of the design that are not visible in the specific view direction.

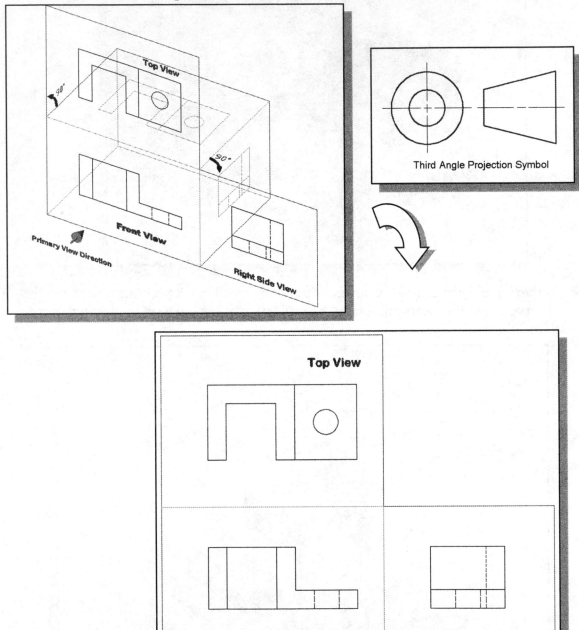

> ➢ Using *Internet Explorer*, open the following **avi** file to view the rotation of the projection planes:
> www.SDCpublications.com/downloads/978-1-63057-240-2

## The 3D Adjuster Model and 3rd angle projection

❖ The **Adjuster3rdAngle.SLDPRT** model file is available on the SDC Publications website.

1. Launch your internet browser, such as the *MS Internet Explorer* or *Mozilla Firefox* web browsers.

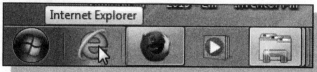

2. Download the SOLIDWORKS part file using the following *URL address*: http://www.SDCpublications.com/downloads/978-1-63057-240-2.

3. Select the **SOLIDWORKS** option on the *Start* menu or select the **SOLIDWORKS** icon on the desktop to start SOLIDWORKS. The SOLIDWORKS main window will appear.

4. Select the **Open** icon with a single click of the left-mouse-button on the *Menu Bar* toolbar.

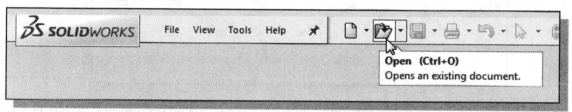

5. Select the downloaded file and click **Open** as shown.

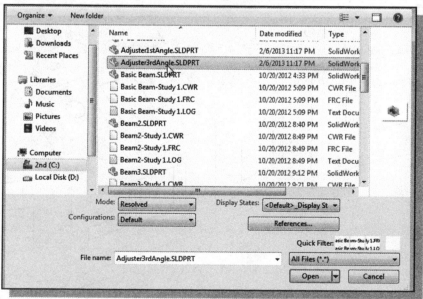

• On your own, use the real-time dynamic rotation feature and examine the relations of the 2D views, projection planes and the 3D object.

## The Glass Box and the Six Principal Views

Considering the third angle projection described in the previous section further, we find that the object can be entirely surrounded by a set of six planes, a Glass box. On these planes, views can be obtained of the object as it is seen from the top, front, right side, left side, bottom, and rear.

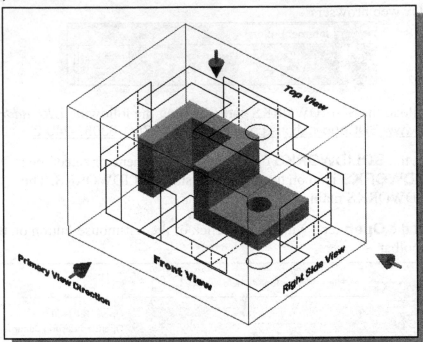

❖ Consider how the six sides of the glass box are being opened up into one plane. The front is the primary plane, and the other sides are hinged and rotated into position.

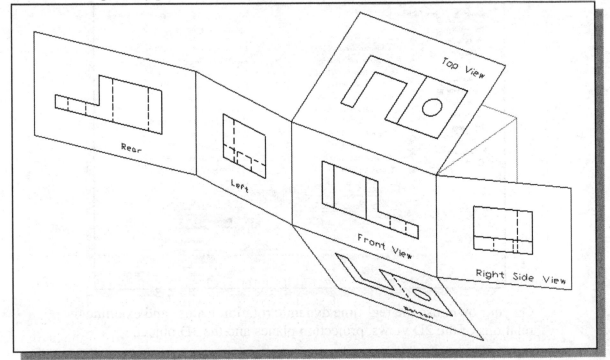

❖ In actual work, there is rarely an occasion when all six principal views are needed on one drawing, but no matter how many are required, their relative positions need to be maintained. These six views are known as the **six principal views**. In performing orthographic projection, each of the 2D views shows only two of the three dimensions (**height**, **width**, and **depth**) of the 3D object.

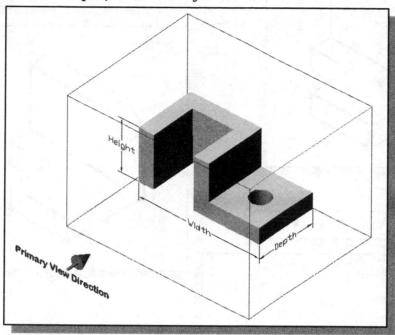

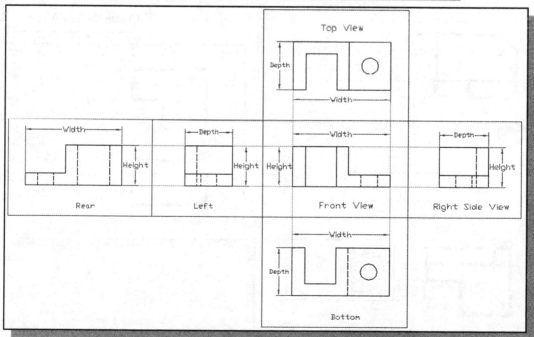

❖ The most common combination selected from the six possible views consists of the *top*, *front* and *right-side* views. The selection of the *Primary View* is the first step in creating a *multi-view drawing*. The Primary View should typically be (1) the view which describes the main features of the design and (2) considerations of the layout of the other 2D views to properly describe the design.

## General Procedure: 3rd Angle Orthographic Projection

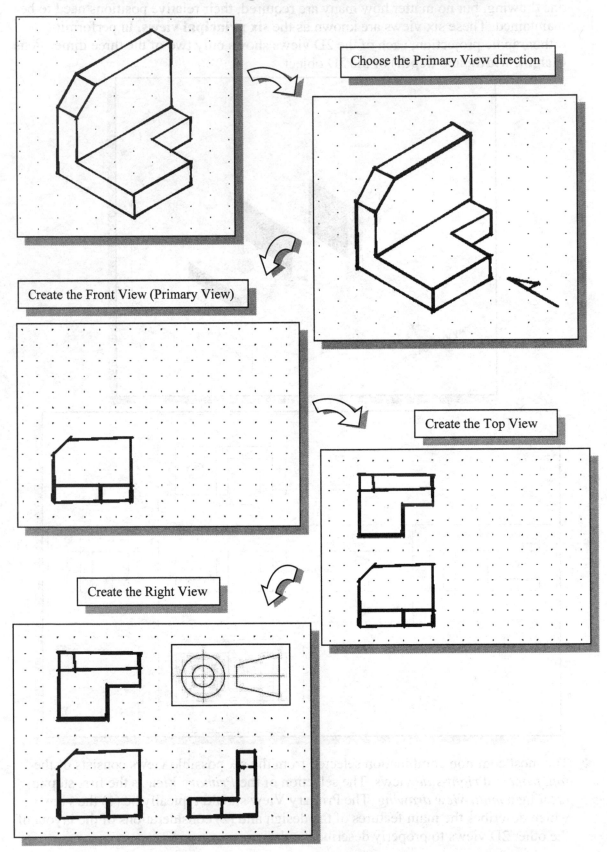

Choose the Primary View direction

Create the Front View (Primary View)

Create the Top View

Create the Right View

# Example 2: 3rd Angle Orthographic Projection

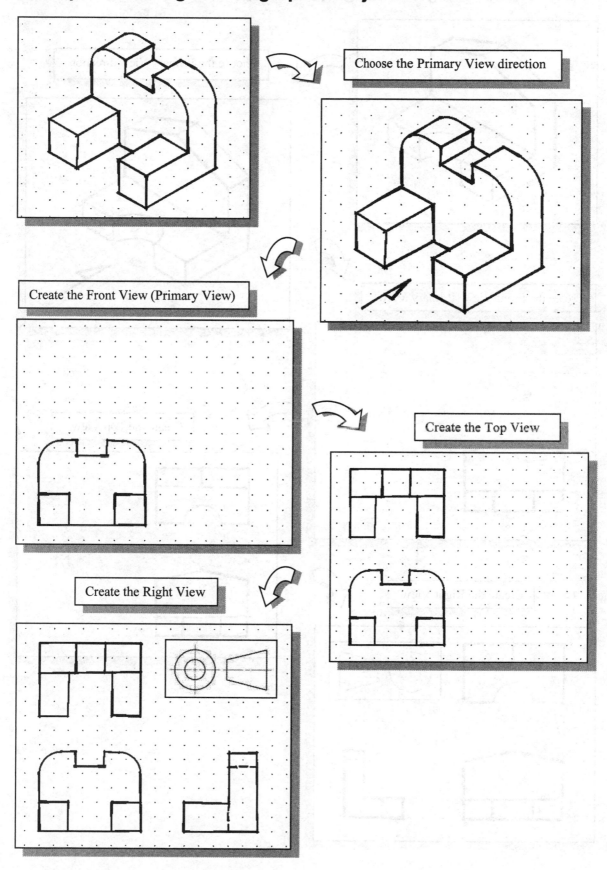

Choose the Primary View direction

Create the Front View (Primary View)

Create the Top View

Create the Right View

# Example 3: 3rd Angle Orthographic Projection

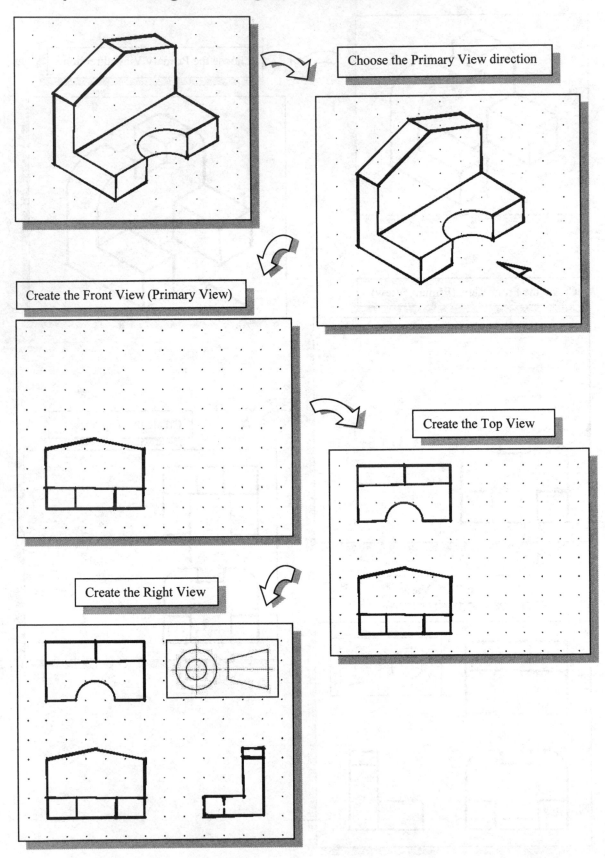

Choose the Primary View direction

Create the Front View (Primary View)

Create the Top View

Create the Right View

## Chapter 7 - 3<sup>rd</sup> Angle Orthographic Sketching Exercise 1:

Using the grid to estimate the size of the parts, create the standard three views by freehand rough sketching.
(Note that 3D Models of chapter examples and exercises are available at http://www.SDCpublications.com/downloads/978-1-63057-240-2.)

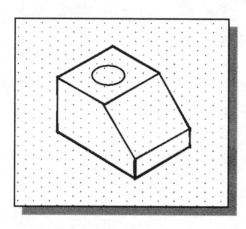

Name: _____    Date: _____

**Notes:**

## Chapter 7 - 3$^{rd}$ Angle Orthographic Sketching Exercise 2:

Using the grid to estimate the size of the parts, create the standard three views by freehand rough sketching.
(Note that 3D Models of chapter examples and exercises are available at http://www.SDCpublications.com/downloads/978-1-63057-240-2.)

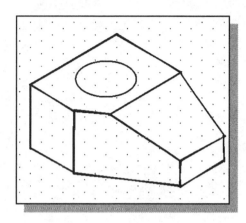

Name: _____        Date: _____

**Notes:**

## Chapter 7 - 3<sup>rd</sup> Angle Orthographic Sketching Exercise 3:

Using the grid to estimate the size of the parts, create the standard three views by freehand rough sketching.
(Note that 3D Models of chapter examples and exercises are available at http://www.SDCpublications.com/downloads/978-1-63057-240-2.)

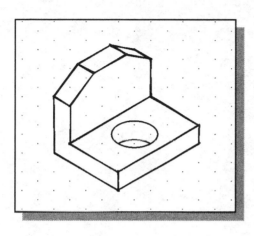

Name: _____        Date: _____

**Notes:**

## Chapter 7 - 3<sup>rd</sup> Angle Orthographic Sketching Exercise 4:

Using the grid to estimate the size of the parts, create the standard three views by freehand rough sketching.
(Note that 3D Models of chapter examples and exercises are available at http://www.SDCpublications.com/downloads/978-1-63057-240-2.)

Name: _____     Date: _____

**Notes:**

## Chapter 7 - 3rd Angle Orthographic Sketching Exercise 5:

Using the grid to estimate the size of the parts, create the standard three views by freehand rough sketching.
(Note that 3D Models of chapter examples and exercises are available at http://www.SDCpublications.com/downloads/978-1-63057-240-2.)

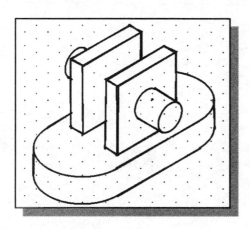

Name: _____     Date: _____

**Notes:**

## Chapter 7 - 3rd Angle Orthographic Sketching Exercise 6:

Using the grid to estimate the size of the parts, create the standard three views by freehand rough sketching.
(Note that 3D Models of chapter examples and exercises are available at http://www.SDCpublications.com/downloads/978-1-63057-240-2.)

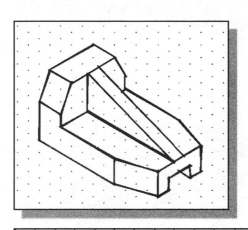

Name: _____    Date: _____

**Notes:**

# Alphabet of Lines

In *Technical Engineering Drawings*, each line has a definite meaning and is drawn in accordance to the line conventions as illustrated in the figure below. Two widths of lines are typically used on drawings; the thick line width should be 0.6 mm and the thin line width should be 0.3 mm.

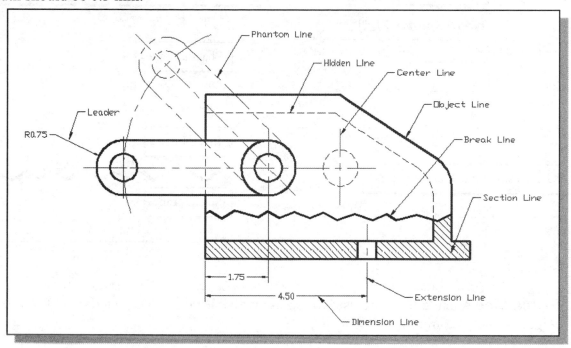

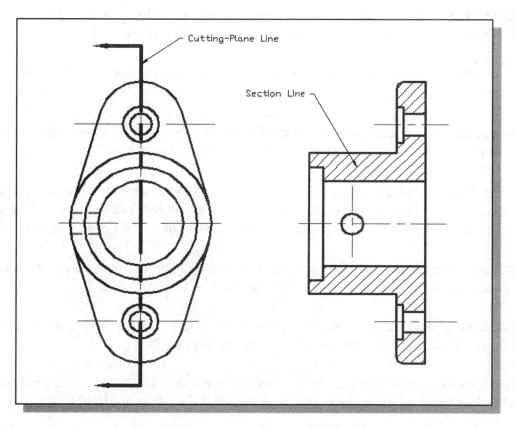

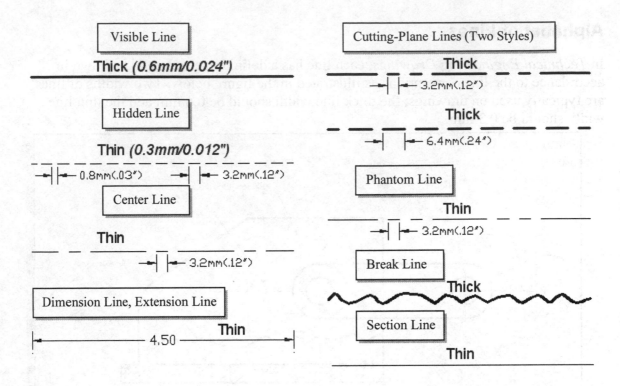

**Visible Line** Visible lines are used to represent visible edges and boundaries. The line weight is thick: 0.6mm/0.024″.

**Hidden Line** Hidden lines are used to represent edges and boundaries that are not visible from the viewing direction. The line weight is thin: 0.3mm/0.012″.

**Center Line** Center lines are used to represent axes of symmetry. The line weight is thin: 0.3mm/0.012″.

**Dimension Line, Extension Line and Leader** Dimension lines are used to show the sizes and locations of objects. The line weight is thin: 0.3mm/0.012″.

**Cutting Plane Lines** Cutting Plane lines are used to represent the location of where an imaginary cut has been made, so that the interior of the object can be viewed. The line weight is thick: 0.6mm/0.024″. (Note that two forms of line types can be used.)

**Phantom Line** Phantom lines are used to represent imaginary features or objects, such as a rotated position of a part. The line weight is thin: 0.3mm/0.012″.

**Break Line** Break lines are used to represent imaginary cuts, so that the interior of the object can be viewed. The line weight is thick: 0.6mm/0.024″.

**Section Line** Section lines are used to represent the regions that have been cut with the *break lines* or *cutting plane lines*. The line weight is thin: 0.3mm/0.012″.

## Precedence of Lines

In multiview drawings, coincidence lines may exist within the same view. For example, hidden features may project lines to coincide with the visible object lines. And center lines may occur where there is a visible or hidden outline.

In creating a multiview drawing, the features of the design are to be represented; therefore, object and hidden lines take precedence over all other lines. And since the visible outline is more important than hidden features, the visible object lines take precedence over hidden lines as shown in the below figure.

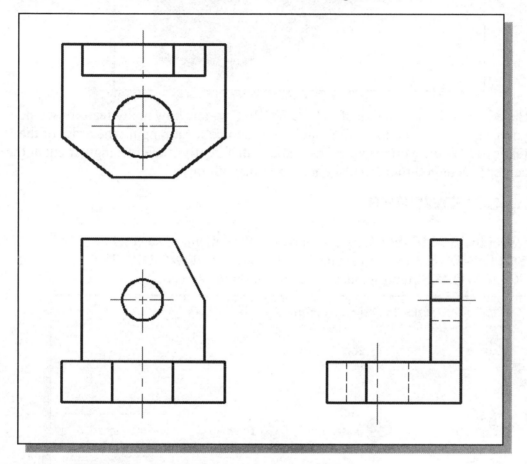

The following list gives the order of precedence of lines:
1.  **Visible object lines**
2.  **Hidden lines**
3.  **Center line or cutting-plane line**
4.  **Break lines**
5.  **Dimension and extension lines**
6.  **Crosshatch/section lines**

In the following sections, the general procedure of creating a 3$^{rd}$ angle three-view orthographic projection from a solid model using SOLIDWORKS is presented.

## The *U-Bracket* Design

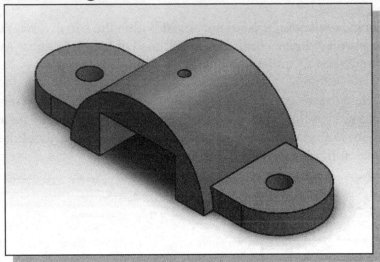

➢ Based on your knowledge of SOLIDWORKS so far, how many features would you use to create the model? What is your choice for arranging the order of the features? Would you organize the features differently if the rectangular cut at the center is changed to a circular shape (Radius: 80mm)?

## Starting SOLIDWORKS

1. Select the **SOLIDWORKS** option on the *Start* menu or select the **SOLIDWORKS** icon on the desktop to start SOLIDWORKS. The SOLIDWORKS main window will appear on the screen.

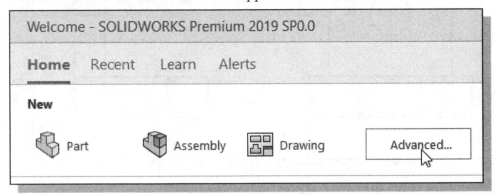

2. In the *Welcome to SOLIDWORKS* dialog box, enter the **Advanced** mode; click once with the left-mouse-button on the **Advanced** icon.

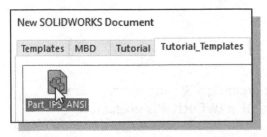

3. Select the the **Part_IPS_ANSI** template under the **Tutorial_Templates** tab. (**NOTE:** You added this tab in Chapter 5. If your template is not available, open the standard Part template, then set the *Drafting Standard* to ANSI and the *Units* to IPS.)

4. Click on the **OK** button to start a new model.

## Applying the BORN Technique

In the previous chapters, we have used a technique of creating solid models known as the **"Base Orphan Reference Node"** (**BORN**) technique. The basic concept of the BORN technique is to use a *Cartesian coordinate system* as the first feature prior to creating any solid features. With the *Cartesian coordinate system* established, we then have three mutually perpendicular datum planes (namely the *XY or 'front'*, *YZ or 'right'*, and *ZX or 'top' planes*) available to use as sketching planes. The three datum planes can also be used as references for dimensions and geometric constructions. Using this technique, the first node in the design tree is called an "orphan," meaning that it has no history to be replayed. The technique of creating the reference geometry in this "base node" is therefore called the "Base Orphan Reference Node" (BORN) technique.

SOLIDWORKS automatically establishes a set of reference geometry when we start a new part, namely a *Cartesian coordinate system* with three work planes, three work axes, and an origin. All subsequent solid features can then use the coordinate system and/or reference geometry as sketching planes.

❖ In the *Feature Manager Design Tree* panel, notice a new part name (Part1) appears with four work features established. The four work features include three *workplanes*, and the *origin*. By default, the three workplanes are aligned to the **world coordinate system** and the origin is aligned to the *origin* of the **world coordinate system**.

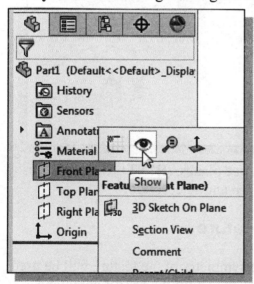

1. Inside the *Feature Manager Design Tree* panel, move the cursor on top of the first work plane—the **Front (XY) Plane**. Notice a rectangle, representing the selected workplane, appears in the graphics window.

2. Inside the *Feature Manager Design Tree* panel, click once with the right-mouse-button on Front (XY) Plane to display the option menu. Click on the **Show** icon to toggle *ON* the display of the plane.

3. On your own, repeat the above steps and toggle *ON* the display of all of the *workplanes* on the screen.

4. On your own, use the *Dynamic Viewing* options (Rotate, Zoom and Pan) to view the planes.

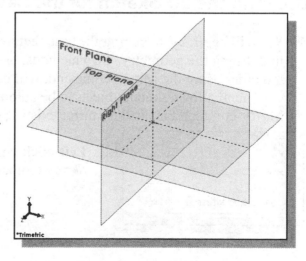

❖ By default, the basic set of planes is aligned to the world coordinate system; the workplanes are the first features of the part. We can now proceed to create solid features referencing the three mutually perpendicular datum planes. We will sketch the first feature, the base feature, in the top (XZ) plane. We will sketch the second feature in the front (XY) plane. Due to the symmetry in the shape of the part, we will use the right (YZ) plane and the front (XY) planes as planes of symmetry for the part. Thus, before starting, we have decided on the orientation and the location of the part in relation to the basic workplanes.

5. Move the cursor inside the graphics area, but away from the planes, and click once with the **left-mouse-button** to make sure no plane is selected.

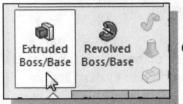

6. Select the **Extruded Boss/Base** button on the *Features* toolbar to create a new extruded feature.

7. In the *Edit Sketch Property-Manager*, the message "*Select a plane on which to create a sketch for the entity*" is displayed. Move the cursor over the edge of the Top Plane in the graphics area. When the Top Plane is highlighted, click once with the **left-mouse-button** to select the **Top (XZ) Plane** as the sketch plane for the new sketch.

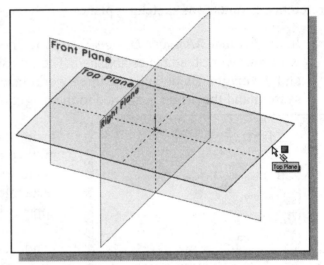

8. On your own, turn *OFF* the visibility of the three planes. (**HINT:** Right-click on each plane in the *Feature Manager Design Tree* and select Hide.)

## Creating the 2D Sketch for the Base Feature

❖ We will begin by sketching the base feature. The symmetry of the feature will be used in defining the geometry. Two methods, one using the Dynamic Mirror command, the other using the Mirror command, will be demonstrated. The Dynamic Mirror command allows sketch objects to be automatically mirrored as they are drawn. The Mirror command is used to mirror existing objects.

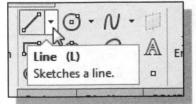

1. **Left-click** on the arrow next to the **Line** button on the *Sketch* toolbar to reveal additional commands.

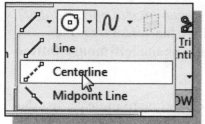

2. Select the **Centerline** command from the pop-up menu.

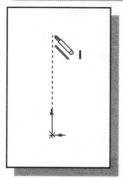

3. Move the cursor to a point directly above the *origin*. The Vertical relation will be automatically highlighted as shown. Click once with **the left-mouse-button** to select the first endpoint of the centerline.

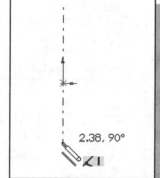

4. Move the cursor to a point directly below the *origin*. The Vertical relation will be highlighted as shown. Click once with the **left-mouse-button** to select the second endpoint of the centerline.

5. Inside the graphics window, right-mouse-click to bring up the **option menu**.

6. In the option menu, choose **Select** to end the Centerline command.

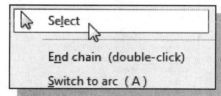

❖ The **Mirror** and **Dynamic Mirror** commands can be used to generate sketch objects with symmetry about a centerline.

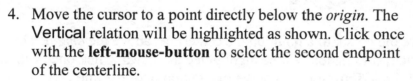

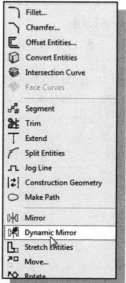

7. Select the **Tools** pull-down menu, and click on **Sketch Tools**. A complete list of sketch tools will appear. (Some of these also appear in the *Sketch* toolbar.)

8. Switch on the **Dynamic Mirror** option as shown.

9. The message *"Please select a sketch line or linear edge to mirror about"* appears in the *Mirror Property Manager*. Move the cursor over the centerline in the graphics area and click once with the **left-mouse-button** to select it as the line to mirror about.

➤ Notice the *hash marks* which appear on the centerline to indicate that the Dynamic Mirror command is active.

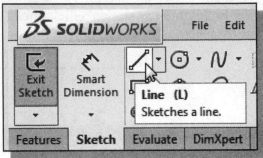

10. Click on the **Line** icon in the *Sketch* toolbar.

11. Move the cursor to a point on the centerline, above the *origin*. The Coincident relation will be highlighted as shown. Click once with the **left-mouse-button** to select the first endpoint of the line.

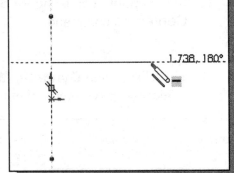

12. Move the cursor to a point to the right, activating the Horizontal relation, as shown. Click once with the **left-mouse-button** to select the second endpoint of the line.

13. Hit the [**Esc**] key once to end the **Line** command.

➤ Notice the mirror image of the line appears to the left of the centerline.

14. On your own, repeat the above steps to create a similar horizontal line below the origin. Be sure that the Horizontal and Vertical relations are active for the second endpoint as shown in the image to the right.

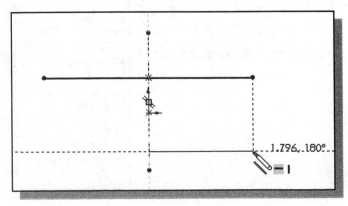

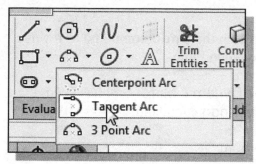

15. **Left-click** on the arrow next to the **Arc** button on the *Sketch* toolbar to reveal additional commands.

16. Select the **Tangent Arc** command from the pop-up menu.

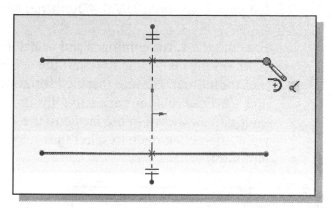

17. Select the right endpoint of the top horizontal line, as shown to the right.

18. Select the right endpoint of the bottom horizontal line, as shown to the right.

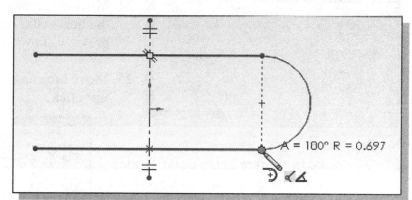

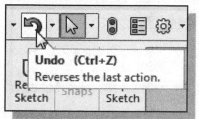

19. Inside the graphics window, right-click to bring up the option menu. In the option menu, select **Select** to end the Tangent Arc command.

❖ SOLIDWORKS also offers the option to *auto-transition* from the Line command to the Tangent Arc command (and vice versa) without activating the Arc command. We will redraw the second line and the tangent arc to demonstrate this utility.

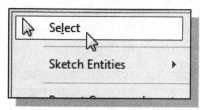

20. Click once with the left mouse button on the **Undo** command on the *Menu Bar* to reverse the last action – the creation of the arc.

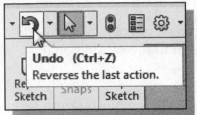

21. Click a second time on the **Undo** command to reverse the creation of the second line.

❖ Notice that the Dynamic Mirror tool has been disabled by the Undo command.

22. On your own, **turn *ON*** the Dynamic Mirror tool.

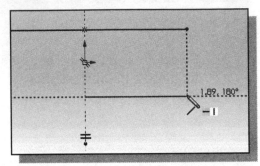

23. Execute the **Line** command and select a point on the centerline to restart the second horizontal line. Be sure that the Horizontal and Vertical relations are active for the endpoint as shown in the image to the right, then click once to select the endpoint.

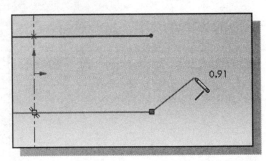

24. Move the cursor away from the endpoint of the horizontal line. **Do not click**. Notice the preview shows another line.

25. Move the cursor back to the endpoint. **Do not click**.

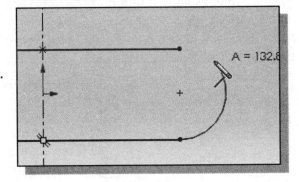

26. Move the cursor away again. Notice the preview now shows a tangent arc.

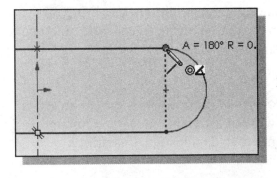

27. Select the right endpoint of the top horizontal line to end the arc.

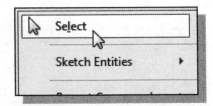

28. Inside the graphics window, right-mouse-click to bring up the option menu. In the option menu, choose **Select** to end the Line/Tangent Arc command.

29. On your own, **turn *OFF*** the Dynamic Mirror option.

❖ We will now constrain the center of the arc to be on a horizontal line with the origin.

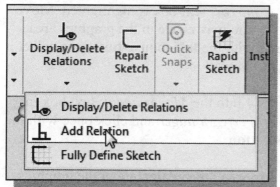

30. **Left-click** on the arrow icon below the **Display/Delete Relations** button on the *Sketch* toolbar to reveal additional sketch relation commands.

31. Select the **Add Relation** command from the pop-up menu. Notice the *Add Relations Property Manager* appears.

32. Pick the **centerpoint** of the tangent arc to the right side by left-clicking once on the geometry.

33. Pick the **origin** of the displayed coordinate system.

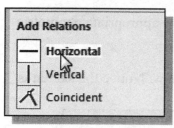

34. Click once with the left-mouse-button on the **Horizontal** icon in the *Add Relations Property Manager* as shown. This activates the **Horizontal** relation.

35. Click the **OK** icon in the *Property Manager*, or hit the **[Esc]** key once, to end the **Add Relations** command.

36. On your own, draw a **circle** as shown in the figure. Make sure the centerpoint for the circle is selected at the center of the tangent arc.

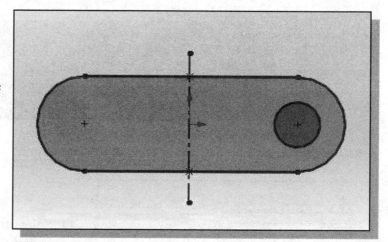

37. Press the **[Esc]** key to ensure no sketch features are selected.

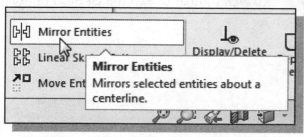

38. Select the **Mirror Entities** command on the *Sketch* toolbar.

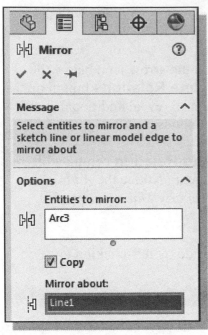

39. In the *Options* panel of the *Mirror Property Manager*, the *Entities to mirror:* text box is highlighted. Move the cursor over the **new circle** in the graphics area and click once with the **left-mouse-button** to select it as the object to mirror.

40. Move the cursor into the *Mirror about:* text box in the *Mirror Property Manager* and click once with the **left-mouse-button**.

41. Move the cursor over the **centerline** in the graphics area and click once with the **left-mouse-button** to select it as the object to mirror about.

42. Select **OK** in the *Mirror Property Manager*.

❖ We have now completed the rough sketch and applied the appropriate geometric relations. Remember the concept is 'shape before size'.

43. On your own, apply the *Smart Dimensions* as shown below. Notice the sketch is now **Fully Defined**.

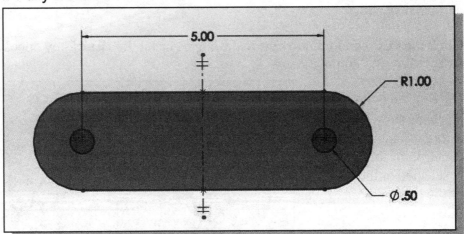

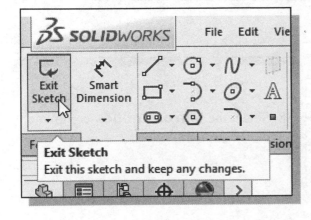

44. Select the **Exit Sketch** icon on the *Sketch* toolbar to exit the **Sketch** option.

## Creating the First Extrude Feature

1. On your own, proceed with the **Boss Extrude** command and create a base with a height of **0.5** inches.

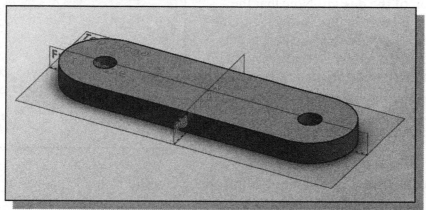

2. On your own, turn the **visibility** of the three major planes back *ON*. (**HINT:** Right-click on each plane in the *Feature Manager Design Tree* and select Show.)

➢ Notice the orientation of the base in relation to the major workplanes of the world coordinate system.

## The Implied Parent/Child Relationships

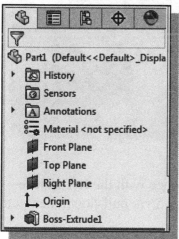

• The *Feature Manager Design Tree* display includes the **Front, Top,** and **Right Planes,** the **Origin** (default work features) and the base feature (**Boss-Extrude1** shown in the figure) we just created. The parent/child relationships were established implicitly when we created the base feature: (1) Top *(XZ) workplane* was selected as the sketch plane; (2) Origin was used as the reference point to align the 2D sketch of the base feature.

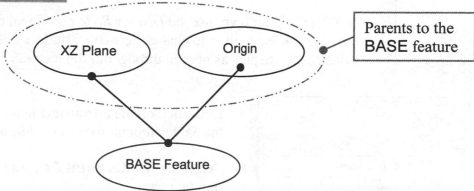

## Creating the Second Solid Feature

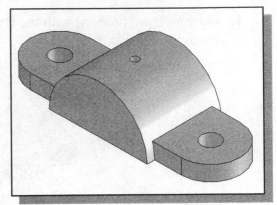

- For the next solid feature, we will create the top section of the design. Note that the center of the base feature is aligned to the **Origin** of the default work features. This was done intentionally so that additional solid features can be created referencing the default work features. For the second solid feature, the Front *(XY) workplane* will be used as the sketch plane.

1. Move the cursor inside the graphics area, but away from the planes, and click once with the **left-mouse-button** to make sure no plane is selected.

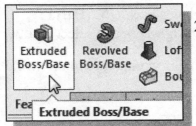

2. In the *Sketch* toolbar select the **Sketch** command by left-clicking once on the icon.

3. In the *Edit Sketch Property Manager*, the message "*Select a plane on which to create a sketch for the entity*" is displayed. Move the cursor over the edge of the Front Plane in the graphics area. When the Front Plane is highlighted, click once with the **left-mouse-button** to select the **Front (XY) Plane** as the sketch plane for the new sketch.

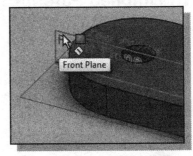

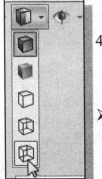

4. Change the display to wireframe by clicking once with the **left-mouse-button** on the **Wireframe** icon on the *Display Style* pull-down menu on the *Heads-up View* toolbar.

➤ NOTE: On your own, use the *Display Style* pull-down menu on the *Heads-up View* toolbar to change between **Shaded With Edges** and **Wireframe** display as needed throughout the lesson.

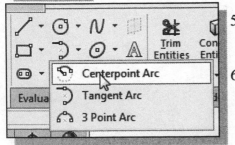

5. **Left-click** on the arrow next to the **Arc** button on the *Sketch* toolbar to reveal additional commands.

6. Select the **Centerpoint Arc** command from the pop-up menu.

7.  Pick the **Origin** as the center location of the new arc.

8.  On your own, create a semi-circle of arbitrary size, with both endpoints aligned to the X-axis, as shown below.

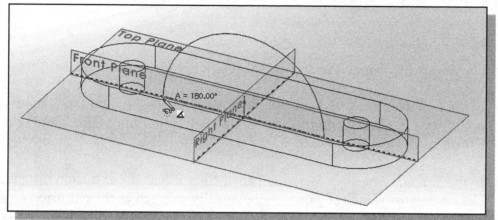

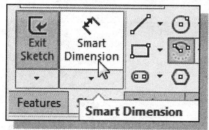

9.  Move the cursor on top of the **Smart Dimension** icon. Left-click once on the icon to activate the Smart Dimension command.

10. On your own, create and adjust the radius dimension of the arc to **1.75**.

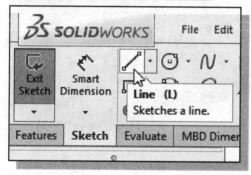

11. Select the **Line** command in the *Sketch* toolbar.

12. Create a line connecting the two endpoints of the arc as shown in the figure below.

13. On your own, add a **Collinar** relation between the **top plane** and the newly created **line**. The sketch should now be Fully Defined.

14. On your own, turn *OFF* the visibility of the workplanes.

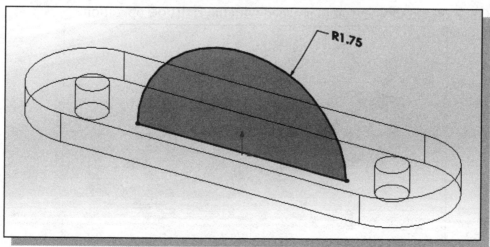

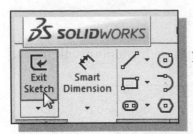

15. Select the **Exit Sketch** icon on the *Sketch* toolbar to exit the **Sketch** option.

16. On your own, change the display style back to '*Shaded with edges*'.

17. In the *Extrude Property Manager* panel, click the drop-down arrow to reveal the options for the *End Condition* (the default end condition is Blind) and select **Mid Plane** as shown.

18. In the *Extrude Property Manager* panel, enter **2.5** as the extrusion distance. Notice that the sketch region is automatically selected as the extrusion profile.

19. Make sure the **Merge result** box is checked.

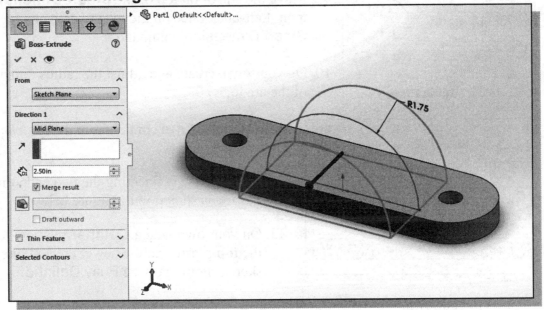

20. Click on the **OK** button to proceed with the **Extrude** operation.

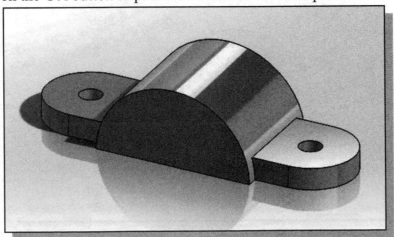

## Creating the First Extruded Cut Feature

- A rectangular cut will be created as the next solid feature.

1. In the *Sketch* toolbar select the **Sketch** command by left-clicking once on the icon.

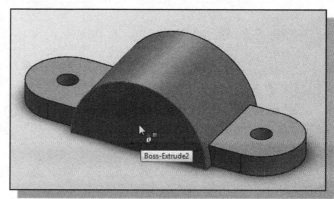

2. In the *Edit Sketch Property Manager*, the message "*Select a plane on which to create a sketch for the entity*" is displayed. Pick the front vertical face of the solid model.

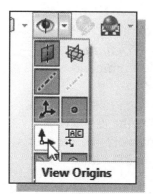

3. We will use the origin as an endpoint for one of the dimensions. Toggle on the **View Origins** option under the *Hide/Show Items* pull-down menu on the *Heads-up View* toolbar.

4. On your own, create a rectangle and apply the dimensions as shown. (**HINT:** To apply the horizontal location dimension, select the **origin** of the world coordinate system as an endpoint. SOLIDWORKS will automatically project the origin onto the workplane.)

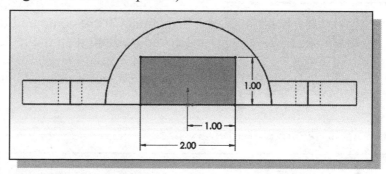

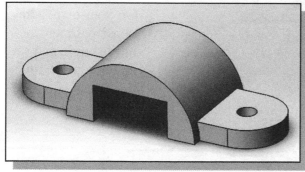

5. On your own, exit the sketch and use the **Extruded Cut** command with the **Through All** end condition option to create a cutout that cuts through the entire 3D solid model as shown.

## Creating the Second Extruded Cut Feature

1. In the *Features* toolbar select the **Extruded Cut** command by left-clicking once on the icon.

2. In the *Edit Sketch Property Manager*, the message "*Select a plane on which to create a sketch for the entity*" is displayed. Select the horizontal face of the last cut feature as the *sketching plane*.

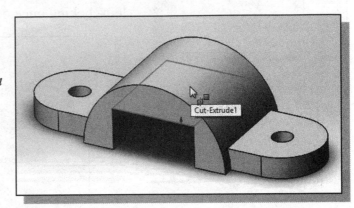

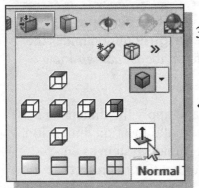

3. Left-click on the **View Orientation** icon on the *Heads-up View* toolbar to reveal the *View Orientation* pull-down menu. Select the **Normal To** command.

❖ The *Normal To* command can be used to switch to a 2D view of a selected surface. Since the surface defining the current sketch plane was pre-selected, this surface is automatically used for the 2D view.

4. Select the **Circle** command by clicking once with the left-mouse-button on the icon in the *Sketch* toolbar.

5. Pick the projected origin to align the center of the new circle.

6. Create a circle of arbitrary size.

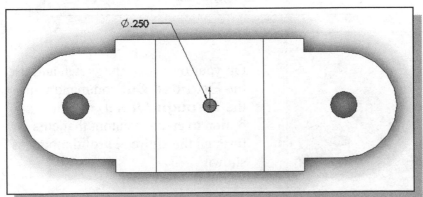

7. On your own, add the size dimension of the circle and set the dimension to **0.25**.

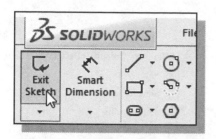

8. Select the **Sketch** icon on the *Sketch* toolbar to exit the Sketch option.

9. In the *Cut-Extrude Property Manager* panel, click the arrow to reveal the pull-down options for the *end condition* (the default end condition is Blind) and select **Through All**.

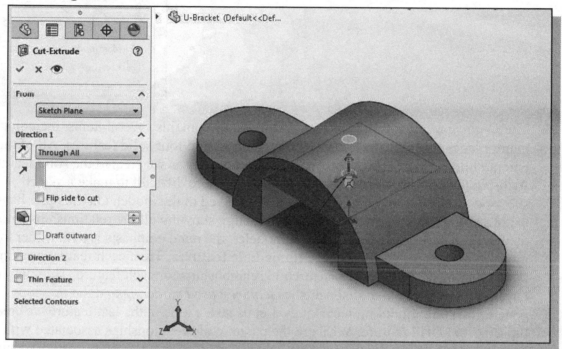

10. Click the **OK** button (green check mark) in the *Cut-Extrude Property Manager* panel.

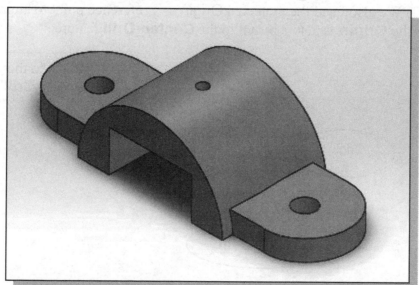

## Examining the Parent/Child Relationships

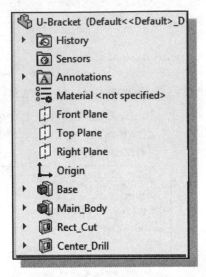

1. On your own, rename the feature names to **Base**, **Main_Body**, **Rect_Cut** and **Center_Drill** as shown in the figure.

2. On your own, save the model using **U-Bracket** as the filename.

❖ The *Feature Manager Design Tree* window now contains several items, including the major planes, the **Origin** (default work features) and four solid features. All of the parent/child relationships were established implicitly as we created the solid features. As more features are created, it becomes much more difficult to make a sketch showing all the parent/child relationships involved in the model. On the other hand, it is not really necessary to have a detailed picture showing all the relationships among the features. In using a feature-based modeler, the main emphasis is to consider the interactions that exist between the **immediate features**. Treat each feature as a unit by itself and be clear on the parent/child relationships for each feature. Thinking in terms of *features* is what distinguishes *feature-based modeling* and the previous generation solid modeling techniques. Let us take a look at the last feature we created, the **Center_Drill** feature. What are the parent/child relationships associated with this feature? (1) Since this is the last feature we created, it is not a parent feature to any other features. (2) Since we used one of the surfaces of the rectangular cutout as the sketching plane, the **Rect_Cut** feature is a parent feature to the **Center_Drill** feature. (3) We also used the projected Origin as a reference point to align the center; therefore, the **Origin** is also a parent to the **Center_Drill** feature.

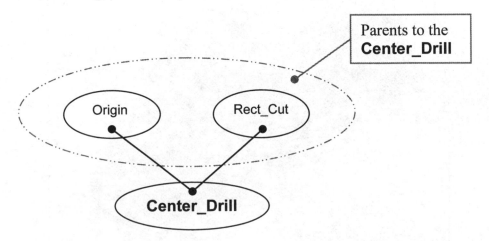

## Modify a Parent Dimension

❖ Any changes to the parent features will affect the child feature. For example, if we modify the height of the **Rect_Cut** feature from 1.0 to 0.75, the depth of the child feature (**Center_Drill** feature) will be affected.

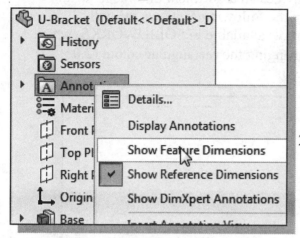

1. In the *Feature Manager Design Tree* window, move the cursor on top of **Annotations** and click once with the **right-mouse-button** to bring up the option menu.

2. In the option menu, select **Show Feature Dimensions**.

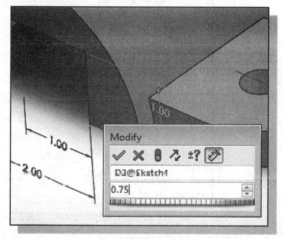

3. Select the **height** dimension (1.0) by double-clicking on the dimension.

4. Enter **0.75** as the new height dimension as shown.

5. Select **OK** in the *Modify* window.

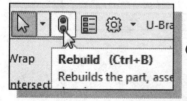

6. Click on the **Rebuild** button in the *Menu Bar* to proceed with updating the solid model.

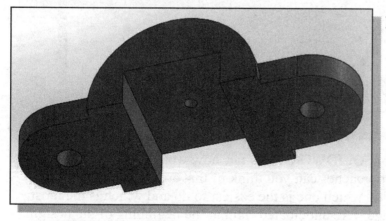

➢ Note that the position of the **Center_Drill** feature is also adjusted as the placement plane is lowered. The drill-hole still goes through the main body of the *U-Bracket* design. The parent/child relationship assures the intent of the design is maintained.

7. On your own, adjust the height of the **Rect_Cut** feature back to **1.0 inch** before proceeding to the next section.

8. Repeat steps 1 and 2 to turn **off** the *Show Feature Dimensions* option.

## A Design Change

➤ Engineering designs usually go through many revisions and changes. For example, a design change may call for a circular cutout instead of the current rectangular cutout feature in our model. SOLIDWORKS provides an assortment of tools to handle design changes quickly and effectively. In the following sections, we will demonstrate some of the more advanced tools available in SOLIDWORKS, which allow us to perform the modification of changing the rectangular cutout (2.0 × 1.0 inch) to a circular cutout (radius: 1.25 inch).

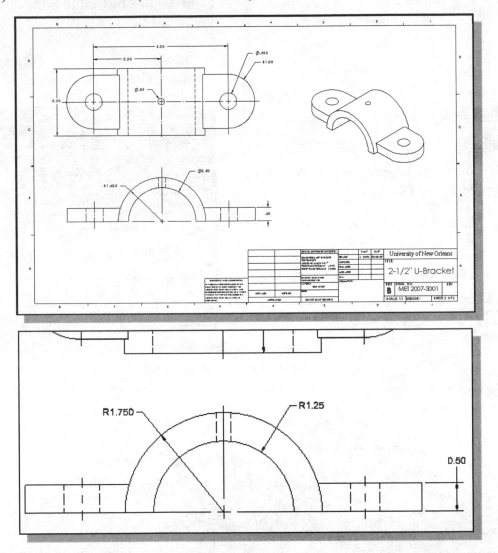

❖ Based on your knowledge of SOLIDWORKS so far, how would you accomplish this modification? What other approaches can you think of that are also feasible? Of the approaches you came up with, which one is the easiest to do and which is the most flexible? If this design change were anticipated right at the beginning of the design process, what would be your choice in arranging the order of the features? You are encouraged to perform the modifications prior to following through the rest of the tutorial.

# Feature Suppression

❖ With SOLIDWORKS, we can take several different approaches to accomplish this modification. We could (1) create a new model, or (2) change the shape of the existing cut feature, or (3) perform **feature suppression** on the rectangular cut feature and add a circular cut feature. The third approach offers the most flexibility and requires the least amount of editing to the existing geometry. **Feature suppression** is a method that enables us to disable a feature while retaining the complete feature information; the feature can be reactivated at any time. Prior to adding the new cut feature, we will first suppress the rectangular cut feature.

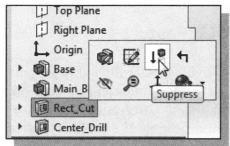

1. Move the cursor inside the *Feature Manager Design Tree* window. Click once with the **left-mouse-button** on top of **Rect_Cut** to bring up the option menu.

2. Pick **Suppress** in the pop-up list.

❖ We have literally *gone back in time*. The Rect_Cut and Center_Drill features have disappeared in the graphics area and both are shaded grey (meaning they are suppressed) in the *Feature Manager Design Tree*. The child feature cannot exist without its parent(s), and any modification to the parent (Rect_Cut) influences the child (Center_Drill).

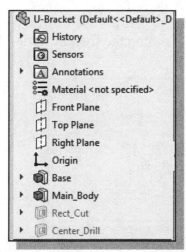

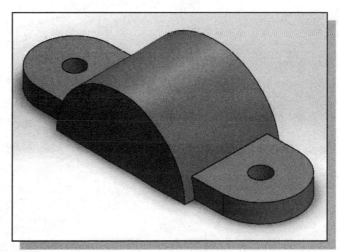

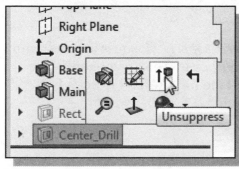

3. Move the cursor inside the *Feature Manager Design Tree* window. Click once with the **right-mouse-button** on top of **Center_Drill** to bring up the option menu.

4. Pick **Unsuppress** in the pop-up list.

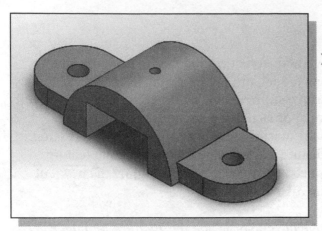

> In the display area and the *Model Tree* window, both the **Rect_Cut** feature and the **Center_Drill** feature are re-activated. The child feature cannot exist without its parent(s); the parent (Rect_Cut) must be activated to enable the child (Center_Drill).

## A Different Approach to the CENTER_DRILL Feature

The main advantage of using the BORN technique is to provide greater flexibility for part modifications and design changes. In this case, the Center_Drill feature can be placed on the Top (XZ) Workplane and therefore not be linked to the Rect_Cut feature. Again, we can take several different approaches to accomplish this modification. We could (1) delete the Center_Drill feature and the corresponding sketch, create a new sketch on the Top (XZ) Workplane, and recreate the Center_Drill feature, or (2) redefine the sketch plane for the existing sketch associated with the Center_Drill feature. We will perform the second option.

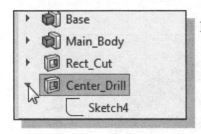

1. Move the cursor inside the *Feature Manager Design Tree* window. Click once with the **left-mouse-button** on the tree to reveal the Sketch defining the Center_Drill feature as shown.

2. Move the cursor inside the *Feature Manager Design Tree* window. Click once with the **left-mouse-button** on top of **Sketch4** to bring up the option menu.

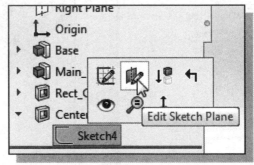

3. Pick **Edit Sketch Plane** in the pop-up menu. This will allow us to redefine the workplane for Sketch4.

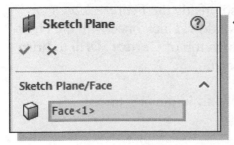

❖ The *Sketch Plane Property Manager* appears, with the *Sketch Plane/Face* text box highlighted. The original sketch plane for Sketch4 (Face<1>) appears in the field.

4. Notice the Part icon of the design tree now appears at the upper left corner of the graphics area. **Left-click** on the box next to the Part icon to reveal the design tree as shown.

5. Move the cursor over the **Top Plane** icon. When the Top Plane is highlighted, click once with the **left-mouse-button** to select the Top (XZ) Plane as the sketch plane for Sketch4.

6. Top Plane now appears in the *Sketch Plane/Face* field of the *Sketch Plane Property Manager*. Click once with the **left-mouse-button** on the **OK** icon to accept the new definition of the sketch plane.

## Suppress the Rect_Cut Feature

❖ Now the **Center_Drill** feature is no longer a child of the **Rect_Cut** feature, any change to the **Rect_Cut** feature does not affect the **Center_Drill** feature.

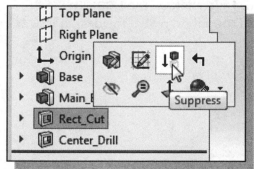

1. Move the cursor inside the *Model Tree* window. Click once with the left-mouse-button on top of **Rect_Cut** to bring up the option menu.

2. Pick **Suppress** in the pop-up menu.

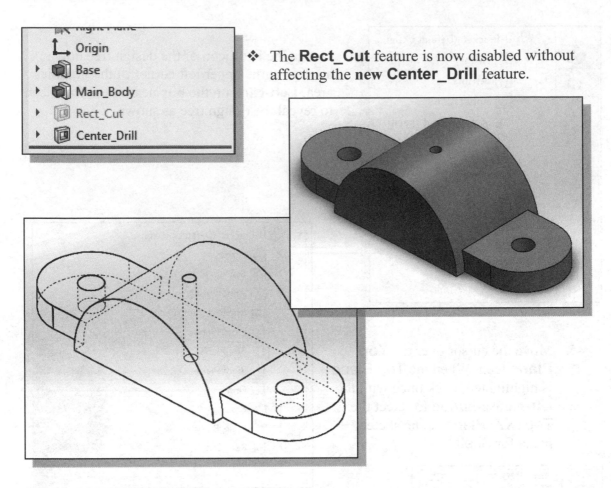

❖ The **Rect_Cut** feature is now disabled without affecting the **new Center_Drill** feature.

## Creating a Circular Extruded Cut Feature

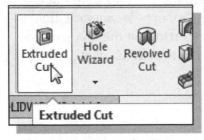

1. In the *Features* toolbar select the **Extruded Cut** command by left-clicking once on the icon.

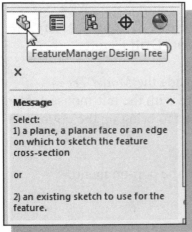

2. In the *Edit Sketch Property Manager*, the message *"Select a plane on which to create a sketch for the entity"* is displayed. **Left-click** on the Feature Manager Design Tree tab to reveal the design tree as shown.

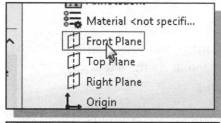

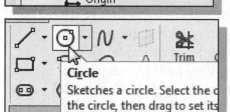

3. Move the cursor over the edge of the **Front Plane** icon in the *Feature Manager Design Tree*. When the Front Plane is highlighted, click once with the **left-mouse-button** to select the Front (XY) Plane as the sketch plane for the new sketch.

4. Select the **Circle** command by clicking once with the left-mouse-button on the icon in the *Sketch* toolbar.

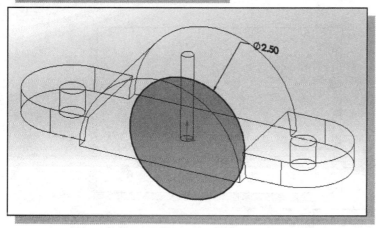

5. Pick the **Origin** to align the center of the new circle.

6. Create a circle of arbitrary size.

7. On your own, create the size dimension of the circle and set the dimension to **2.5**, as shown in the figure.

8. On your own, complete the cut feature using the **Extruded Cut** command. Use the **Through All-Both** option for the *depth*.

9. Rename the feature **Circular_Cut**.

❖ Note that the parents of the Circular_Cut feature are the Front (XY) Plane and the Origin.

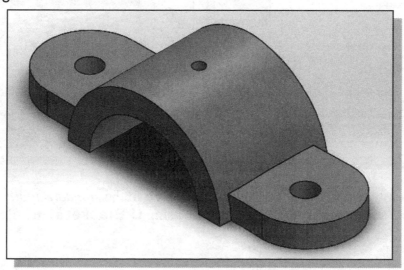

## A Flexible Design Approach

In a typical design process, the initial design will undergo many analyses, testing, reviews and revisions. SOLIDWORKS allows the users to quickly make changes and explore different options of the initial design throughout the design process.

The model we constructed in this chapter contains two distinct design options. The *feature-based parametric modeling* approach enables us to quickly explore design alternatives and we can include different design ideas into the same model. With parametric modeling, designers can concentrate on improving both the design and the design process. The key to successfully using parametric modeling as a design tool lies in understanding and properly controlling the interactions of features, especially the parent/child relations.

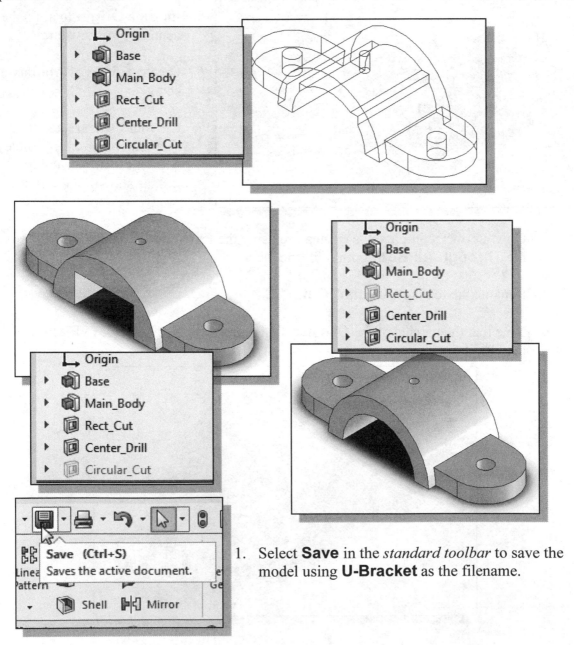

1.  Select **Save** in the *standard toolbar* to save the model using **U-Bracket** as the filename.

# Drawings from Parts and Associative Functionality

With the software/hardware improvements in solid modeling, the importance of two-dimensional drawings is decreasing. Drafting is considered one of the downstream applications of using solid models. In many production facilities, solid models are used to generate machine tool paths for *computer numerical control* (CNC) machines. Solid models are also used in *rapid prototyping* to create 3D physical models out of plastic resins, powdered metal, etc. Ideally, the solid model database should be used directly to generate the final product. However, the majority of applications in most production facilities still require the use of two-dimensional drawings. Using the solid model as the starting point for a design, solid modeling tools can easily create all the necessary two-dimensional views. In this sense, solid modeling tools are making the process of creating two-dimensional drawings more efficient and effective.

SOLIDWORKS provides associative functionality in the different SOLIDWORKS modes. This functionality allows us to change the design at any level, and the system reflects it at all levels automatically. For example, a solid model can be modified in the *Part Modeling* mode and the system automatically reflects that change in the *Drawing* mode. And we can also modify a feature dimension in the *Drawing* mode, and the system automatically updates the solid model in all modes.

In this lesson, the general procedure of creating multi-view drawings is illustrated. The *U_Bracket* design from the last chapter is used to demonstrate the associative functionality between the model and drawing views.

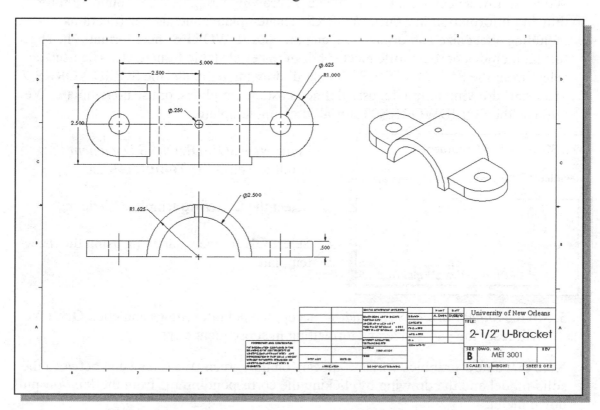

## Drawing Mode

➢ SOLIDWORKS allows us to generate 2D engineering drawings from solid models so that we can plot the drawings to any exact scale on paper. An engineering drawing is a tool that can be used to communicate engineering ideas/designs to manufacturing, purchasing, service, and other departments. Until now we have been working in *Model* mode to create our design in ***full size***. We can arrange our design on a two-dimensional sheet of paper so that the plotted hardcopy is exactly what we want. This two-dimensional sheet of paper is saved in a ***drawing*** file in SOLIDWORKS. We can place borders and title blocks, objects that are less critical to our design, in the *drawing*. In general, each company uses a set of standards for drawing content, based on the type of product and also on established internal processes. The appearance of an engineering drawing varies depending on when, where, and for what purpose it is produced. However, the general procedure for creating an engineering drawing from a solid model is fairly well defined. In SOLIDWORKS, creation of 2D engineering drawings from solid models consists of four basic steps: drawing sheet formatting, creating/positioning views, annotations, and printing/plotting.

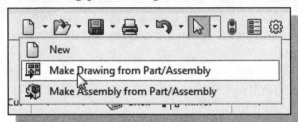

1. Click on the arrow next to the **New** icon on the *Menu Bar* and select **Make Drawing from Part/Assembly** in the pull-down menu.

➢ We must now select a drawing template. A drawing template is the foundation for drawing information in SOLIDWORKS. The template contains specifications including sheet size and orientation, and the SOLIDWORKS sheet format. The sheet format includes borders, title blocks, bill of materials table format, etc. The user can select from the SOLIDWORKS 'Standard' drawing template, the SOLIDWORKS 'Tutorial' drawing template, user-defined custom templates, or use no template. We will use the SOLIDWORKS 'Tutorial' drawing template.

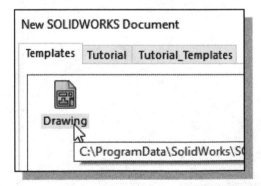

2. In the *New SOLIDWORKS Document* window, select the **Templates** tab.

3. Select the **Drawing** template file icon.

4. Select **OK** to open a drawing using the *draw* template.

5. In the *Sheet Format/Size* window, accept the default settings and click **OK**. (We will reset the paper size and formatting in subsequent steps.)

➢ Note that a new graphics window appears on the screen. We can switch between the solid model and the drawing by clicking the corresponding file from the *Window* pull-down menu.

➢ In the graphics window, SOLIDWORKS displays a default drawing sheet that includes a title block. The title block also indicates the paper size being used.

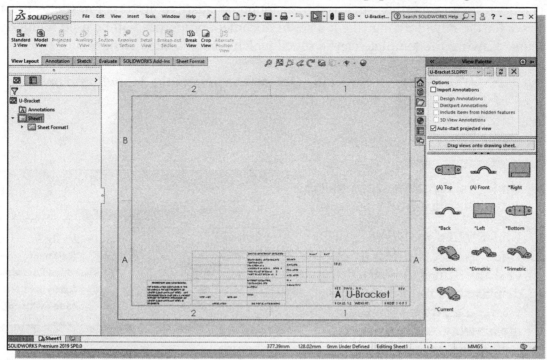

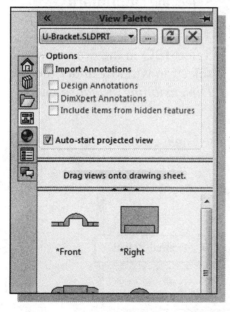

➢ The **View Palette** appears in the task pane. This can be used to insert drawing views. We will first set the document properties and sheet properties for the drawing.

➢ In the *Feature Manager Design Tree* area, the **Drawing** icon is displayed at the top, which indicates that we have switched to **Drawing** mode. The default filename is the filename of the solid part file from which we created the drawing – U-Bracket. *Sheet1* is the current drawing sheet that is displayed in the graphics window.

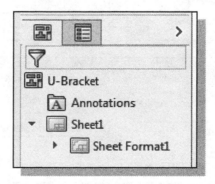

## Setting Document Properties

➤ The document properties, such as dimensioning standard, units, text fonts, etc., must be set for the new drawing file. If a set of document properties will be used regularly, a new drawing file template can be saved which contains these settings.

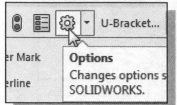

1. Select the **Options** icon from the *Menu Bar* to open the options dialog box.

2. Under the **System Options** tab, select **Display Style** as shown.

3. Under '*Display style for new views*', select **Hidden lines visible**. This will be the default display style for orthographic views (e.g., front view) added to the drawing.

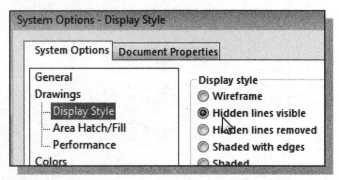

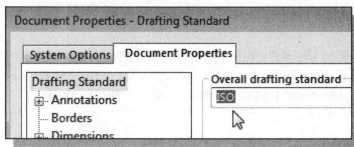

4. Select the **Document Properties** tab.

5. Select **ISO** in the drop-down selection window under the *Dimensioning standard* panel to reset.

6. Select **IPS (inch, pound, second)** under the *Unit system* options.

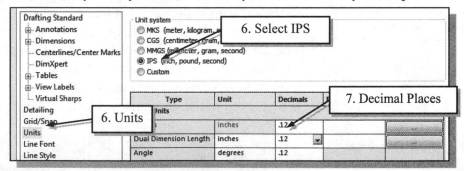

7. Confirm the degree of accuracy with which the different units will be displayed to 2 decimal places.

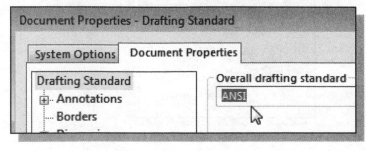

8. Set to use the ANSI standard and click **OK** to accept the settings.

## Setting Sheet Properties using the Pre-Defined Sheet Formats

➢ A set of properties are also set for each sheet in the drawing file. These settings include sheet size, sheet orientation, scale, and projection type.

1. Move the cursor on **Sheet1** in the model tree manager and click once with the **right-mouse-button** to open the pop-up menu.

2. Select **Properties** as shown to edit the properties for Sheet1.

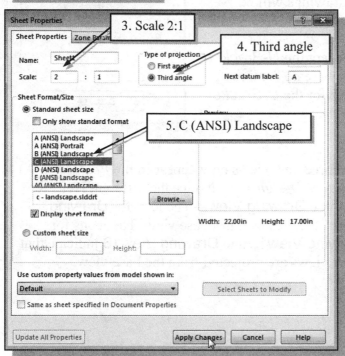

3. In the *Sheet Properties* window, set the scale to **2:1** as shown.

4. Set the *Type of projection* to **Third angle**.

5. Set the *Standard sheet size* to **C (ANSI) Landscape**.

6. Click **Apply changes** in the *Sheet Properties* window to accept the selected settings.

7. Press the **[F]** key to scale the new sheet to fit the display area.

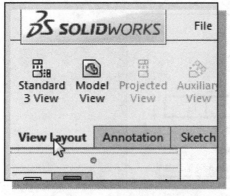

8. Click on the **View Layout tab** to display the Layout related commands as shown.

## Creating Three Standard Views

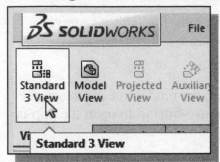

1. Select the **Standard 3 View** button on the *Drawing* toolbar.

➢ The *Standard 3 View Property Manager* appears. In the *Open documents* field of the *Part/Assembly to Insert* option window, the **U-Bracket** part file appears and is highlighted. The *U-Bracket* model is the only model opened. By default, all of the 2D drawings will be generated from this model file.

2. Left-click on the **OK** icon to accept the *U-Bracket* part file as the document used to create the three views. The three standard views for the *U-Bracket* appear on the active sheet.

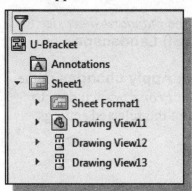

➢ The three new views now appear in the *Feature Manager Design Tree*. Notice the different icons for the three **Drawing Views**. The icon for **Drawing View11** indicates it is a base view. The icons for **Drawing View12** and **Drawing View13** indicate that these views are projected from the base view.

## Repositioning Views

1. Move the cursor on top of the front view and watch for the **four-arrow Move symbol** as the cursor is near the border indicating the view can be dragged to a new location as shown in the figure.

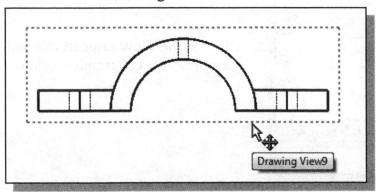

2. Press and hold down the left-mouse-button and reposition the view to a new location.

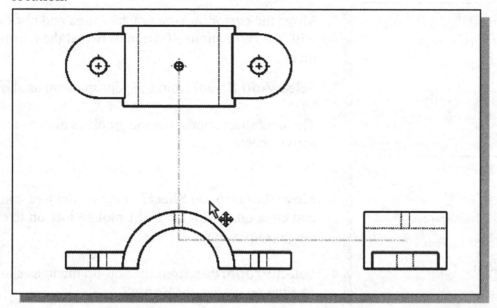

3. On your own, reposition the views we have created so far. Note that the top view can be repositioned only in the vertical direction. The top view remains aligned to the base view, the front view. Similarly, the right-side view can be repositioned only in the horizontal direction.

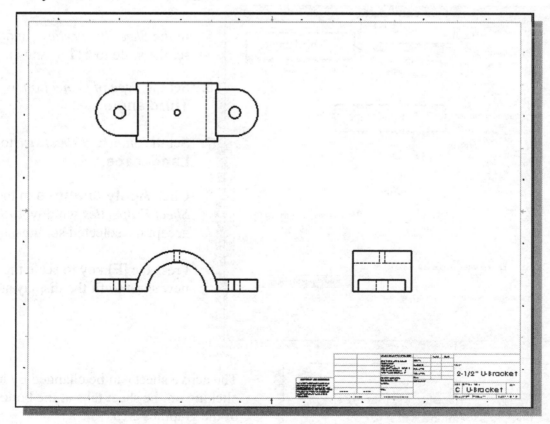

## Adding a New Sheet

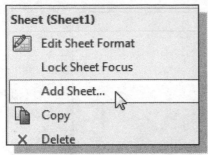

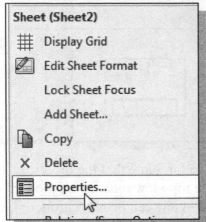

1. Move the cursor into the graphics area and click once with the **right-mouse-button** to reveal the pop-up menu.

2. Select **Add Sheet** from the pop-up menu as shown.

➢ The new sheet appears in the graphics area as the active sheet.

3. Move the cursor on **Sheet2** in the model tree manager and click once with the **right-mouse-button** to open the pop-up menu.

4. Select **Properties** from the pop-up menu as shown to edit the properties for **Sheet2**.

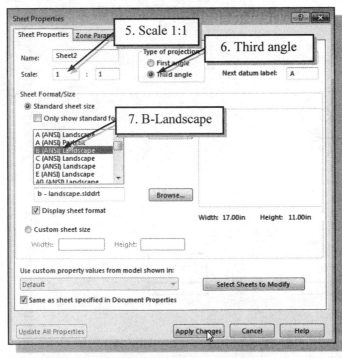

5. In the *Sheet Properties* window, set the scale to **1:1** as shown.

6. Set the *Type of projection* to **Third angle**.

7. Set the *Standard sheet size* to **B-Landscape**.

8. Click **Apply changes** in the *Sheet Properties* window to accept the selected settings.

9. Press the **[F]** key to scale the new sheet to fit the display area.

➢ The active sheet can be changed by left-clicking on the sheet tabs at the bottom of the graphics area.

## Adding a Base View

❖ In SOLIDWORKS *Drawing* mode, the first drawing view we create is called a **base view**. A *base view* is the primary view in the drawing; other views can be derived from this view. When creating a *base view*, SOLIDWORKS allows us to specify the view to be shown. By default, SOLIDWORKS will treat the *world XY plane* as the front view of the solid model. Note that there can be more than one *base view* in a drawing.

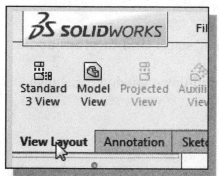

1. Make sure Sheet2 is the active drawing sheet.

2. Click on the **View Layout tab** to display the Layout related commands as shown.

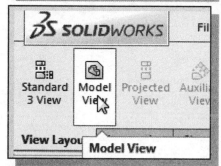

3. Click on the **Model View** icon on the *Drawing* toolbar.

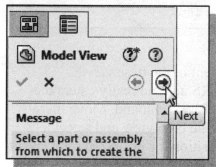

4. The *Model View Property Manager* appears with the U-Bracket part file selected as the part from which to create the base view. Click the **Next** arrow as shown to proceed with defining the base view.

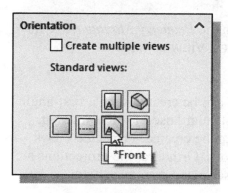

5. In the *Orientation* panel of the *Model View Property Manager*, select the **Front View** for the *Orientation*, as shown. (Make sure the *Create multiple views* box is **unchecked**.)

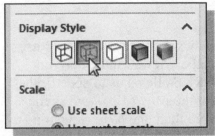

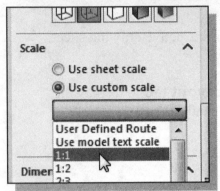

6. **Scroll down** in the *Property Manager* to the *Display Style* panel. Set the display style to *Hidden Lines Visible* if necessary. The default display setting can be defined in the *Document Settings*.

7. Under the *Scale* option, select **Use custom scale**.

8. Select **1:1** from the pull-down window, as shown.

9. Move the cursor inside the graphics area and left-click to place the **base** view near the lower left corner of the drawing sheet as shown below.

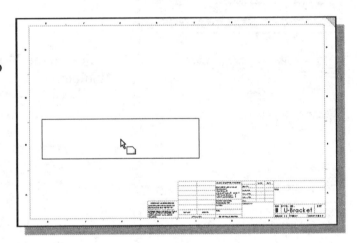

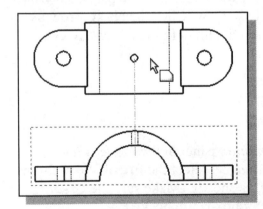

10. Upon placement of the base view, SOLIDWORKS automatically enters the **Projected View** command. Move the cursor above the base view and select a location to position the projected top view of the model.

11. Click **OK** in the *Property Manager* to exit the Projected View command.

❖ In SOLIDWORKS *Drawing* mode, **projected views** can be created with a first-angle or third-angle projection, depending on the drafting standard used for the drawing. We must have a base view before a projected view can be created. Projected views can be orthographic projections or isometric projections. Orthographic projections are aligned to the base view and inherit the base view's scale and display settings.

➤ **NOTE:** In your drawing, the inserted views may include *center marks*, *centerlines*, and/or *dimensions*. If so, these were added with the **auto-insert** option. To access these options, click **Options** (*Standard* toolbar), select the **Document Properties** tab, and then select **Detailing**. This lesson was done with only the **Center marks-holes-part** auto-insert option selected. Other center marks, centerlines, and dimensions will be added individually.

➤ Notice the hidden lines are visible. The dash lengths and spaces can be adjusted in the document properties.

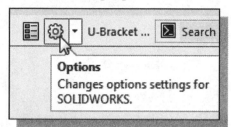

12. Select the **Options** icon from the *Menu Bar* to open the options window.

13. Select the **Document Properties** tab.

14. Select **Line Style**.

15. Select the **Dashed** line style.

16. Adjust the *Line length and spacing values* to, for example, **A,0.75,-0.25**. (This will scale the dash length by 0.75 and the space by 0.25.)

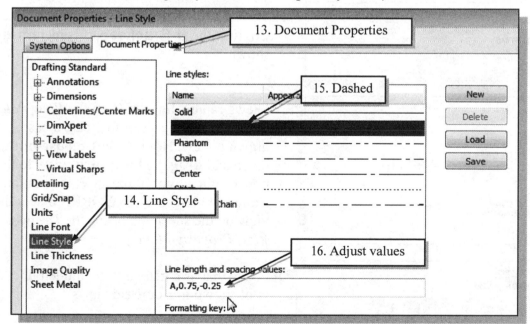

17. Left-click on the **OK** button to accept the settings.

18. Press the **[F]** key to fit the drawing to the graphics viewing area.

## Adding an Isometric View using the View Palette

❖ We will add an isometric view. The isometric view could be added as a *Projected View* based on the existing *Front View*. We will add the isometric view as a second base view. It can be added using the Model View command as was used for the *Front View*. An alternate method to add a base view is to use the **View Palette**. We will use this option.

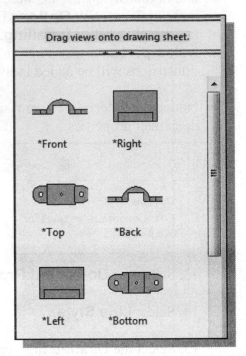

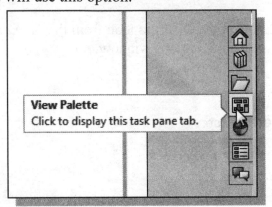

1. Click on the **View Palette** icon to the right of the graphics area.

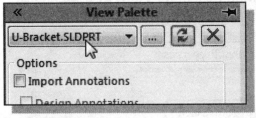

2. The *View Palette* appears in the task pane. Confirm the **U-Bracket** part is identified as shown.

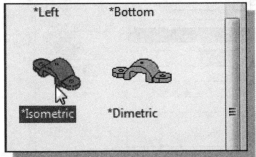

3. Views can be dragged from the *View Palette* onto the drawing sheet to create a base drawing view. Hold the **left-mouse-button** down and **drag** the **Isometric** view to the upper right corner of the sheet to position the isometric view. SOLIDWORKS places the view on the sheet and opens the *Drawing View Property Manager*.

➤ Note that as a general rule, hidden lines are not shown in isometric views.

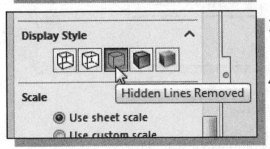

4. In the *Drawing View Property Manager*, select the **Hidden Lines Removed** option for the *Display Style*.

5.  Left-click on the **OK** icon in the *Drawing View Property Manager* to accept the view.

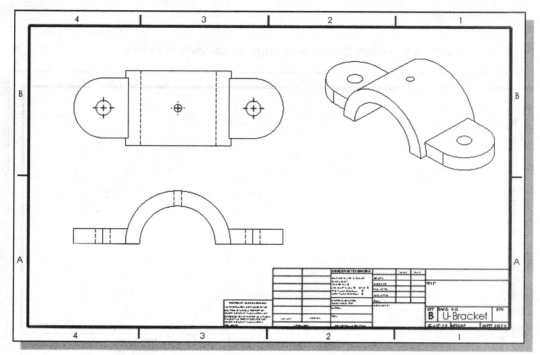

## Adjusting the View Scale

❖  SOLIDWORKS allows us to easily adjust the drawing view settings. Simply selecting a drawing view will open the corresponding *Property Manager*.

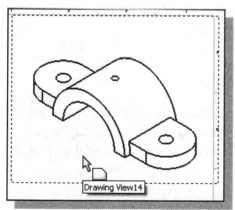

1.  Move the cursor on top of the isometric view and watch for the box around the entire view indicating the view is selectable as shown in the figure. **Left-mouse-click** once to bring up the *Property Manager*.

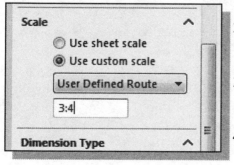

2.  Under the *Scale* options in the *Drawing View Property Manager*, select **Use custom scale**.

3.  Select **User Defined** from the pull-down options (scroll to the top of the list), as shown.

4.  Enter **3:4** as the user defined scale.

5.  Left-click on the **OK** icon in the *Drawing View Property Manager* to accept the new settings.

6.  If necessary, reposition the views to appear as shown below.

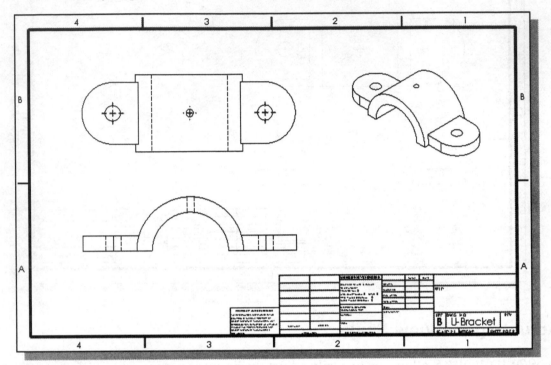

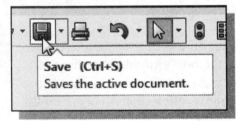

7.  Click Save to save the drawing file.

➢  Note that we have completed the creation of the necessary 2D orthographic views of the U-Bracket design for a multiview engineering drawing. In the next chapter, we will look into the general procedure in providing detailed descriptions of the created 2D views to document the design.

## Questions:

1. Explain what an orthographic view is and why it is important to engineering graphics.

2. What is the purpose of creating 2D orthographic views?

3. What information is missing in the front view? Top view? Right view?

4. Which dimension stays the same in between the Top view and the front view?

5. How do you decide which views are necessary for a multi-view drawing?

6. Explain the creation of 2D views using the 1$^{st}$ Angle Projection method.

7. Explain the creation of 2D views using the 3$^{rd}$ Angle Projection method.

8. How do you modify the sheet size in SOLIDWORKS?

## Exercises: (Unless otherwise specified, dimensions are in inches.)

1. **Swivel Yoke**

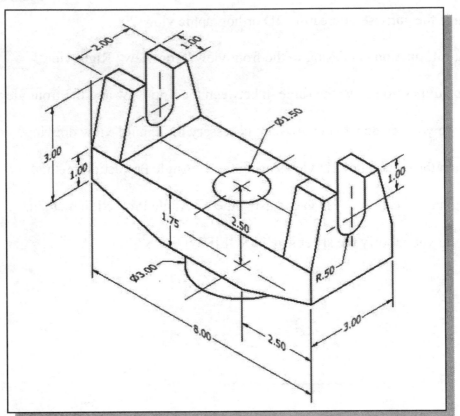

2. **Angle Bracket**

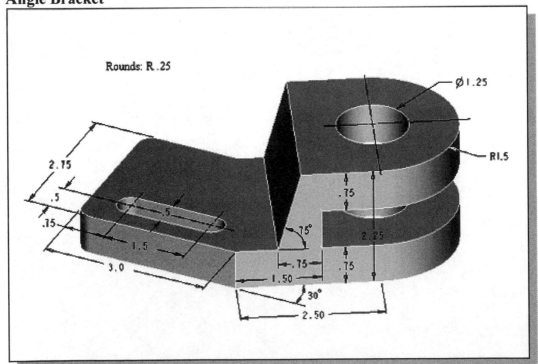

3.  **Connecting Rod**

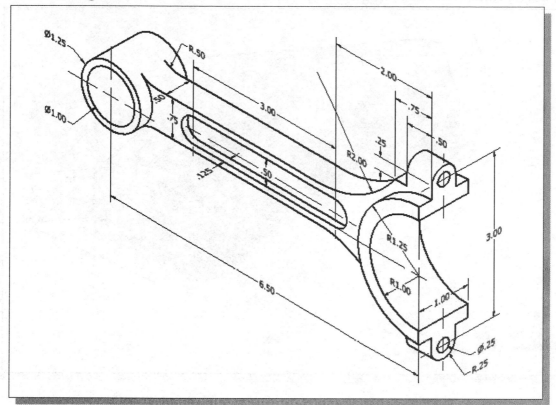

4.  **Tube Hanger**

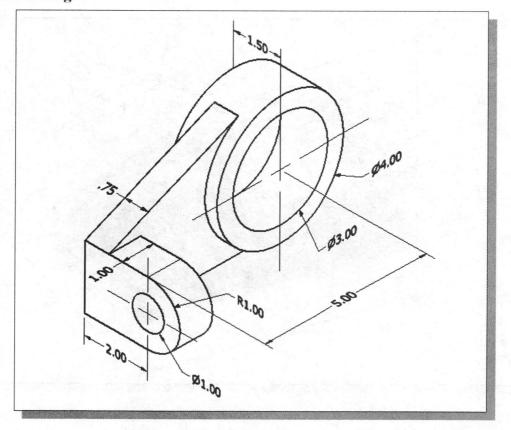

5.  **Angle Latch** (Dimensions are in Millimeters.)

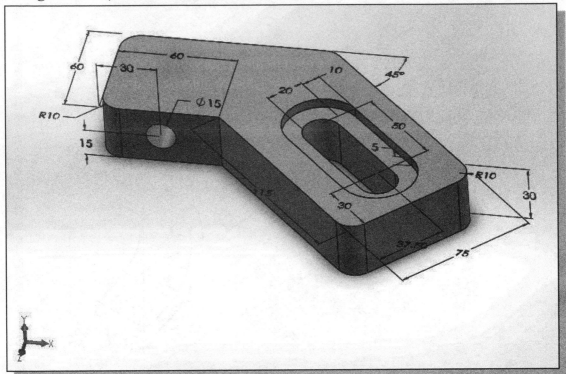

6.  **Inclined Lift**

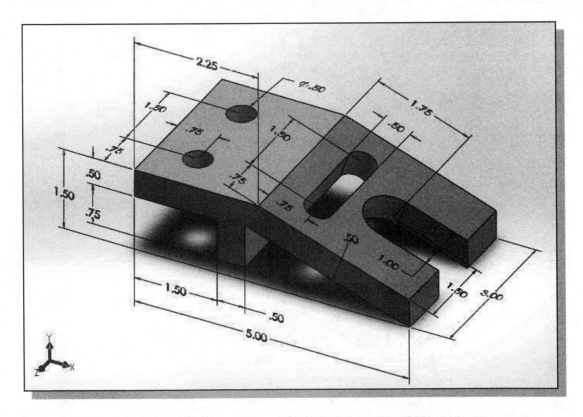

7. **Pivot Lock** (Dimensions are in inches. The circular features in the design are all aligned to the two centers at the base.)

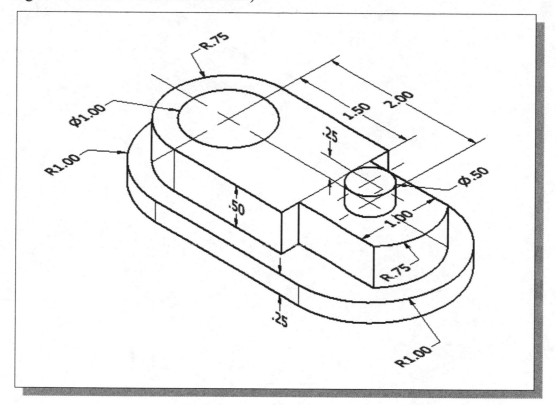

**Notes:**

# Chapter 8
# Dimensioning and Notes

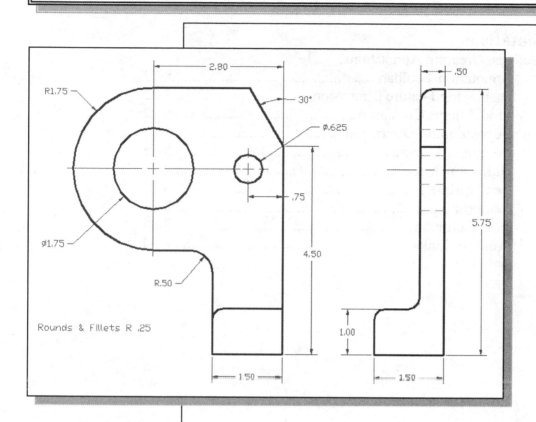

## Learning Objectives

- ♦ **Understand Dimensioning Nomenclature and Basics**
- ♦ **Create Drawing Layouts from Solid Models**
- ♦ **Understand Associative Functionality**
- ♦ **Use the default Sheet Formats**
- ♦ **Arrange and Manage 2D Views in Drawing Mode**
- ♦ **Display and Hide Feature Dimensions**
- ♦ **Create Reference Dimensions**
- ♦ **Understand *Edit Sheet* and *Edit Sheet Format* Modes**

## Certified SOLIDWORKS User Examination Objectives Coverage

**Certified Associate Reference Guide**

### Annotations

Objectives: Creating Annotations.

Annotation Toolbar ........................................................8-22
Displaying Feature Dimensions ..............................8-22
Model Items Command ............................................8-22
Repositioning Dimensions .......................................8-23
Reference Dimensions .............................................8-25
Hide Dimensions ......................................................8-25
Center Mark .............................................................8-27
Centerline ................................................................8-30
Note Command .........................................................8-34
Property Links .........................................................8-34

## Introduction

In engineering graphics, a detailed drawing is a drawing consisting of the necessary orthographic projections, complete with all the necessary dimensions, notes, and specifications needed to manufacture the design. The dimensions put on the drawing are not necessarily the same ones used to create the drawing but are those required for the proper functioning of the part, as well as those needed to manufacture the design. It is therefore important, prior to dimensioning the drawing, to study and understand the design's functional requirements; then consider the manufacturing processes that are to be performed by the pattern-maker, die-maker, machinist, etc., in determining the dimensions that best give the information. In engineering graphics, a **detailed drawing** is a drawing consisting of the necessary orthographic projections, complete with all the necessary dimensions, notes, and specifications needed to manufacture the design.

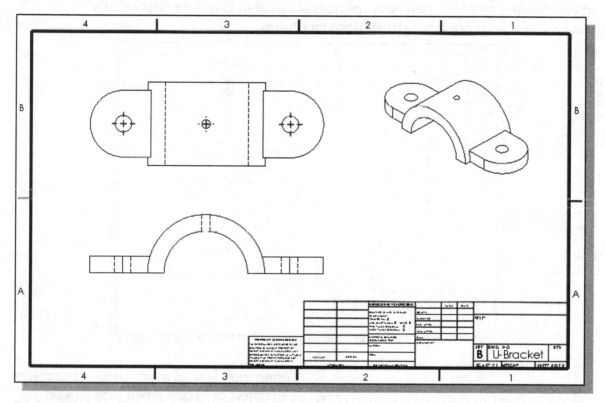

Considerable experience and judgment is required for accurate size and shape description. Dimensioning a design correctly requires conformance to many rules. For example, detail drawings should contain only those dimensions that are necessary to make the design. Dimensions for the same feature of the design should be given only once in the same drawing. Nothing should be left to chance or guesswork on a drawing. Drawings should be dimensioned to avoid any possibility of questions. Dimensions should be carefully positioned, preferably near the profile of the feature being dimensioned. The designer and the CAD operator should also be as familiar as possible with materials, methods of manufacturing and shop processes.

## Dimensioning Standards and Basic Terminology

The first step to learning to dimension is a thorough knowledge of the terminology and elements required for dimensions and notes on engineering drawings. Various industrial standards exist for the different industrial sectors and countries; the appropriate standards must be used for the particular type of drawing that is being produced. Some commonly used standards include **ANSI** (American National Standard Institute), **BSI** (British Standards Association), **DIN** (Deutsches Institut fur Normüng), **ISO** (International Standardization Organization), and **JIS** (Japanese Industry Standards). In this text, the discussions of dimensioning techniques are based primarily on the standards of the *American National Standard Institute* (ANSI/ASME Y14.5); other common sense practices are also illustrated.

In engineering graphics, two basic methods are used to give a description of a measurement on a drawing: **Dimensions** and **Notes**.

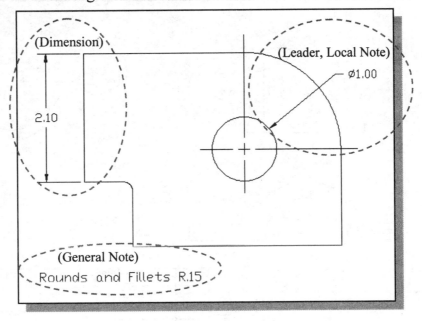

A dimension is used to give the measurement of an object or between two locations on a drawing. The **numerical value** gives the actual measurement; the **dimension lines** and the **arrowheads** indicate the direction in which the measurement applies. **Extension lines** are used to indicate the locations where the dimension is associated.

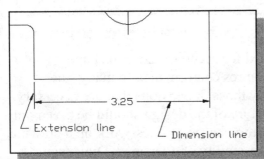

Two types of notes are generally used in engineering drawings: **Local Notes** and **General Notes**. A *Local Note* is done with a leader and an arrowhead referring the statement of the note to the proper place on the drawing. A *General Note* refers to descriptions applying to the design as a whole and it is given without a leader.

## Selection and Placement of Dimensions and Notes

One of the most important considerations to assure proper manufacturing and functionality of a design is the selection of dimensions to be given in the drawing. The selection should always be based upon the **functionality of the part**, as well as the **production requirements in the shop**. The method of manufacturing can affect the type of detailed dimensions and notes given in the drawing. It would be very beneficial to have a good knowledge of the different shop practices to give concise information on an engineering drawing.

After the dimensions and notes to be given have been selected, the next step is the actual placement of the dimensions and notes on the drawing. In reading a drawing, it is natural to look for the dimensions of a given feature wherever that feature appears most characteristic. One of the views of an object will usually describe the shape of some detailed feature better than with other views. Placing the dimensions in those views will promote clarity and ease of reading. This practice is known as the **Contour Rule**.

The two methods of positioning dimension texts on a dimension line are the **aligned** system and the **unidirectional** system. The *unidirectional* system is more widely used as it is easier to apply, and easier to read, the text horizontally. The texts are oriented to be read from the **bottom** of the drawing with the *unidirectional* system. The *unidirectional* system originated in the automotive and aircraft industries and is also referred to as the "horizontal system."

For the *aligned* system, the texts are oriented to be aligned to the dimension line. The texts are oriented to be read from the **bottom or right side** of the drawing with the *aligned* system.

**General Notes** must be oriented horizontally and be read from the bottom of the drawing in both systems.

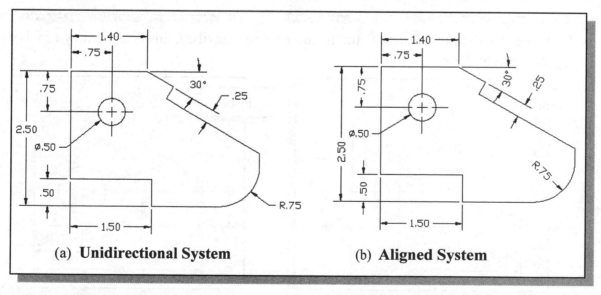

(a) **Unidirectional System**          (b) **Aligned System**

The following rules need to be followed for clearness and legibility when applying dimensions and/or notes to a design:

1.  Always dimension **features**, not individual geometric entities. Consider the dimensions related to the **sizes** and **locations** of each feature.

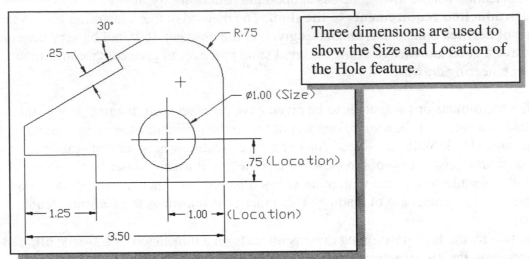

> Three dimensions are used to show the Size and Location of the Hole feature.

2.  Choose the view that best describes the feature to place the associated dimensions; this is known as the **Contour Rule**.

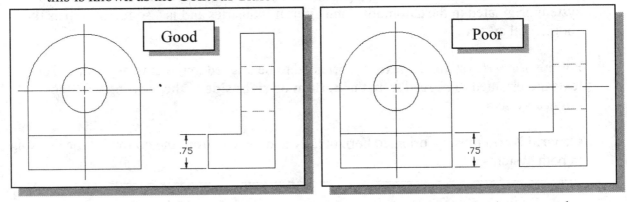

3.  Place the dimensions next to the features being described; this is also known as the **Proximity Rule**.

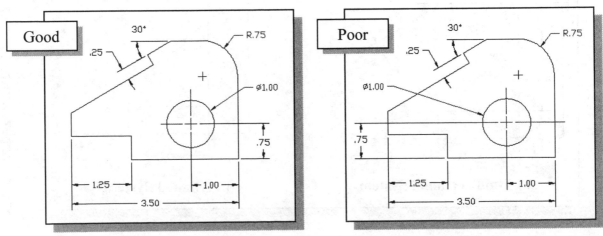

4.  Dimensions should be placed outside the view if possible; only place dimensions on the inside if they add clarity, simplicity, and ease of reading.

5.  Dimensions common to two views are generally placed in between the views.

6.  Center lines may also be used as extension lines; therefore, there should be **no gap** in between an extension line and a center line.

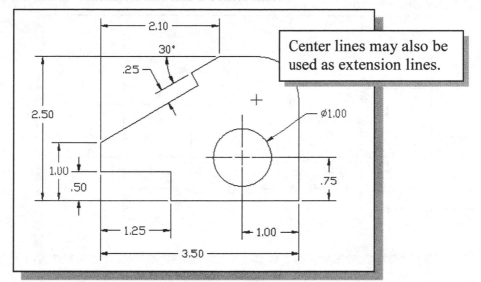

Center lines may also be used as extension lines.

7.  Dimensions should be applied to one view only. When dimensions are placed between views, the extension lines should be **drawn from one view**, not from both views.

8.  Center lines are also used to indicate the **symmetry** of shapes and frequently eliminate the need for a positioning dimension. They should extend about 6 mm (0.25 in.) beyond the shape for which they indicate symmetry, unless they are carried further to serve as extension lines. Center lines should not be continued between views.

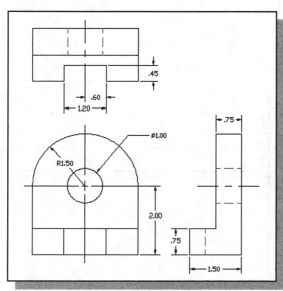

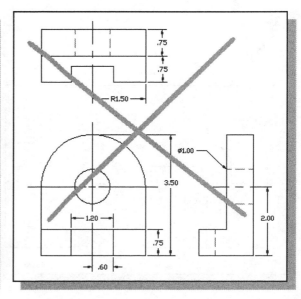

9.  Dimensions should be placed only on the view that shows the measurement in its **true length**.

10. Always place a shorter dimension line inside a longer one to avoid crossing dimension lines with the extension lines of other dimensions. Thus, an overall dimension, the maximum size of a part in a given direction, will be placed outside all other dimensions.

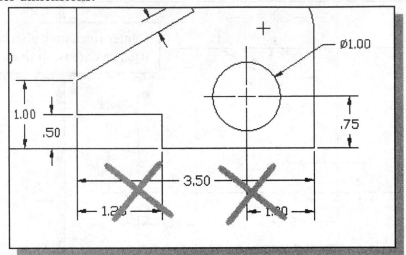

11. As a general rule, local notes (leaders) and general notes are created and placed after the regular dimensions.

12. Dimensions should never be crowded. If the space is small and crowded, an enlarged view or a partial view may be used.

13. Do not dimension to hidden lines; if necessary, create sectional views to convert the hidden features into visible features.

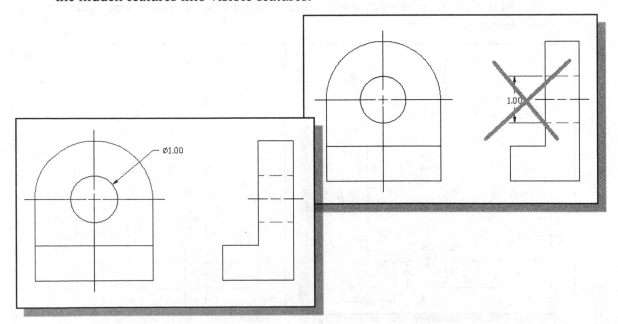

14. The spacing of dimension lines should be uniform throughout the drawing. Dimension lines should be spaced, in general, **10 mm** (0.4 inches) away from the outlines of the view. This applies to a single dimension or to the first dimension of several in a series. The space between subsequent dimension lines should be at least **6 mm** (0.25 inches) and uniformly spaced.

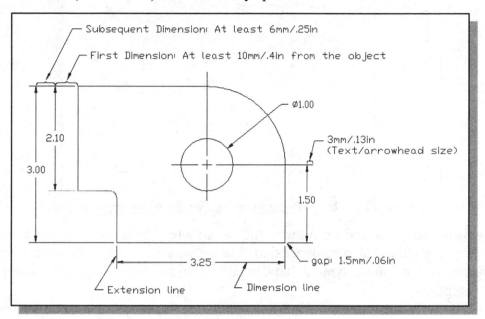

15. The **Overall Width, Height**, and **Depth** of the part should generally be shown in a drawing unless curved contours are present.

16. Never allow the crossing of dimension lines.

17. Dimension texts should be placed midway between the arrowheads, except when several parallel dimensions are present, where the texts should be staggered.

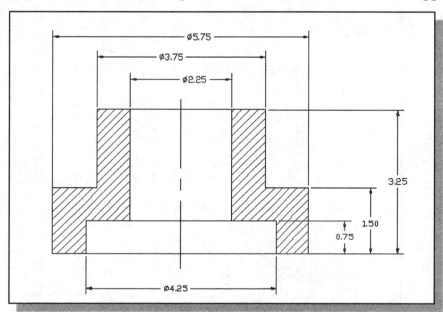

18. A leader should always be placed to allow its extension to pass through the **center** of all round features.

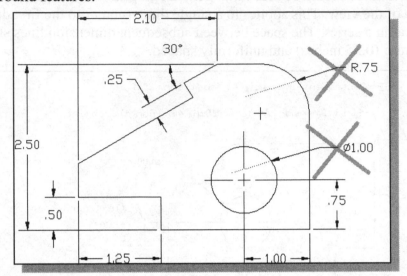

19. **Symbols** are preferred for features such as counter-bores, countersinks and spot faces; they should also be dimensioned using a leader. Each of these features has a special dimensioning symbol that can be used to show (a) Shape (b) Diameter and (c) Depth.

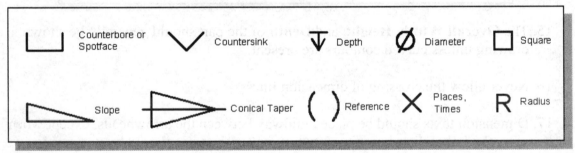

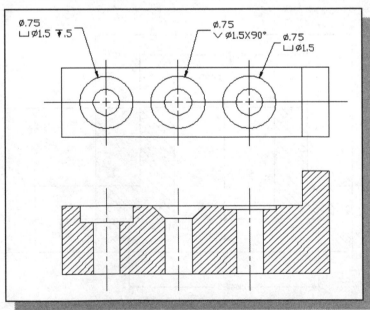

20. For leaders: avoid crossing leaders, avoid long leaders, avoid leaders in a horizontal or vertical direction, and avoid leaders parallel to adjacent dimension lines, extension lines, or section lines.

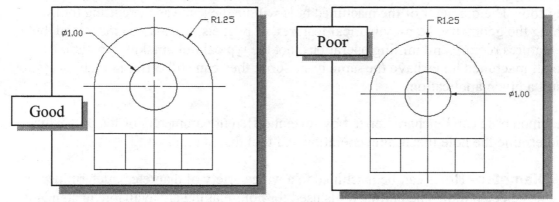

21. When dimensioning a circular feature, full cylinders (holes and bosses) must always be measured by their diameters. Arcs should always be shown by specifying a radius value. If a leader is used, the leader should meet the surface at an angle between 30° to 60° if possible.

22. Dual dimensions may be used, but they must be consistent and clearly noted. The dual-dimensioning can be shown by placing the alternate-unit's values below or next to the primary-unit's dimension, as shown in the figure.

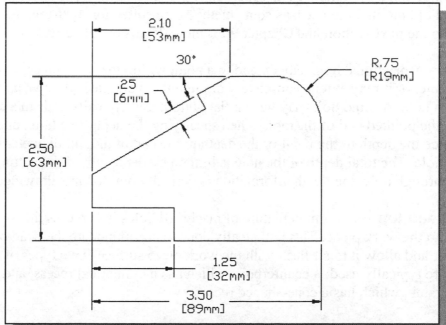

23. Never pass through a dimension text with any kind of line.

24. In general, *Metric* system is given in millimeters and they are generally rounded to the nearest whole number. English system is given in inches and they are typically shown with two decimal places. When a metric dimension is less than a millimeter, a zero is placed before the decimal point, but no zero is placed before a decimal point when inches are used.

## Machined Holes

In machining, a hole is a cylindrical feature that is cut from the work piece by a rotating cutting tool. The diameter of the machined hole will be the same as the cutting tool with matching the geometry. Non-cylindrical features, or pockets, can also be machined, but these features require end milling operations, not the typical hole-making operations. While all machined holes have the same basic form, they can still differ in many ways to best suit a given application.

A machined hole can be characterized by several different parameters or features which will determine the hole machining operation and tool that is required.

**Diameter** - Holes can be machined in a wide variety of diameters, determined by the selected tool. The cutting tools used for hole-making are available in standard sizes that can be as small as 0.0019 inches and as large as 3 inches. Several standards exist, including fractional sizes, letter sizes, number sizes, and metric sizes. A custom tool can be created to machine a non-standard diameter, but it is generally more cost effective to use the standard sized tool.

**Tolerance** - In any machining operation, the precision of a cut can be affected by several factors, including the sharpness of the tool, any vibration of the tool, or the buildup of chips of material. The specified tolerance of a hole will determine the method of hole-making used, as some methods are suited for tight-tolerance holes. Refer to the next section and Chapter 9 for more details on tolerances.

**Depth** - A machined hole may extend to a point within the work piece, known as a blind hole, or it may extend completely through the work piece, known as a through hole. A blind hole may have a flat bottom, but typically ends in a point due to the pointed end of the tool. When specifying the depth of a hole, one may reference the depth to the point or the depth to the end of the full diameter portion of the hole. The total depth of the hole is limited by the length of the cutting tool. For a through hole, the depth information is typically omitted in a drawing.

**Recessed top** - A common feature of machined holes is to recess the top of the hole into the work piece. This is typically done to accommodate the head of a fastener and allow it to sit flush with the work piece surface. Two types of recessed holes are typically used: a counterbore, which has a cylindrical recess, and a countersink, which has a cone-shaped recess.

**Threads** - Threaded holes are machined to accommodate a threaded fastener and are typically specified by their outer diameter and pitch. The pitch is a measure of the spacing between threads and may be expressed in the English standard, as the number of threads per inch (TPI), or in the metric standard, as the distance in between threads. For more details on threads, refer to Chapter 13 of this text.

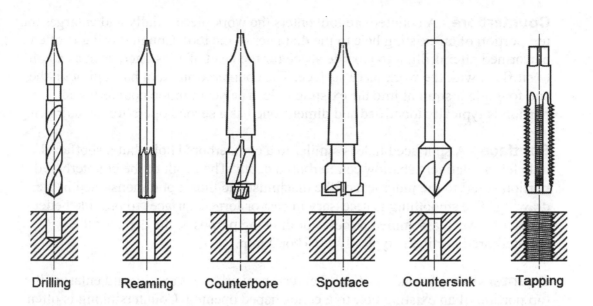

Drilling     Reaming     Counterbore     Spotface     Countersink     Tapping

**Drilling** - A drill bit enters the work piece in the axial direction and cuts a blind hole or a through hole with a diameter equal to that of the tool. A drill bit is a multi-point tool and typically has a pointed end. A twist drill is the most commonly used, but other types of drill bits, such as center drill, spot drill, or tap drill can also be used to start a hole that will be completed by additional operation.

**Reaming** - A reamer enters the work piece in the axial direction and enlarges an existing hole to the diameter of the tool. A reamer is a multi-point tool that has many flutes, which may be straight or in a helix. Reaming removes a minimal amount of material and is often performed after drilling to obtain both a more accurate diameter and a smoother internal finish.

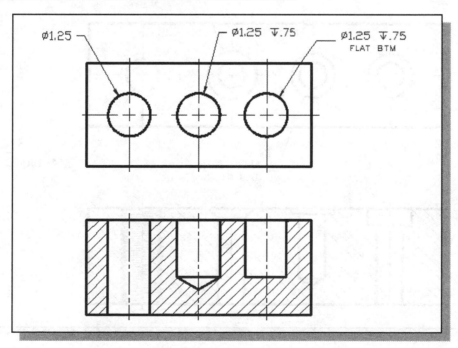

**Counterbore** - A counterbore tool enters the work piece axially and enlarges the top portion of an existing hole to the diameter of the tool. Counterboring is often performed after drilling to provide space for the head of a fastener, such as a bolt, to sit flush with the work piece surface. The counterboring tool has a pilot on the end to guide it straight into the existing hole. The depth of the counterbored portion is typically identified and dimensioned as a second operation on a drawing.

**Spotface** - A spotfaced hole is similar to a counterbored hole, but a spotfaced hole has a relatively shallow counterbored depth. The depth of the counterbored portion is left to the judgment of the machinist and thus, not dimensioned in the drawing. The smoothing is necessary in cast or forged surfaces to provide better contact between assembled pieces. For drawing purposes, the depth of the counterbored portion is typically 1/16″ or 2 mm.

**Countersink** - A countersink tool enters the work piece axially and enlarges the top portion of an existing hole to a cone-shaped opening. Countersinking is often performed after drilling to provide space for the head of a fastener, such as a screw, to sit flush with the work piece surface. Common included angles for a countersink include 60, 82, 90, 100, 118, and 120 degrees.

**Tapping** - A tap enters the work piece in the axial direction and cuts internal threads inside an existing hole. The existing hole is typically drilled with a drill size that will accommodate the desired tap. The tap is selected based on the major diameter and pitch of the threaded hole. Threads may be cut to a specified depth inside the hole (bottom tap) or the complete depth of a through hole (through tap).

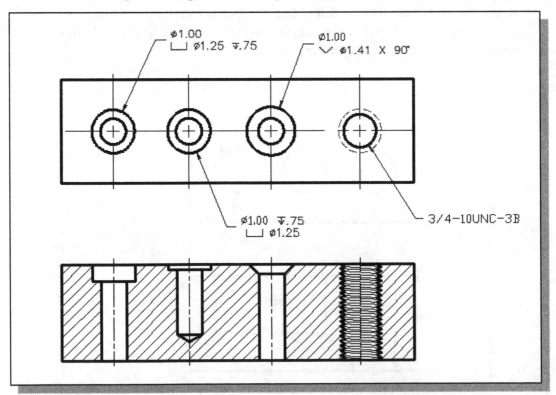

# Baseline and Chain Dimensioning

Frequently the need arises to dimension a feature which involves a series of dimensions. Two methods should be considered for this type of situation: **Baseline dimensioning** and **Chain dimensioning**. The baseline dimensioning method creates dimensions by measuring from a common baseline. The chain dimensioning method creates a series or a chain of dimensions that are placed one after another. It is important to always consider the functionality and the manufacturing process of the specific features to be dimensioned. Note that both methods can be used to dimension different features within the same drawing as shown in the below figure.

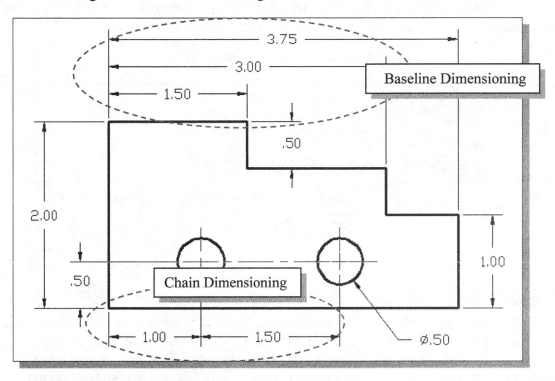

- **Baseline dimensioning:** Used when the **location** of features must be controlled from a common reference point or plane.

- **Chain dimensioning:** Used when **tolerances** between adjacent features is more important than the overall tolerance of the feature. For example, dimensioning mating parts, hole patterns, slots, etc. Note that more in-depth discussions on tolerances are presented in the next section and also in the next chapter.

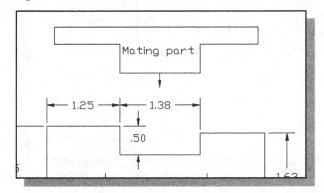

## Dimensioning and Tolerance Accumulation

Drawings with dimensions and notes often serve as construction documents to ensure the proper functioning of a design. When dimensioning a drawing, it is therefore critical to consider the *precision* required for the part. *Precision* is the degree of accuracy required during manufacturing. Different machining methods will produce different degrees of precisions. However, one should realize it is nearly impossible to produce any dimension to an absolute, accurate measurement; some variation must be allowed in manufacturing. **Tolerance** is the allowable variation for any given size and provides a practical means to achieve the precision necessary in a design. With **tolerancing**, each dimension is allowed to vary within a specified zone. The general machining tolerances of some of the more commonly used manufacturing processes are listed in the figure below.

| International Tolerance Grade (IT) | | | | | | | |
|---|---|---|---|---|---|---|---|
| 4 | 5 | 6 | 7 | 8 | 9 | 10 | 11 |
| Lapping or Honing | | | | | | | |
| | Cylindrical Grinding | | | | | | |
| | Surface Grinding | | | | | | |
| | Diamond Turning or Boring | | | | | | |
| | Broaching | | | | | | |
| | Powder Metal-sizes | | | | | | |
| | | Reaming | | | | | |
| | | | Turning | | | | |
| | | | Powder Metal-sintered | | | | |
| | | | Boring | | | | |
| | | | | | | Milling, Drilling | |
| | | | | | | Planing & Shaping | |
| | | | | | | Punching | |
| | | | | | | | Die Casting |

As it was discussed in the previous sections, the selections and placements of dimensions and notes can have a drastic effect on the manufacturing of the designs. Furthermore, a poorly dimensioned drawing can create problems that may invalidate the design intent and/or become very costly to the company. One of the undesirable results in poorly dimensioned drawings is **Tolerance Accumulation**. *Tolerance Accumulation* is the effect of cumulative tolerances caused by the relationship of tolerances associated with relating dimensions. As an example, let's examine the *tolerance accumulation* that can occur with two mating parts, as shown in the below figure, using a standard tolerance of ± 0.01 for all dimensions.

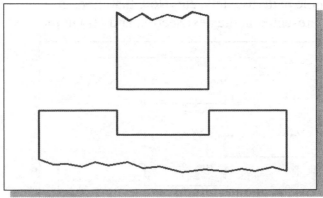

## (1) Tolerance Accumulation - Baseline Dimensioning

Consider the use of the *Baseline dimensioning* approach, as described in the previous sections, to dimension the parts. At first glance, it may appear that the tolerance of the overall size is maintained and thus this approach is a better approach in dimensioning the parts!?

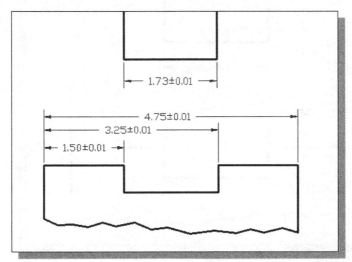

Now let's examine the size of the center slot, which is intended to fit the other part:

**Upper limit** of the slot size: 3.26-1.49= 1.77
**Lower limit** of the slot size: 3.24-1.51= 1.73

The tolerance of the center slot has a different range of tolerance than the standard ± 0.01. With this result, the two parts will not fit as well as the tolerance has been loosened.

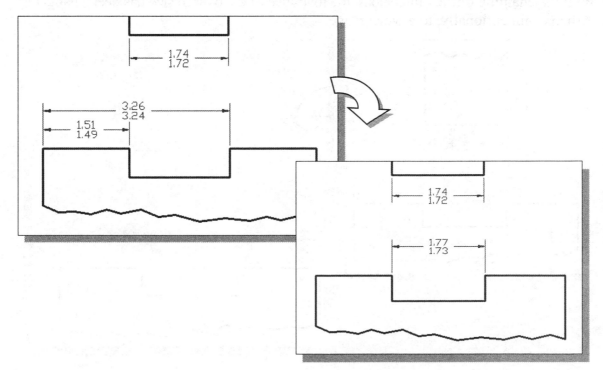

## (2) Tolerance Accumulation - Chain Dimensioning

Now consider the use of the *chain dimensioning* approach, as described in the previous sections, to dimension the parts.

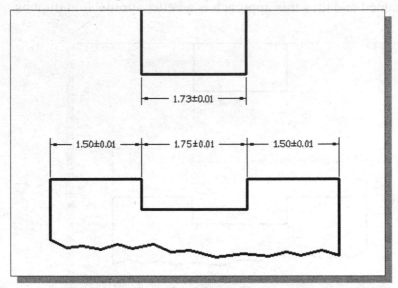

The *tolerance accumulation* occurred since the overall size is determined with the three dimensions that are chained together. Considering the upper and lower limits of the overall size of the bottom part:

**Upper limit** of the overall size: 1.51+1.76+1.51= 4.78
**Lower limit** of the overall size: 1.49+1.74+1.49= 4.72

With the chaining of the dimensions, the tolerance of the overall size has been changed, perhaps unintentionally, to a wider range.

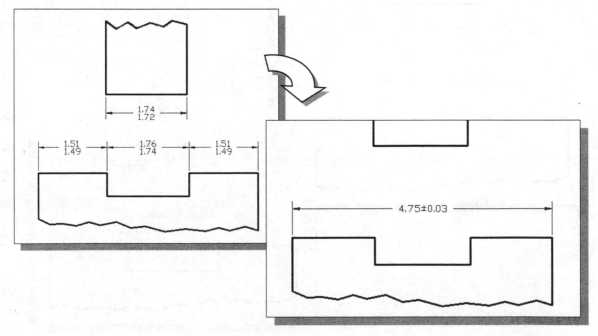

## (3) Avoid Tolerance Accumulation – Dimensioning Features

The effect of the *tolerance accumulation* is usually undesirable. We could be using fairly precise and expensive manufacturing processes but resulting in lower quality products simply due to the poor selections and placements of dimensions.

To avoid *tolerance accumulation,* properly dimensioning a drawing is critical. The first rule of dimensioning states that one should always dimension **features**, not individual geometric entities. Always consider the dimensions related to the **sizes** and **locations** of each feature. One should also consider the associated **design intent** of the **feature**, the **functionality** of the design and the related **manufacturing processes**.

For our example, one should consider the center slot as a feature, where the other part will be fit into, and therefore the **size** and **location** of the feature should be placed as shown in the figure below.

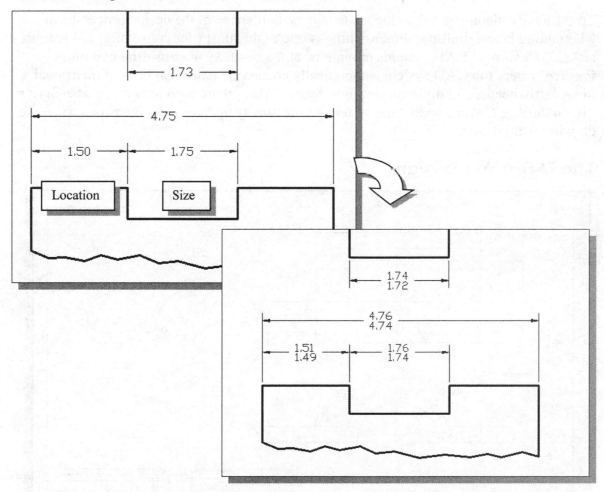

> ➤ More detailed discussions on tolerances are covered in the next chapter; however, the effects of *tolerance accumulation* should be avoided even during the initial selection and placement of dimensions. Note that there are also other options, generally covered under the topics of **Geometric Dimensioning and Tolerancing (GD&T),** to assure the accuracy of designs through productions.

## Dimensioning Tools in SOLIDWORKS

Thus far, the use of **SOLIDWORKS** to define the *shape* of designs has been illustrated. In this chapter, the general procedures to convey the *size* definitions of designs using **SOLIDWORKS** are discussed. The *tools of size description* are known as *dimensions* and *notes*.

Considerable experience and judgment are required for accurate size description. As it was outlined in the previous sections, detail drawings should contain only those dimensions that are necessary to make the design. Dimensions for the same feature of the design should be given only once in the same drawing. Nothing should be left to chance or guesswork on a drawing. Drawings should be dimensioned to avoid any possibility of questions. Dimensions should be carefully positioned, preferably near the profile of the feature being dimensioned. The designer and the CAD operator should be as familiar as possible with materials, methods of manufacturing, and shop processes.

Traditionally, detailing a drawing is the biggest bottleneck of the design process; and when doing board drafting, dimensioning is one of the most time consuming and tedious tasks. Today, most CAD systems provide what is known as an **auto-dimensioning feature**, where the CAD system automatically creates the extension lines, dimensional lines, arrowheads, and dimension values. Most CAD systems also provide an **associative dimensioning feature** so that the system automatically updates the dimensions when the drawing is modified.

## The *U-Bracket* Design

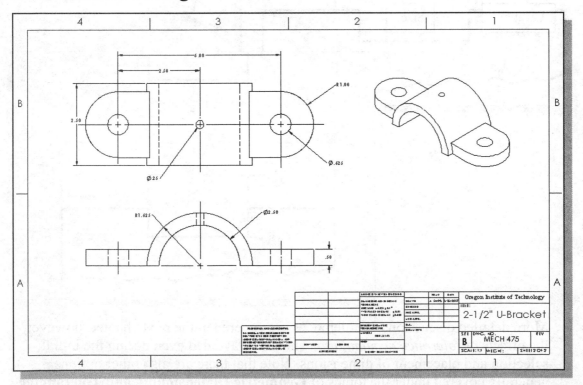

## Start SOLIDWORKS

1. Select the **SOLIDWORKS** option on the *Start* menu or select the **SOLIDWORKS** icon on the desktop to start SOLIDWORKS. The SOLIDWORKS main window will appear on the screen.

2. In the **Recent Document** section, select the **U-Bracket.SLDDRW** file to open it.

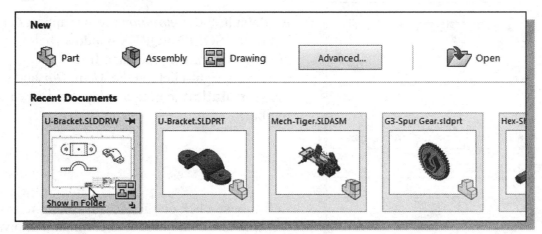

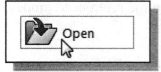

3. If the **U-Bracket.SLDDRW** file isn't shown in the recent document section, use the **Open** to locate the file.

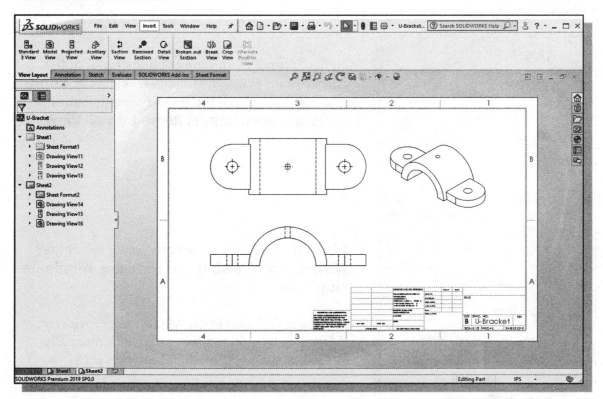

## Display Feature Dimensions

- By default, feature dimensions are not displayed in 2D views in SOLIDWORKS. The dimensions used to create the part can be imported into the drawing using the **Model Items** command. We can also add dimensions to the drawing manually using the **Smart Dimension** command. The **Model Items** command appears on the *Annotation* toolbar.

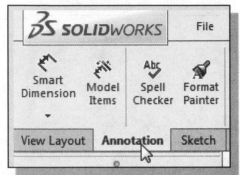

1. By default, the *Annotation* toolbar appears at the left of the SOLIDWORKS window (below or next to the *Drawing* toolbar). If it is not displayed, right-click on the *Menu Bar* and select the **Annotation** icon to display the *Annotation* toolbar.

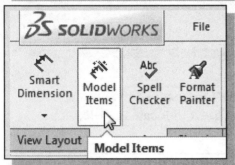

2. Left-mouse-click on the **Model Items** icon on the *Annotation* toolbar. This command allows us to import dimensions from the referenced model into selected drawing views.

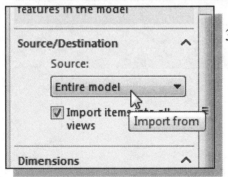

3. In the *Source/Destination* options panel of the *Model Items Property Manager*, select the **Entire model** option from the *Import from* pull-down menu and check **Import items into all views** as shown.

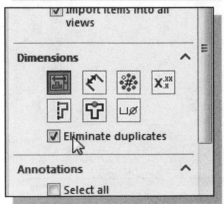

4. Under the *Dimensions* options panel, select the **Marked for drawing** icon and check **Eliminate duplicates** as shown.

5. Click the **OK** icon in the *Model Items Property Manager*. Notice dimensions are automatically placed on the front and top drawing views.

## Repositioning, Appearance, and Hiding of Feature Dimensions

1.  Move the cursor on top of the width dimension text **5.00** and watch for when the dimension text becomes highlighted as shown in the figure, indicating the dimension is selectable.

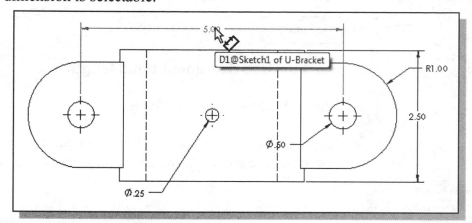

2.  Reposition the dimension by using the left-mouse-button to drag the dimension text to a new location. (**NOTE:** SOLIDWORKS may execute snaps to align the text automatically. To negate these, hold down the **[Alt]** key while dragging the text and it will move freely.)

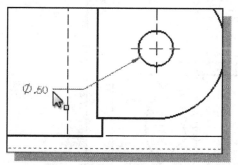

3.  Move the cursor on top of the diameter dimension **0.50** and click once with the left-mouse-button to select the dimension.

4.  Left-click on the arrow at the end of the leader line and drag to lengthen the bent leader line as shown.

❖  The default bent leader length can be adjusted to improve the appearance of the current view. This is done using the **Document Settings**. There are many document settings, including dimension settings, line style, etc., which can be adjusted to achieve the desired drawing display.

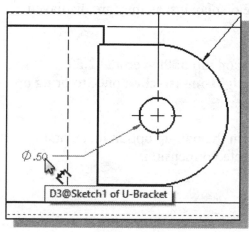

5.  On your own, change the **Bent leader length** to **0.2 in**. (Select the **Options** icon from the *Menu Bar* to open the *Options* window. Select the **Document Properties** tab. Select **Dimensions**. In the **Leader length:** field, enter **0.2 in**. Click **OK**.)

6.  Move the cursor on top of the diameter dimension **0.50,** and click once with the left-mouse-button to select the dimension.

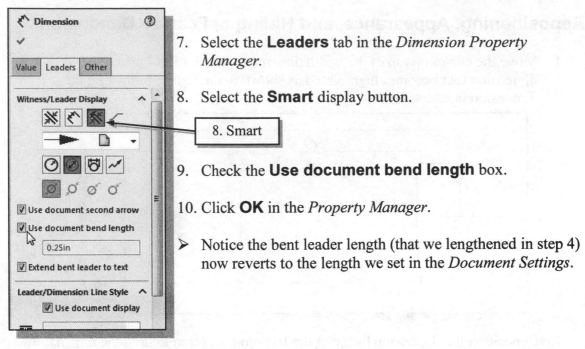

7.  Select the **Leaders** tab in the *Dimension Property Manager*.

8.  Select the **Smart** display button.

8. Smart

9.  Check the **Use document bend length** box.

10. Click **OK** in the *Property Manager*.

➢  Notice the bent leader length (that we lengthened in step 4) now reverts to the length we set in the *Document Settings*.

11. On your own, reposition the dimensions displayed in the *top* view as shown. (Note the 2.5 vertical dimension is intentionally placed to the right with the extension lines crossing the other dimensions.)

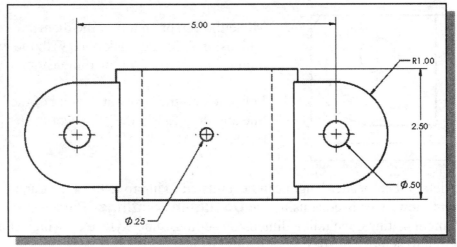

➢  Any feature dimensions can be removed from the display just as they are displayed.

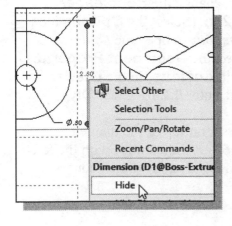

12. Move the cursor on top of the vertical **2.50** dimension and **right-mouse-click** once to bring up the option menu.

13. Select **Hide** from the pop-up option menu and remove the selected dimension.

## Adding Additional Dimensions – Reference Dimensions

- Besides displaying the **feature dimensions**, dimensions used to create the features, we can also add additional **reference dimensions** in the drawing. *Feature dimensions* are used to control the geometry, whereas *reference dimensions* are controlled by the existing geometry. In the drawing layout, therefore, we can ***add*** or ***delete*** *reference dimensions*, but we can only ***hide*** the *feature dimensions*. One should try to use as many *feature dimensions* as possible and add *reference dimensions* only if necessary. It is also more effective to use *feature dimensions* in the drawing layout since they are created when the model was built. Note that additional *Drawing* mode entities, such as lines and arcs, can be added to drawing views. Before *Drawing* mode entities can be used in a reference dimension, they must be associated to a *drawing view*.

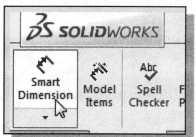

1. Left-mouse-click on the **Smart Dimension** icon on the *Annotation* toolbar.

2. In the top view, select the top horizontal line as the first entity to dimension.

3. In the top view, select the bottom horizontal line as the second entity to dimension.

4. Place the **2.50** vertical dimension to the left side of the top view as shown.

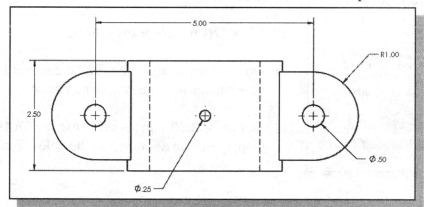

5. On your own, add the **2.50** horizontal dimension and arrange the display of the top view to look like the figure below.

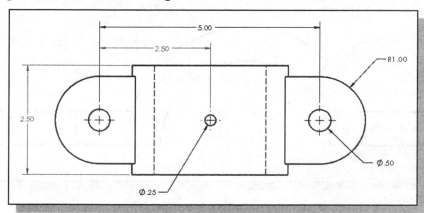

6. In the front view, left click on the **R1.75** dimension and move it to the left hand side.

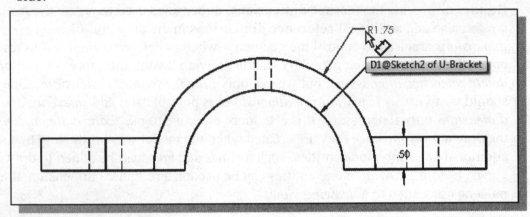

7. In the *Dimension Property Manager*, select the **Leaders** tab.

8. In the *Witness/Leader Display* panel, select the **Inside** option as shown.

9. Switch **On** the Arc extension line option if necessary.

10. Click **OK** to accept the setting.

11. On your own, select the Ø2.50 dimension and experiment with the available options.

12. Set the Ø2.50 dimension to use the **Outside** display option as shown in the below figure.

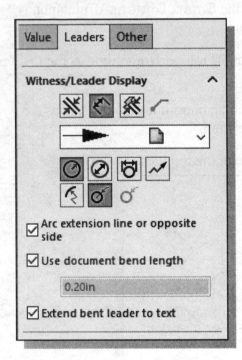

13. On your own, adjust the front view dimension display to appear as in the figure below.

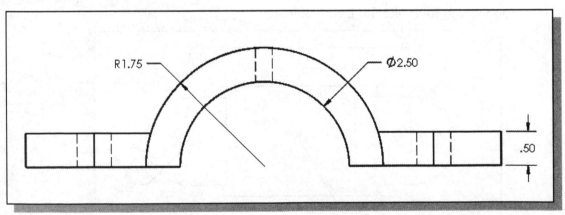

## Tangent Edge Display

❖ Another important setting for drawing views is the way tangent edges – where a curved surface meets tangent to a flat surface – are represented. In the figure above, the tangent edges (the vertical lines coincident with the centerlines of the two hidden holes) are visible.

1. Press the **[Esc]** key a few times to ensure no objects are selected.

2. Move the cursor over the *front view* in the graphics area and **right-click** to open the pop-up option menu.

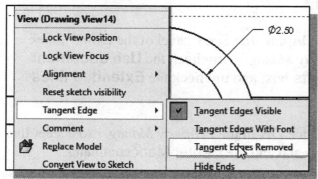

3. Select **Tangent Edge**, then **Tangent Edges Removed** from the pop-up menu, as shown.

4. Click **OK** in the *Property-Manager* or press the **[Esc]** key to accept the setting.

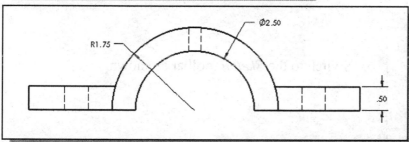

❖ Notice the tangent edges are no longer visible.

5. On your own, remove the tangent edges from the isometric view.

## Adding Center Marks, Center Lines, and Sketch Objects

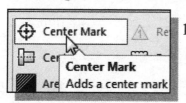

1. Click on the **Center Mark** button in the *Annotation* toolbar.

2. Click on the larger arc in the front view to add the center mark.

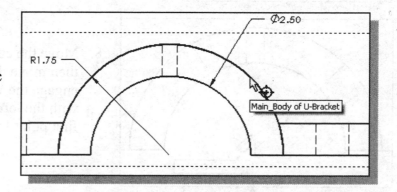

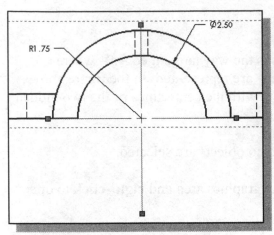

3.  A preview of the center mark appears in the graphics area and the *Center Mark Property Manager* is opened.

❖  The default center mark included extended lines in all four directions. Since the feature is semicircular, we will turn *OFF* the default extended lines and keep only the center mark.

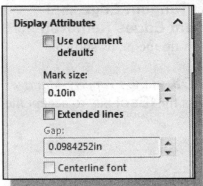

4.  In the *Display Attributes* panel of the *Center Mark Property Manager*, **uncheck** the **Use document defaults** box, and **uncheck** the **Extended lines** box.

5.  Left-Click **OK** in the *Property Manager* to accept the settings and end the **Center Mark** command.

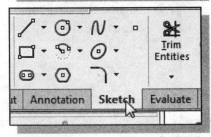

6.  Switch to the *Sketch* toolbar as shown.

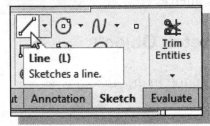

7.  Left-click on the **Line** button on the *Sketch* toolbar.

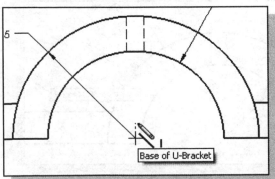

8.  Move the cursor over the center mark, then move the cursor above the center to engage the **Vertical** relation. Click once with the **left-mouse-button** to locate the first point for the line.

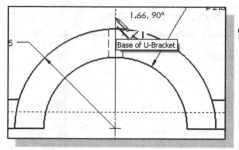

9. Move the cursor vertically to locate the endpoint for the line. Click once with the **left-mouse-button** to locate the endpoint.

10. Press the **[Esc]** key once to end the Line command.

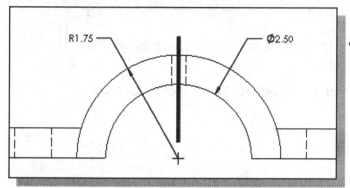

• The properties of the sketched line might not be what is desired. We will adjust the line properties through the *Line Format* toolbar.

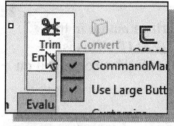

11. Move the cursor on top of any icon on the Command Manager and click once with the **left-mouse-button** to display the *Option List*.

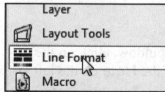

12. Click **Line Format** to display the *Line Format* toolbar.

13. Left-click on the **Sketched Line** we just created.

14. Click on the **Line Thickness** icon and select **Default** as the new line thickness.

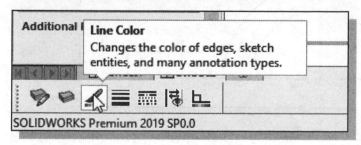

15. On your own, adjust the **Line Color** to black and uncheck the *Default* option.

16. Press the **[Esc]** key once to de-select the sketched line.

17. On your own, sketch the two horizontal lines as shown in the figure.

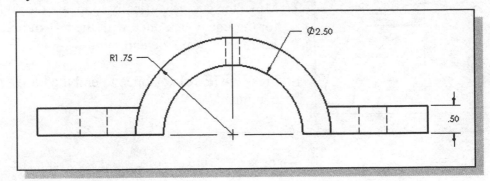

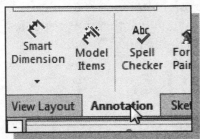

18. Switch to the **Annotation** toolbar as shown.

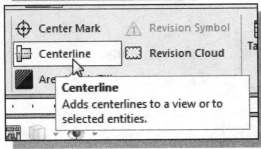

19. Click on the **Centerline** button in the *Annotation* toolbar. (**NOTE:** This is different from the Centerline command on the *Sketch* toolbar.)

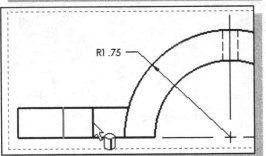

20. Inside the graphics window, click on the two hidden edges of the **Hole** on the left as shown in the figure.

- Note the centerline on the right side of the front view is created automatically as shown below.

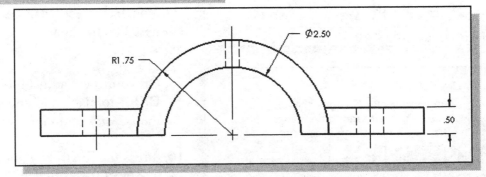

21. Left-click the **OK** icon in the *Centerline Property Manager* or press the **[Esc]** key to end the Centerline command.

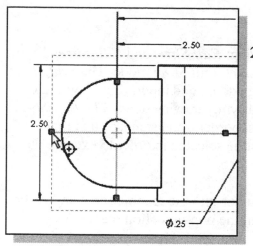

22. Add the three center marks if they were not added automatically. Drag the endpoint of the *Extended Lines* to desired locations.

23. On your own, adjust your settings and endpoint locations until the top view and front view appear as in the figure below.

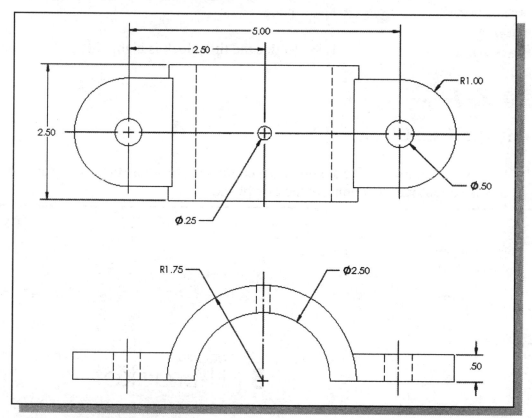

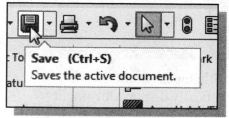

24. Click on the **Save** icon in the *Menu Bar* as shown.

25. If the SOLIDWORKS window appears asking to *'Update the drawing?'* select **Yes**.

## Edit Sheet vs. Edit Sheet Format

❖ There are two modes in a SOLIDWORKS drawing: *Edit Sheet* and *Edit Sheet Format*. To this point we have been operating in the *Edit Sheet* mode. The *Edit Sheet* mode is used to make detail drawings, including adding and modifying views, adding and modifying dimensions, and adding and modifying drawing notes. *The Edit Sheet Format* mode is used to add and edit title blocks, borders, and standard text that appear in every drawing. The correct mode must be selected in order to execute the corresponding commands.

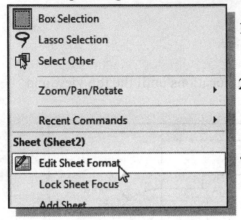

1. Move the cursor into the graphics area and **right-click** to open the pop-up option menu.

2. Select **Edit Sheet Format** from the pop-up option menu to change to *Edit Sheet Format* mode.

❖ Notice the title block is highlighted.

## Modify the Title Block

1. On your own, use the **Zoom** and **Pan** commands to adjust the display as shown; this is so that we can complete the title block.

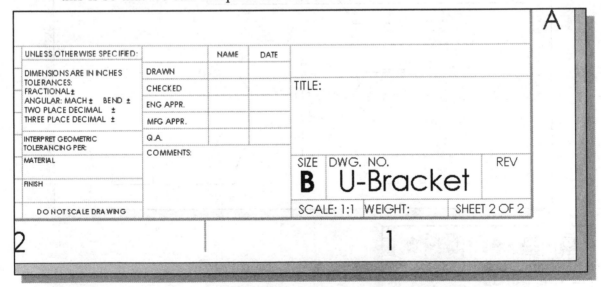

❖ There are many notes added as entries in the default title block. Some are simple **Text** notes; some include **Property Links** to automatically display document properties. In the following steps, we will learn to (1) edit existing notes, (2) insert new notes, and (3) control **Property Links**. We will first edit a simple text note.

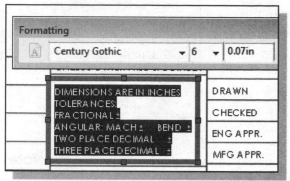

2. Zoom in on the tolerance specification box. This note contains only simple text entries. **Double-click** on the text with the **left-mouse-button** to enter the text editor mode.

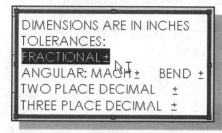

3. Highlight the FRACTIONAL ± line as shown by clicking and dragging with the left-mouse-button.

4. Press the [**Delete**] key twice to delete the line.

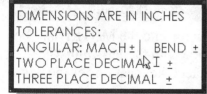

5. Move the cursor to a location to the right of the MACH ± text, and click once with the **left-mouse-button** to select a location to enter text.

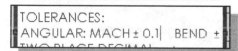

6. Enter **0.1**.

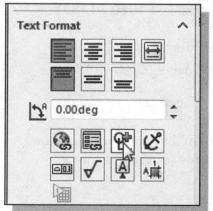

7. In the *Text Format* panel of the *Note Property Manager*, select the **Add Symbol** icon.

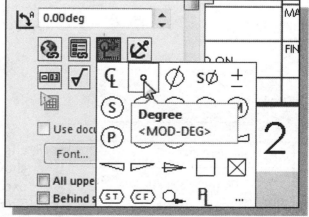

8. In the *Symbols* pop-up list, select **Degree** to insert a degree symbol.

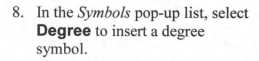

9.  On your own, continue editing the text in the tolerance box until it appears as shown.

10. Select the **OK** icon in the *Note Property Manager* to accept the change to the note.

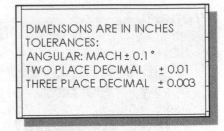

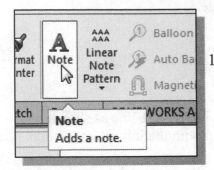

11. We will now insert a new note. Left-click on the **Note** icon on the *Annotation* toolbar.

12. Select the **No Leader** option in the *Leader* panel of the *Note Property Manager*.

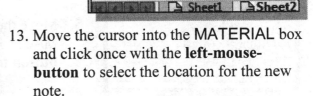

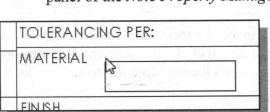

13. Move the cursor into the MATERIAL box and click once with the **left-mouse-button** to select the location for the new note.

14. Enter the text **1060 ALLOY**.

15. **Highlight** the text **1060 ALLOY**.

16. Change the font size to **8** in the *Formatting* window as shown.

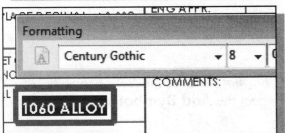

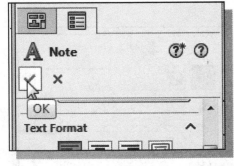

17. Select the **OK** icon in the *Note Property Manager* or press the **[Esc]** key to accept the note.

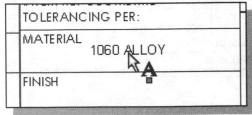

18. Click and drag on the new note to locate it as shown. (**NOTE:** SOLIDWORKS may execute snaps to align the text automatically. To negate these, hold down the **[Alt]** key while dragging the text and it will move freely.)

## Property Links

➤ SOLIDWORKS allows notes to be linked to properties for automatic display. In SOLIDWORKS, these properties are classified as system-defined *System Properties* and user-defined *Custom Properties*. Each system or custom property has a property name. There are system and custom properties associated with each SOLIDWORKS **Part, Assembly,** or **Drawing** file. Each **Property Link** contains (1) the definition of the source and (2) the property name. The source can be (1) the current document, (2) a model specified in a sheet, (3) a model in a specified drawing view, or (4) a component to which the annotation is attached. Several linked properties are automatically included in the default SOLIDWORKS sheet formats.

1. Move the cursor over the note in the center of the **DWG. NO.** box as shown. A letter **A** appears indicating a note. A pop-up dialog box appears which reads *$PRP: "SW-File Name"*. This is the definition of a **Property Link**. The prefix *$PRP:* defines the source as the current document. The Property Name is *SW-File Name*. This is an example of a **System Property.** Based on this Property Link, the filename – **U-Bracket** – of the current file is displayed.

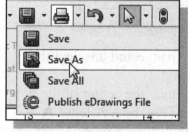

2. We will change the file name. Select **Save As** from the *Menu Bar* pull-down menu.

3. Enter **MECH 475** as the new *File name*.

4. Enter **U-Bracket Detail** for the *Description* and click **Save**.

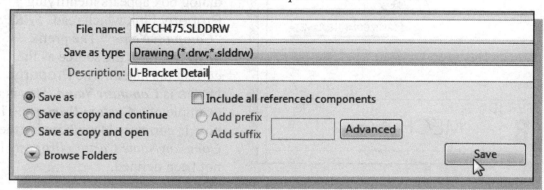

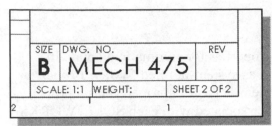

5.  Notice the new file name is now displayed in the **DWG. NO.** box. Click once with the left-mouse-button on the **MECH 475** note to open the *Note Property Manager*.

6.  Click the **Font** button in the *Text Format* panel of the *Note Property Manager*.

7.  In the *Choose Font* pop-up window, select **Points** under the *Height* option.

8.  Select **16** as the font size.

9.  Click **OK** in the *Choose Font* pop-up window.

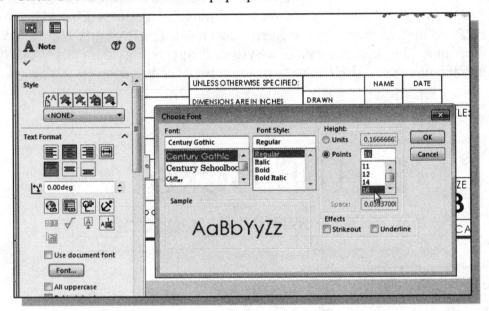

10. In the *Note Property Manager*, click **OK** to accept the change and unselect the note.

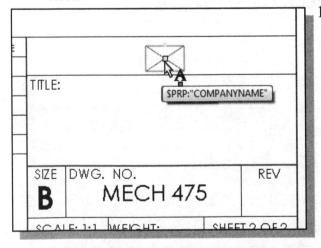

11. Left-click near the center of the box above the **Title:** box. A letter **A** appears indicating a note. A pop-up dialog box appears identifying a **Property Link** which reads *$PRP:* "*CompanyName.*" The prefix *$PRP:* defines the source as the current document. The **Property Name** is *CompanyName*. This is an example of a **Custom Property**. The note is currently blank because the *CompanyName Custom Property* has not been defined.

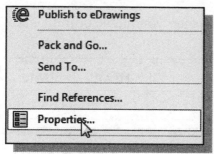

12. Select **Properties** from the *File* **pull-down menu** on the *Menu Bar*.

13. In the *Summary Information* window select the **Custom** tab. Notice the *Custom Property* **Description** appears in the table. We defined this property when we saved the drawing file.

14. The cell below the cell containing Description reads **<Type a new property>**. Move the cursor over this cell and click once with the **left-mouse-button** as shown.

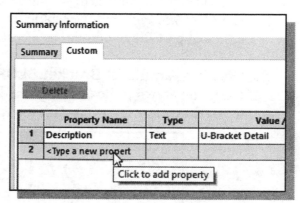

15. Type **CompanyName** in the cell as the new *Property Name*. (Note: If pull-down menu selection options appear, select **CompanyName**.)

16. Enter **your company name** to appear in the title block in the **Value / Text Expression** cell as shown.

| | Property Name | Type | Value / Text Expression | | Evaluated Value | | ⌒ |
|---|---|---|---|---|---|---|---|
| 1 | Description | Text | U-Bracket Detail | | U-Bracket Detail | | |
| 2 | CompanyName | Text | Oregon Institute of Technology | ⌄ | | | |
| 3 | <Type a new propert | | | | | | |

17. Click **OK** to accept the user-defined property.

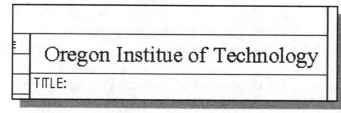

18. Notice the entry now automatically appears in the title block. On your own, adjust the font size to fit in the box (see steps 5-9 above).

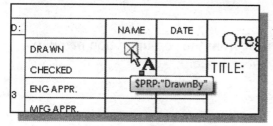

19. Move the cursor over the box next to **Drawn** as shown. A note containing a link to the *Custom Property* **DrawnBy** is contained in this box.

20. On your own, go to the *File Properties* window and enter your name or initials as the entry for the **DrawnBy** *Custom Property*.

21. On your own, determine the *Property Link* in the **DrawnDate** box. Enter the date for the *Custom Property*.

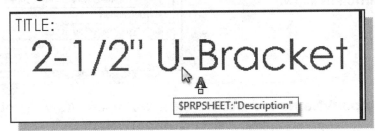

| | NAME | DATE |
|---|---|---|
| DRAWN | A. Smith | 1/15/2019 |

22. Left-click near the center of the **Title** box. A letter **A** appears indicating a note. A pop-up dialog box appears identifying a *Property Link* which reads *$PRPSHEET: "Description."* The prefix *$PRPSHEET:* defines the source as the model (part file) appearing in the sheet. The *Property Name* is **Description**. This is a **Custom Property**. We can define this description when we save the U-Bracket part file or add this as a custom property.

23. Open the part file (**U-Bracket.SLDPRT**). Follow the **Save As** procedure as we did in the drawing file to change the Description property (P. 8-35). Then return to the drawing window, and hit the **Rebuild** button to update the drawing.

TITLE:

## 2-1/2" U-Bracket

$PRPSHEET:"Description"

24. On your own investigate the rest of the title block to determine where other *Property Links* appear.

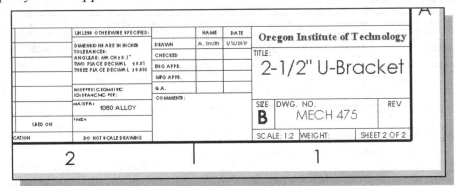

25. Press the **[F]** key to scale the view to fit the graphics area.

26. Move the cursor into the graphics area and **right-click** to open the pop-up option menu.

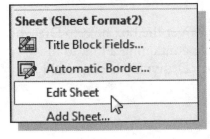

27. Select **Edit Sheet** from the pop-up option menu to change to *Edit Sheet* mode.

## Associative Functionality – Modifying Feature Dimensions

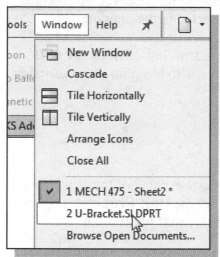

- *Associative functionality* allows us to change the design at any level, and the system reflects the changes at all levels automatically.

1. Select the **U-Bracket** part window from the *Window* pull-down menu to switch to *Part Modeling* mode.

2. In the *Feature Manager Design Tree*, **right-click** on **Annotations** to open the option menu.

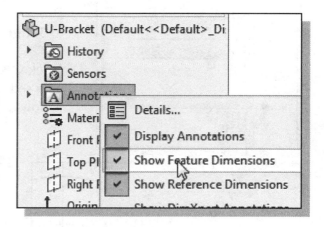

3. Select **Show Feature Dimensions** in the pop-up option menu.

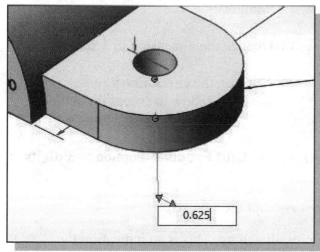

4. Double-click on the diameter dimensions (**0. 50**) of the hole on the base feature as shown in the figure. (Change the display to *Wireframe* mode if necessary.)

5. In the *Modify* dialog box, enter **0.625** as the new diameter dimension.

6. Click on the **OK** button to accept the new setting.

7. Select the **Rebuild** icon on the *Menu Bar*. Notice both holes are increased to 0.625 due to the **Mirror** relation.

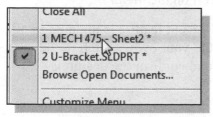

8. Select the **MECH 475- Sheet2** drawing window from the *Window* pull-down menu to switch to *Drawing* mode.

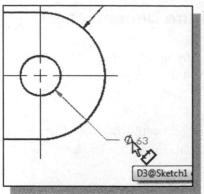

➢ Notice the change made to the model – the changed hole diameter – is reflected in the drawing.

9. Inside the graphics window, left-click on the **0.63** dimension in the *top* view to bring up the *Dimension Property Manager*.

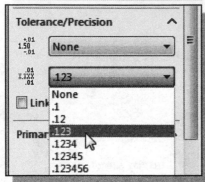

10. In the *Tolerance/Precision* panel of the *Dimension Property Manager*, set the **Unit Precision** option to **3 digits after the decimal point** as shown.

11. Click **OK** in the *Property Manager* to accept the new setting. Notice the accurate dimension of **.625** is now displayed.

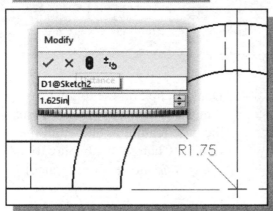

12. Inside the graphics window, **double-click** with the **left-mouse-button** on the **R 1.75** dimension in the *front* view to bring up the *Modify* pop-up option window.

13. Change the dimension to **1.625**.

14. Click on the **check mark** button in the *Modify* window to accept the setting.

15. In the *Dimension Property Manager*, set the **Unit Precision** option to **3 digits after the decimal point**.

16. Click **OK** in the *Property Manager* to accept the new setting.

➢ Notice the views become shaded with cross-hatching. This is an indication by SOLIDWORKS that changes have been made and a rebuild is needed to update the drawing and model.

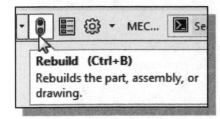

17. Select the **Rebuild** icon on the *Menu Bar*.

❖ Note the geometry of the cut feature is updated in all views automatically. On your own, switch to the *Part Modeling* mode and confirm the design is updated as well.

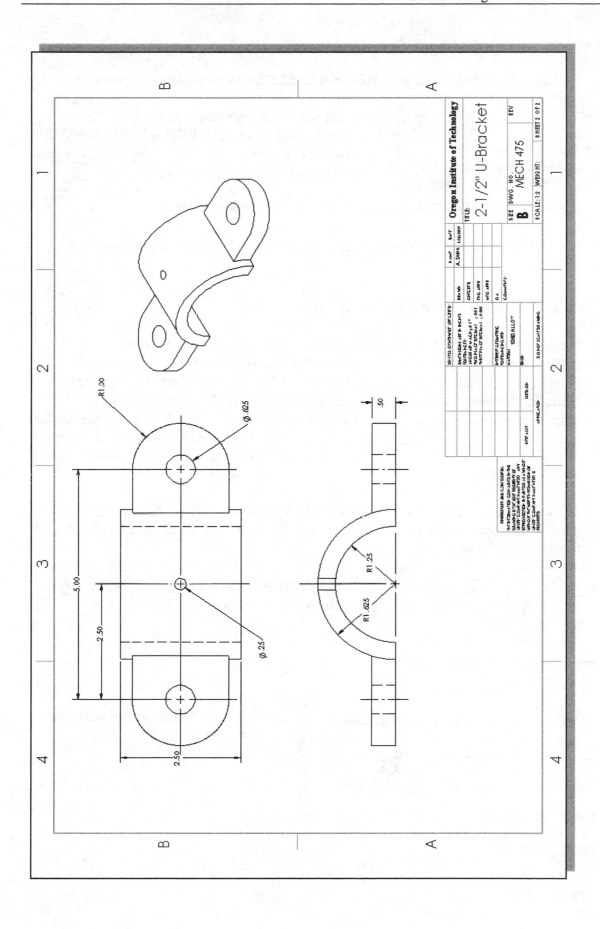

Drawing title block contents:

**Oregon Institute of Technology**

TITLE: 2-1/2" U-Bracket

DWG. NO.: MECH 475

SIZE: B

SCALE: 1:2    SHEET 2 OF 2

DRAWN: A. Smith    DATE: 1/01/2001

MATERIAL: 1060 ALLOY

Dimensions on drawing:
- R1.00
- Ø.625
- Ø.25
- 5.00
- 2.50
- 2.50
- R1.25
- R1.625
- .50

Title block standard notes:
- UNLESS OTHERWISE SPECIFIED
- DIMENSIONS ARE IN INCHES
- TOLERANCES
- DO NOT SCALE DRAWING
- PROPRIETARY AND CONFIDENTIAL

## Saving the Drawing File

1.  Go to *Drawing* mode with **MECH 475.SLDDRW** as the active document.

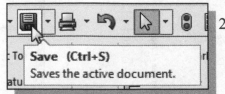

2.  Select the **Save** icon on the *Menu Bar*. This command will save the **MECH 475.SLDDRW** file. This file includes the document properties, the sheet format, and the drawing views.

3.  In the *Save Modified Documents* window, select **Save All**.

4.  If the SOLIDWORKS window appears asking to *'Update the drawing?'* select **Yes**.

## Create a B size Drawing Template

❖ In SOLIDWORKS, each new drawing is created from a template. During the installation of SOLIDWORKS, a default drafting standard was selected which sets the default template used to create drawings. We can use this template or another predefined template, modify one of the predefined templates, or create our own templates to enforce drafting standards and other conventions. A drawing file can be used as a template; a drawing file becomes a template when it is saved as a template (*.drwdot) file. Once the template is saved, we can create a new drawing file using the new template.

❖ We will first create a custom SOLIDWORKS drawing template file with the document properties (e.g., ANSI dimension standard, English units, etc.) and sheet format (size B paper size, etc.).

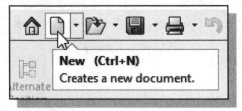

1.  Click **New** in the *Quick access toolbar* to start a **New file**.

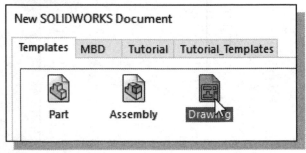

2.  Select **Drawing** in the Templates tab as shown.

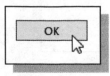

3.  Click **OK** to proceed.

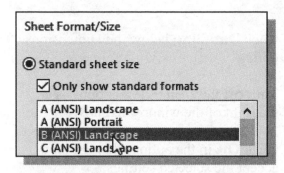

4. Select B(ANSI) Landscape in the Standard sheet size option as shown.

5. Click **OK** to proceed.

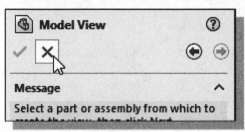

6. Click **Cancel** to exit the Insert Model view command.

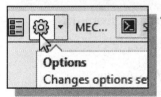

7. Select the **Options** icon, and then select the **Document Properties** tab.

8. Click **Units** as shown.

9. Select **IPS (inch, pound, second)** as the *Unit system* option.

10. Verify that **.12** is selected in the *Decimals* spin boxes.

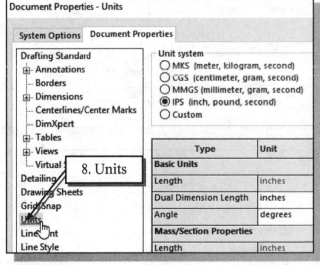

11. Click **Dimensions** as shown.

12. Verify that **.12** is selected in the *Primary precision* spin box.

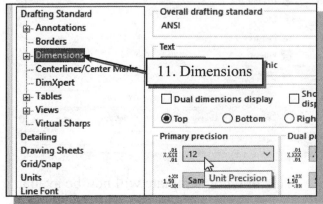

13. Click **Detailing** below the Dimensions option.

14. Turn **off** the *Center Marks* options at the right of the window as shown.

15. Click **OK** in the options dialog box to accept the selected settings.

16. Select **Save As** from the *File* pull-down menu as shown.

17. Click in the *Save as type:* entry box and select **Drawing Templates** from the pull-down options.

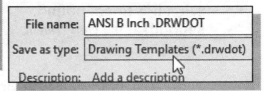

| File name: | ANSI B Inch .DRWDOT |
|---|---|
| Save as type: | Drawing Templates (*.drwdot) |
| Description: | Add a description |

18. The folder selection will automatically change to the default Templates folder. Use the browser to change the folder selection to the **Tutorial Templates** folder you created in Chapter 5.

19. Enter **ANSI-B-Inch** for the *File name* and enter **ANSI B Inch** for the *Description*.

20. Select **Save**. If any of the SOLIDWORKS pop-up window appears, click **OK**.

❖ This drawing template will now be available for use to create future drawings.

## Questions:

1.  What does *associative functionality* within SOLIDWORKS allow us to do?

2.  How do we move a view on the *Drawing Sheet*?

3.  How do we display feature/model dimensions in the *Drawing* mode?

4.  What is the difference between a *feature dimension* and a *reference dimension*?

5.  How do we reposition dimensions?

6.  What is a *base view*?

7.  Identify and describe the following commands:

    (a)

    (b)

    (c)

    (d)

## Exercises:

1. **Shaft Guide** (Design is symmetrical and dimensions are in inches.)

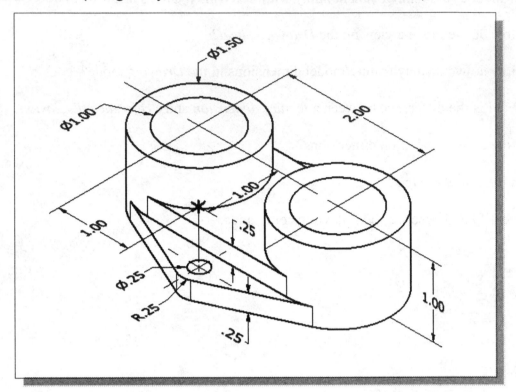

2. **Shaft Guide** (Dimensions are in inches.)

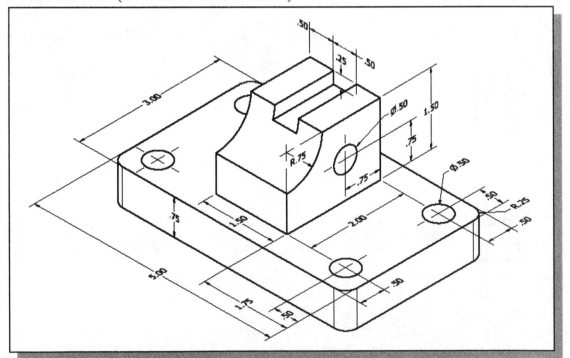

3. **Cylinder Support** (Dimensions are in inches.)

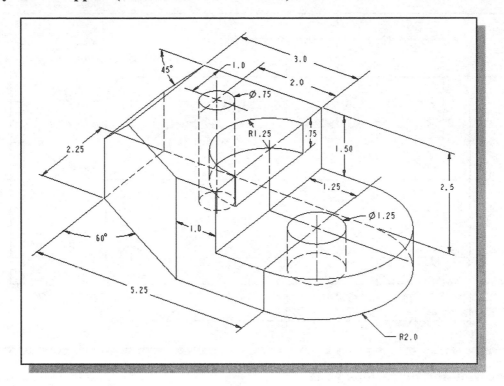

4. **Swivel Base** (Dimensions are in inches.)

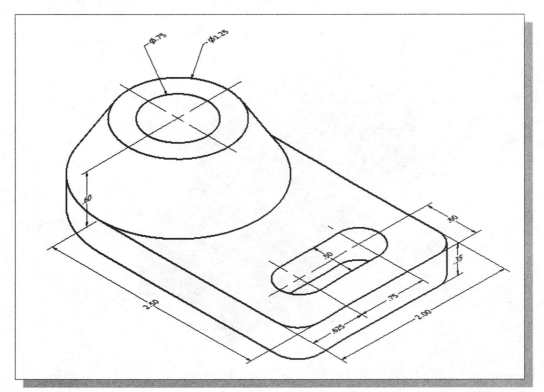

5. **Block Base** (Dimensions are in inches. Plate Thickness: 0.25)

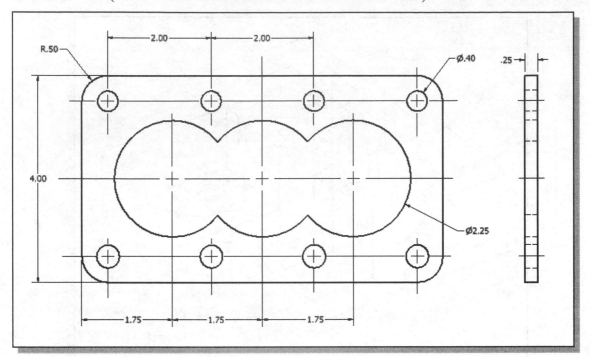

6. **Slide Mount** (Dimensions are in inches.)

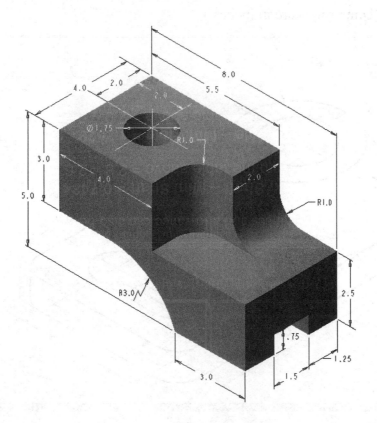

# Chapter 9
# Tolerancing and Fits

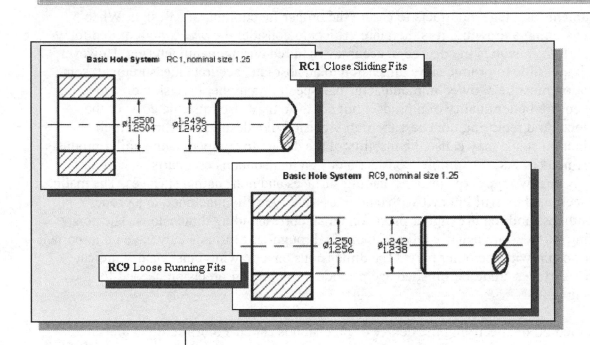

## Learning Objectives

- ◆ **Understand Tolerancing Nomenclature**
- ◆ **The Tolerancing Designation Method**
- ◆ **Understand the Basics of ANSI Standard Fits**
- ◆ **Use of the ISO Metric Standard Fits**
- ◆ **Use the Open Files SOLIDWORKS Option**
- ◆ **Set up the Tolerancing Option in SOLIDWORKS**

# Precision and Tolerance

In the manufacturing of any product, quality and cost are always the two primary considerations. Drawings with dimensions and notes often serve as construction documents and legal contracts to ensure the proper functioning of a design. When dimensioning a drawing, it is therefore critical to consider the *precision* required for the part. *Precision* is the degree of accuracy required during manufacturing. However, it is impossible to produce any dimension to an absolute, accurate measurement; some variation must be allowed in manufacturing. Specifying higher precision on a drawing may ensure better quality of a product but doing so may also raise the costs of the product. And requiring unnecessary high precision of a design, resulting in high production costs, may cause the inability of the design to compete with similar products in the market. As an example, consider a design that contains cast parts. A cast part usually has two types of surfaces: mating surfaces and non-mating surfaces. The mating surfaces, as they will interact with other parts, are typically machined to a proper smoothness and require higher precision on all corresponding dimensions. The non-mating surfaces are usually left in the original rough-cast form as they have no important relationship with the other parts. The dimensions on a drawing must clearly indicate which surfaces are to be finished and provide the degree of precision needed for the finishing.

The method of specifying the degree of precision is called **Tolerancing**. **Tolerance** is the allowable variation for any given size and provides a practical means to achieve the precision necessary in a design. Tolerancing also ensures interchangeability in manufacturing; parts can be made by different companies in different locations and still maintain the proper functioning of the design. With tolerancing, each dimension is allowed to vary within a specified zone. By assigning as large a tolerance as possible, without interfering with the functionality of a design, the production costs can be reduced. The smaller the tolerance zone specified, the more expensive it is to manufacture.

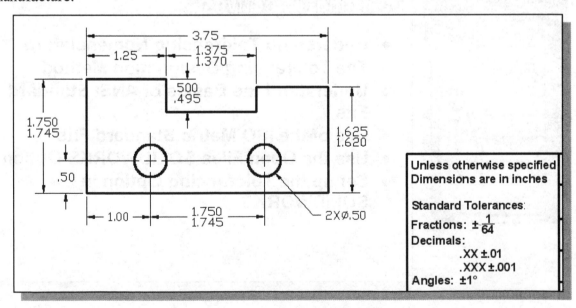

# Methods of Specifying Tolerances – English System

Three methods are commonly used in specifying tolerances: **Limits**, **Unilateral tolerances** and **Bilateral tolerances**. Note that the unilateral and bilateral methods employ the use of a **Base Dimension** in specifying the variation range.

- Limits – Specifies the range of size that a part may be.
  This is the preferred method as approved by ANSI/ASME Y14.5M; the maximum and minimum limits of size and locations are specified. The maximum value is placed above the minimum value. In single-line form, a dash is used to separate the two values.

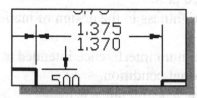 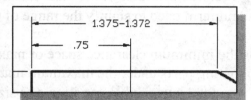

- Bilateral – Variation of size is in both directions.
  This method uses a basic size and is followed by a plus-and-minus expression of tolerance. The bilateral tolerances method allows variations in both directions from the basic size. If an equal variation in both directions is desired, the combined plus and minus symbol is used with a single value.

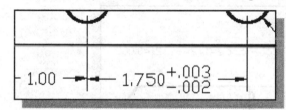 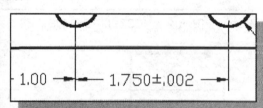

- Unilateral – Variation of size is in only one direction.
  This method uses a basic size and is followed by a plus-or-minus expression of tolerance. The unilateral tolerances method allows variations only in one direction from the basic size.

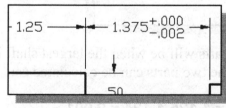 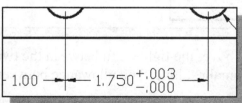

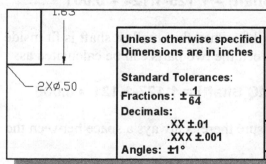

Unless otherwise specified
Dimensions are in inches

Standard Tolerances:
Fractions: $\pm\frac{1}{64}$
Decimals:
.XX $\pm$.01
.XXX $\pm$.001
Angles: $\pm$1°

- General tolerances – General tolerances are typically covered in the title block; the general tolerances are applied to the dimensions in which tolerances are not given.

## Nomenclature

The terms used in tolerancing and dimensioning should be clearly understood before more detailed discussions are studied. The following are definitions as defined in the ANSI/ASME Y 14.5M standard.

- **Nominal size** is the designation used for the purpose of general identification.
- **Basic size, or basic dimension,** is the theoretical size from which limits of size are derived. It is the size from which limits are determined for the size or location of a feature in a design.
- **Actual size** is the measured size of the manufactured part.
- **Fit** is the general term used to signify the range of tightness in the design of mating parts.
- **Allowance** is the minimum clearance space or maximum interference intended between two mating parts under the maximum material condition.
- **Tolerance** is the total permissible variation of a size. The tolerance is the difference between the limits of size.
- **Maximum Material Condition (MMC)** is the size of the part when it consists of the most material.
- **Least Material Condition (LMC)** is the size of the part when it consists of the least material.

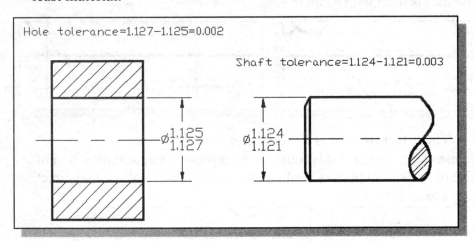

In the figure above, the tightest fit between the two parts will be when the largest shaft is fit inside the smallest hole. The allowance between the two parts can be calculated as:

**Allowance = (MMC Hole)-(MMC Shaft) = 1.125-1.124 = 0.001**

The loosest fit between the above two parts will be when the smallest shaft is fit inside the largest hole. The maximum clearance between the two parts can be calculated as:

**Max. Clearance = (LMC Hole)-(LMC Shaft) = 1.127-1.121 = 0.006**

➤ Note that the above dimensions will assure there is always a space between the two mating parts.

## Example 9.1

The size limits of two mating parts are as shown in the below figure. Determine the following items, as described in the previous page: nominal size, tolerance of the shaft, tolerance of the hole, allowance and maximum clearance between the parts.

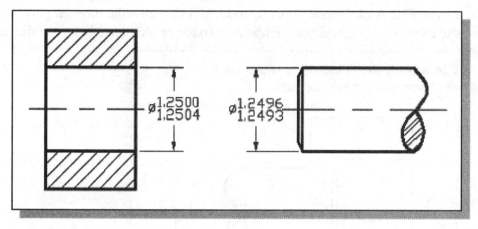

**Solution:**

**Nominal size = 1.25**

**Tolerance of the shaft = Max. Shaft - Min. Shaft**
     **= 1.2496-1.2493 = 0.0003**

**Tolerance of the hole = Max. Hole - Min. Hole**
     **= 1.2504-1.2500 = 0.0004**

**Allowance = Min. Hole (MMC Hole) - Max. Shaft (MMC Shaft)**
     **= 1.2500-1.2496 = 0.0004**

**Max. Clearance = Max. Hole (LMC Hole) - Min. Shaft (LMC Shaft)**
     **= 1.2504-1.2493 = 0.0011**

**Exercise:** Given the size limits of two mating parts, determine the following items: nominal size, tolerance of the shaft, tolerance of the hole, allowance and maximum clearance between the parts.

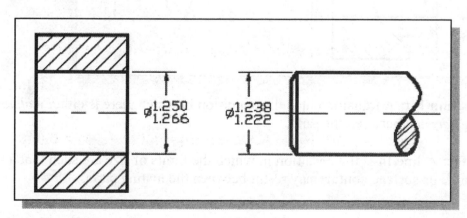

## Fits between Mating Parts

Fit is the general term used to signify the range of tightness in the design of mating parts. In ANSI/ASME Y 14.5M, four general types of fits are designated for mating parts.

- **Clearance Fit:** A clearance fit is the condition in which the internal part is smaller than the external part; and always leaves a space or clearance between the parts.

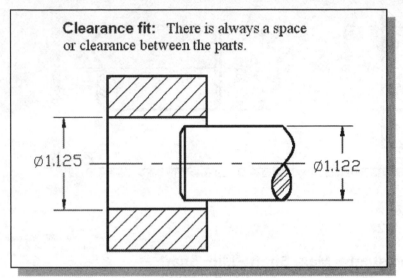

- **Interference Fit:** An interference fit is the condition in which the internal part is larger than the external part, and there is always an interference between the parts.

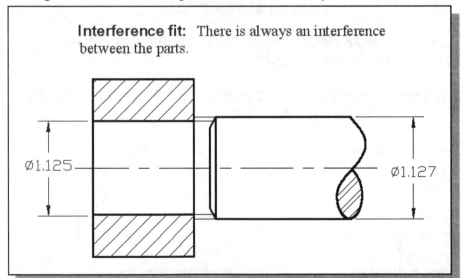

- **Transition Fit:** A transition fit is the condition in which there is either a *clearance* or *interference* between the parts.

- **Line Fit:** A line fit is the condition in which the limits of size are such that a clearance or surface contact may result between the mating parts.

## Selective Assembly

By specifying the proper allowances and tolerances, mating parts can be completely interchangeable. But sometimes the fit desired may require very small allowances and tolerances, and the production cost may become very high. In such cases, either manual or computer-controlled selective assembly is often used. The manufactured parts are then graded as small, medium and large based on the actual sizes. In this way, very satisfactory fits may be achieved at a much lower cost compared to manufacturing all parts to very accurate dimensions. Interference and transition fits often require the use of selective assembly to get the desirable interference or clearance.

## Basic Hole and Basic Shaft Systems

In manufacturing, standard tools (such as reamers, drills) are often used to produce holes. In the **basic hole system**, the **minimum size of the hole** is taken as a base, and an allowance is assigned to derive the necessary size limits for both parts. The hole can often be made with a standard size tool and the shaft can easily be machined to any size desired.

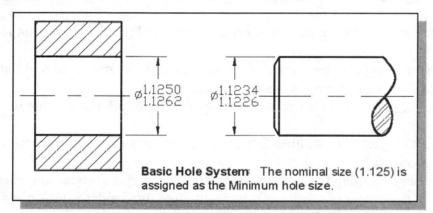

**Basic Hole System**: The nominal size (1.125) is assigned as the Minimum hole size.

On the other hand, in some branches of manufacturing industry, where it is necessary to manufacture the shafts using standard sizes, the **basic shaft system** is often used. This system should only be used when there is a special need for it. For example, if a shaft is to be assembled with several other parts using different fits. In the **basic shaft system**, the **maximum shaft** is taken as the basic size, and all size limits are derived from this basic size.

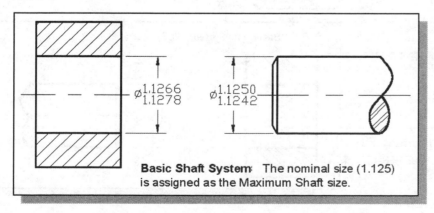

**Basic Shaft System**: The nominal size (1.125) is assigned as the Maximum Shaft size.

# American National Standard Limits and Fits – Inches

The standard fits designated by the American National Standards Institute are as follows:

- **RC**  Running or Sliding Clearance Fits (See appendix A for the complete table.)
- **LC**  Locational Clearance Fits
- **LT**  Transition Clearance or Interference Fits
- **LN**  Locational Interference Fits
- **FN**  Force or Shrink Fits

These letter symbols are used in conjunction with numbers for the class of fit. For example, **RC5** represents a class 5 Running or Sliding Clearance Fits. The limits of size for both of the mating parts are given in the standard.

1.  **Running or Sliding Clearance Fits:** RC1 through RC9
    These fits provide limits of size for mating parts that require sliding and running performance, with suitable lubrication allowance.
    RC1, **close sliding fits,** are for mating parts to be assembled without much play in between.
    RC2, **sliding fits,** are for mating parts that need to move and turn easily but not run freely.
    RC3, **precision running fits,** are for precision work running at low speed and light journal pressure. This class provides the closest running fits.
    RC4, **close running fits,** are for moderate running speed and journal pressure in accurate machinery.
    RC5, RC6 and RC7, **medium running fits,** are for running at higher speed and/or higher journal pressure.
    RC8 and RC9, **loose running fits,** are for general purpose running condition where wide commercial tolerances are sufficient; typically used for standard stock parts.

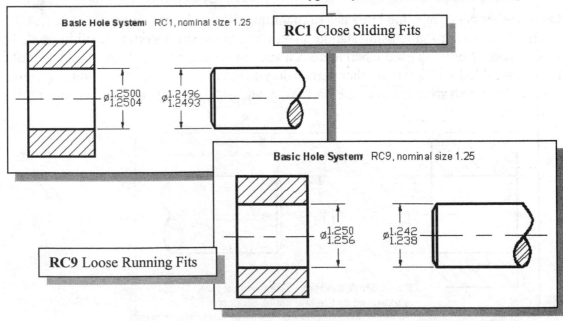

## 2. Locational Clearance Fits: LC1 through LC11

These fits provide limits of size for mating parts that are normally stationary and can be freely assembled and disassembled. They run from snug fits for parts requiring accuracy location, through the medium clearance fits for parts (such as spigots) to the looser fastener fits where freedom of assembly is of prime importance.

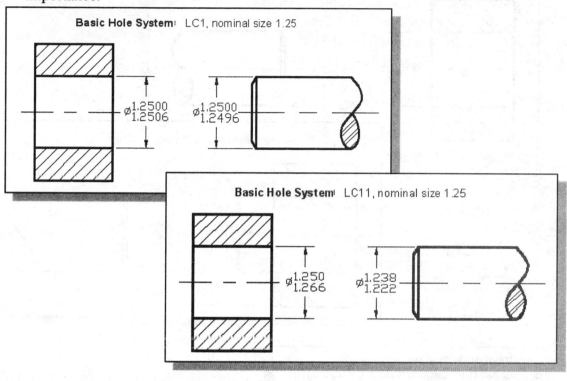

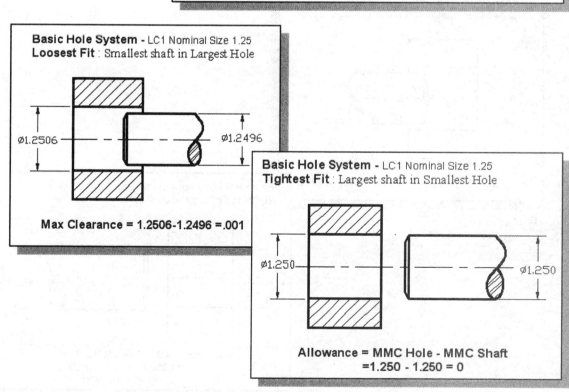

3.  **Transition Clearance or Interference Fits:** LT1 through LT6
    These fits provide limits of size for mating parts requiring accuracy location, but either a small amount of clearance or interference is permissible.

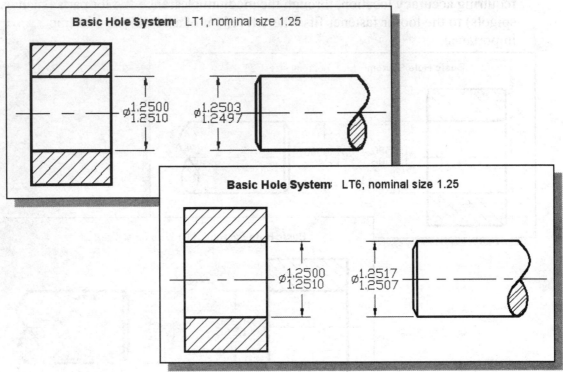

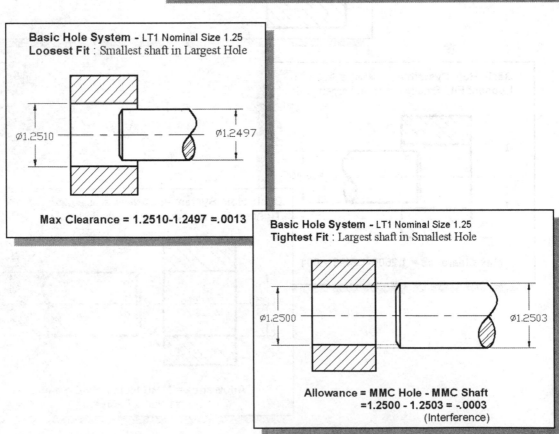

4. **Locational Interference Fits:** LN1 through LN3

These fits provide limits of size for mating parts requiring accuracy location, and for parts needing rigidity and alignment. Such fits are not for parts that transmit frictional loads from one part to another by the tightness of the fit.

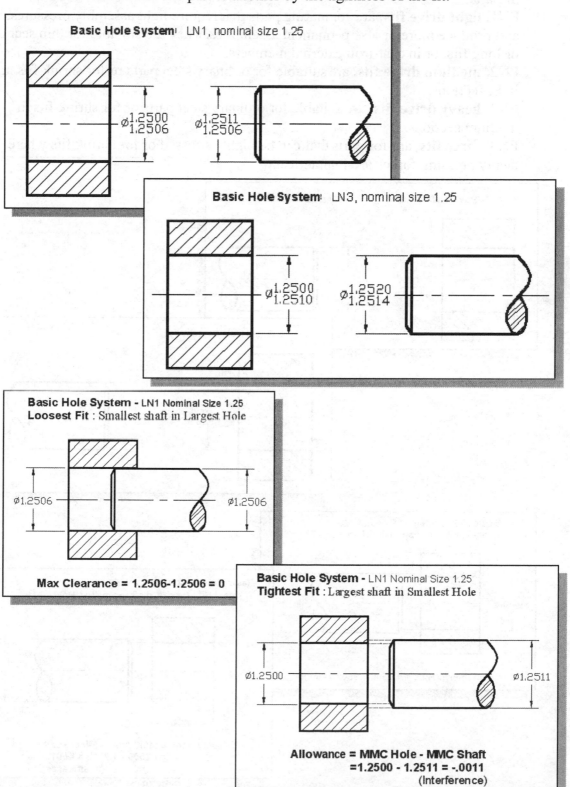

**Basic Hole System** LN1, nominal size 1.25

$\phi$1.2500 / 1.2506

$\phi$1.2511 / 1.2506

**Basic Hole System** LN3, nominal size 1.25

$\phi$1.2500 / 1.2510

$\phi$1.2520 / 1.2514

**Basic Hole System** - LN1 Nominal Size 1.25
**Loosest Fit** : Smallest shaft in Largest Hole

$\phi$1.2506

$\phi$1.2506

**Max Clearance = 1.2506-1.2506 = 0**

**Basic Hole System** - LN1 Nominal Size 1.25
**Tightest Fit** : Largest shaft in Smallest Hole

$\phi$1.2500

$\phi$1.2511

**Allowance = MMC Hole - MMC Shaft**
**=1.2500 - 1.2511 = -.0011**
(Interference)

5.   **Force or Shrink Fits:** FN1 through FN5
These fits provide limits of size for mating parts requiring constant bore pressure
in between the parts. The interference amount varies almost directly with the size
of parts.

FN1, **light drive fits,** are for mating parts that require light assembly pressures,
and produce more or less permanent assemblies. They are suitable for thin sections
or long fits, or in cast-iron external members.

FN2, **medium drive fits,** are suitable for ordinary steel parts or for shrink fits in
light sections.

FN3, **heavy drive fits,** are suitable for ordinary steel parts or for shrink fits in
medium sections.

FN4, **force fits,** are for parts that can be highly stressed or for shrink fits where
heavy pressing forces are impractical.

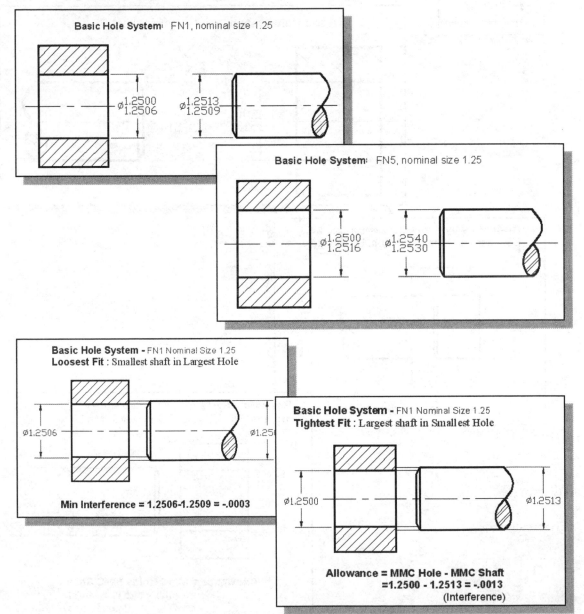

**Example 9.2 Basic Hole System**

A design requires the use of two mating parts with a nominal size of **0.75** inches. The shaft is to run with moderate speed but with a fairly heavy journal pressure. The fit class chosen is **Basic Hole System RC6**. Determine the size limits of the two mating parts.

From Appendix A, the RC6 fit for 0.75 inch nominal size:

| Nominal size Range | Limits of Clearance | RC6 Standard Limits – Basic Hole | |
|---|---|---|---|
| | | Hole | Shaft |
| 0.71 - 1.19 | 1.6 | +2.0 | -1.6 |
| | 4.8 | -0.0 | -2.8 |

Using the *Basic Hole System*, the *BASIC* size 0.75 is designated as the minimum hole size. And the maximum hole size is 0.75+0.002 = 0.752 inches.

The maximum shaft size is 0.75-0.0016 = 0.7484 inches
The minimum shaft size is 0.75-0.0028 = 0.7472 inches

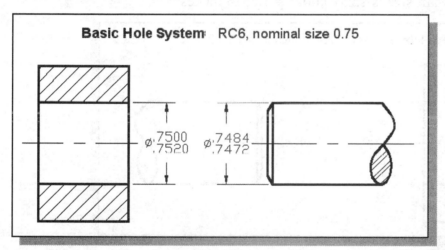

Basic Hole System  RC6, nominal size 0.75

➤  Note the important tolerances and allowance can also be obtained from the table:

**Tolerance of the shaft = Max. Shaft – Min. Shaft = 0.7484 – 0.7472 = 0.0012**
This can also be calculated from the table above = 2.8 – 2.6 = **1.2 (thousandths)**

**Tolerance of the hole = Max. Hole – Min. Hole = 1.2520 – 1.2500 = 0.002**
This can also be calculated from the table above = 2.0 – 0.0 = **2.0 (thousandths)**

**Allowance = Min. Hole (MMC Hole) – Max. Shaft (MMC Shaft)**
         **= 0.7500 – 0.7484 = 0.0016**
This is listed as the minimum clearance in the table above: **+1.6 (thousandths)**

**Max. Clearance = Max. Hole (LMC Hole) – Min. Shaft (LMC Shaft)**
         **= 0.7520 – 0.7472 = 0.0048**
This is listed as the maximum clearance in the table above: **+4.8 (thousandths)**

**Example 9.3 Basic Hole System**

A design requires the use of two mating parts with a nominal size of **1.25** inches. The shaft and hub is to be fastened permanently using a drive fit. The fit class chosen is **Basic Hole System FN4**. Determine the size limits of the two mating parts.

From ANSI/ASME B4.1 Standard, the FN4 fit for 1.25 inches nominal size:

| Nominal size Range | Limits of Interference | FN4 Standard Limits – Basic Hole | |
|---|---|---|---|
| | | Hole | Shaft |
| 1.19 - 1.97 | 1.5<br>3.1 | +1.0<br>-0.0 | +3.1<br>+2.5 |

Using the *Basic Hole System*, the *BASIC* size 1.25 is designated as the minimum hole size. And the maximum hole size is 1.25+0.001 = 1.251 inches.

The maximum shaft size is 1.25+0.0031 = 1.2531 inches
The minimum shaft size is 1.25+0.0025 = 1.2525 inches

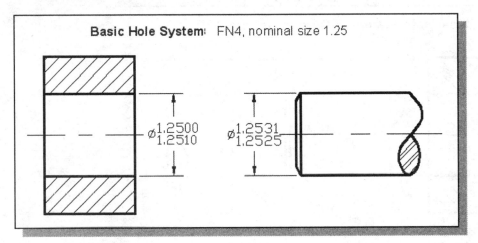

> Note the important tolerances and allowance can also be obtained from the table:

**Tolerance of the shaft = Max. Shaft – Min. Shaft  = 1.2531 – 1.2525 = 0.0006**
This can also be calculated from the table above = **3.1 – 2.5 = 0.6 (thousandths)**

**Tolerance of the hole = Max. Hole – Min. Hole = 1.2510 – 1.2500 = 0.001**
This can also be calculated from the table above = **1.0 – 0.0 = 1.0 (thousandths)**

**Allowance = Min. Hole (MMC Hole) – Max. Shaft (MMC Shaft)**
           **= 1.2500 – 1.2531 = -0.0031**
This is listed as the maximum interference in the table above: **3.1 (thousandths)**

**Min. interference = Max. Hole (LMC Hole) – Min. Shaft (LMC Shaft)**
              **= 1.2510 – 1.2525 = -0.0015**
This is listed as the minimum interference in the table above: **1.5 (thousandths)**

**Example 9.4 Basic Shaft System**

Use the **Basic Shaft System** from example 9.2. A design requires the use of two mating parts with a nominal size of **0.75** inches. The shaft is to run with moderate speed but with a fairly heavy journal pressure. The fit class chosen is **Basic Shaft System RC6**. Determine the size limits of the two mating parts.

From Appendix A, the Basic Hole System RC6 fit for 0.75 inch nominal size:

| Nominal size Range | Limits of Clearance | RC6 Standard Limits – Basic Hole | |
|---|---|---|---|
| | | Hole | Shaft |
| 0.71 - 1.19 | 1.6 | +2.0 | -1.6 |
| | 4.8 | -0.0 | -2.8 |

Using the Basic Shaft System, the BASIC size 0.75 is designated as the MMC of the shaft, the maximum shaft size. The conversion of limits to the Basic shaft System is performed by adding 1.6 (thousandths) to the four size limits as shown:

| Nominal size Range | Limits of Clearance | RC6 Standard Limits – Basic Shaft | |
|---|---|---|---|
| | | Hole | Shaft |
| 0.71 - 1.19 | 1.6 | +3.6 | -0.0 |
| | 4.8 | +1.6 | -1.2 |

Therefore, the maximum shaft size is 0.75
The minimum shaft size is 0.75-0.0012 = 0.7488 inches.
The maximum hole size is 0.75+0.0036 = 0.7536 inches
The minimum hole size is 0.75+0.0016 = 0.7516 inches

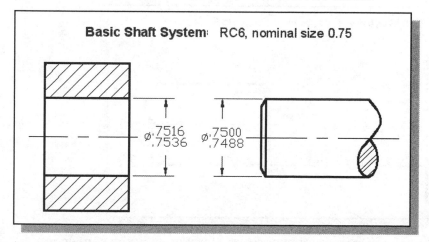

Basic Shaft System RC6, nominal size 0.75

➤ Note that all the tolerances and allowance are maintained in the conversion.

**Allowance = Min. Hole – Max. Shaft = 0.7517 – 0.7500 = 0.0016**
This is listed as the minimum clearance in the table above: **1.6 (thousandths)**

**Max. Clearance = Max. Hole – Min. Shaft = 0.7537 – 0.7488 = 0.0048**
This is listed as the maximum clearance in the table above: **4.8 (thousandths)**

**Example 9.5 Basic Shaft System**

Use the **Basic Shaft System** from example 9.3. A design requires the use of two mating parts with a nominal size of **1.25** inches. The shaft and hub is to be fastened permanently using a drive fit. The fit class chosen is **Basic Shaft System FN4**. Determine the size limits of the two mating parts.

From ANSI/ASME B4.1 Standard, the FN4 fit for 1.25 inches nominal size:

| Nominal size Range | Limits of Interference | FN4 Standard Limits – Basic Hole | |
|---|---|---|---|
| | | Hole | Shaft |
| 1.19 - 1.97 | 1.5 | +1.0 | +3.1 |
| | 3.1 | -0.0 | +2.5 |

With the *Basic Shaft System*, the *BASIC* size 1.25 is designated as the MMC of the shaft, the maximum shaft size. The conversion of limits to the Basic shaft System is performed by subtracting 3.1 (thousandths) from the four size limits as shown:

| Nominal size Range | Limits of Interference | FN4 Standard Limits – Basic Shaft | |
|---|---|---|---|
| | | Hole | Shaft |
| 1.19 - 1.97 | 1.5 | -2.1 | -0.0 |
| | 3.1 | -3.1 | -0.6 |

Therefore, the maximum shaft size is 1.25
The minimum shaft size is 1.25-0.0006 = 1.2494 inches.
The maximum hole size is 1.25-0.0021 = 1.2479 inches
The minimum hole size is 1.25-0.0031 = 1.2469 inches

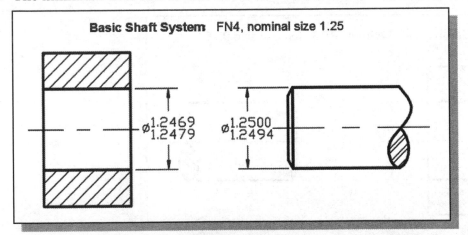

> Note that all the tolerances and allowance are maintained in the conversion.

**Allowance = Min. Hole – Max. Shaft = 1.2469 – 1.2500 = -0.0031**
This is listed as the maximum interference in the table above: **3.1 (thousandths)**

**Min. Interference = Max. Hole – Min. Shaft = 1.2479 – 1.2494 = -0.0015**
This is listed as the minimum interference in the table above: **1.5 (thousandths)**

## Tolerancing – Metric System

The concepts described in previous sections on the English tolerancing system are also applicable to the Metric System. The commonly used system of preferred metric tolerancing and fits are outlined by the International Organization for Standardization (ISO). The system is specified for holes and shafts, but it is also adaptable to fits between features of parallel surfaces.

- **Basic size** is the theoretical size from which limits or deviations of size are derived.

- **Deviation** is the difference between the basic size and the hole or shaft size. This is equivalent to the word "Tolerance" in the English system.

- **Upper Deviation** is the difference between the basic size and the permitted maximum size. This is equivalent to "Maximum Tolerance" in the English system.

- **Lower Deviation** is the difference between the basic size and the permitted minimum size. This is equivalent to "Minimum Tolerance" in the English system.

- **Fundamental Deviation** is the deviation closest to the basic size. This is equivalent to "Minimum Allowance" in the English system.

- **Tolerance** is the difference between the permissible variations of a size. The tolerance is the difference between the limits of size.

- **International tolerance grade (IT)** is a set of tolerances that varies according to the basic size and provides a uniform level of accuracy within the grade. There are 18 IT grades- IT01, IT0, and IT1 through IT16.

  - ❖ IT01 through IT7 is typically used for measuring tools; IT5 through IT11 are used for Fits and IT12 through IT16 for large manufacturing tolerances.

  - ❖ The following is a list of IT grade related to machining processes:

| International Tolerance Grade (IT) | | | | | | | |
|---|---|---|---|---|---|---|---|
| **4** | **5** | **6** | **7** | **8** | **9** | **10** | **11** |
| Lapping or Honing | | | | | | | |
| | Cylindrical Grinding | | | | | | |
| | Surface Grinding | | | | | | |
| | Diamond Turning or Boring | | | | | | |
| | Broaching | | | | | | |
| | Powder Metal-sizes | | | | | | |
| | | Reaming | | | | | |
| | | | Turning | | | | |
| | | | Powder Metal-sintered | | | | |
| | | | Boring | | | | |
| | | | | | | Milling, Drilling | |
| | | | | | | Planing & Shaping | |
| | | | | | | Punching | |
| | | | | | | | Die Casting |

## Metric Tolerances and Fits Designation

Metric tolerances and fits are specified by using the fundamental deviation symbol along with the IT grade number. An upper-case **H** is used for the fundamental deviation for the hole tolerance using the **Hole Basis system,** where a lower case **h** is used to specify the use of the **Shaft Basis System**.

1. **Hole:  30H8**

    Basic size = 30 mm
    Hole size is based on hole basis system using IT8 for tolerance.

2. **Shaft:  30f7**

    Basic size = 30 mm
    Shaft size is based on hole basis system using IT7 for tolerance.

3. **Fit:  30H8/f7**

    Basic size = 30 mm
    Hole basis system with IT8 hole tolerance and IT7 for shaft tolerance.

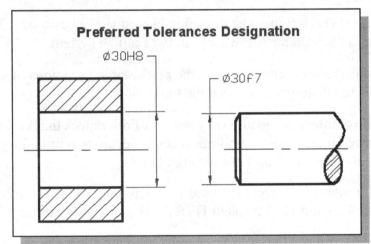

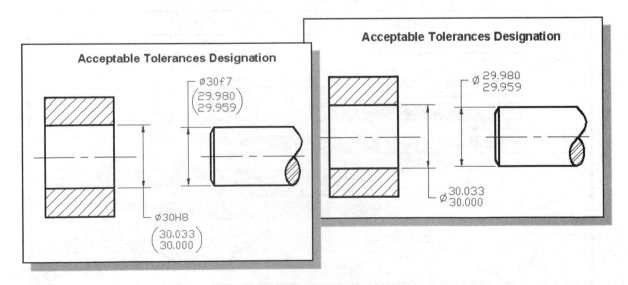

# Preferred ISO Metric Fits

The symbols for either the hole-basis or shaft-basis preferred fits are given in the table below. These are the preferred fits; select the fits from the table whenever possible. See **Appendix B** for the complete *ISO Metric Fits* tables.

- The **Hole-Basis** system is a system in which the basic size is the minimum size of the hole. The *hole-basis* system is the preferred method; the fundamental deviation is specified by the uppercase **H**.

- The **Shaft-Basis** system is a system in which the basic size is the maximum size of the shaft. The fundamental deviation of the *shaft-basis* system is specified by the lowercase **h**.

| ISO Preferred Metric Fits | | | |
|---|---|---|---|
| | **Hole Basis** | **Shaft Basis** | **Description** |
| **Clearance Fits** | H11/c11 | C11/h11 | *Loose running* fit for wide commercial tolerance or allowances. |
| | H9/d9 | D9/h9 | *Free running* fit not for use when accuracy is essential, but good for large temperature variation, high running speed, or heavy journal pressure. |
| | H8/f7 | F8/h7 | *Close running* fit for running on accurate machines running at moderate speeds and journal pressure. |
| | H7/g6 | G7/h6 | *Sliding* fit not intended to run freely, but to move and turn freely and locate accurately. |
| **Transition Fits** | H7/h6 | H7/h6 | *Locational clearance* fit provides snug fit for locating stationary parts but can be freely assembled and disassembled. |
| | H7/k6 | K7/h6 | *Locational transition* fit for accurate location, a compromise between clearance and interference. |
| | H7/n6 | N7/h6 | *Locational transition* fit for accurate location where interference is permissible. |
| | H7/p6 | P7/h6 | *Locational interference* fit for parts requiring rigidity and alignment with prime accuracy but without special bore pressure. |
| **Interference Fits** | H7/s6 | S7/h6 | *Medium drive* fit for ordinary steel parts or shrink fits on light sections, the tightest fit usable with cast iron. |
| | H7/u6 | U7/h6 | *Force* fit suitable for parts which can be highly stressed or for shrink fits where the heavy pressing forces required are impractical. |

### Example 9.6 Metric Hole Basis System

A design requires the use of two mating parts with a nominal size of **25** millimeters. The shaft and hub is to be fastened permanently using a drive fit. The fit class chosen is **Hole-Basis System H7/s6**. Determine the size limits of the two mating parts.

From Appendix B, the H7/s6 fit for 25 millimeters size:

| Basic size | H7/s6 Medium Drive –Hole Basis | | |
|:---:|:---:|:---:|:---:|
| | Hole H7 | Shaft s6 | Fit |
| 25 | 25.021 | 25.048 | -0.014 |
| | 25.000 | 25.035 | -0.048 |

Using the *Hole Basis System*, the *BASIC* size 25 is designated as the minimum hole size. And the maximum hole size is 25.021 millimeters.
The maximum shaft size is 25.048 millimeters
The minimum shaft size is 25.035 millimeters

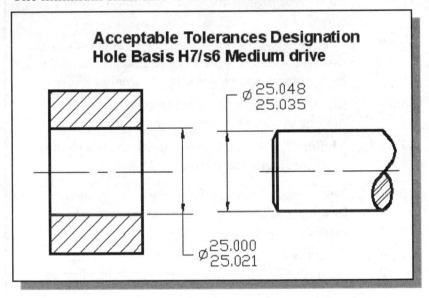

**Tolerance of the shaft = Max. Shaft – Min. Shaft = 25.048 – 25.035 = 0.013**

**Tolerance of the hole = Max. Hole – Min. Hole = 25.021 – 25.000 = 0.021**

**Allowance = Min. Hole (MMC Hole) – Max. Shaft (MMC Shaft)**
       **= 25.000 – 25.048 = -0.0048**
This is listed as the maximum interference in the table above: **-0.048**

**Min. interference = Max. Hole (LMC Hole) – Min. Shaft (LMC Shaft)**
       **= 25.021 – 25.035 = -0.0014**
This is listed as the minimum interference in the table above: **-0.0014**

**Example 9.7 Shaft Basis System**

A design requires the use of two mating parts with a nominal size of **25** millimeters. The shaft is to run with moderate speed but with a fairly heavy journal pressure. The fit class chosen is **Shaft Basis System F8/h7**. Determine the size limits of the two mating parts.

From Appendix B, the Shaft Basis System F8/h7 fit for 25 millimeters size:

| Basic size | F8/h7 Medium Drive –Shaft Basis | | |
|:---:|:---:|:---:|:---:|
| | Hole F8 | Shaft h7 | Fit |
| 25 | 25.053 | 25.000 | 0.074 |
| | 25.020 | 24.979 | 0.020 |

Therefore, the maximum shaft size is 25
The minimum shaft size is 24.979
The maximum hole size is 25.053
The minimum hole size is 25.020

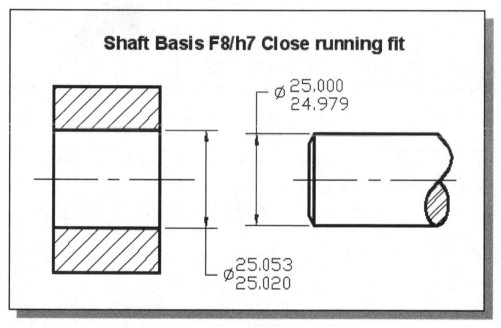

**Shaft Basis F8/h7 Close running fit**

**Tolerance of the shaft = Max. Shaft – Min. Shaft  = 25.000 – 24.979 = 0.021**

**Tolerance of the hole = Max. Hole – Min. Hole = 25.053 – 25.020 = 0.023**

**Allowance = Min. Hole – Max. Shaft = 25.020 – 25.000 = 0.020**
This is listed as the minimum clearance in the table above: **0.020**

**Max. Clearance = Max. Hole – Min. Shaft = 25.053 – 24.979 = 0.074**
This is listed as the maximum clearance in the table above: **0.074**

## Updating the U-Bracket Drawing

1. Select the **SOLIDWORKS 2019** option on the *Program* menu or select the **SOLIDWORKS 2019** icon on the *Desktop*.

2. In the **Recent Document** section, select the **MECH475.SLDDRW** file to open it.

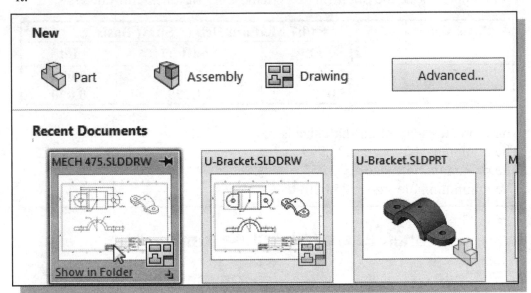

3. Use the **Open** button to locate the file if it is not shown in the recent document section.

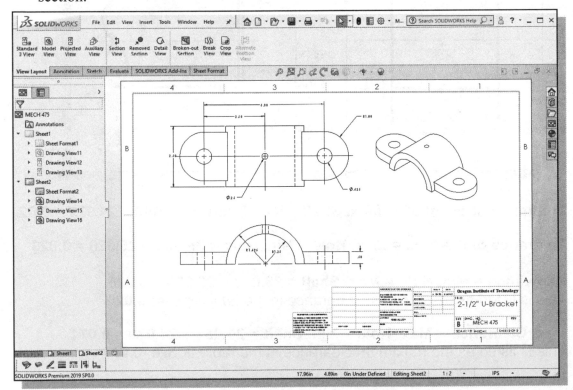

## Determining the Tolerances Required

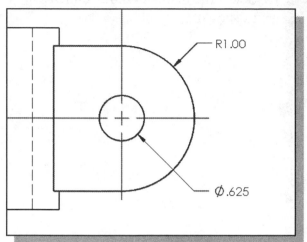

The *Bracket* design requires one set of tolerances: the nominal size of the **0.625** inch holes.

### RC2 - Nominal size of 0.625

A pin is to be placed through the diameter 0.625 hole. The fit class chosen is **Basic Hole System RC2**. We will first determine the size limits of the two mating parts.

From Appendix A, the RC2 fit for 0.625 inch nominal size:

| Nominal size Range | Limits of Clearance | RC2 Standard Limits – Basic Hole | |
|---|---|---|---|
| | | Hole | Shaft |
| 0.40 - 0.71 | 0.25<br>0.95 | +0.4<br>-0.0 | -0.25<br>-0.55 |

Using the *Basic Hole System*, the *BASIC* size 0.625 is designated as the minimum hole size.

The maximum hole size is 0.625+0.0004 = 0.6254 inches
The minimum hole size is 0.625-0.00= 0.62500 inches

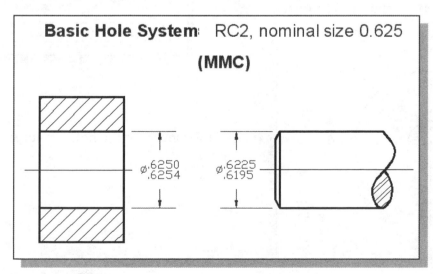

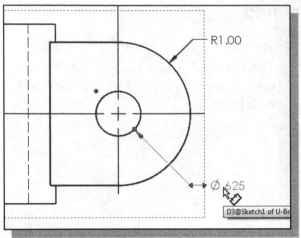

1.  Select the diameter **0.625** dimension as shown.

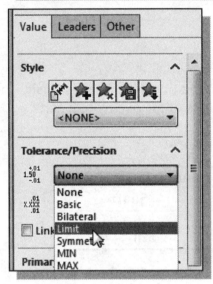

2.  In the Property Manager, select Limit as the style of Tolerance as shown.

    •   Note that the other styles are also available.

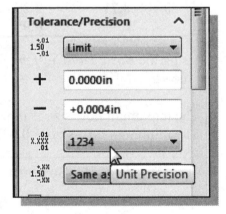

3.  Enter +0.0004 in the lower bound box and set to display 4 digits after the decimal point as shown. Note that we are setting the limits using the MMC (Maximum material condition) approach.

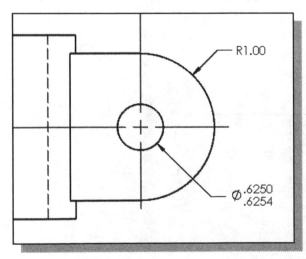

➤   The tolerances for the hole are set as shown.

## Questions:

1.  Why are **Tolerances** important to a technical drawing?

2.  Using the **Basic Shaft system**, calculate the size limits of a nominal size of 1.5 inches.

    Use the following FN4 fit table for 1.5 inches nominal size (numbers are in thousandth of an inch):

| Nominal size Range | Limits of Interference | FN4 Standard Limits – Basic Hole | |
|---|---|---|---|
| | | Hole | Shaft |
| 1.19 - 1.97 | 1.5 | +1.0 | +3.1 |
| | 3.1 | -0.0 | +2.5 |

3.  Using the **Shaft Basis D9/h9** fits, calculate the size limits of a nominal size of 30 mm.

4.  What is the procedure to set tolerances in SOLIDWORKS?

5.  Explain the following terms:

    **(a) Limits**

    **(b) MMC**

    **(c) Tolerances**

    **(d) Basic Hole System**

    **(e) Basic Shaft System**

6.  Given the dimensions as shown in the below figure, determine the tolerances of the two parts, and the allowance between the parts.

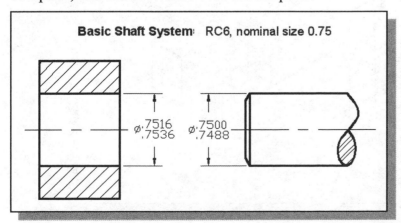

**Exercises:** (Create the drawings with the following Tolerances.)

1. **Shaft Guide** (Dimensions are in inches.)
   Fits: Diameter 0.25, Basic Hole system RC1, Diameter 1.00, Basic Hole system RC6

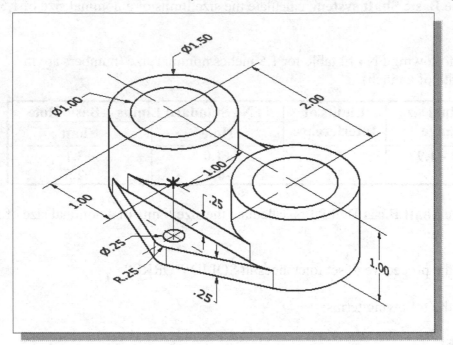

2. **Pivot Lock** (Dimensions are in inches.)
   Fits: Diameter 0.5, Basic Hole system RC2, Diameter 1.00, Basic Hole system RC5

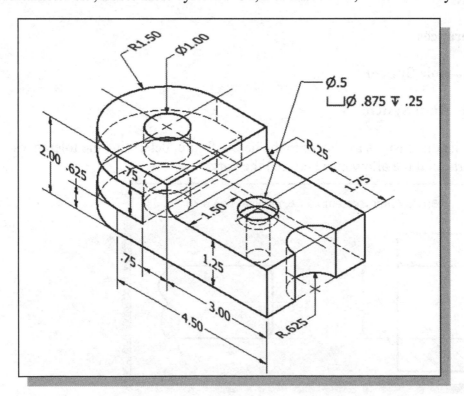

3. **U-Bracket** (Dimensions are in millimeters.) Fits: Diameter 40, H7/g6.

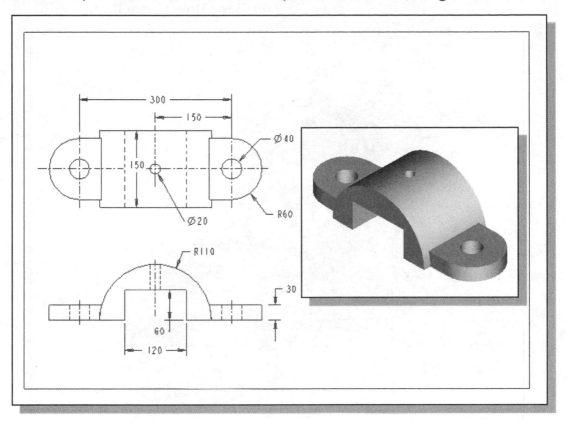

4. **Cylinder Support** (Inches.)  Fits: Diameter 1.50, Basic Hole system RC8.

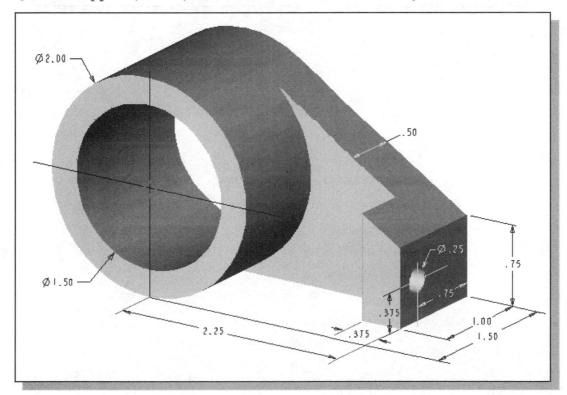

5. **Idler Mount** (Inches.)  Fits: Diameter 3.00, Basic Shaft system FN3.
                                             Diameter 2.00, Basic Hole system RC3.

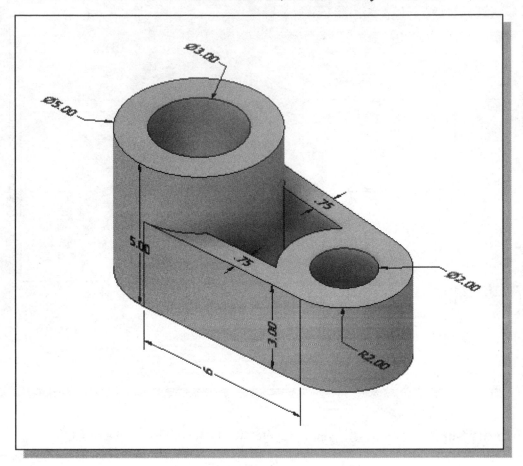

# Chapter 10
# Pictorials and Sketching

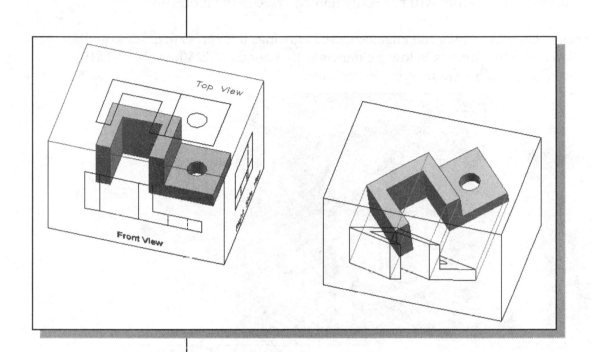

## Learning Objectives

♦ **Understand the Importance of Freehand Sketching**

♦ **Understand the Terminology Used in Pictorial Drawings**

♦ **Understand the Basics of the Following Projection Methods: Axonometric, Oblique and Perspective**

♦ **Be Able to Create Freehand 3D Pictorials**

## Engineering Drawings, Pictorials and Sketching

One of the best ways to communicate your ideas is through the use of a picture or a drawing. This is especially true for engineers and designers. Without the ability to communicate well, engineers and designers will not be able to function in a team environment and therefore will have only limited value in the profession.

For many centuries, artists and engineers used drawings to express their ideas and inventions. The two figures below are drawings by Leonardo da Vinci (1452-1519) illustrating some of his engineering inventions.

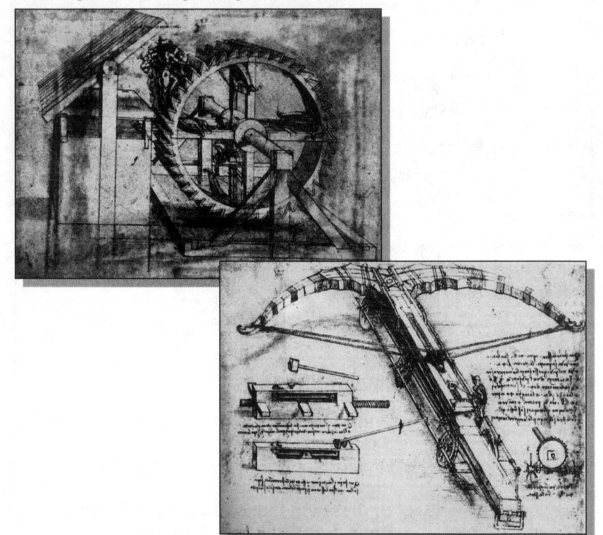

Engineering design is a process to create and transform ideas and concepts into a product definition that meets the desired objective. The engineering design process typically involves three stages: (1) Ideation/conceptual design stage: this is the beginning of an engineering design process, where basic ideas and concepts take shape. (2) Design development stage: the basic ideas are elaborated and further developed. During this stage, prototypes and testing are commonly used to ensure the developed design meets the desired objective. (3) Refine and finalize design stage: This stage of the design process is the last stage of the design process, where the finer details of the design are further refined. Detailed information on the finalized design is documented to assure the design is ready for production.

Two types of drawings are generally associated with the three stages of the engineering process: (1) Freehand Sketches and (2) Detailed Engineering Drawings.

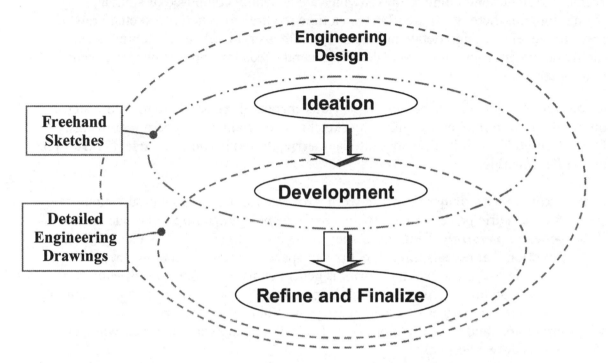

Freehand sketches are generally used in the beginning stages of a design process: (1) to quickly record designer's ideas and help in formulating different possibilities, (2) to communicate the designer's basic ideas with others and (3) to develop and elaborate further the designer's ideas/concepts.

During the initial design stage, an engineer will generally picture the ideas in his/her head as three-dimensional images. The ability to think visually, specifically three-dimensional visualization, is one of the most essential skills for an engineer/designer. And freehand sketching is considered one of the most powerful methods to help develop visualization skills.

Detailed engineering drawings are generally created during the second and third stages of a design process. The detailed engineering drawings are used to help refine and finalize the design and also to document the finalized design for production. Engineering drawings typically require the use of drawing instruments, from compasses to computers, to bring precision to the drawings.

Freehand Sketches and Detailed Engineering Drawings are essential communication tools for engineers. By using the established conventions, such as perspective and isometric drawings, engineers/designers are able to quickly convey their design ideas to others.

The ability to sketch ideas is absolutely essential to engineers. The ability to sketch is helpful, not just to communicate with others, but also to work out details in ideas and to identify any potential problems. Freehand sketching requires only simple tools, a pencil and a piece of paper, and can be accomplished almost anywhere and anytime. Creating freehand sketches does not require any artistic ability. Detailed engineering drawing is employed only for those ideas deserving a permanent record.

Freehand sketches and engineering drawings are generally composed of similar information, but there is a tradeoff between time required to generate a sketch/drawing versus the level of design detail and accuracy. In industry, freehand sketching is used to quickly document rough ideas and identify general needs for improvement in a team environment.

Besides the 2D views described in the previous chapter, there are three main divisions commonly used in freehand engineering sketches and detailed engineering drawings: (1) **Axonometric**, with its divisions into **isometric**, **dimetric** and **trimetric**; (2) **Oblique**; and (3) **Perspective**.

1. **Axonometric projection**: The word *Axonometric* means "to measure along axes." Axonometric projection is a special *orthographic projection* technique used to generate *pictorials*. **Pictorials** show a 2D image of an object as viewed from a direction that reveals three directions of space. In the figure below, the adjuster model is rotated so that a *pictorial* is generated using *orthographic projection* (projection lines perpendicular to the projection plane) as described in Chapter 4. There are three types of axonometric projections: isometric projection, dimetric projection, and trimetric projection. Typically, in an axonometric drawing, one axis is drawn vertically.

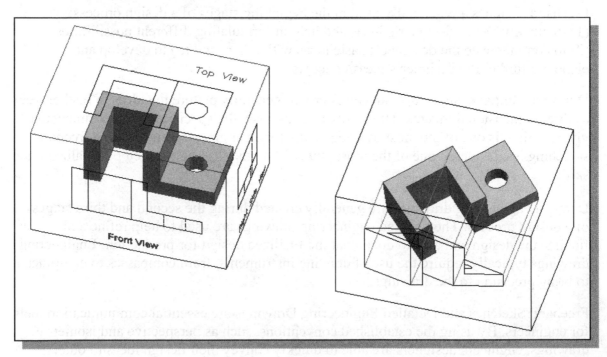

In **isometric projections**, the direction of viewing is such that the three axes of space appear equally foreshortened, and therefore the angles between the axes are equal. In **dimetric projections**, the directions of viewing are such that two of the three axes of space appear equally foreshortened. In **trimetric projections**, the direction of viewing is such that the three axes of space appear unequally foreshortened.

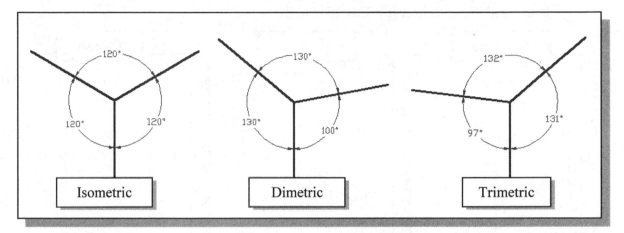

Isometric projection is perhaps the most widely used for pictorials in engineering graphics, mainly because isometric views are the most convenient to draw. Note that the different projection options described here are not particularly critical in freehand sketching as the emphasis is generally placed on the proportions of the design, not the precision measurements. The general procedure to constructing isometric views is illustrated in the following sections.

2.  **Oblique Projection** represents a simple technique of keeping the front face of an object parallel to the projection plane and still reveals three directions of space. An **orthographic projection** is a parallel projection in which the projection lines are perpendicular to the plane of projection. An **oblique projection** is one in which the projection lines are other than perpendicular to the plane of projection.

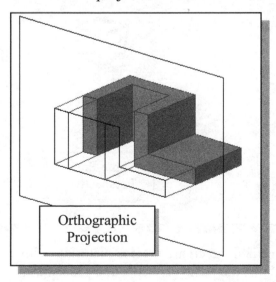

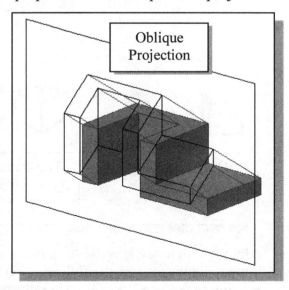

In an oblique drawing, geometry that are parallel to the frontal plane of projection are drawn true size and shape. This is the main advantage of the oblique drawing over the axonometric drawings. The three axes of the oblique sketch are drawn horizontal, vertical, and the third axis can be at any convenient angle (typically between 30 and 60 degrees). The proportional scale along the $3^{rd}$ axis is typically a scale anywhere between ½ and 1. If the scale is ½, then it is a **Cabinet** oblique. If the scale is 1, then it is a **Cavalier** oblique.

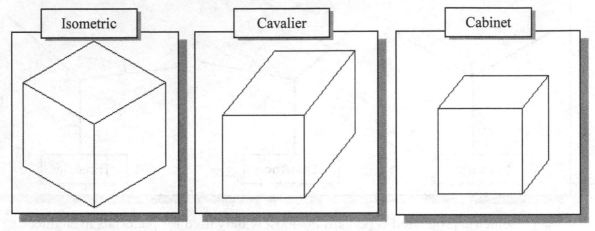

3. **Perspective Projection** adds realism to the three-dimensional pictorial representation; a perspective drawing represents an object as it appears to an observer; objects that are closer to the observer will appear larger to the observer. The key to the perspective projection is that parallel edges converge to a single point, known as the **vanishing point**. If there is just one vanishing point, then it is called a one-point perspective. If two sets of parallel edge lines converge to their respective vanishing points, then it is called a two-point perspective. There is also the case of a three-point perspective in which all three sets of parallel lines converge to their respective vanishing points.

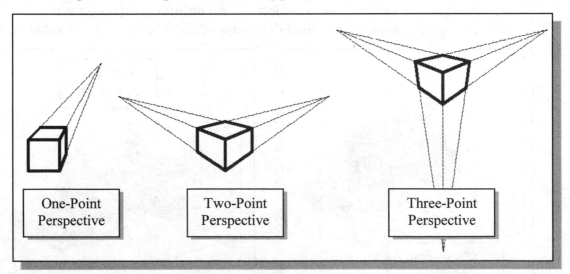

- Although there are specific techniques available to create precise pictorials with known dimensions, in the following sections, the basic concepts and procedures relating to freehand sketching are illustrated.

## Isometric Sketching

Isometric drawings are generally done with one axis aligned to the vertical direction. A **regular isometric** is when the viewpoint is looking down on the top of the object, and a **reversed isometric** is when the viewpoint is looking up on the bottom of the object.

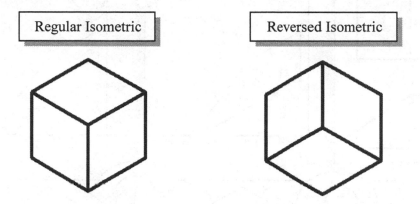

Two commonly used approaches in creating isometric sketches are (1) the **enclosing box** method and (2) the **adjacent surface** method. The enclosing box method begins with the construction of an isometric box showing the overall size of the object. The visible portions of the individual 2D-views are then constructed on the corresponding sides of the box. Adjustments of the locations of surfaces are then made, by moving the edges, to complete the isometric sketch.

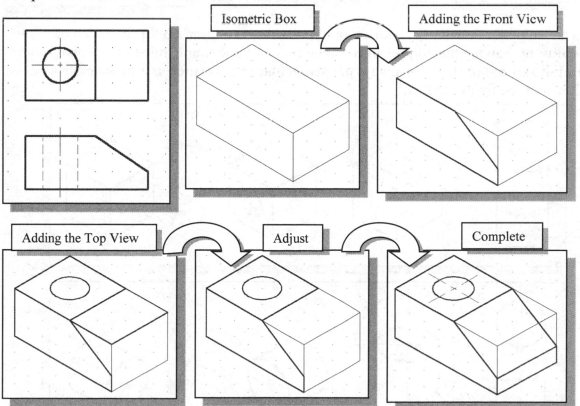

The adjacent surface method begins with one side of the isometric drawing, again with the visible portion of the corresponding 2D-view. The isometric sketch is completed by identifying and adding the adjacent surfaces.

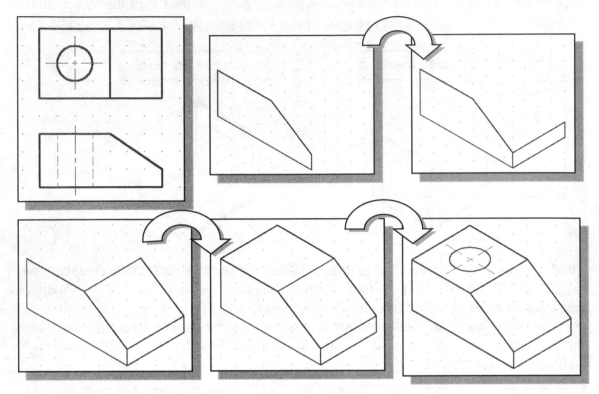

In an isometric drawing, cylindrical or circular shapes appear as ellipses. It can be confusing drawing the ellipses in an isometric view; one simple rule to remember is the **major axis** of the ellipse is always **perpendicular** to the **center axis** of the cylinder as shown in the figures below.

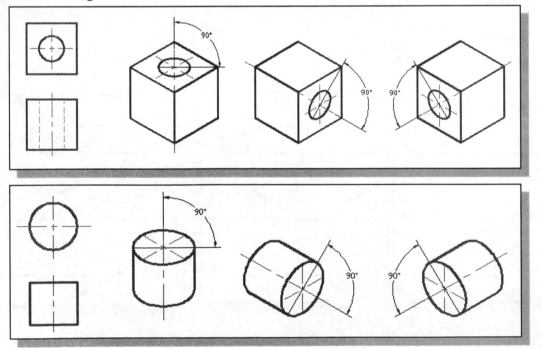

## Chapter 10 - Isometric Sketching Exercise 1:

Given the Orthographic Top view and Front view, create the isometric view.

(Note that 3D Models of chapter examples and exercises are available at www.SDCpublications.com/downloads/978-1-63057-240-2.)

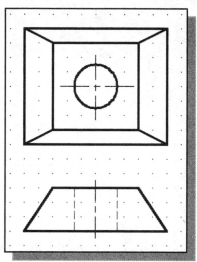

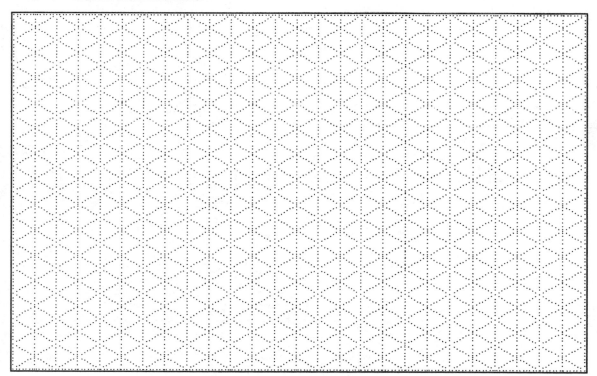

Name: _____     Date: _____

**Notes:**

## Chapter 10 - Isometric Sketching Exercise 2:

Given the Orthographic Top view and Front view, create the isometric view.

(Note that 3D Models of chapter examples and exercises are available at www.SDCpublications.com/downloads/978-1-63057-240-2.)

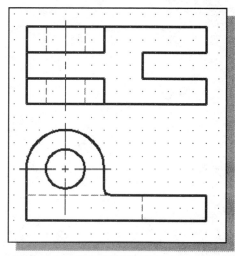

Name: _____    Date: _____

**Notes:**

## Chapter 10 - Isometric Sketching Exercise 3:

Given the Orthographic Top view and Front view, create the isometric view.

(Note that 3D Models of chapter examples and exercises are available at www.SDCpublications.com/downloads/978-1-63057-240-2.)

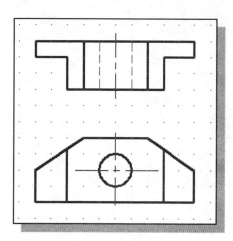

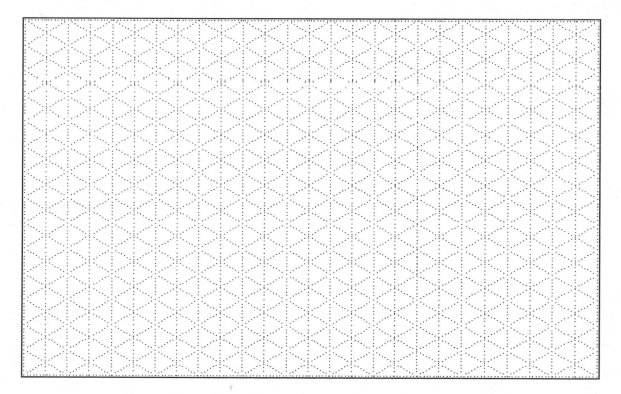

Name: _____     Date: _____

**Notes:**

## Chapter 10 - Isometric Sketching Exercise 4:

Given the Orthographic Top view and Front view, create the isometric view.

(Note that 3D Models of chapter examples and exercises are available at www.SDCpublications.com/downloads/978-1-63057-240-2.)

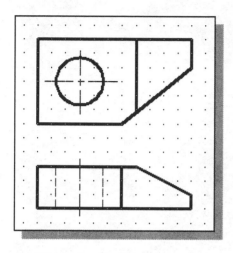

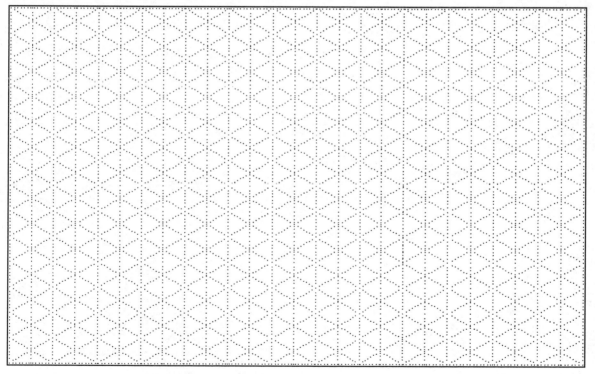

Name: _____      Date: _____

**Notes:**

## Chapter 10 - Isometric Sketching Exercise 5:

Given the Orthographic Top view, Front view, and Side view create the isometric view.

(Note that 3D Models of chapter examples and exercises are available at www.SDCpublications.com/downloads/978-1-63057-240-2.)

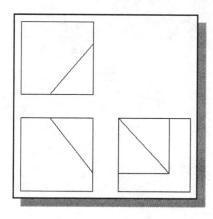

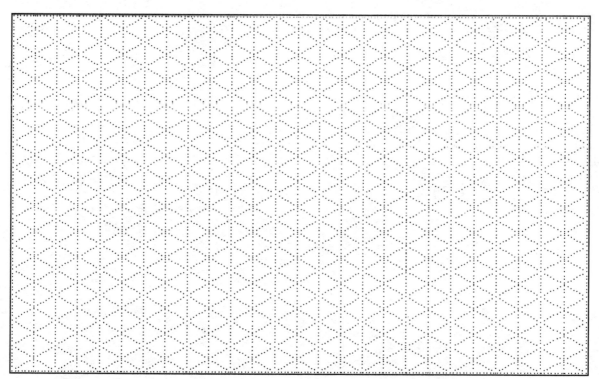

Name: _____    Date: _____

**Notes:**

## Chapter 10 - Isometric Sketching Exercise 6:

Given the Orthographic Top view, Front view, and Side view create the isometric view.

   (Note that 3D Models of chapter examples and exercises are available at www.SDCpublications.com/downloads/978-1-63057-240-2.)

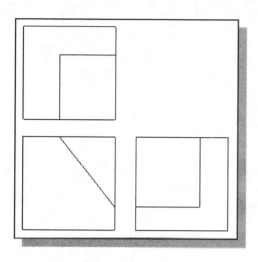

Name: _____     Date: _____

**Notes:**

## Chapter 10 - Isometric Sketching Exercise 7:

Given the Orthographic Top view, Front view, and Side view create the isometric view.

(Note that 3D Models of chapter examples and exercises are available at www.SDCpublications.com/downloads/978-1-63057-240-2.)

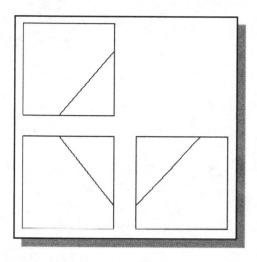

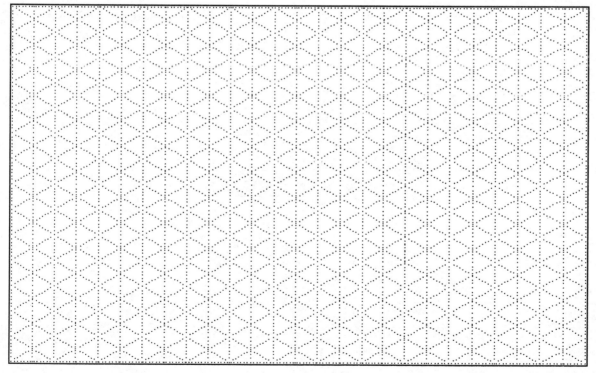

Name: _____     Date: _____

**Notes:**

## Chapter 10 - Isometric Sketching Exercise 8:

Given the Orthographic Top view, Front view, and Side view create the isometric view.

(Note that 3D Models of chapter examples and exercises are available at www.SDCpublications.com/downloads/978-1-63057-240-2.)

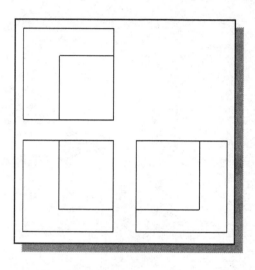

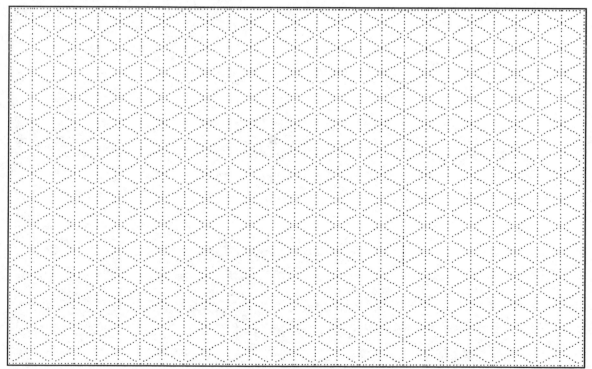

Name: _____    Date: _____

**Notes:**

## Chapter 10 - Isometric Sketching Exercise 9:

Given the Orthographic Top view, Front view, and Side view create the isometric view.

(Note that 3D Models of chapter examples and exercises are available at www.SDCpublications.com/downloads/978-1-63057-240-2.)

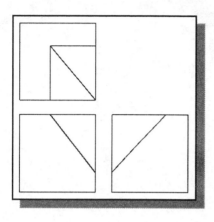

Name: _____     Date: _____

**Notes:**

## Chapter 10 - Isometric Sketching Exercise 10:

Given the Orthographic Top view, Front view, and Side view create the isometric view.

   (Note that 3D Models of chapter examples and exercises are available at www.SDCpublications.com/downloads/978-1-63057-240-2.)

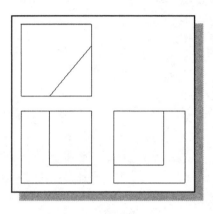

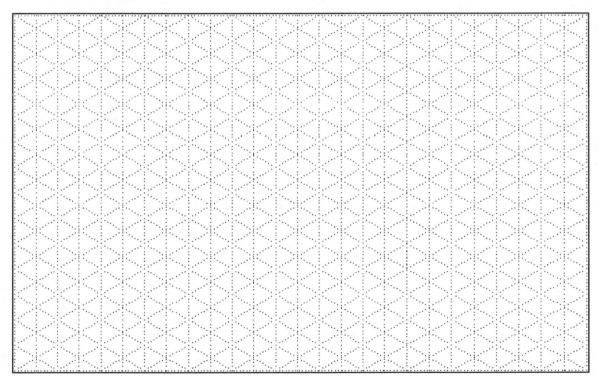

Name: _____      Date: _____

**Notes:**

## Oblique Sketching

Keeping the geometry that is parallel to the frontal plane true to size and shape is the main advantage of the oblique drawing over the axonometric drawings. Unlike isometric drawings, circular shapes that are paralleled to the frontal view will remain as circles in oblique drawings. Generally speaking, an oblique drawing can be created very quickly by using a 2D view as the starting point. For designs with most of the circular shapes in one direction, an oblique sketch is the ideal choice over the other pictorial methods.

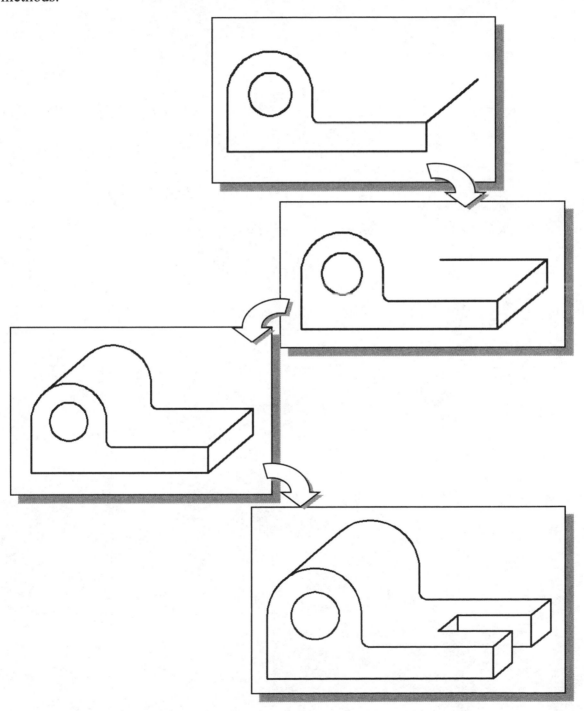

**Notes:**

## Chapter 10 -  Oblique Sketching Exercise 1:

Given the Orthographic Top view and Front view, create the oblique view.

(Note that 3D Models of chapter examples and exercises are available at www.SDCpublications.com/downloads/978-1-63057-240-2.)

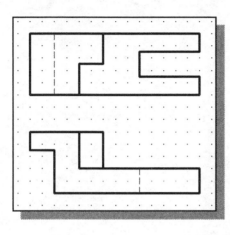

Name: _____    Date: _____

**Notes:**

## Chapter 10 - Oblique Sketching Exercise 2:

Given the Orthographic Top view and Front view, create the oblique view.

(Note that 3D Models of chapter examples and exercises are available at www.SDCpublications.com/downloads/978-1-63057-240-2.)

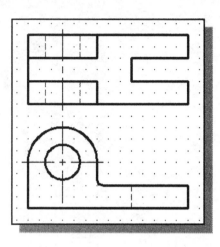

Name: _____    Date: _____

**Notes:**

## Chapter 10 - Oblique Sketching Exercise 3:

Given the Orthographic Top view and Front view, create the oblique view.

(Note that 3D Models of chapter examples and exercises are available at www.SDCpublications.com/downloads/978-1-63057-240-2.)

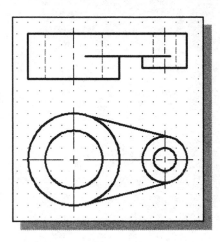

Name: _____          Date: _____

**Notes:**

## Chapter 10 - Oblique Sketching Exercise 4:

Given the Orthographic Top view and Front view, create the oblique view.

(Note that 3D Models of chapter examples and exercises are available at www.SDCpublications.com/downloads/978-1-63057-240-2.)

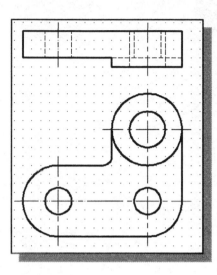

Name: _____    Date: _____

**Notes:**

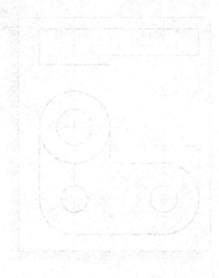

## Chapter 10 - Oblique Sketching Exercise 5:

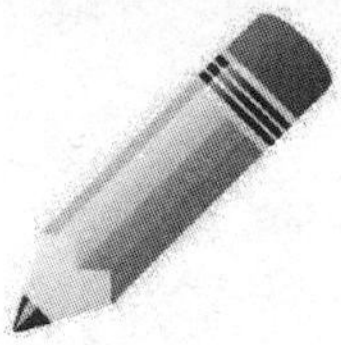

Given the Orthographic Top view and Front view, create the oblique view.

(Note that 3D Models of chapter examples and exercises are available at www.SDCpublications.com/downloads/978-1-63057-240-2.)

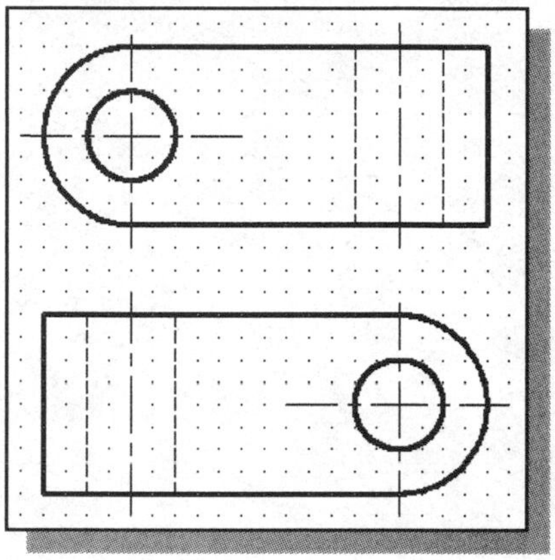

Name: ______________________ Date: ______________

**Notes:**

## Chapter 10 - Oblique Sketching Exercise 6:

Given the Orthographic Top view and Front view, create the oblique view.

(Note that 3D Models of chapter examples and exercises are available at www.SDCpublications.com/downloads/978-1-63057-240-2.)

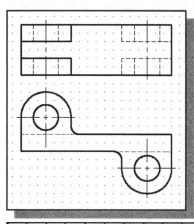

Name: _____        Date: _____

**Notes:**

# Perspective Sketching

A perspective drawing represents an object as it appears to an observer; objects that are closer to the observer will appear larger to the observer. The key to the perspective projection is that parallel edges converge to a single point known as the **vanishing point**. The vanishing point represents the position where projection lines converge.

The selection of the locations of the vanishing points, which is the first step in creating a perspective sketch, will affect the looks of the resulting images.

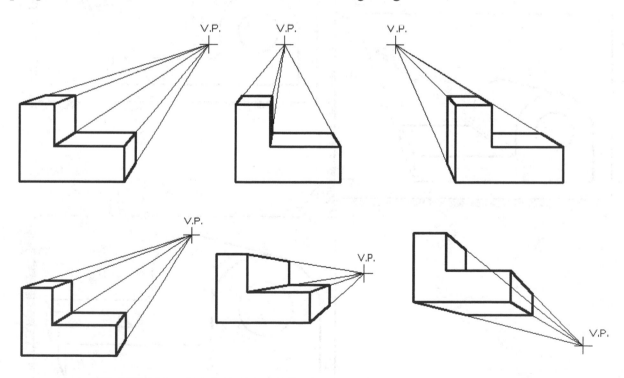

## One-Point Perspective

One-point perspective is commonly used because of its simplicity. The first step in creating a one-point perspective is to sketch the front face of the object just as in oblique sketching, followed by selecting the position for the vanishing point. For Mechanical designs, the vanishing point is usually placed above and to the right of the picture. The use of construction lines can be helpful in locating the edges of the object and to complete the sketch.

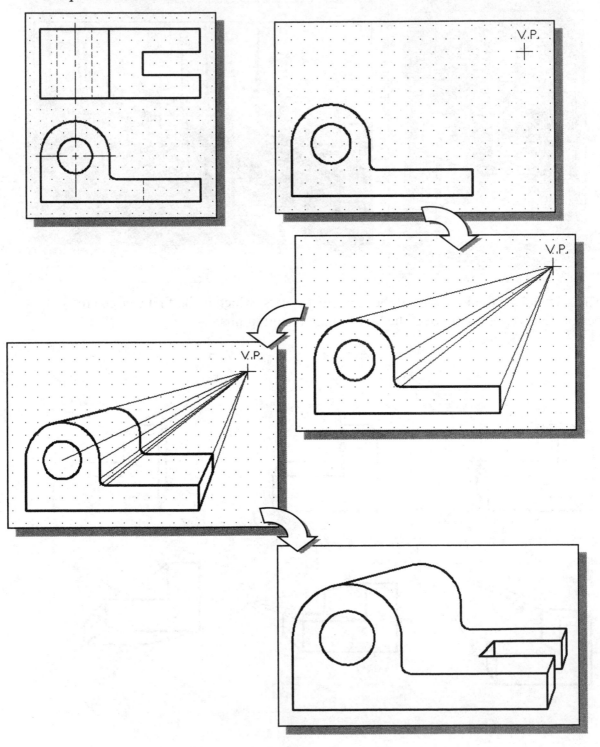

# Two-Point Perspective

Two-point perspective is perhaps the most popular of all perspective methods. The use of the two vanishing points gives very true to life images. The first step in creating a two-point perspective is to select the locations for the two vanishing points, followed by sketching an enclosing box to show the outline of the object. The use of construction lines can be very helpful in locating the edges of the object and to complete the sketch.

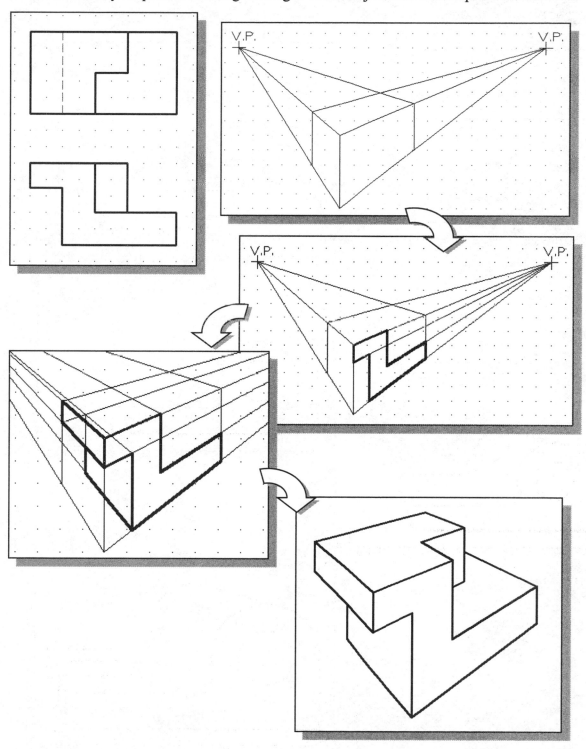

**Notes:**

## Chapter 10 - Perspective Sketching Exercise 1:

Given the Orthographic Top view and Front view, create one-point or two-point perspective views.
(Note that 3D Models of chapter examples and exercises are available at www.SDCpublications.com/downloads/978-1-63057-240-2.)

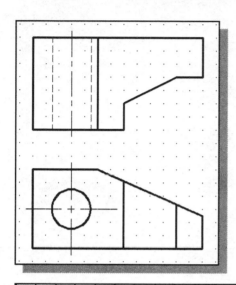

Name: _____    Date: _____

**Notes:**

## Chapter 10 - Perspective Sketching Exercise 2:

Given the Orthographic Top view and Front view, create one-point or two-point perspective views.
(Note that 3D Models of chapter examples and exercises are available at www.SDCpublications.com/downloads/978-1-63057-240-2.)

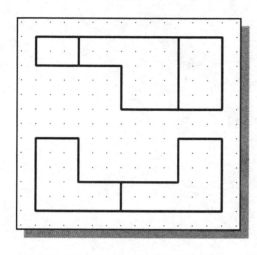

Name: _____    Date: _____

**Notes:**

## Chapter 10 - Perspective Sketching Exercise 3:

Given the Orthographic Top view and Front view, create one-point or two-point perspective views.
(Note that 3D Models of chapter examples and exercises are available at www.SDCpublications.com/downloads/978-1-63057-240-2.)

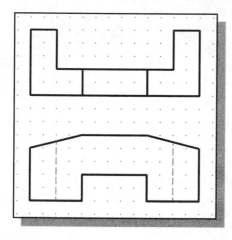

Name: _____    Date: _____

**Notes:**

## Chapter 10 - Perspective Sketching Exercise 4:

Given the Orthographic Top view and Front view, create one-point or two-point perspective views.
(Note that 3D Models of chapter examples and exercises are available at www.SDCpublications.com/downloads/978-1-63057-240-2.)

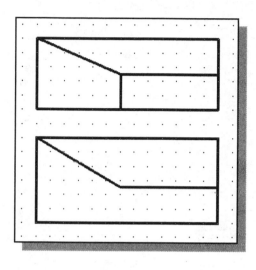

Name: _____     Date: _____

**Notes:**

## Chapter 10 - Perspective Sketching Exercise 5:

Given the Orthographic Top view and Front view, create one-point or two-point perspective views.
(Note that 3D Models of chapter examples and exercises are available at www.SDCpublications.com/downloads/978-1-63057-240-2.)

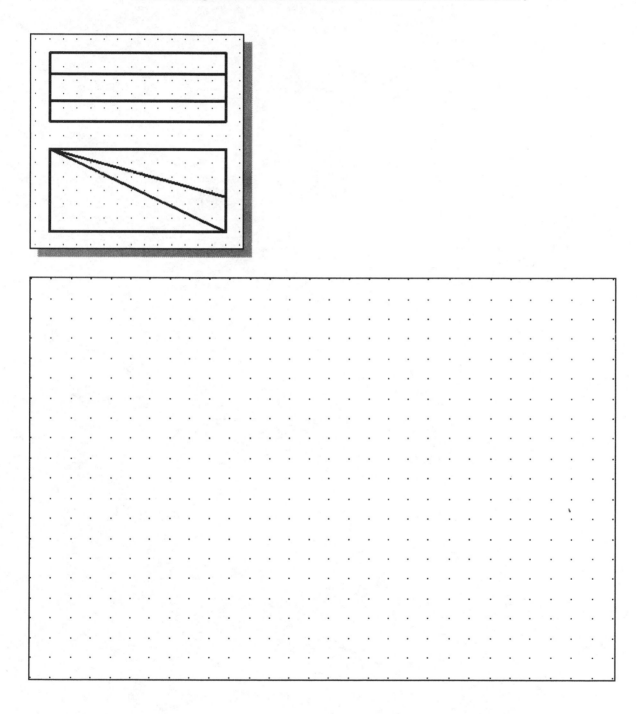

Name: _____    Date: _____

**Notes:**

## Chapter 10 - Perspective Sketching Exercise 6:

Given the Orthographic Top view and Front view, create one-point or two-point perspective views.
(Note that 3D Models of chapter examples and exercises are available at www.SDCpublications.com/downloads/978-1-63057-240-2.)

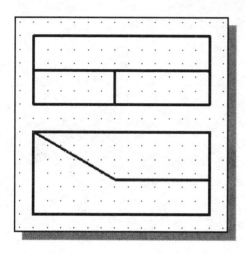

Name: _____    Date: _____

**Notes:**

## Questions:

1.  What are the three types of *Axonometric* projection?

2.  Describe the differences between an *Isometric* drawing and a *Trimetric* drawing.

3.  What is the main advantage of *Oblique* projection over the *isometric* projection?

4.  Describe the differences between a one-point perspective and a two-point perspective.

5.  Which pictorial methods maintain true size and shape of geometry on the frontal plane?

6.  What is a vanishing point in a perspective drawing?

7.  What is a *Cabinet Oblique*?

8.  What is the angle between the three axes in an *isometric* drawing?

9.  In an *Axonometric* drawing, are the projection lines perpendicular to the projection plane?

10. A cylindrical feature, in a frontal plane, will remain a circle in which pictorial methods?

11. In an *Oblique* drawing, are the projection lines perpendicular to the projection plane?

12. Create freehand pictorial sketches of:
    *   Your desk
    *   Your computer
    *   One corner of your room
    *   The tallest building in your area

## Exercises:

**Complete the missing views.** (Hint: Create a pictorial sketch as an aid in reading the views.)

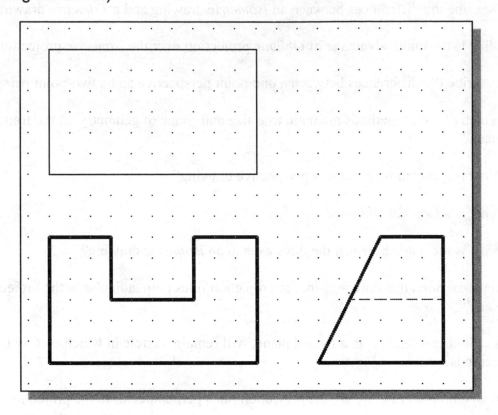

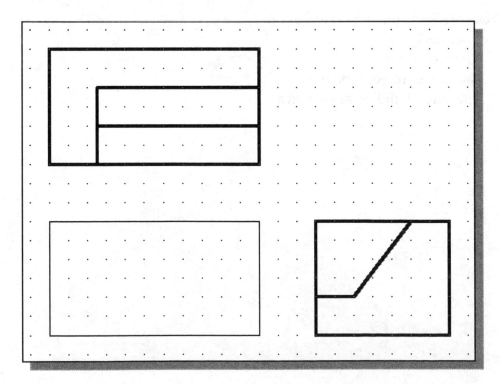

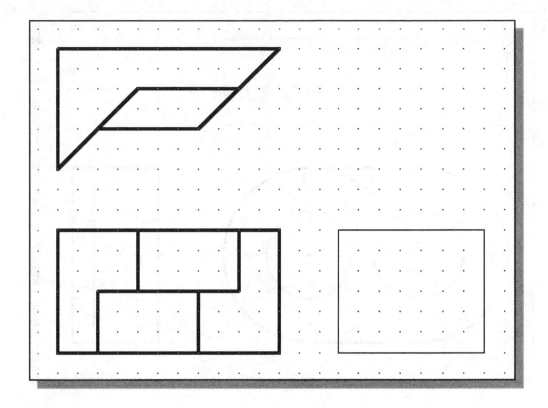

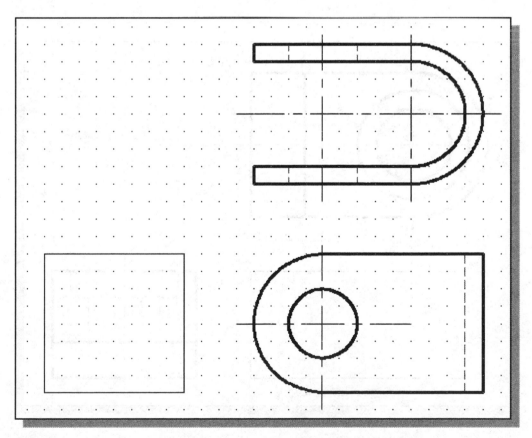

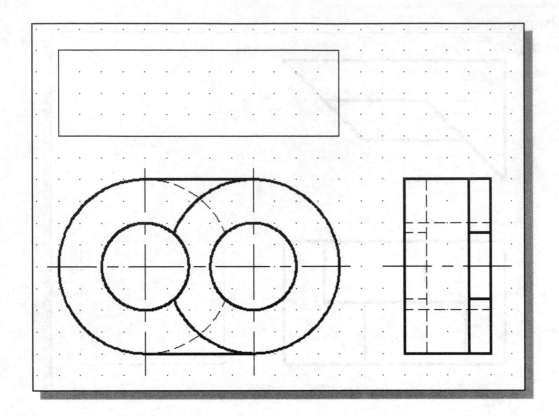

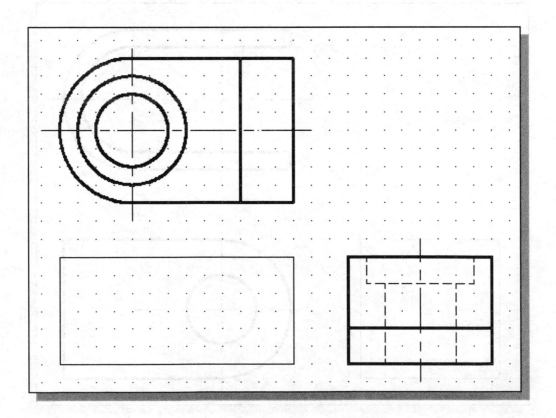

# Chapter 11
# Section Views & Symmetrical Features in Designs

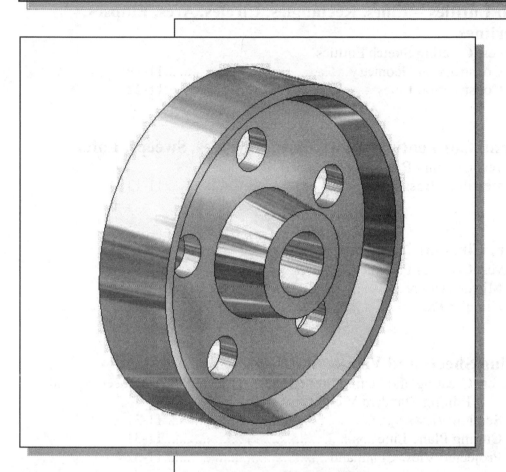

## Learning Objectives

- ♦ **Create Revolved Features**
- ♦ **Use the Mirror Feature Command**
- ♦ **Create New Borders and Title Blocks**
- ♦ **Create Circular Patterns**
- ♦ **Create and Modify Linear Dimensions**
- ♦ **Use Associative Functionality**
- ♦ **Identify Symmetrical Features in Designs**
- ♦ **Create a Section View in a Drawing**

## Certified SOLIDWORKS Associate Exam Objectives Coverage

### Sketch Entities – Lines, Rectangles, Circles, Arcs, Ellipses, Centerlines
Objectives: Creating Sketch Entities.
>        Construction Geometry.................................................11-19
>        Construction Lines......................................................11-22

### Boss and Cut Features – Extrudes, Revolves, Sweeps, Lofts
Objectives: Creating Basic Swept Shapes.
>        Revolved Boss/Base....................................................11-13

### Linear, Circular, and Fill Patterns
Objectives: Creating Patterned Features.
>        Mirror Feature..............................................................11-17
>        Circular Pattern...........................................................11-24

### Drawing Sheets and Views
Objectives: Creating and Setting Properties for Drawing Sheets; Inserting and Editing Standard Views.
>        Section Views ..............................................................11-31
>        Cutting Plane Line .......................................................11-31
>        Projected View Command ...........................................11-32

### Annotations
Objectives: Creating Annotations.
>        Center Mark, Circular Option ....................................11-37

## Introduction

In the previous chapters, we have explored the basic CAD methods of creating orthographic views. By carefully selecting a limited number of views, the external features of most complicated designs can be fully described. However, we are frequently confronted with the necessity of showing the interiors of parts, or when parts are assembled, that cannot be shown clearly by means of hidden lines.

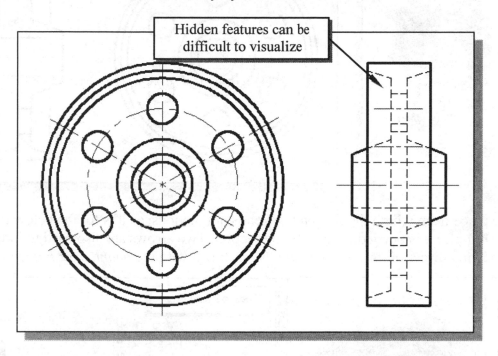

Hidden features can be difficult to visualize

In cases of this kind, to aid in describing the object, one or more views are drawn to show the object as if a portion of the object had been cut away to reveal the interior.

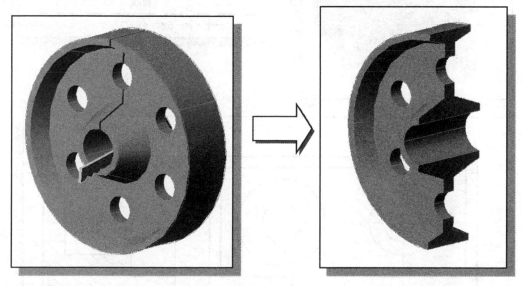

This type of convention is called a **section**, which is defined as an imaginary cut made through an object to expose the interior of a part. This kind of cutaway view is known as a **section view**.

In a *section view*, the place from which the section is taken must be identifiable on the drawing. If the place from which the section is taken is obvious, as it is for the figure below, no further description is needed.

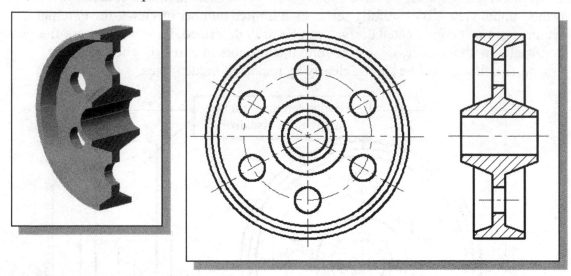

If the place from which the section is taken is not obvious, as it is for the below figure, a **cutting plane** is needed to identify the section. Two arrows are also used to indicate the viewing direction. A *cutting plane line* is drawn with the *phantom* or *hidden* line.

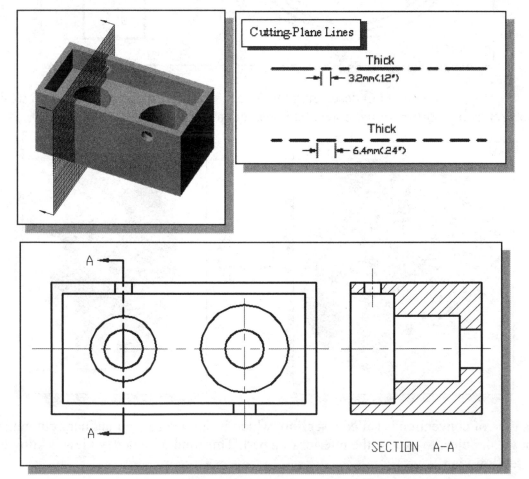

# General Rules of Section Views

Section views are used to make a part drawing more understandable, showing the internal details of the part. Since the sectioned drawing is showing the internal features there is generally no need to show hidden lines.

A section view still follows the general rules of any view in a multiview drawing. Cutting planes may be labeled at their endpoints to clarify the association of the cut and the section views, especially when multiple cutting plane lines are used. When using multiple cutting planes, each sectioned drawing is drawn as if the other cutting plane lines do not exist. The cutting plane line takes precedence over center lines and, remember, cutting plane lines may be omitted when their location is obvious.

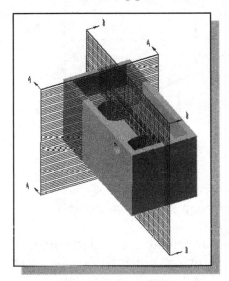

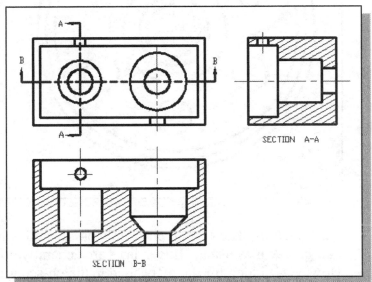

**Section lines** or **cross-hatch lines** are drawn where the cutting plane passes through the object. This is done as if a saw was used to cut the part and then section lines are used to represent the cutting marks left by the saw blade. Different materials are represented by the use of different section line types. The line type for iron may be used as the general section line type for any material.

Cast iron, Malleable
iron and general use
for all materials

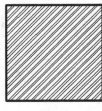

Steel

Bronze, Brass and
composition Materials

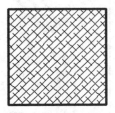

Magnesium, aluminum
and aluminum alloys

Section lines should not be parallel or perpendicular to object lines. Section lines are generally drawn at 45°, 30° or 60°, unless there are conflicts with other rules. Section lines should be oriented at different angles for different parts.

## Section Drawing Types

- **Full Section:** The cutting plane passes completely through the part as a single flat plane.

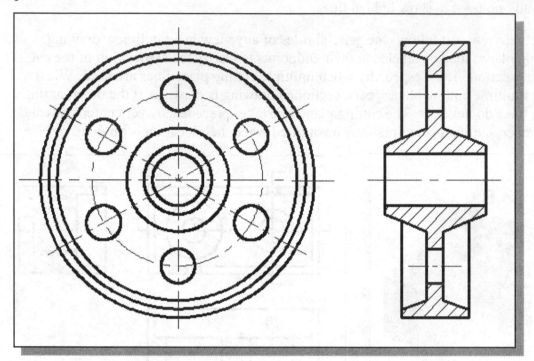

- **Half Section:** The cutting plane only passes half way through the part and the other half is drawn as usual. Hidden lines are generally not shown on either half of the part. However, hidden lines may be used in the un-sectioned half if necessary for clarity. Half section is mostly used on cylindrical parts, and a center line is used to separate the two halves.

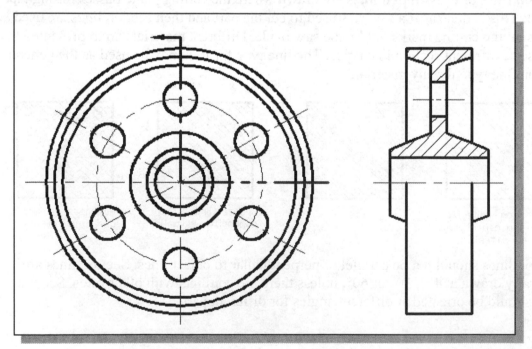

- **Offset Section:** It is often difficult to interpret a multiview drawing when there are multiple features on the object. An **offset section** allows the cutting plane to pass through several places by **"offsetting"** or **bending** the cutting plane. The offsets or bends in the cutting plane are all 90° and are *not* shown in the section view.

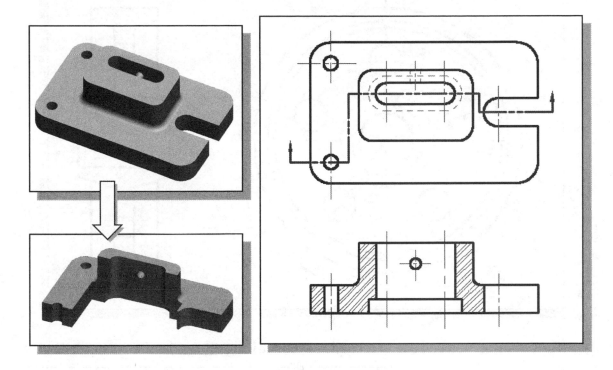

- **Broken-Out Section:** It is quite often that a full section is not necessary, but only a partial section is needed to show the interior details of a part. In a **broken-out section**, only a portion of the view is sectioned and a jagged break-line is used to divide the sectioned and un-sectioned portion of the view.

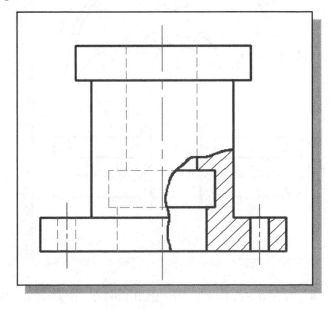

- **Aligned Section:** To include certain angled features in a section view, the cutting plane may be bent to pass through those features to their true radial position.

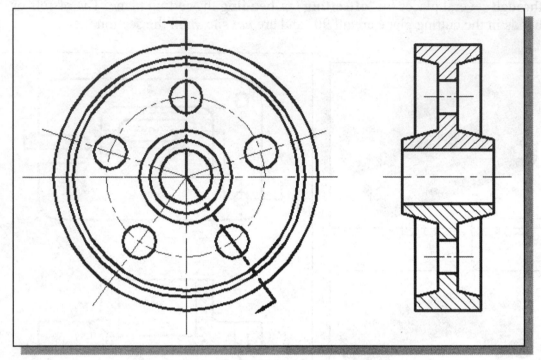

- **Half Views:** When space is limited, a symmetrical part can be shown as a half view.

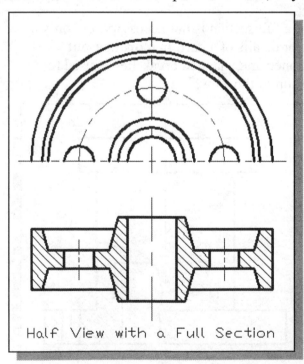

Half View with a Full Section

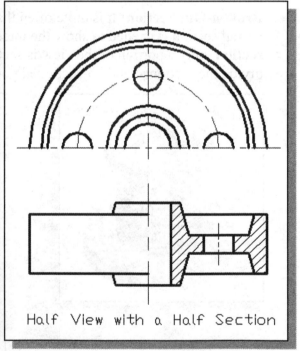

Half View with a Half Section

- **Thin Sections:** For thin parts, typically parts that are less than 3mm thick, such as sheet metal parts and gaskets, section lines are ineffective and therefore should be omitted. Thin sections are generally shown in solid black without section lines.

- **Revolved Section:** A cross section of the part is revolved 90° and superimposed on the drawing.

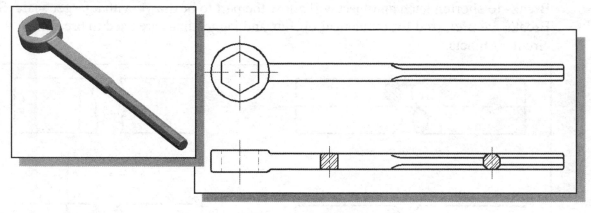

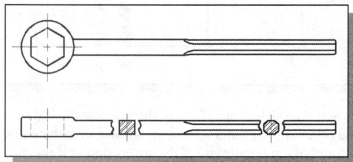

- The object lines adjacent to a revolved section may be broken out with short break lines. The true shape of a revolved section is retained regardless of the object lines in the view.

- **Removed Section:** Similar to the revolved section except that the sectioned drawing is not superimposed on the drawing but placed adjacent to it. The removed section may also be drawn at a different scale. Very useful for detailing small parts or features.

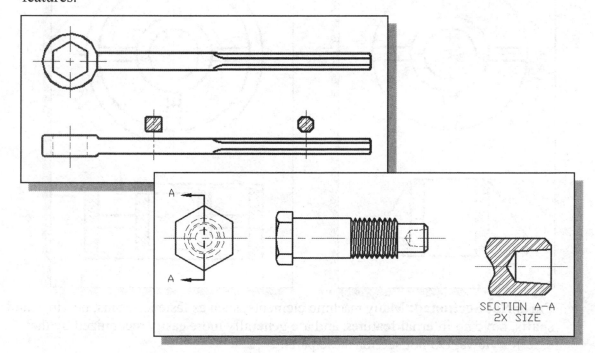

- **Conventional Breaks:** In making the details of parts that are long and with a uniform cross section, it is rarely necessary to draw its whole length. Using the Conventional Breaks to shorten such an object will allow the part to be drawn with a larger scale. S-Breaks are preferred for cylindrical objects and jagged lines are used to break non-circular objects.

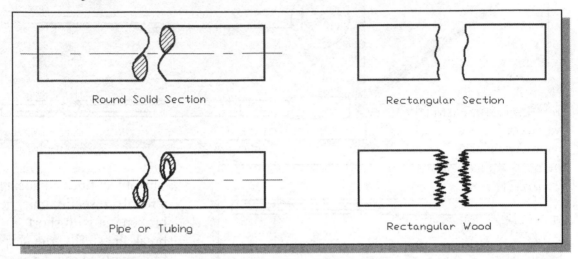

- **Ribs and Webs in Sections:** To avoid a misleading effect of thickness and solidity, ribs, webs, spokes, gear teeth and other similar parts are drawn without crosshatching when the cutting plane passes through them lengthwise. Ribs are sectioned if the cutting plane passes through them at other orientations.

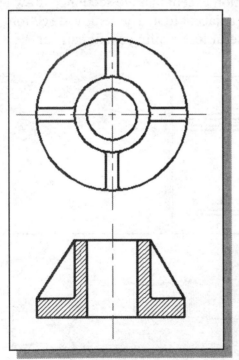
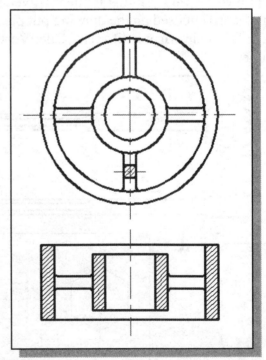

- **Parts Not Sectioned:** Many machine elements, such as fasteners, pins, bearings and shafts, have no internal features, and are generally more easily recognized by their exterior views. Do not section these parts.

## Section Views in SOLIDWORKS

In parametric modeling, it is important to identify and determine the features that exist in the design. *Feature-based parametric modeling* enables us to build complex designs by working on smaller and simpler units. This approach simplifies the modeling process and allows us to concentrate on the characteristics of the design. Symmetry is an important characteristic that is often seen in designs. Symmetrical features can be easily accomplished by the assortments of tools that are available in feature-based modeling systems, such as SOLIDWORKS.

The modeling technique of extruding two-dimensional sketches along a straight line to form three-dimensional features, as illustrated in the previous chapters, is an effective way to construct solid models. For designs that involve cylindrical shapes, shapes that are symmetrical about an axis, revolving two-dimensional sketches about an axis can form the needed three-dimensional features. In solid modeling, this type of feature is called a ***revolved feature***.

In SOLIDWORKS, besides using the **Revolved Boss/Base** command to create revolved features, several options are also available to handle symmetrical features. For example, we can create multiple identical copies of symmetrical features with the **Circular Pattern** command, or create mirror images of models using the **Mirror** command. We can also use *construction geometry* to assist the construction of more complex features. In this lesson, the construction and modeling techniques of these more advanced options are illustrated.

## A Revolved Design: *PULLEY*

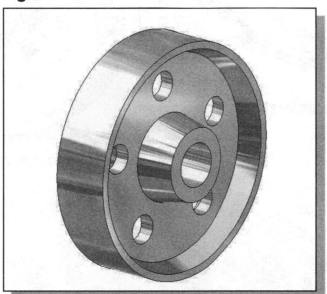

❖ Based on your knowledge of SOLIDWORKS, how many features would you use to create the design? Which feature would you choose as the **base feature** of the model? Identify the symmetrical features in the design and consider other possibilities in creating the design. You are encouraged to create the model on your own prior to following through the tutorial.

# Modeling Strategy – A Revolved Design

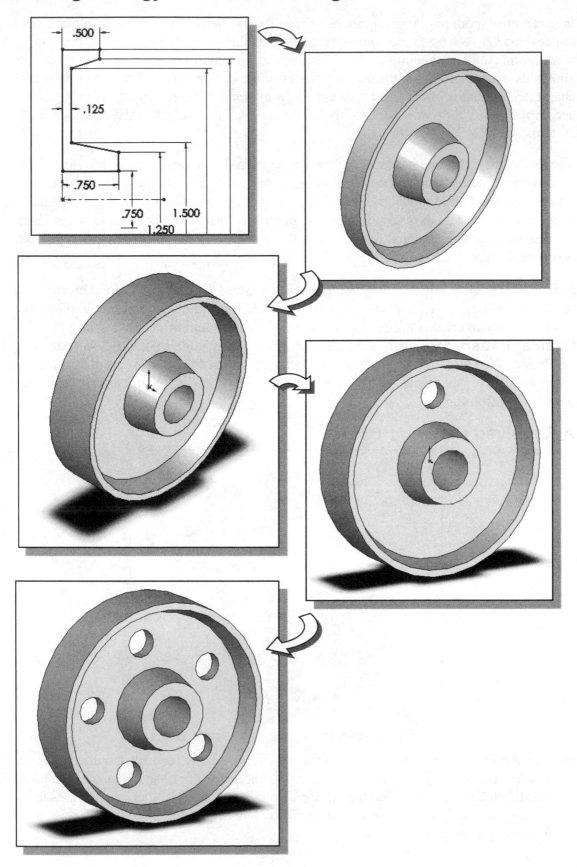

## Starting SOLIDWORKS

1.  Select the **SOLIDWORKS** option on the *Start* menu or select the **SOLIDWORKS** icon on the desktop to start SOLIDWORKS. The SOLIDWORKS main window will appear on the screen.

2.  In the *Welcome to SOLIDWORKS* dialog box, enter the **Advanced** mode by clicking once with the left-mouse-button on the **Advanced** icon.

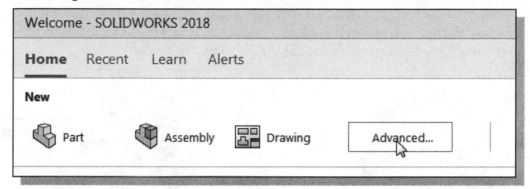

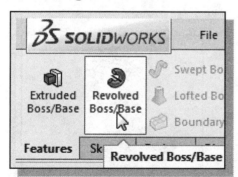

3.  Select the **Tutorial_Templates** tab. (Refer to Chapter 5 on the setup of template files.)

4.  Select the **Part_IPS_ANSI** template as shown.

5.  Click on the **OK** button to open a new document using the selected template.

## Creating the Base Feature

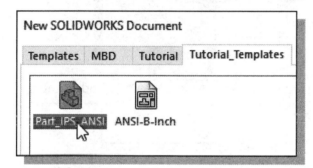

1.  Select the **Revolved Boss/Base** button on the *Features* toolbar to create a new feature.

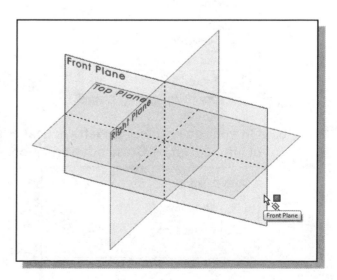

2.  Move the cursor over the edge of the Front Plane in the graphics area. When the Front Plane is highlighted, click once with the **left-mouse-button** to select the Front (XY) Plane as the sketch plane for the new sketch.

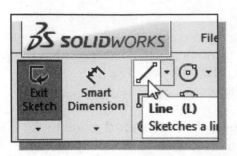

3.  Click on the **Line** icon in the *Sketch* toolbar.

4.  Create a closed-region sketch, above the origin, as shown below. (Note that the *Pulley* design is symmetrical about a horizontal axis as well as a vertical axis, which allows us to simplify the 2D sketch as shown below.)

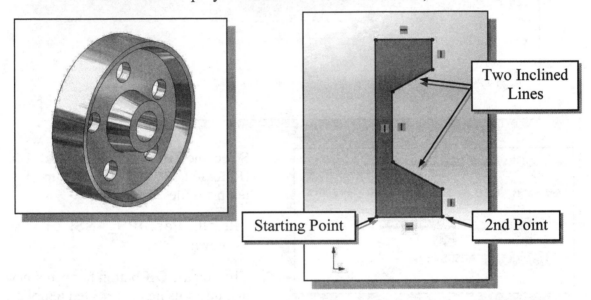

5.  We want the left vertical edge to be aligned with the Y-axis. Select the **Add Relation** command from the *Display/Delete Relations* pop-up menu on the *Sketch* toolbar.

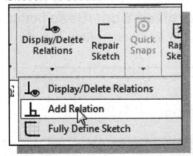

6.  On your own, select the **left vertical edge** and the **origin** and add a **Coincident** relation.

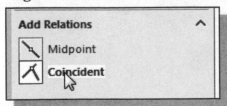

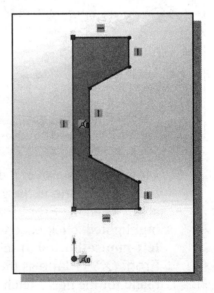

➢ The X-axis is an axis of symmetry. It will serve as the rotation axis for the revolve feature, and we will use it to apply dimensions. A centerline will be created along the X-axis for these purposes.

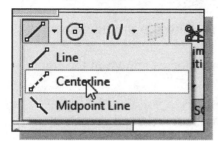

7. Select the **Centerline** command from the *Line* pop-up menu on the *Sketch* toolbar.

8. Left-click on the **origin** as the first endpoint of the centerline.

9. Move the cursor to the right and left-click to create a **horizontal centerline** along the X-axis.

10. Press the **[Esc]** key to exit the Centerline command.

11. Select the **Smart Dimension** command in the *Sketch* toolbar.

12. Pick the **centerline** as the first entity to dimension as shown in the figure below.

13. Select the **bottom horizontal line** as the second object to dimension.

14. Locate the dimension **below the centerline**. This will automatically create a diameter dimension bisected by the centerline.

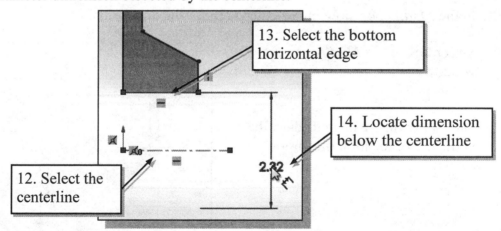

13. Select the bottom horizontal edge

14. Locate dimension below the centerline

12. Select the centerline

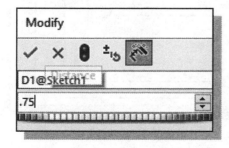

15. In the *Modify* pop-up window, enter **.75** as the dimension value.

16. Select **OK** in the *Modify* pop-up window.

- **To create a dimension that will account for the symmetrical nature of the design: pick the centerline (axis of symmetry), pick the entity, and then place the dimension on the side of the centerline opposite to the entity.**

17. Pick the **centerline** as the first entity to dimension as shown in the figure below.

18. Select the **corner point** as the second object to dimension as shown in the figure below.

19. Locate the dimension **below the centerline**. This will automatically create a diameter dimension bisected by the centerline.

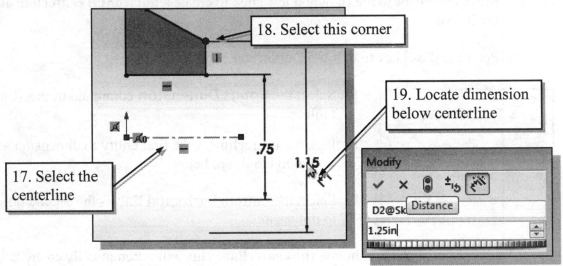

18. Select this corner

19. Locate dimension below centerline

17. Select the centerline

.75

1.15

Modify

D2@Sk Distance

1.25in

20. In the *Modify* pop-up window enter **1.25** for the dimension.

21. Select **OK** in the *Modify* pop-up window.

22. On your own, create and adjust the vertical size/location dimensions as shown.

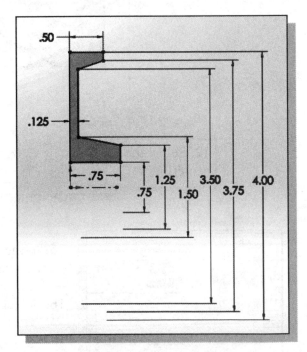

## Creating the Revolved Feature

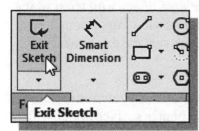

1. Select the **Exit Sketch** icon on the *Sketch* toolbar to exit the **Sketch** option.

2. The *Revolve Property Manager* appears. In the *Revolve Parameters* panel, the centerline is automatically selected as the revolve axis. The default revolution is **360°**. Click the **OK** icon to accept these parameters and create the revolved feature.

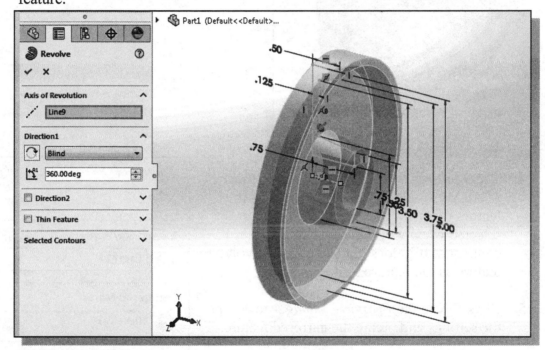

## Mirroring Features

- In SOLIDWORKS, features can be mirrored to create and maintain complex symmetrical features. We can mirror a feature about a work plane or a specified surface. We can create a mirrored feature while maintaining the original parametric definitions, which can be quite useful in creating symmetrical features. For example, we can create one quadrant of a feature, then mirror it twice to create a solid with four identical quadrants.

1. In the *Features* toolbar, select the **Mirror** command as shown.

2. The *Mirror Property Manager* appears. The *Mirror Face/Plane* panel is highlighted and the message "*Select a plane or planar face about which to mirror...*" appears in the *Status Bar* area. Use the dynamic viewing options to rotate/position the model so that we are viewing the back surface as shown.

3. Select the surface as shown in the figure as the planar surface about which to mirror. The selection appears in the *Mirror Face/Plane* panel.

4. The **Features to Mirror** panel in the *Property Manager* should contain **Revolve1** as the feature to mirror because this was pre-selected when we entered the **Mirror** command. If it does not, select the **Revolve1** feature as the feature to mirror.

5. Click **OK** in the *Property Manager* to accept the settings and create the mirrored feature.

6. On your own, return to an isometric view.

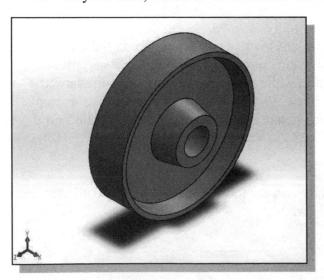

➤ Now is a good time to save the model. It is a good habit to save your model periodically, just in case something might go wrong while you are working on it. You should also save the model after you have completed any major constructions.

7. On your own, save the model as a *part* file, entering **Pulley** for the filename and **4″ Pulley** for the Description.

## Creating an Extruded Cut Feature using Construction Geometry

- In SOLIDWORKS, we can also use **construction geometry** to help define, constrain, and dimension the required geometry. **Construction geometry** can be lines, arcs, and circles that are used to line up or define other geometry but are not themselves used as the shape geometry of the model. When profiling the rough sketch, SOLIDWORKS will separate the construction geometry from the other entities and treat them as construction entities. Construction geometry can be dimensioned and constrained just like any other profile geometry. When the profile is turned into a 3D feature, the construction geometry remains in the sketch definition but does not show in the 3D model. Using construction geometry in profiles may mean fewer constraints and dimensions are needed to control the size and shape of geometric sketches. In SOLIDWORKS, any sketch entity can be converted to construction geometry. Points and centerlines are always construction entities. We will illustrate the use of the construction geometry to create a cut feature.

- The *Pulley* design requires the placement of five identical holes on the base solid. Instead of creating the five holes one at a time, we can simplify the creation of these holes by using the **Circular Pattern** command, which allows us to create duplicate features. Prior to using the Circular Pattern command, we will first create a *pattern leader*, which is a regular extruded feature.

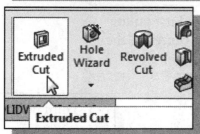

1. In the *Features* toolbar select the **Extruded Cut** command by left-clicking once on the icon.

2. Use the *Design Tree* in the graphics area to select the **Right** (YZ) **Plane** as shown.

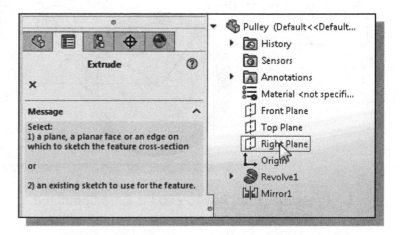

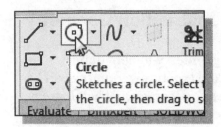

3.  Select the **Circle** command by clicking once with the left-mouse-button on the icon in the *Sketch* toolbar.

4.  Create a circle of arbitrary size as shown.

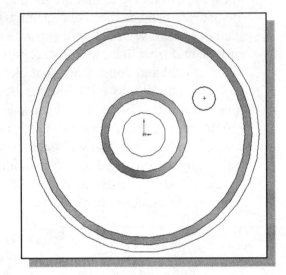

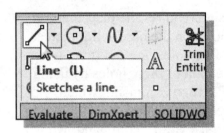

5.  Select the **Line** command in the *Sketch* toolbar.

6.  Create 2 *lines* by left-clicking on (1) the center of the circle we just created, (2) the *origin*, and (3) a point to the right of the origin creating a horizontal line, as shown below.

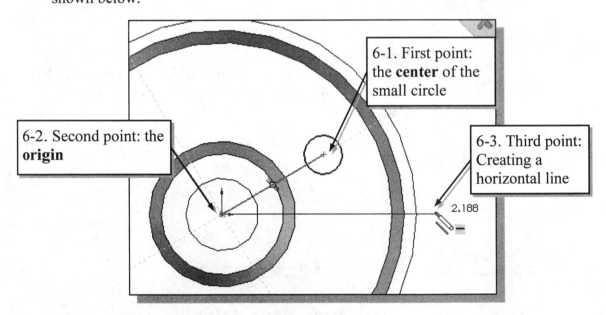

6-1. First point: the **center** of the small circle

6-2. Second point: the **origin**

6-3. Third point: Creating a horizontal line

2.188

7.  Press the **[Esc]** key to exit the **Line** command.

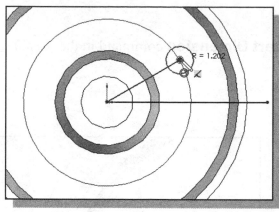

8.  Select the **Circle** command on the *Sketch* toolbar.

9.  Place the center at the ***origin***.

10. Pick the center of the small circle to set the size of the circle as shown.

11. Press the **[Esc]** key to exit the *Circle* command.

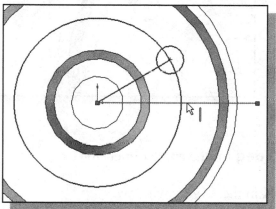

12. In the graphics area, left-click on the ***horizontal line*** to select it.

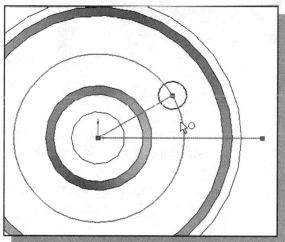

13. Press and hold down the **[Ctrl]** key and left-click on the **lines** and **the last circle** drawn. The two lines and the circle are all highlighted. All three appear as selected entities (*Arc2, Line1, Line2*) in the *Properties Property Manager*.

14. **Check** the box next to the **For construction** option in the options panel of the *Properties Property Manager*.

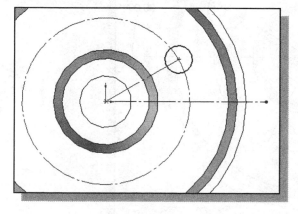

15. Click **OK** in the *Property Manager* to accept the setting. Notice the entities have been converted to construction geometry and appear with the same line style as centerlines.

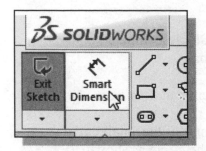

16. Select the **Smart Dimension** command in the *Sketch* toolbar.

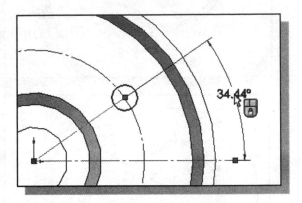

17. Pick the **horizontal construction line** as the first entity to dimension.

18. Select the **angled construction line** as the second entity to dimension.

19. Place the dimension text to the right of the model as shown.

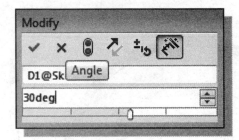

20. Enter **30 deg** for the angle dimension.

21. Click **OK** in the *Modify* window.

➢ Note that the location of the small circle is adjusted as the location of the construction line is adjusted by the *angle dimension* we created.

22. On your own, create the two diameter dimensions, **.5** and **2.5**, as shown below.

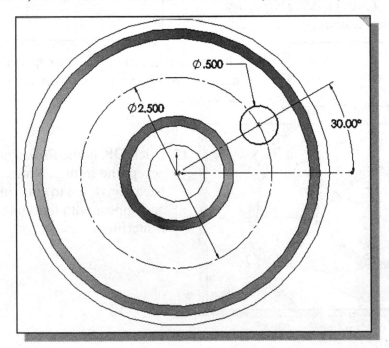

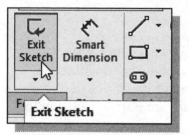

23. Select the **Exit Sketch** icon on the *Sketch* toolbar to exit the Sketch option.

24. On your own, complete the **Extruded Cut** feature. Use the **Through All - Both** option for the distance option.

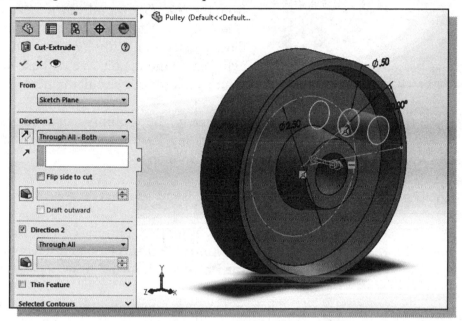

25. Click **OK** in the *Property Manager* to create the extruded cut feature.

26. On your own, adjust the angle dimension applied to the construction line of the cut feature to **90** and observe the effect of the adjustment.

## Circular Pattern

➢ In SOLIDWORKS, existing features can be easily duplicated. The **Linear Pattern** and **Circular Pattern** commands allow us to create rectangular and polar arrays of features. The patterned features are parametrically linked to the original feature; any modifications to the original feature are also reflected in the arrayed features.

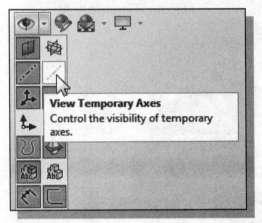

1. Select the **View Temporary Axes** option from the *Hide/Show* pull-down menu on the *Heads-up View* toolbar. Temporary axes are those created implicitly by cones and cylinders in the model. We will use the central axis for our circular pattern.

➢ Alternately, the visibility can be toggled *ON* by selecting **Temporary Axes** on the *View* pull-down menu.

2. Press the **[Esc]** key once to ensure no feature is pre-selected.

3. In the *Features* toolbar, left-click on the **arrow** below the *Linear Pattern* icon and select the **Circular Pattern** command from pop-up menu.

4. Click on the Direction 1 selection box as shown.

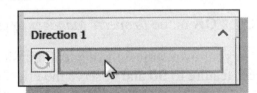

5. The message *"Select edge or axis for direction reference..."* is displayed in the *Status Bar* area. SOLIDWORKS expects us to select an axis to pattern about. Select the **axis** at the center of the *Pulley* as shown. Notice the axis (Axis<1>) appears in the *Pattern Axis* box in the *Parameters* panel of the *Circular Pattern Property Manager*.

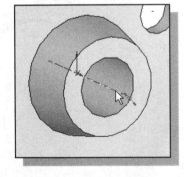

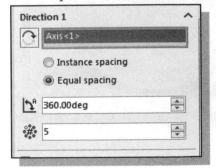

6. In the *Parameters* panel, check the **Equal Spacing** box, enter **5** in the *Number of Instances* box, and enter **360** in the *Angle* box as shown.

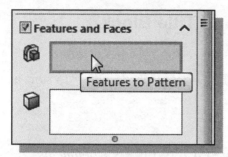

7. The *Features to Pattern* text box is now activated for us to begin the selection of features to be patterned.

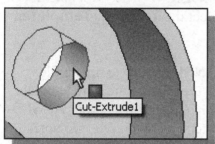

8. Select the **circular cut feature** when it is highlighted as shown.

9. Select **OK** in the *Circular Pattern Property Manager* to accept the settings and create the circular pattern.

10. On your own, turn *OFF* the visibility of the **Temporary Axes** and the **Origins**. (See Step 1 above.)

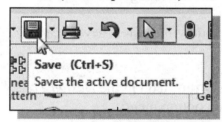

11. Select **Save** in the *Menu Bar*; we can also use the **"Ctrl+S"** combination (press down the [Ctrl] key and hit the [S] key once) to save the part as *Pulley*.

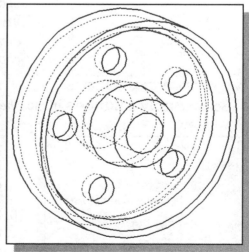

## Drawing Mode – Defining a New Border and Title Block

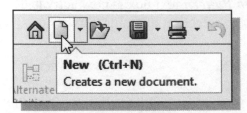

1. Click **New** in the *Quick access toolbar* to start a **New file**.

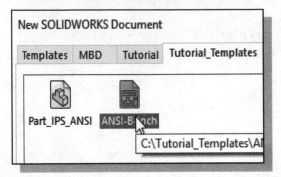

2. In the *New* SOLIDWORKS *Document* window, select the **Tutorial_Templates** tab.

3. Select the **ANSI-B-Inch** template file icon. This is the custom drawing template you created in Chapter 8. (If your ANSI-B-Inch template is not available, see note below.)

4. Select **OK** in the *New* SOLIDWORKS *Document* window to open the new drawing file. Select **Cancel** to exit the *Insert Model View* option.

❖ **NOTE:** If the ANSI-B-Inch template is not available: (1) open the default **Drawing** template; (2) set document properties following instructions on page 8-43.

➢ We will use the *Document Properties* in this template but will create a new *Sheet Format*.

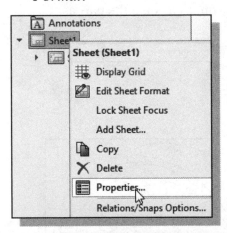

5. Right-click on the **Sheet** icon in the *Design Tree* to reveal the pop-up option menu.

6. Select **Properties** from the menu.

7. Set the scale to **1:1** and select **A (ANSI) Landscape** for the *Sheet Format/Size* in the *Sheet Properties* window.

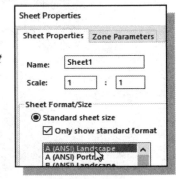

8. Select **Apply Changes** in the *Sheet Properties* window.

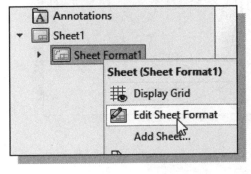

9. Move the cursor on top of Sheet1 and **right-click** to open the pop-up option menu.

10. Select **Edit Sheet Format** from the pop-up option menu to change to *Edit Sheet Format* mode. (Notice this method could also be used to access the *Sheet Properties*.)

11. Inside the graphics area, click and drag to create a selection window picking all the entities in the title block, leaving only the border.

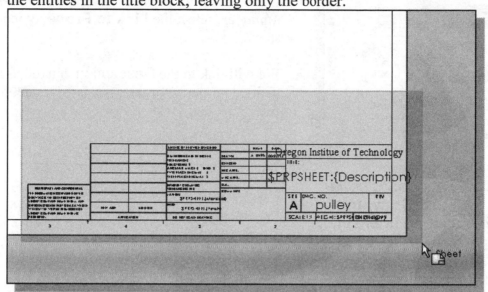

12. Press the **[Delete]** key to delete these entities.

13. On your own, create a title block using the **Line** and **Smart Dimension** commands in the *Sketch* toolbar and the **Note** command in the *Annotation* toolbar.

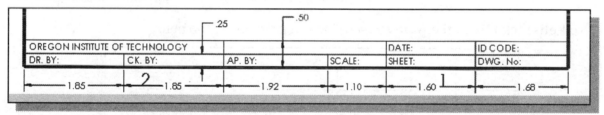

➢ We will now add a **Note** with a **Property Link**.

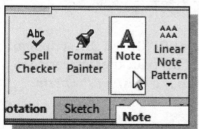

14. Left-click on the **Note** icon on the *Annotation* toolbar.

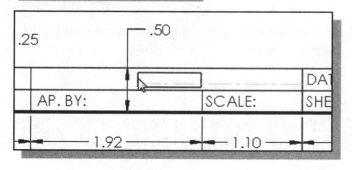

15. Select a location inside the central empty box in the new title block as shown.

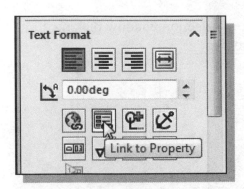

16. In the *Text Format* panel of the *Note Property Manager*, select the **Link to Property** icon as shown.

➢ We will link to the Description entered as a user-defined property for the associated model part file.

17. In the *Link to Property* window, set to *Model found here* and select **Drawing view specified in sheet properties**, as shown.

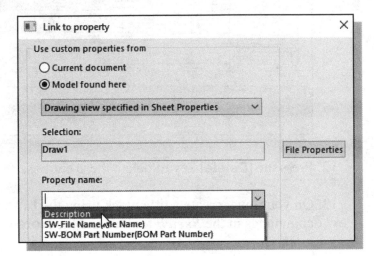

18. In the *Property name* list, select **Description** from the pull-down menu as shown.

19. Left-click **OK** in the *Link to Property* window to accept the selection.

20. Left-click **OK** in the *Note Property Manager* to create the note.

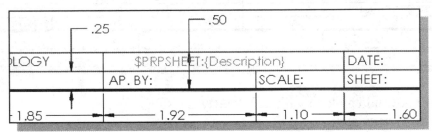

➢ Notice the Property Link appears in the title block as **$PRPSHEET:{Description}**. The prefix **$PRPSHEET:** defines the source as the model (part file) appearing in the sheet. The Property Name is **Description**, which is a *Custom Property*.

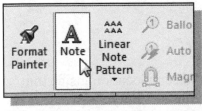

21. On your own, **exit** the *Note* command.

22. Move cursor on top of any icon in the *Command Manager* area.

23. **Right-click** once to open the pop-up option menu and switch on the Line Format toolbar as shown.

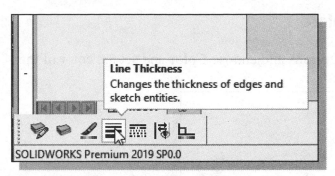

- The **Line Format** toolbar appears near the bottom of the SOLIDWORKS main window.

24. Pre-select the two horizontal lines we just created.

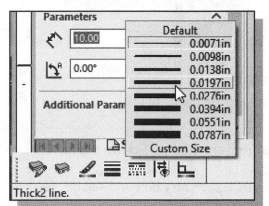

25. Set the line thickness to Thick2 line thickness, which is 0.0197 in.

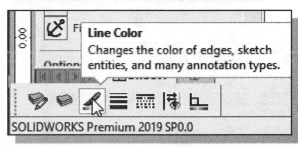

26. On your own, also adjust the color of the lines.

27. On your own, repeat the above steps and adjust the thickness and color of the vertical lines we just created.

28. On your own, delete the dimensions you created to set up the title block.

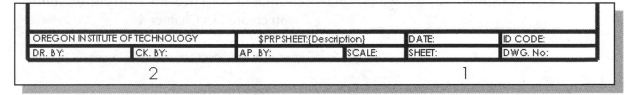

29. Move the cursor into the graphics area and **right-click** to open the pop-up option menu.

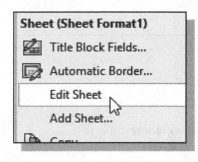

30. Select **Edit Sheet** from the pop-up option menu to change to *Edit Sheet* mode.

31. Select the **Rebuild** icon on the *Menu Bar*.

## Create a New Drawing Template

1.  On your own, use the **System Options** and choose **Color** and set the color of the **Inactive Entities** to *Black* as shown.

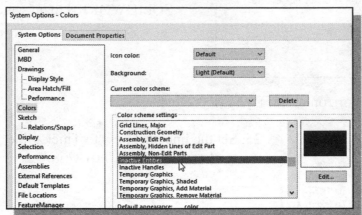

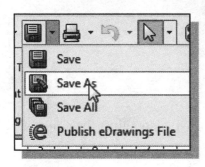

2.  Select **Save as** from the *File* pull-down menu.

3.  Click in the *Save as type:* entry box and select **Drawing Template** from the pull-down options.

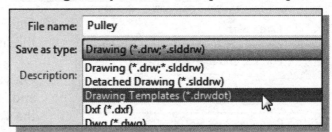

4.  The folder selection will automatically change to the default Templates folder. Use the browser to change the folder selection to the **Tutorial Templates** folder you created in Chapter 4.

5.  Enter **A-Custom** for the *File name*.

6.  Enter **ANSI A Inch Custom** for the *Description*.

7.  Click **Save**.

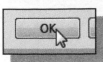

8.  If the SOLIDWORKS warning window appears, click **OK**.

## Creating Views

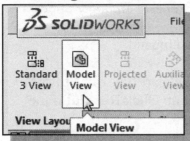

1.  Click on the **Model View** icon in the *View Layout* toolbar.

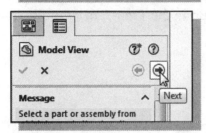

2.  The *Model View Property Manager* appears with the *Pulley* part file selected as the part from which to create the base view. Click the **Next** arrow as shown to proceed with defining the base view.

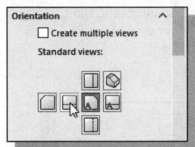

3.  In the *Model View Property Manager*, select the **Left View** for the *Orientation*, as shown.  (Make sure the *Create multiple views* box is unchecked.)

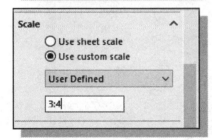

4.  Under the *Scale* options in the *Drawing View Property Manager*, select **Use custom scale**.

5.  Select **User Defined** from the pull-down options, as shown. (You may have to use the scroll bar to locate the User Defined option.)

6.  Enter **3:4** as the user defined scale.

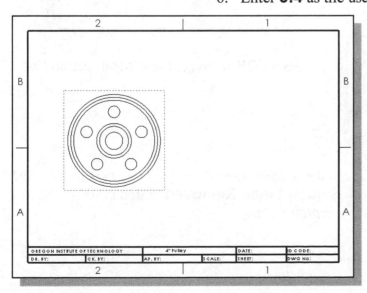

7.  Move the cursor inside the graphics area and place the **base view** toward the left side of the border as shown.

➢ When the base view is placed, SOLIDWORKS automatically executes the **Projected View** command and the *Projected View Property Manager* appears.

8. Press the **[Esc]** key once to exit the **Projected View** command.

➢ Notice the **Description** we entered for the *Pulley* model part file – **4″ Pulley** – appears in the title block due to the **Property Link**. (Hint: Click **Rebuild** if the text does not appear automatically.)

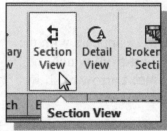

9. Select the **Section View** icon in the *View Layout* toolbar.

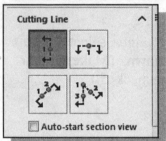

10. Verify that the **Vertical** option is selected in the *Cutting Line* panel of the *Section View* Property Manager.

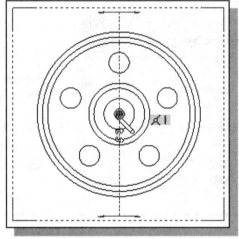

11. Inside the graphics window, align the cursor to the center of the base view and click the **left mouse button** to create the vertical cutting plane line as shown.

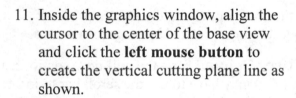

12. Select **OK** to create the vertical section line.

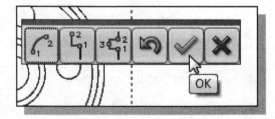

13. In the *Section View Property Manager*, confirm the **Hidden Lines Removed** option is set for the *Display Style*.

14. Next, SOLIDWORKS expects us to place the projected section. Select a location that is toward the right side of the base view as shown in the figure and click once with the **left mouse button**.

15. Select **OK** in the *Property Manager* to create the *Section View*.

16. Press the **[Esc]** key to ensure that no view or objects are selected.

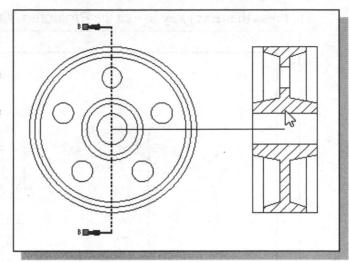

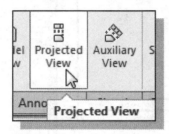

➤ We will add an isometric view using the **Projected View** command.

17. Select the **Projected View** icon from the toolbar.

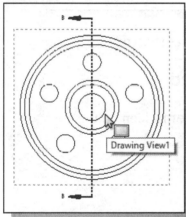

18. **Left-click** to select the **base view** for the projection as shown.

19. In the *Projected View Property Manager*, select the **Hidden Lines Removed** option for the *Display Style*.

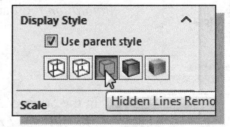

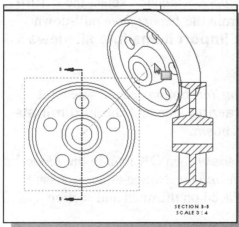

20. Place the projected isometric view by moving the cursor to the right and above the base view as shown and clicking once with the **left-mouse-button**. (**NOTE:** When it is created, the projected isometric view is automatically aligned along a line at a 45° angle from the base view. We can relocate the isometric view after it is created.)

21. Press the **[Esc]** key to exit the Projected View command.

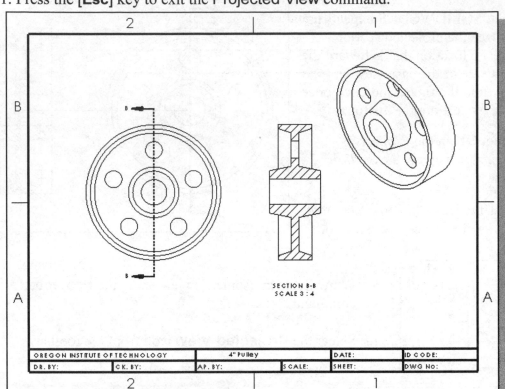

> On your own, reposition the views so that they appear as shown.

## Retrieve Dimensions – Model Items Command

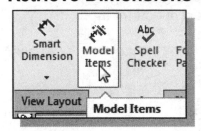

1. Left-click on the **Model Items** icon on the *Annotation* toolbar.

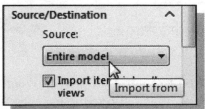

2. In the *Source/ Destination* options panel of the *Model Items Property Manager*, select the **Entire model** option from the *Import from* pull-down menu and check **Import items into all views** as shown.

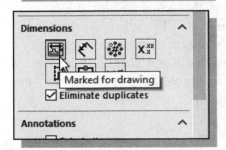

3. Under the *Dimensions* options panel, select the **Marked for drawing** icon and check **Eliminate duplicates** as shown.

4. Click the **main view** then **OK** icon in the *Model Items Property Manager*. Notice dimensions are automatically placed on the main and section drawing views.

5.  On your own, adjust the drawing to appear as shown below. You will have to reposition some dimensions, and hide some dimensions. You may also have to reposition the views.

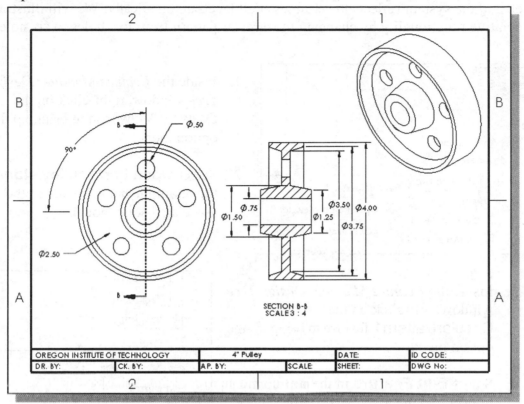

## Save the Drawing File

1.  Select **Save as** from the *File* pull-down menu.

2.  Select SOLIDWORKS **Drawing** (*.slddrw) as the file type.

3.  Use the browser to select the directory. Save the drawing file in the same directory as the part file.

4.  Enter **Pulley** for the *File name* and **Pulley Detail** for the *Description*.

5.  Click **Save**.

6.  If the SOLIDWORKS *warning* window appears, select **Save All**.

## Associative Functionality – A Design Change

*Associative functionality* within SOLIDWORKS allows us to change the design at any level, and the system reflects the changes at all levels automatically. We will illustrate the associative functionality by changing the circular pattern from five holes to six holes.

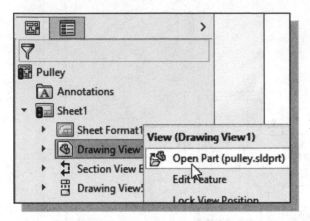

1. Inside the *Feature Manager Design Tree* window, right-click on the **Drawing View** icon to bring up the option menu.

2. Select **Open Part (pulley.sldprt)** in the pop-up menu to switch to the associated solid model.

3. Inside the *Feature Manager Design Tree* window, left-click on the **CircularPattern1** feature to bring up the option menu.

4. Select **Edit Feature** in the pop-up menu to bring up the associated feature option.

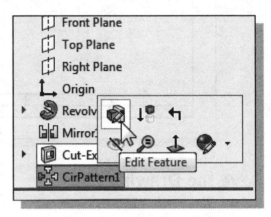

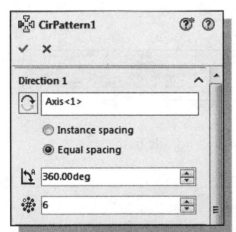

5. In the *Circular Pattern Property Manager*, change the number to **6** as shown.

6. Click on the **OK** button to accept the setting.

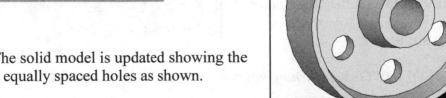

❖ The solid model is updated showing the 6 equally spaced holes as shown.

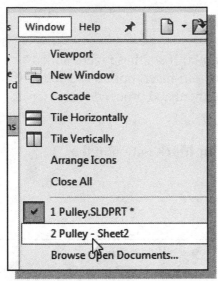

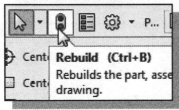

7.  Switch back to the *Pulley* drawing by selecting it in the *Window* pull-down menu. (**NOTE:** The * appearing next to the filename indicates that this file has been modified but not saved. When you save the file, this * will no longer appear.)

8.  If the SOLIDWORKS warning window appears, select **Yes** to update the drawing sheet.

9.  When the drawing window opens, some views may be shaded, indicating that they have not been updated to include the changes to the model. Select the **Rebuild** icon on the *Menu Bar* to update the views.

❖  Notice, in the *Pulley* drawing, the circular pattern is also updated automatically in all views.

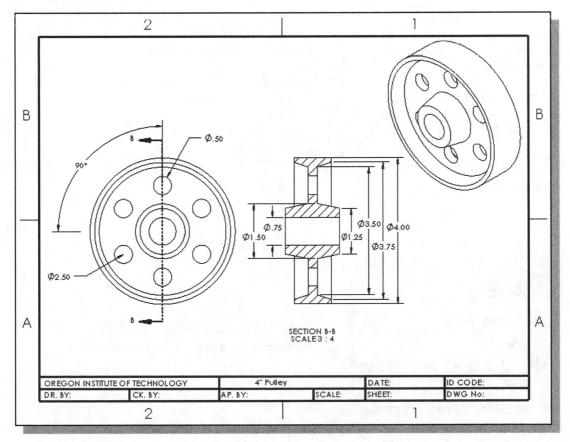

## Adding Centerlines to the Pattern Feature

In your drawing, the inserted views may include *center marks*. If so, these were added with the **auto-insert** option. This lesson was done with no auto-insert options selected. If your drawing already has the center marks, you can delete them and proceed with Step 1, or select them and proceed to Step 6.

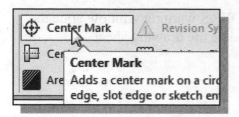

1. Click on the **Center Mark** button in the *Annotation* toolbar.

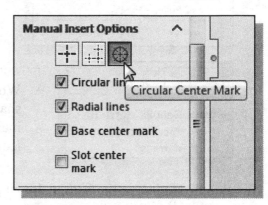

2. In the *Manual Insert Options* panel of the *Center Mark Property Manager*, activate the **Circular Center Mark** option, and **check** the **Circular lines**, **Radial lines**, and **Base center mark** option boxes.

➢ The **Centered Pattern** option allows us to add centerlines to a patterned feature.

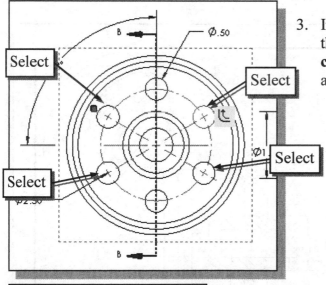

3. Inside the graphics area, hold down the **[Ctrl]** key and select the **four circular edges** of the pattern feature as shown.

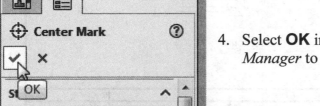

4. Select **OK** in the *Center Mark Property Manager* to create the center marks.

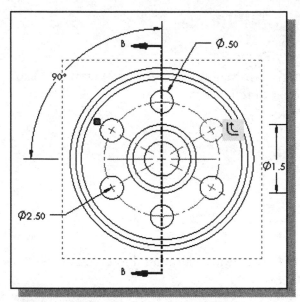

5. Notice the base circle center mark interferes with the cutting plane line. **Select** any of the newly created center marks or the associated centerlines to reopen the *Property Manager*.

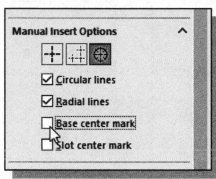

6. In the *Manual Insert Options* panel of the *Center Mark Property Manager*, **uncheck** the **Base center mark** option box.

7. Select **OK** in the *Property Manager* to accept the new setting.

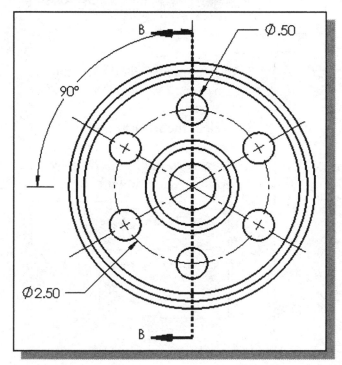

8. On your own, extend the segments of the centerlines so that they appear as shown.

## Completing the Drawing

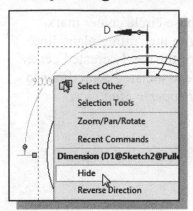

1. Select the angle dimension in the Primary view and use the Hide option to remove it.

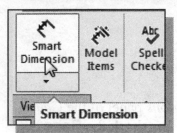

2. Use the **Smart Dimension** command to create the **60°** dimension as shown.

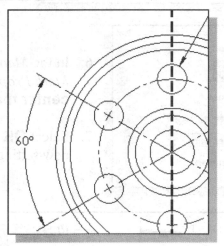

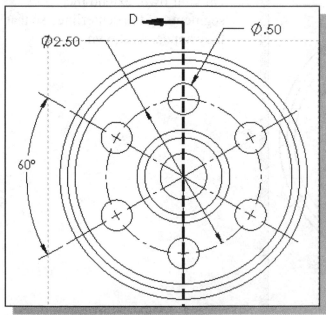

3. Select the **Ø2.50** dimension to open the *Property Manager*.

4. Select the **Inside** option for the arrow placement under the **Leaders** tab in the *Property Manager* as shown below.

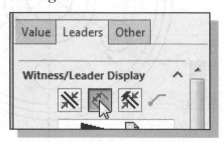

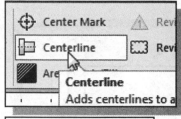

5. Click on the **Centerline** button in the *Annotation* toolbar.

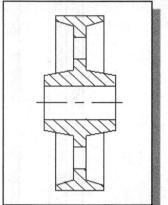

6. Inside the graphics window, click on the two edges of the *Section View* to create a centerline through the view as shown in the figure.

7. Repeat the above step and create the centerlines through the other two holes.

8. On your own, complete the title block and complete the drawing to appear as shown on the next page.

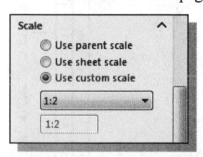

9. Select the **Isometric view**.

10. In the *Property Manager*, set the *Scale* of the view to **Use custom scale** and choose **1:2** as shown.

- Note the default view scale is set to **Use Parent Scale**.

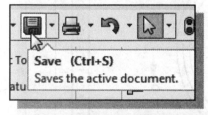

11. Select the **Save** icon on the *Menu Bar*. This command will save the **Pulley.SLDDRW** file. This file includes the document properties, the sheet format, and the drawing views.

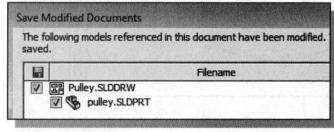

➢ Because the **Pulley** model part file was not saved after the design change, the *Save Modified Documents* window will appear.

12. Select **Save All** to save the Pulley.SLDDRW drawing file and the Pulley.SLDPRT model part file.

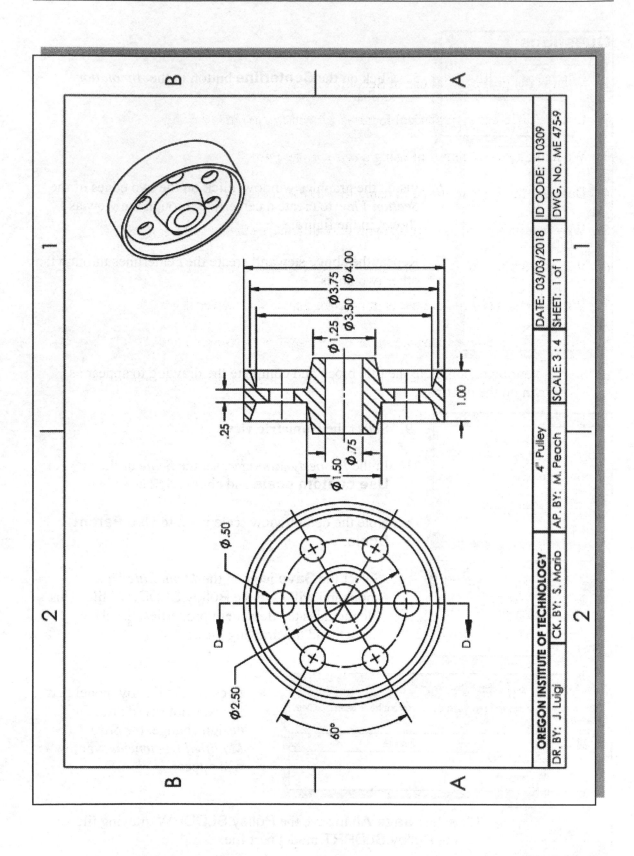

OREGON INSTITUTE OF TECHNOLOGY

4" Pulley

| DR. BY: J. Luigi | CK. BY: S. Mario | AP. BY: M. Peach | SCALE: 3 : 4 |
|---|---|---|---|
| DATE: 03/03/2018 | | ID CODE: 110309 | |
| | | DWG. No. ME 475-9 | |
| | | SHEET: 1 of 1 | |

## Questions:

1.  What is the purpose and use of Section Views?

2.  List the different symmetrical features created in the *Pulley* design.

3.  What are the advantages of using a *drawing template*?

4.  Describe the steps required in using the **Mirror** command.

5.  Why is it important to identify symmetrical features in designs?

6.  When and why should we use the **Circular Pattern** command?

7.  What are the required elements in order to generate a sectional view?

8.  How do we create a linear *diameter dimension* for a revolved feature?

9.  What is the difference between *construction geometry* and *normal geometry*?

10. Identify and describe the following commands:

    (a)

    (b)

    (c)

    (d)

**Exercises:** (Unless otherwise specified, all dimensions are in inches.)

1. **Circular Spacer**

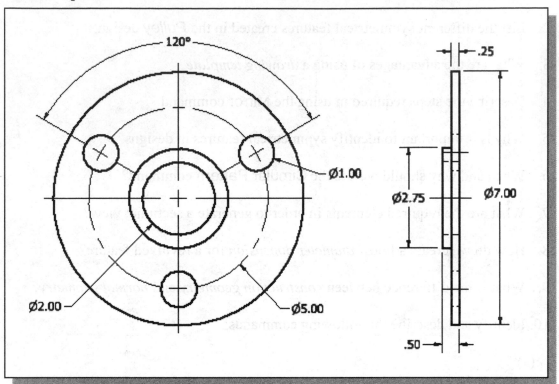

2. **Ratchet Plate** (thickness: 0.125 inch)

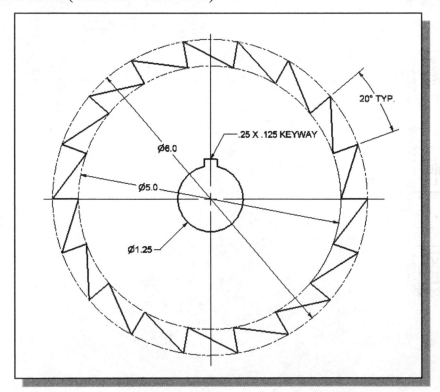

### 3.  Geneva Wheel

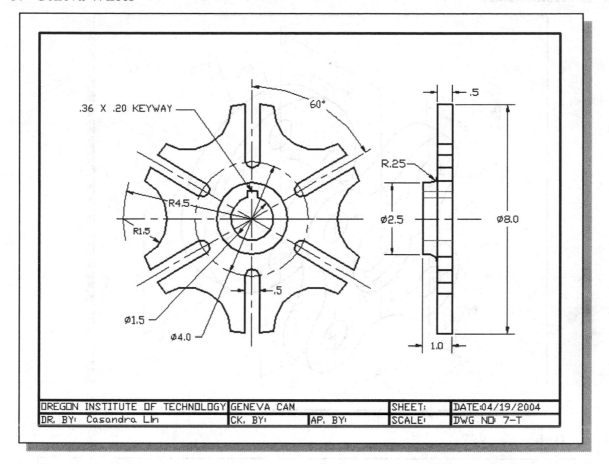

| | | | |
|---|---|---|---|
| OREGON INSTITUTE OF TECHNOLOGY | GENEVA CAM | | SHEET: | DATE:04/19/2004 |
| DR. BY: Casandra Lin | CK. BY: | AP. BY: | SCALE: | DWG NO: 7-T |

### 4.  Support Mount

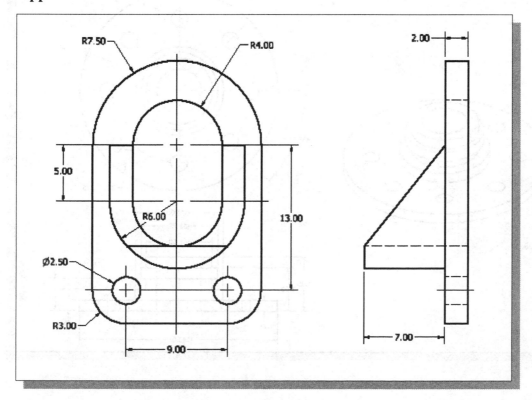

### 5. Holder Base

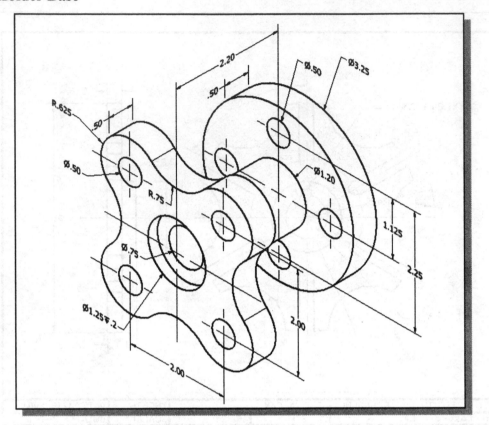

### 6. Hub

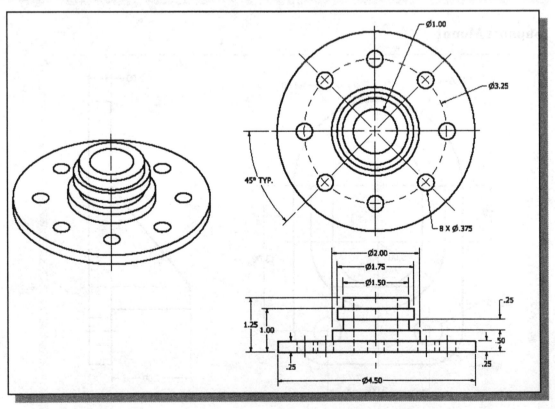

7.  **Valve Knob** (Dimensions are in mm, height: 6 mm)

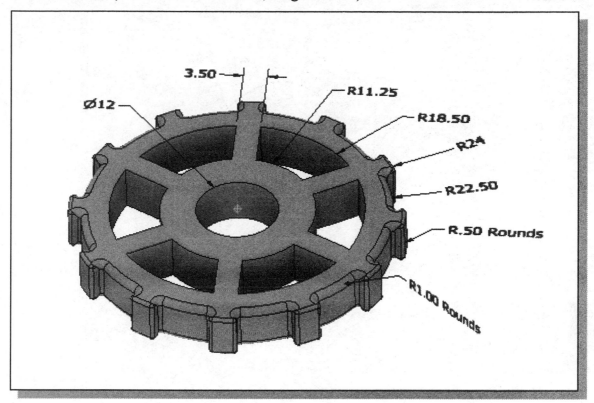

**Notes:**

# Chapter 12
# Auxiliary Views and Reference Geometry

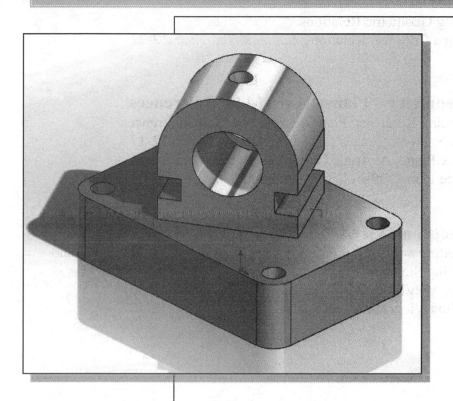

## Learning Objectives

- ♦ **Understand the Principles of Creating Auxiliary Views**
- ♦ **Understand the Concepts and the Use of Reference Geometry**
- ♦ **Use the Different Options to Create Reference Geometry**
- ♦ **Create Auxiliary Views in 2D Drawing Mode**
- ♦ **Create and Adjust Centerlines**
- ♦ **Create Shaded Images in the 2D Drawing mode**

## Certified SOLIDWORKS Associate Exam Objectives Coverage

### Sketch Relations
Objectives: Using Geometric Relations.
          Applying a Collinear Relation ....................................12-22

### Reference Geometry – Planes, Axis, Mate References
Objectives: Creating Reference Planes, Axes, and Mate References.
          Reference Axis.............................................................12-17
          Reference Plane, At Angle Option............................12-16
          Reference Plane, Offset Distance Option ..................12-26

### Drawing Sheets and Views
Objectives: Creating and Setting Properties for Drawing Sheets; Inserting and
          Editing Standard Views.
          Auxiliary Views.........................................................12-30
          Controlling View and Sheet Scales...........................12-40

Certified Associate Reference Guide

## Introduction

An important rule concerning multiview drawings is to draw enough views to accurately describe the design. This usually requires two or three of the regular views, such as a front view, a top view and/or a side view.

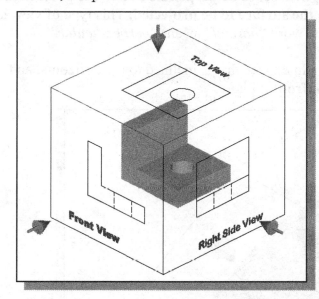

In the left figure, the L-shaped object is placed with the surfaces parallel to the principal planes of projection. The top, front and right side views show the true shape of the different surfaces of the object. Note especially that planes of projection are parallel to the top, front and right side of object.

Based on the principle of orthographic projection, it is clear that a plane surface is shown in true shape when the direction of view is perpendicular to the surface.

Many designs have features located on inclined surfaces that are not parallel to the regular planes of projection. To truly describe the feature, the true shape of the feature must be shown using an **auxiliary view**. An *auxiliary view* has a line of sight that is perpendicular to the inclined surface, as viewed looking directly at the inclined surface. An *auxiliary view* is a supplementary view that can be constructed from any of the regular views. A primary *auxiliary view* is projected onto a plane that is perpendicular to one of the principal planes of projection and is inclined to the other two. A secondary *auxiliary view* is projected onto a plane that is inclined to all three principal planes of projection. In the figures below, the use of the standard views does not show the true shape of the upper feature of the design; the use of an auxiliary view provided the true shape of the feature and also eliminated the need for front and right side views.

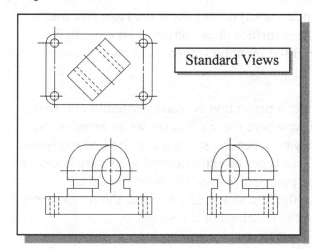

Standard Views

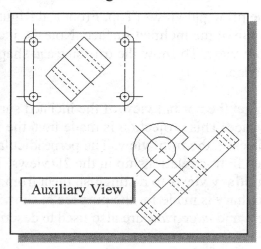

Auxiliary View

## Normal View of an Inclined Surface

No matter what the position of a surface may be, the fundamentals of projecting a normal view of the surface remain the same: **The projection plane is placed parallel to the surface to be projected. The line of sight is set to be perpendicular to the projection plane and therefore perpendicular to the surface to be projected.** This type of view is known as **normal view**. In geometry, the word "*normal*" means "*perpendicular.*"

In the figure below, the design has an inclined face that is inclined to the horizontal and profile planes and **perpendicular to the frontal plane**.

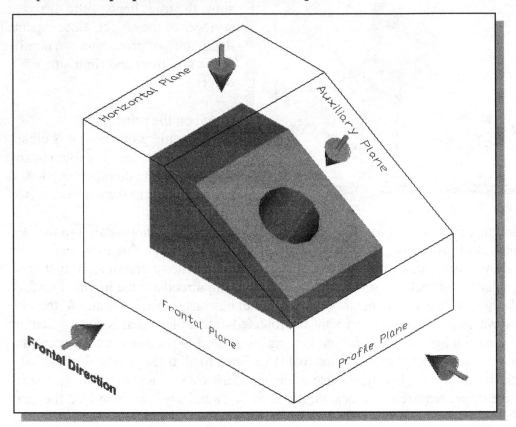

The principal views (Top, Front and Right side views) do not show the **true size and shape** of the inclined surface. Note the inclined surface does appear as an edge in the front view. To show the true size and shape of the inclined surface, a *normal* view is needed.

To get the normal view of the inclined surface, a projection is made perpendicular to the surface. This projection is made from the view where the surface shows as an edge--in this case, the front view. The perpendicularity between the surface and the line of sight is seen in true relationship in the 2D views. These types of extra normal views are known as **auxiliary views** to distinguish them from the principal views. However, since an auxiliary is made for the purpose of showing the true shape of a surface, the terms *normal view* and *edge view* are also used to describe the relations of the views.

An auxiliary is constructed following the rules of orthographic projection. An auxiliary view is aligned to the associated views--in this case, the front view. The line of sight is perpendicular to the edge view of the surface, as shown in the below figures.

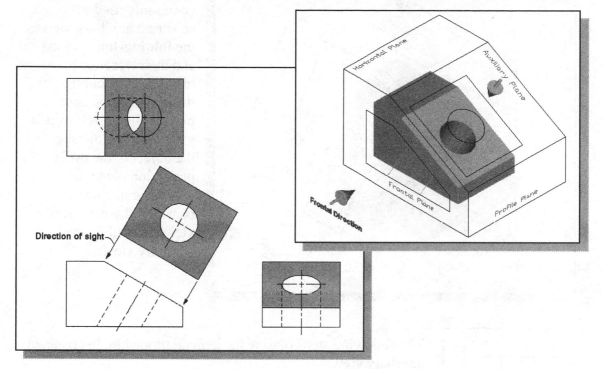

The orientation of the inclined surface may be different, but the direction of the normal view remains the same as shown in the below figures.

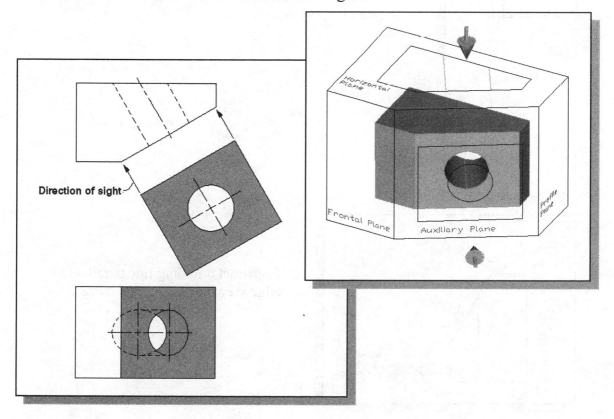

## Construction Method I – Folding Line Method

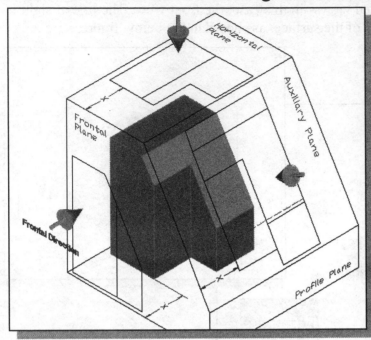

Two methods are commonly used to construct auxiliary views: the **folding-line** method and the **reference plane** method. The folding-line method uses the concept of placing the object inside a glass-box; the distances of the object to the different projection planes are used as measurements to construct the necessary views, including the auxiliary views.

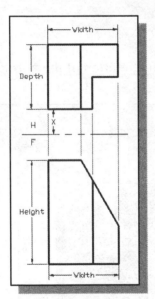

The following steps outline the general procedure to create an auxiliary view:

1.  Construct the necessary principal views; in this case, the front and top views.

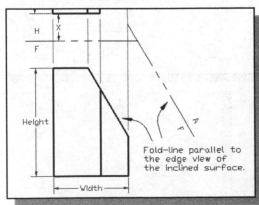

2.  Construct a folding line parallel to the edge view of the inclined surface.

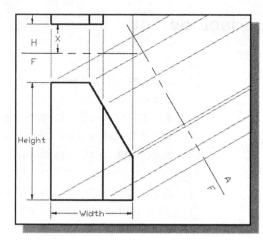

3.  Construct the projection lines perpendicular to the edge view of the inclined surface, and also perpendicular to the folding line.

4.  Use the corresponding distances, X, and the depth of the object from the principal views to construct the inclined surface.

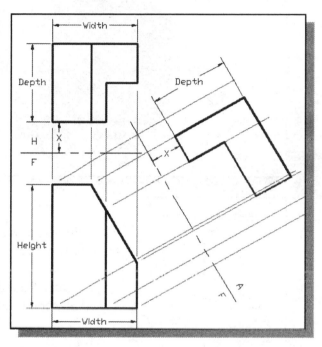

5.  Complete the auxiliary view by following the principles of orthographic projection.

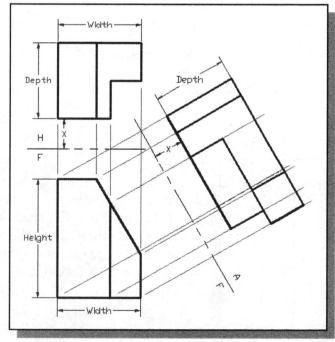

# Construction Method II – Reference Plane Method

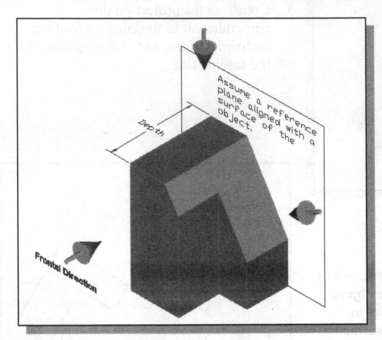

The **Reference plane method** uses the concept of placing a reference plane aligned with a surface of the object that is perpendicular to the inclined surface. The reference plane is typically a flat surface or a plane which runs through the center of the object. The distances of the individual corner to the reference plane are then used as measurements to construct the auxiliary view.

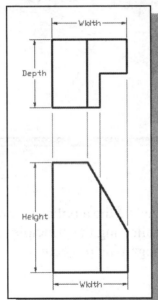

The following steps outline the general procedure to create an auxiliary view:

1.  Construct the necessary principal views; in this case, the front and top views.

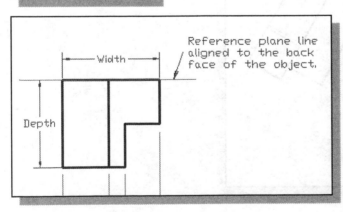

2.  Construct a reference plane line aligned to a flat face and perpendicular to the inclined surface of the object--in this case, the back face of the object.

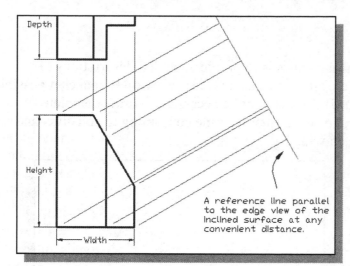

3. Construct a reference line parallel to the edge view of the inclined surface and construct the projection lines perpendicular to the edge view of the inclined surface.

4. Using the corresponding distances, such as the depth of the object, construct the inclined surface.

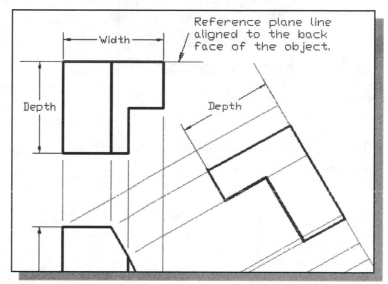

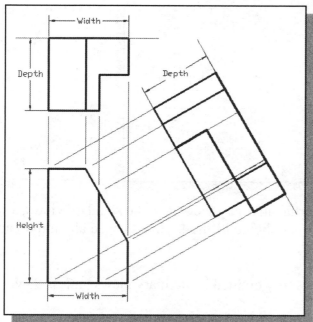

5. Complete the auxiliary view by following the principles of orthographic projection.

## Partial Views

The primary purpose of using auxiliary views is to provide detailed descriptions of features that are on inclined surfaces. The use of an auxiliary view often makes it possible to omit one or more standard views. But it may not be necessary to draw complete auxiliary views, as the completeness of detail may be time consuming to construct and add nothing to the clearness of the drawing.

 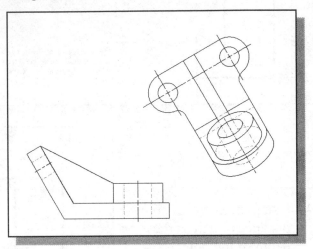

In these cases, partial views are often sufficient, and the resulting drawings are much simpler and easier to read.

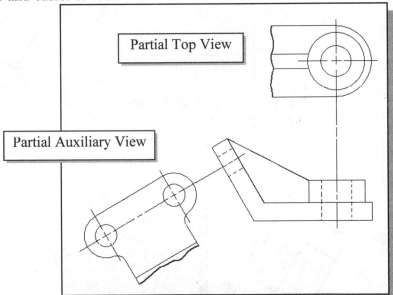

- To clarify the relationship of partial auxiliary views and the standard views, a center line or a few projection lines should be used. This is especially important when the partial views are small.

- In practice, hidden lines are generally omitted in auxiliary views, unless they provide clearness to the drawings.

# Reference Geometry in SOLIDWORKS

*Feature-based parametric modeling* is a cumulative process. The relationships that we define between features determine how a feature reacts when other features are changed. Because of this interaction, certain features must, by necessity, precede others. A new feature can use previously defined features to define information such as size, shape, location and orientation. SOLIDWORKS provides several tools to automate this process. *Reference geometry* can be thought of as user-definable datum, which are updated with the part geometry. We can create planes, axes, points, or local coordinate systems that do not already exist. Reference geometry can also be used to align features or to orient parts in an assembly. In this chapter, the use of the **Offset** option and the **Angled** option to create new reference planes, surfaces that do not already exist, are illustrated. By creating parametric reference geometry, the established feature interactions in the CAD database assure the capturing of the design intent. The default planes, which are aligned to the origin of the coordinate system, can be used to assist the construction of the more complex geometric features.

## Auxiliary Views in 2D Drawings

An important rule concerning multiview drawings is to draw enough views to accurately describe the design. This usually requires two or three of the regular views, such as a front view, a top view and/or a side view. However, many designs have features located on inclined surfaces that are not parallel to the regular planes of projection. To truly describe the feature, the true shape of the feature must be shown using an **auxiliary view**. An *auxiliary view* has a line of sight that is perpendicular to the inclined surface, as viewed looking directly at the inclined surface. An *auxiliary view* is a supplementary view that can be constructed from any of the regular views. Using the solid model as the starting point for a design, auxiliary views can be easily created in 2D drawings. In this chapter, the general procedure of creating auxiliary views in 2D drawings from solid models is illustrated.

## The *Rod-Guide* Design

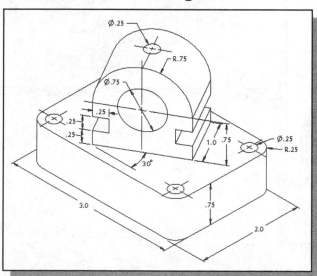

❖ Based on your knowledge of SOLIDWORKS so far, how would you create this design? What are the more difficult features involved in the design? Take a few minutes to consider a modeling strategy and do preliminary planning by sketching on a piece of paper. You are also encouraged to create the design on your own prior to following through the tutorial.

## Modeling Strategy

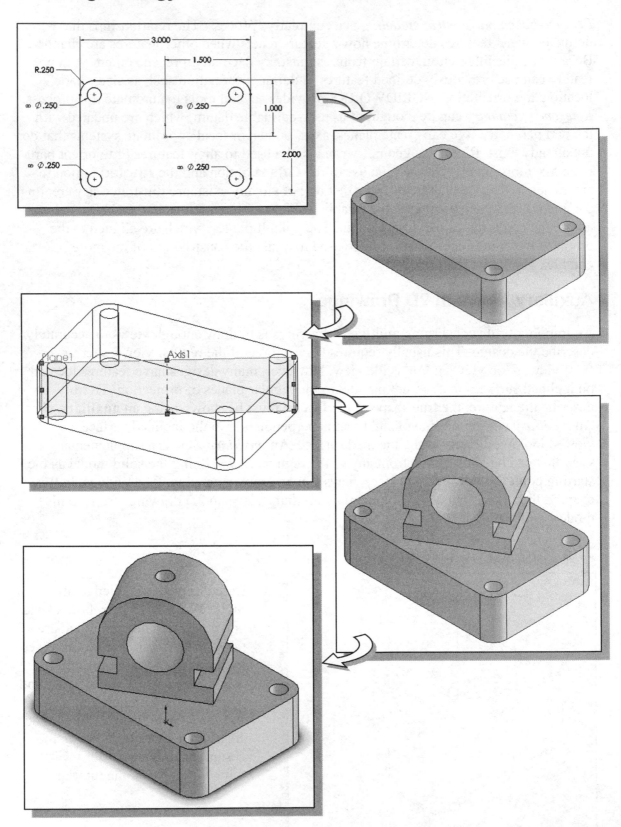

## Starting SOLIDWORKS

1. Select the **SOLIDWORKS** option on the *Start* menu or select the **SOLIDWORKS** icon on the desktop. The SOLIDWORKS main window will appear on the screen.

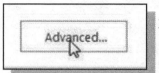

2. In the *Welcome to SOLIDWORKS* dialog box, enter the **Advanced** mode by clicking once with the left-mouse-button on the **Advanced** icon.

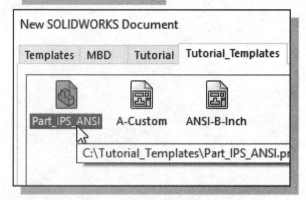

3. Select the **Tutorial_Templates** tab. (Refer to Chapter 5 on the setup of template files.)

4. Select the **Part_IPS_ANSI** template as shown.

5. Click on the **OK** button to open a new document.

## Applying the BORN Technique

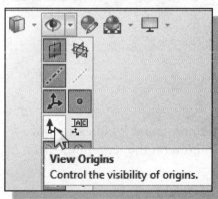

1. Select the **Hide/Show Items** icon on the *Heads-up View* toolbar to reveal the pull-down menu. In the pull-down menu, toggle on the **View Planes** and **View Origins** buttons.

2. In the *Feature Manager Design Tree* window, select the **Front**, **Top**, and **Right Planes** by holding down the **[Ctrl]** key and clicking with the left-mouse-button.

3. In the option menu, click on **Show** to toggle *ON* the display of the planes.

4. On your own, use the viewing options (**Rotate**, **Zoom** and **Pan**) to view the work features established.

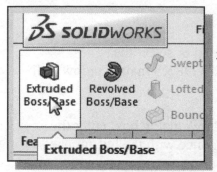

5.  Select the **Extruded Boss/Base** button on the *Features* toolbar to create a new extruded feature.

6.  In the *Edit Sketch Property-Manager*, the message "*Select a plane on which to create a sketch for the entity*" is displayed. Move the cursor over the edge of the **Top Plane** in the graphics area. When the Top Plane is highlighted, click once with the **left-mouse-button** to select the Top (XZ) Plane as the sketch plane for the new sketch.

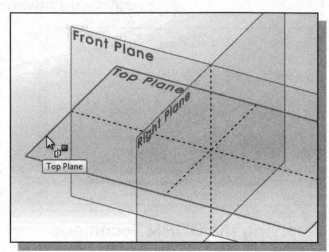

7.  On your own, **hide** the three work planes.

## Creating the Base Feature

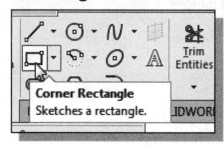

1.  Select the **Rectangle** command by clicking once with the left-mouse-button on the icon in the *Sketch* toolbar.

2.  Create a rectangle of arbitrary size with the origin inside the rectangle as shown.

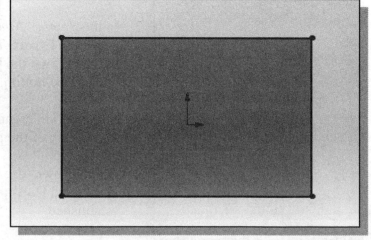

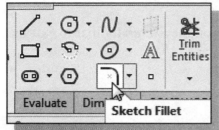

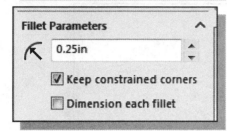

3. Click on the **Sketch Fillet** icon in the *Sketch* toolbar.

4. Enter **0.25 in** for the fillet radius in the *Fillet Parameters* panel of the *Sketch Fillet Property-Manager*.

5. Create four rounded corners as shown. Each fillet can be created by moving the cursor into the graphics area and (1) clicking on the corner, or (2) clicking on the two lines forming the corner.

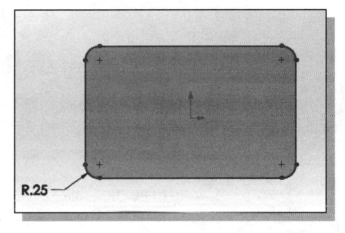

6. Select **OK** in the *Property Manager* or press the **[Esc]** key to exit the Sketch Fillet command.

7. On your own, create four circles of the same diameter and with the centers aligned to the centers of arcs. Also create and modify the six dimensions as shown in the figure below. (Hint: Add relations and establish equations.) Note that these dimensions fix the center of the base at the origin.

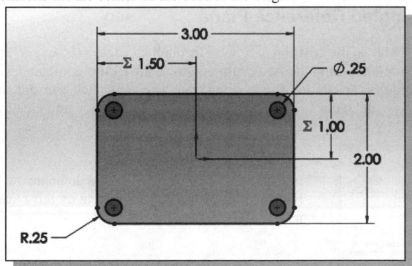

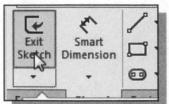

8. Select the **Exit Sketch** icon on the *Sketch* toolbar to exit the Sketch option.

9. Create a **Blind** extrusion with a distance of **0.75** in.

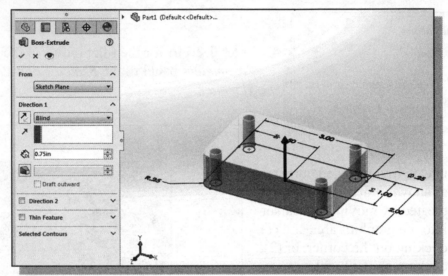

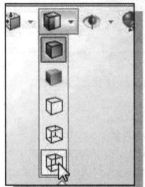

10. Click **OK** in the *Extrude Property Manager* to accept the settings and create the extrusion.

11. In the *Display Style* pull-down menu on the *Heads-up View* toolbar, select the **Wireframe** option to set the display mode to *Wireframe*.

## Create an Angled Reference Plane

SOLIDWORKS allows the creation of a plane through an edge, axis, or sketch line at an angle to a face or plane, using the **At Angle** option of the **Plane** command. Using the **At Angle** option requires selection of an existing edge, axis, or sketch line and an existing face or plane. We will create a reference axis along the Y-direction and create a plane that can be visualized by rotating the Front Plane about the reference (Y) axis.

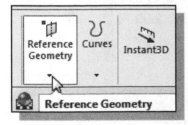

1. In the *Features* toolbar, select the down arrow below the **Reference Geometry** command by left-clicking the icon.

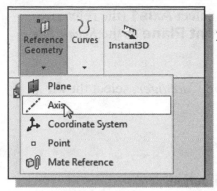

2.  In the pull-down option menu of the **Reference Geometry** command, select the **Axis** option.

3.  In the *Selections* panel of the *Axis Property Manager*, select the **Two Planes** option. (We will define the Y-axis as the intersection of the Front and Right Planes.)

4.  Select the **Front Plane** and **Right Plane** in the *Design Tree* appearing in the graphics area.

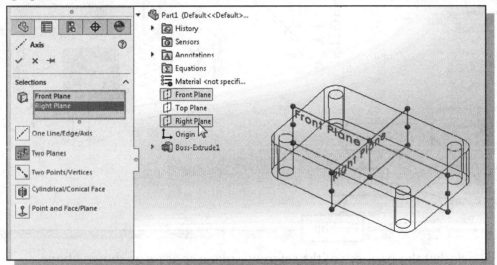

5.  Click **OK** in the *Property Manager* to create the new reference axis.

6.  Press the **[Esc]** key to make sure no objects are pre-selected.

➤  Notice the reference axis, named **Axis1**, has been created. It is aligned with the Y-axis and passes through the center of the base feature.

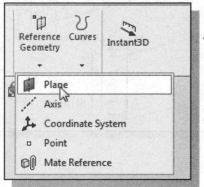

7.  Select the **Reference Geometry** command from the *Features* toolbar, and select the **Plane** option from the pull-down menu.

➤  Notice the message "*Select references and constraints*" appears and the *First Reference* window is highlighted.

8. In the *Design Tree* appearing in the graphics area, select **Axis1** (the reference axis we just created) as the *First Reference* and **Front Plane** as the *Second Reference*.

9. In the *Second Reference* panel of the *Plane Property Manager*, select the **At Angle** option button.

10. Enter **30 deg** as the angle in the *Property Manager*.

11. If necessary, check the **Flip** box to set the plane direction.

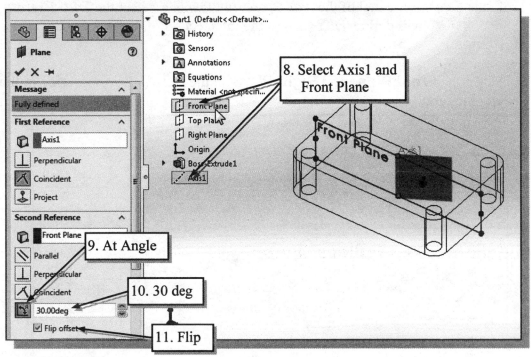

12. Check that the preview of the reference plane appears as shown above.

13. Click **OK** in the *Property Manager* to create the new reference plane.

❖ Note that the *angle* is measured relative to the selected reference plane, the **Front (XY) Plane**.

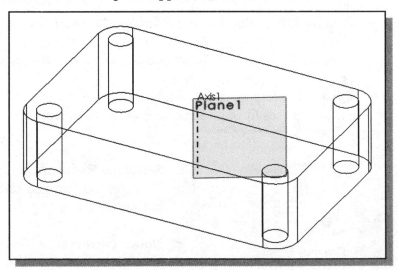

## Creating an Extruded Feature on the Reference Plane

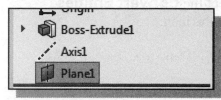

1. Confirm the reference plane Plane1 is pre-selected as shown.

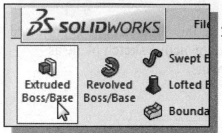

2. In the *Features* toolbar, select the **Extruded Boss/Base** command by clicking once with the left-mouse-button on the icon.

- Note the pre-selected angled reference plane, **Plane1**, is automatically used as the sketch plane.

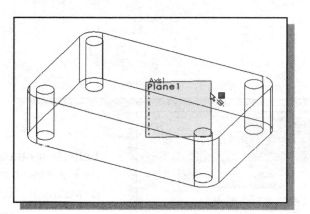

## Using the Convert Entities Option

❖ Projected geometry is another type of *reference geometry*. The Convert Entities tool can be used to project entities from previously defined sketches or features onto the sketch plane. The position of the projected geometry is fixed to the feature from which it was projected. We can use the Convert Entities tool to project geometry from a sketch or feature onto the active sketch plane.

Typical uses of the Convert Entities command include:
- Project a silhouette of a 3D feature onto the sketch plane for use in a 2D profile.
- Project a sketch from a feature onto the sketch plane so that the projected sketch can be used to constrain a new sketch.

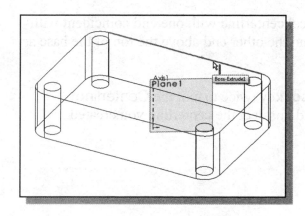

1. Move the cursor over the top rear edge of the base feature. When the edge is highlighted, click once with the **left-mouse-button** to select the edge.

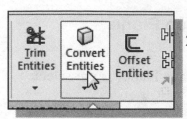

2.  In the *Sketch* toolbar, select the **Convert Entities** command by left-clicking the icon.

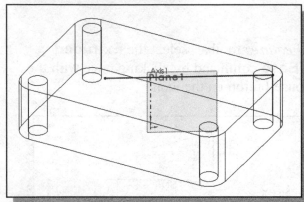

❖   Notice the selected edge is projected onto the active sketch plane as shown.

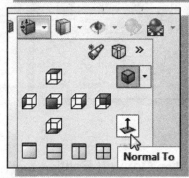

3.  Left-click on the **View Orientation** icon on the *Heads-up View* toolbar to reveal the *View Orientation* pull-down menu. Select the **Normal To** command.

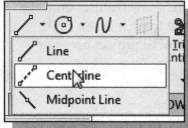

4.  **Left-click** on the arrow next to the **Line** button on the *Sketch* toolbar to reveal additional commands.

5.  Select the **Centerline** command from the pop-up menu.

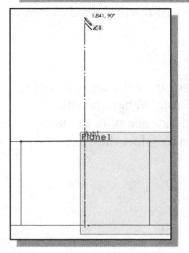

6.  Draw a vertical centerline with one end coincident with the **origin** and the other end above the top of the base as shown.

7.  Press the **[Esc]** key twice to exit the Centerline command and unselect the centerline you created.

❖ The **Dynamic Mirror** command will be used to generate sketch objects with symmetry about a centerline.

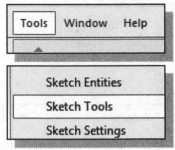

8.  Select the **Tools** pull-down menu, then click on **Sketch Tools**. A complete list of sketch tools will appear. (Some of these also appear in the *Sketch* toolbar.)

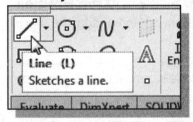

9.  Select **Dynamic Mirror** as shown.

10. The message "*Please select a sketch line or linear edge to mirror about*" appears in the *Mirror Property Manager*. Move the cursor over the **centerline** in the graphics area and click once with the **left-mouse-button** to select it as the line to mirror about.

➢ Notice the hash marks which appear on the centerline to indicate that the Dynamic Mirror command is active.

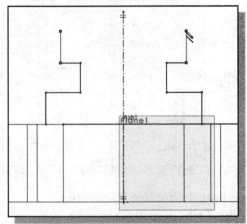

11. Click on the **Line** icon in the *Sketch* toolbar.

12. On your own, create a rough sketch using the projected edge as the bottom line as shown in the figure. (Note that all edges are either horizontal or vertical.)

13. Press the **[Esc]** key to exit the Line command.

14. On your own, **turn *OFF*** the Dynamic Mirror command.

15. Press the **[Esc]** key twice to ensure no entities are selected.

16. Select the **Add Relation** command from the *Display/Delete Relations* pop-up menu on the *Sketch* toolbar.

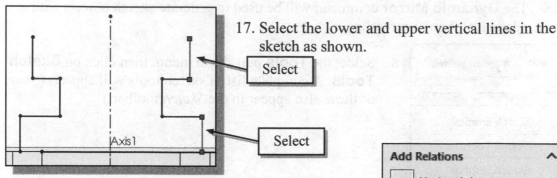

17. Select the lower and upper vertical lines in the sketch as shown.

18. Select the **Collinear** option in the *Add Relations* panel of the *Add Relations Property Manager*.

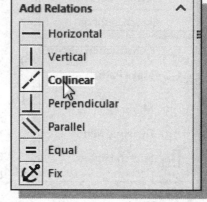

19. Repeat the above steps and align the two vertical lines on the left.

20. Press the **[Esc]** key to exit the Add Relations command.

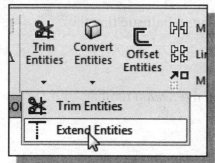

21. **Left-click** on the **arrow** below the **Trim Entities** button on the *Sketch* toolbar to reveal additional commands.

22. Select the **Extend Entities** command from the pop-up menu.

23. Move the cursor near the left endpoint of the horizontal line we created with the Convert Entities command. When the extended line appears, click once with the **left-mouse-button** to extend the horizontal line to meet the bottom of the vertical line as shown.

24. Press the **[Esc]** key once to end the Extend command.

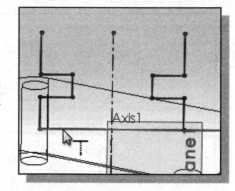

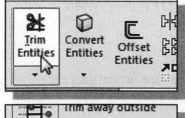

25. We will next trim the bottom horizontal line to the inclined line. Select the **Trim Entities** icon on the *Sketch* toolbar.

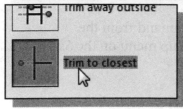

26. Select the **Trim to closest** option in the *Trim Property Manager*.

27. Move the cursor near the right endpoint of the horizontal line we created with the **Convert Entities** command. SOLIDWORKS will automatically display the possible result of the selection. When the preview appears, click once with the **left-mouse-button** to trim the horizontal line to meet the bottom of the vertical line as shown.

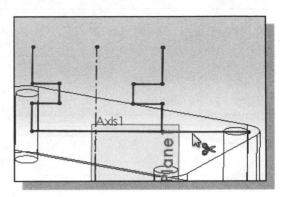

28. Press the [**Esc**] key once to end the Trim command.

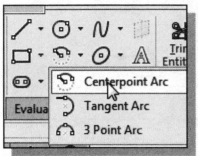

29. **Left-click** on the arrow next to the **Arc** button on the *Sketch* toolbar and select the **Centerpoint Arc** command from the pop-up menu.

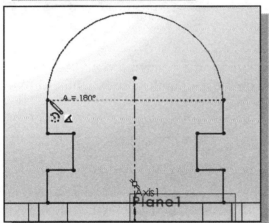

30. On your own, create the arc with the centerpoint coincident with the centerline and aligned with the top endpoints of the vertical lines as shown.

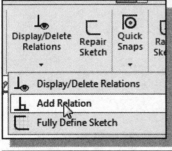

31. Use the **Add Relation** command to create a **Tangent** relation between the arc and the upper right vertical line.

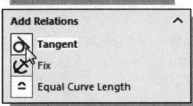

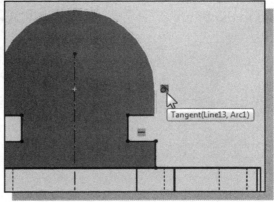

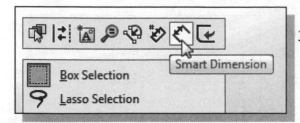

32. Inside the graphics window, click once with the right-mouse-button to display the option menu. Select the **Smart Dimension** option in the pop-up menu.

33. On your own, create the Dimensions as shown.

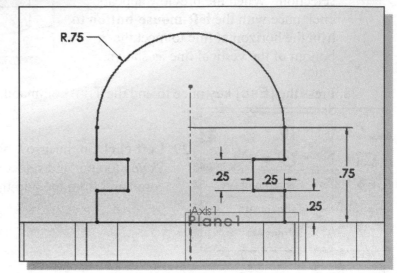

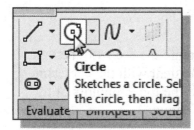

34. Select the **Circle** command by clicking once with the left-mouse-button on the icon in the *Sketch* toolbar.

35. On your own, add a **0.75** inch circle concentric with the arc.

36. On your own, complete the sketch as shown in the figure.

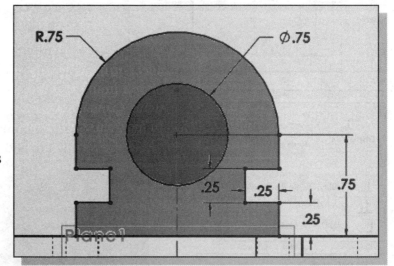

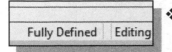

❖ The *Status Bar* should now read *Fully Defined* because the geometric and dimensional constraints fully define the sketch.

## Completing the Solid Feature

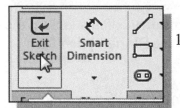

1. Select the **Exit Sketch** icon on the *Sketch* toolbar to exit the Sketch option.

2. Select the **Midplane** option.

3. Enter **1.0 inch** for the extrusion distance.

4. Select the **Contour** as shown and confirm the **Merge result** option is on.

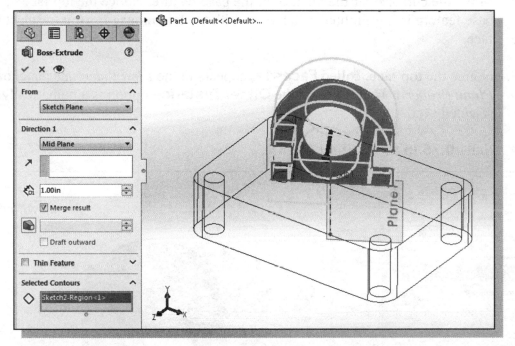

5. Click **OK** in the *Property Manager* to accept the settings and create the extrusion.

6. Using the *Heads-up View* toolbar, set the display style to **Shaded with edges**.

7. On your own, turn **off** the visibility of the angled reference plane.

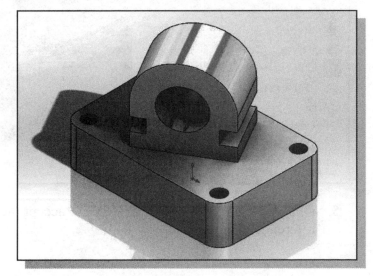

## Create an Offset Reference Plane

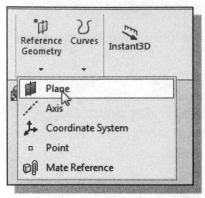

1. Select the **Reference Geometry** command from the *Features* toolbar, and select the **Plane** option from the pull-down menu.

❖ In the *Status Bar* area, the message: "*Select valid entities to define a plane (plane, face, edge, line, or point*" is displayed). SOLIDWORKS expects us to select any existing geometry, which will be used as a reference to create the new work plane.

2. Move the cursor over the top face of the base feature. Notice the top face of the base feature is highlighted. Click once with the **left-mouse-button** to select the face.

3. Notice the top face, called **Face<1>**, appears in the *First Reference* panel of the *Plane Property Manager*, and the **Offset Distance** option is automatically selected.

4. Enter **0.75 in** as the offset distance.

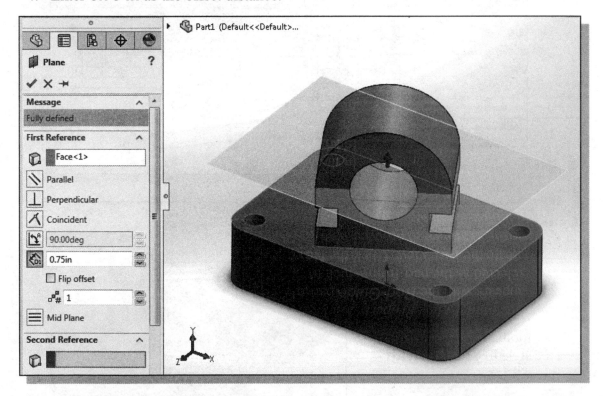

5. Click **OK** in the *Property Manager* to accept the settings and create the new reference plane.

## Creating another Extruded Feature using the Reference Plane

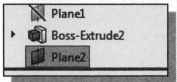

1. Confirm the reference plane Plane2 is pre-selected as shown.

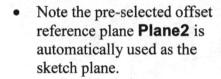

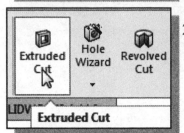

2. In the *Features* toolbar, select the **Extruded Cut** command by clicking once with the left-mouse-button on the icon.

- Note the pre-selected offset reference plane **Plane2** is automatically used as the sketch plane.

3. In the *Display Style* pull-down menu on the *Heads-up View* toolbar, select the **Wireframe** option to set the display mode to *Wireframe*.

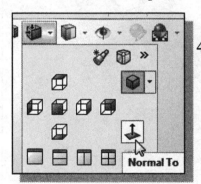

4. In the *View Orientation* pull-down menu on the *Heads-up View* toolbar, select the **Normal To** command.

5. On your own, create a **Ø 0.25** inch circle aligned to the origin as shown in the figure.

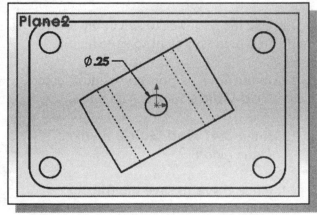

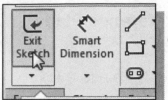

6. Select the **Exit Sketch** icon on the *Sketch* toolbar to exit the sketch.

7. On your own, create the **Extruded Cut** feature. Use the **Through All** option and click the Reverse Direction button if necessary.

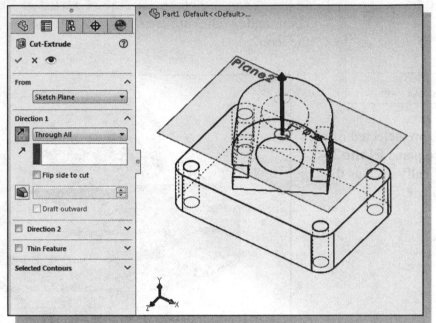

8. Click **OK** in the *Property Manager* to create the extruded cut feature.

9. On your own, hide all planes, hide Axis1, and set the display to **Shaded With Edges**.

10. Select the **Save As** option from the *File* pull-down menu.

11. Use the *Browser* to select the folder in which you want to save the file.

12. In the *Save As* pop-up window, enter **Rod-Guide** for the *File name*.

13. Enter **3/4″ Rod Guide** for the *Description*.

14. Click **Save** to save the file.

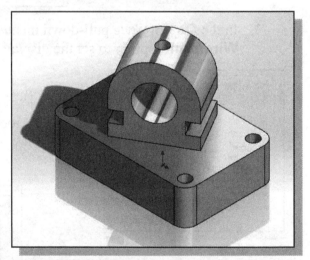

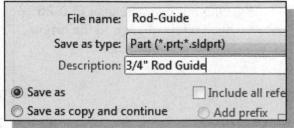

## Starting a New 2D Drawing and Adding a Base View

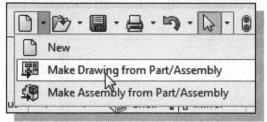

1. Click on the arrow next to the **New** icon on the *Menu Bar* and select **Make Drawing from Part/Assembly** in the pull-down menu.

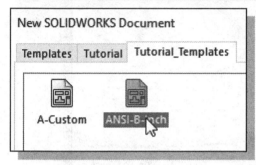

2. In the *New* SOLIDWORKS *Document* window, select the **Tutorial_Templates** tab.

3. Select the **ANSI-B-Inch** template file icon. This is the custom B size drawing template you created in Chapter 8.

4. Select **OK** in the *New* SOLIDWORKS *Document* window to open the new drawing file.

➢ The new drawing file opens using the **ANSI-B-Inch.SLDDOT** template. The B-size drawing sheet appears. The *View Palette* appears in the *task pane*.

❖ **NOTE:** If your ANSI-B-Inch template is not available: (1) open the default Drawing template; (2) set document properties following instructions on page 8-43.

❖ In SOLIDWORKS, the first drawing view we create is called a **base view**. A *base view* is the primary view in the drawing; other views can be derived from this view. In SOLIDWORKS, a base view can be added using the **Model View** command on the *Drawing* toolbar, or using the **View Palette**. By default, SOLIDWORKS will treat the *world XY plane* as the front view of the solid model. Note that there can be more than one *base view* in a drawing.

5. If the *View Palette* is collapsed, select the **View Palette** tab at the right of the graphics area.

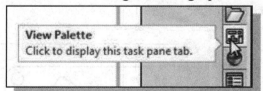

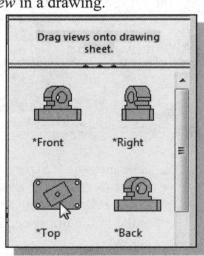

6. Select the **Top** view in the *View Palette* by clicking on the icon with the **left-mouse-button** as shown.

7.  Hold down the **left-mouse-button** and **drag** the top view from the *View Palette* into the graphics window and place the **base view** near the upper left corner of the graphics window as shown below.

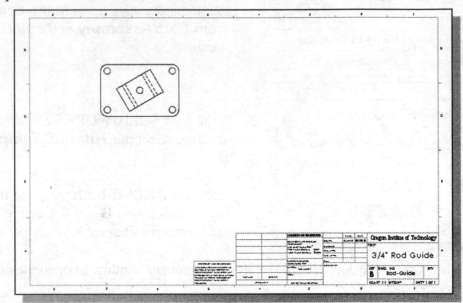

➤  When the base view is placed, SOLIDWORKS automatically executes the **Projected View** command and the *Projected View Property Manager* appears.

8.  Press the **[Esc]** key to exit the **Projected View** command.

## Creating an Auxiliary View

❖  In SOLIDWORKS *Drawing* mode, the **Projected View** command is used to create standard views such as the *top* view, *front* view or *isometric* view. For non-standard views, the **Auxiliary View** command is used. *Auxiliary views* are created using orthographic projections. Orthographic projections are aligned to the base view and inherit the base view's scale and display settings.

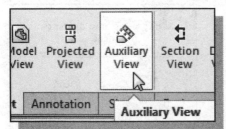

1.  Click on the **Auxiliary View** button in the *View Layout* toolbar.

2.  The message "*Please select a Reference Edge to continue*" appears in the *Auxiliary View Property Manager*. Pick the front edge of the upper section of the model as shown.

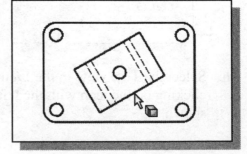

❖ The orthographic projection direction will be perpendicular to the selected edge.

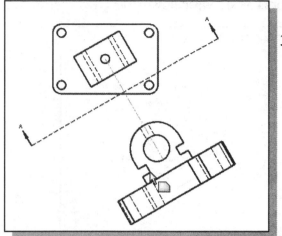

3. Move the cursor below the base view and select a location to position the auxiliary view of the model as shown. Click once with the left-mouse-button to place the view.

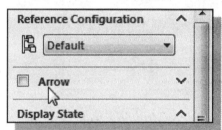

4. In the *Auxiliary View Property Manager* **uncheck** the box next to the **Arrow** option. This will turn *OFF* the display of the direction arrow.

5. Click **OK** to accept the settings and exit the Auxiliary View command.

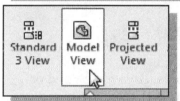

6. Click on the **Model View** icon *Drawing* toolbar.

7. The *Model View Property Manager* appears with the *Rod-Guide* part file selected as the part from which to create the base view. Click the **Next** arrow as shown to proceed with defining the base view.

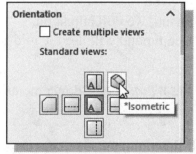

8. In the *Model View Property Manager*, select the **Isometric** view for the *Orientation*, as shown.

➢ In general, hidden lines are not used in isometric views.

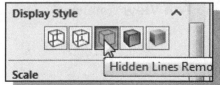

9. In the *Model View Property Manager*, select the **Hidden Lines Removed** option for the *Display Style*.

10. Move the cursor toward the upper right corner of the sheet, and select a location to position the isometric view of the model as shown below.

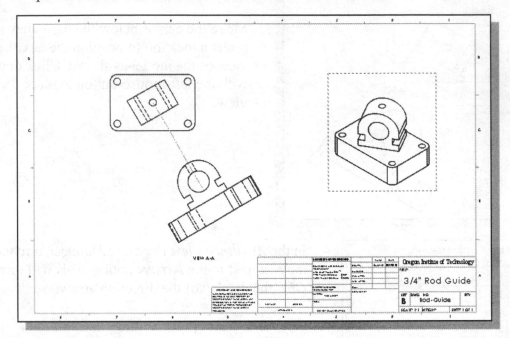

11. Left-click on the **OK** icon in the *Drawing View Property Manager* to accept the view.

## Displaying Feature Dimensions

➢ In your drawing, the inserted views may include *center marks*, *centerlines*, and/or *dimensions*. If so, these were added with the **auto-insert** option. To access these options, click **Options** (*Standard* toolbar), select the **Document Properties** tab, and then select **Detailing**. This lesson was done with no auto-insert options selected. The center marks, centerlines, and dimensions will be added individually. The dimensions used to create the part can be imported into the drawing using the **Model Items** command. We can also add dimensions to the drawing manually using the **Smart Dimension** command.

➢ Before we apply dimensions using the Model Items command, we will adjust the size/position of the *Reference Plane 1*. The Model Items command will use the mid-line of the plane to locate the angle dimension.

1. Select the ***Rod-Guide.SLDPRT*** part window from the *Window* pull-down menu to switch to *Part Modeling* mode.

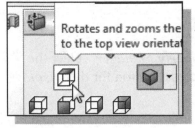

2. In the *View Orientation* pull-down menu on the *Heads-up View* toolbar, select the **Top View** command.

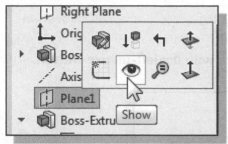

3. On your own, toggle *ON* the visibility of **Plane1**.

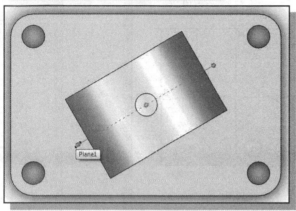

4. Using the left-mouse-button, **click and drag** the end of **Plane1** and position it such that the **mid-point** of the plane coincides with the center of the hole. This point will be used as the reference point for applying the angle dimension when the Model Items command is executed.

5. On your own, switch back to the **2D Drawing** file.

6. Press the **[Esc]** key to make sure no items are selected.

7. Switch to the *Annotation* toolbar in the *Command Manager*.

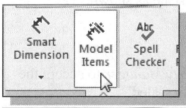

8. Left-mouse-click on the **Model Items** icon on the *Annotation* toolbar.

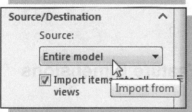

9. In the *Source/Destination* options panel of the *Model Items Property Manager*, select the **Entire model** option from the *Import from* pull-down menu and check **Import items into all views** as shown.

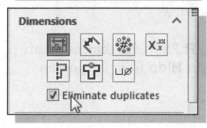

10. Under the *Dimensions* options panel, select the **Marked for drawing** icon. (**NOTE:** This option may be selected by default.) Check **Eliminate duplicates** as shown.

11. Click on the top view and then the **OK** icon in the *Model Items Property Manager*. Notice dimensions are automatically placed on the top and auxiliary drawing views.

## Adjusting the View Scale

1.  Move the cursor over the top view and click once with the **left-mouse-button**. The *Drawing View1 Property Manager* appears.

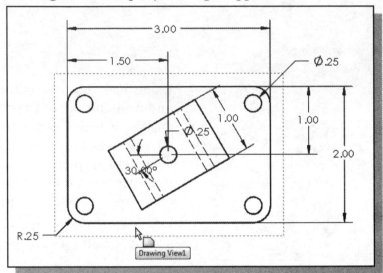

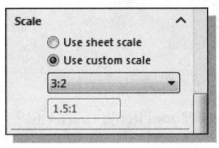

2.  Under the *Scale* options in the *Drawing View Property Manager*, select **Use custom scale**.

3.  Select **3:2** as custom scale as shown.

4.  Left-click on the **OK** icon in the *Drawing View Property Manager* to accept the new settings. Notice the auxiliary view is automatically scaled.

5.  If necessary, reposition the views so they do not overlap.

## Repositioning, Appearance, and Hiding of Feature Dimensions

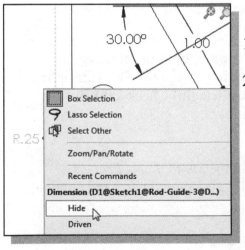

1.  Zoom in on the top view.

2.  **Right-click** on the **R.25** *radius* dimension of the round, and select **Hide** from the pop-up option menu.

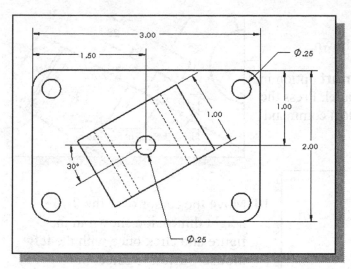

3. On your own, reposition the dimensions to appear as shown in the figure.

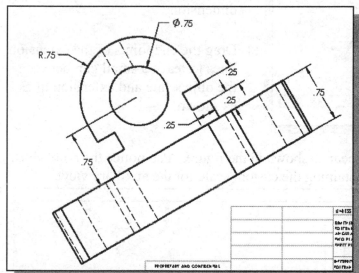

4. **Pan** to the auxiliary view. On your own, hide and reposition dimensions as shown in the figure.

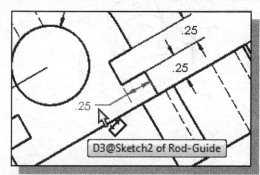

❖ Note the appearance of dimensions can be adjusted independently.

5. Move the cursor over the **0.25** dimension shown in the figure and click once with the **left-mouse-button** to select the dimension. The *Dimension Property Manager* will appear.

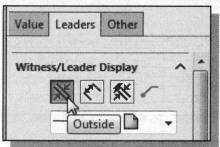

6. In the *Dimension Property Manager*, select the **Leaders** tab.

7. In the *Witness/Leader Display* panel, select the **Outside** option as shown. This will result in the arrows being located inside the extension lines.

8. Reposition the dimension as shown.

9. On your own, return to the **Smart** option in the *Witness/Leader Display* panel. Press the **[Esc]** key to exit the Dimension command and unselect the dimension.

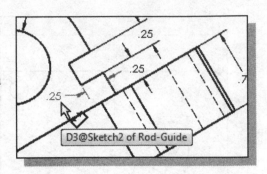

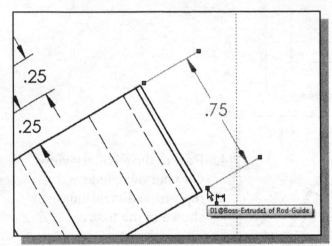

10. Move the cursor over the **0.75** height dimension shown in the figure and click once with the **left-mouse-button** to select the dimension.

11. Drag the endpoints of the extension lines to leave a small gap between the object line and extension lines as shown.

12. Adjust the dimensions to appear as shown in the figure. Also notice the note which was automatically added containing the custom scale for the auxiliary view.

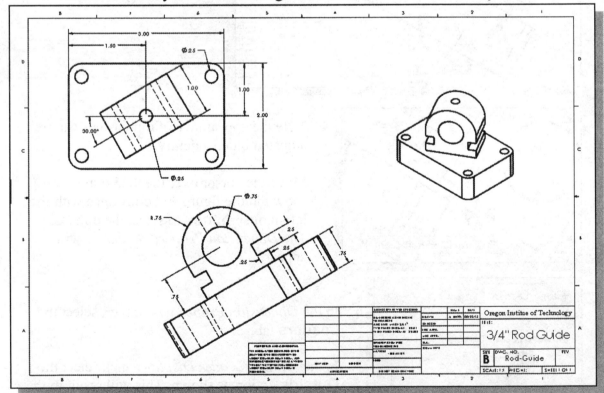

## Tangent Edge Display

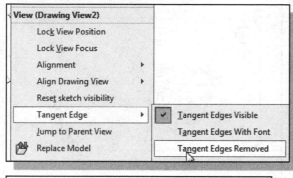

1. Move the cursor over the *auxiliary view* and click once with the **right-mouse-button**.

2. In the pop-up option menu, select **Tangent Edge**, then **Tangent Edges Removed**.

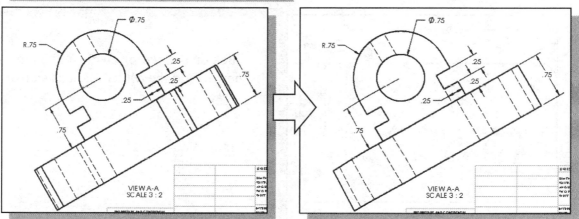

## Adding Center Marks and Center Lines

➤ In your drawing, the inserted views may include *center marks*. If so, these were added with the **auto-insert** option. This lesson was done with no auto-insert options selected. The center marks will be added individually. If your drawing already has the center marks, you can delete them and proceed with the lesson, or select them and adjust the *Display Attributes* in the *Center Mark Property Manager* as described below.

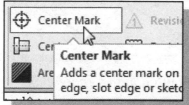

1. Click on the **Center Mark** button in the *Annotation* toolbar.

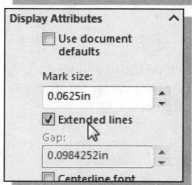

2. In the *Display Attributes* panel of the *Center Mark Property Manager*, **uncheck** the **Use document defaults** box, enter **0.0625** for the Mark size, and **check** the Extended lines box.

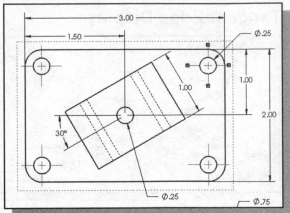

3. Click on the four arcs in the top view to add the center marks as shown.

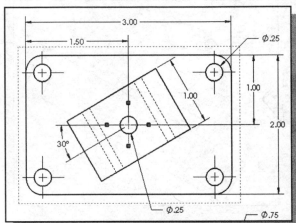

4. Click on the **center circle** in the top view to add the center mark as shown.

5. Click on the **Arc** in the Auxiliary view to add the center mark as shown.

6. Click **OK** in the *Property Manager* to exit the **Center Mark** command.

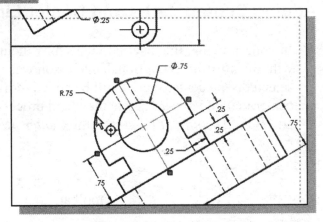

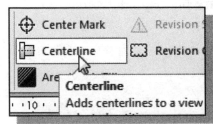

7. Click on the **Centerline** button in the *Annotation* toolbar.

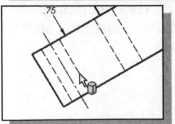

8. Inside the graphics window, click on the two hidden edges of one of the **Drill** features and create a center line in the auxiliary view as shown in the figure.

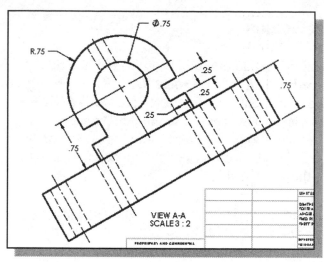

VIEW A-A
SCALE 3 : 2

9. On your own, repeat the above step and create additional centerlines as shown.

10. Click **OK** in the *Property Manager* to exit the CenterLine command.

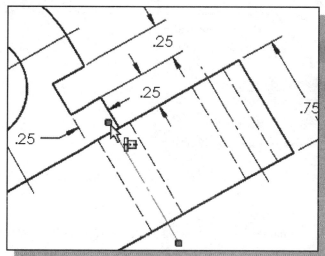

11. Select the centerline that intercepts with the dimension and move the endpoint of the line by **clicking and dragging** as shown.

12. On your own, adjust extension line endpoints, by **clicking and dragging**, to appear approximately as shown in the figure.

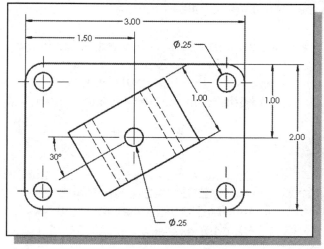

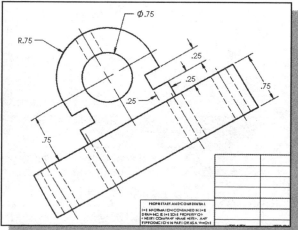

## Controlling the View and Sheet Scales

➤ The scale for the Sheet is set at 1:1. This setting automatically appears in the title block due to the Property Link described as *$PRP:"SW-Sheet Scale."* The prefix *$PRP:* defines the source as the current document. The Property Name is *SW-Sheet Scale*. This is a *System Property*. Each view can be set to use the default sheet scale or to use a custom scale defined for the view. When we created the top and auxiliary views, a custom scale of 3:2 was created. For the isometric view the default sheet scale of 1:1 was used. We will change the sheet scale to 3:2.

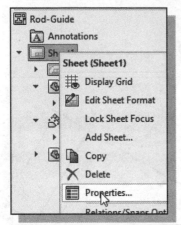

1. Move the cursor over **Sheet1** in the *Feature Manager Design Tree* and **right-click** to open the option menu.

2. Select **Properties**.

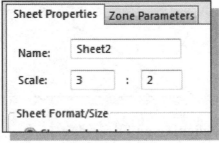

3. Enter **3:2** as the Scale in the *Sheet Properties* window.

4. Click **OK** in the *Sheet Properties* window, and choose Scale the annotations' position.

➤ Notice the scale note in the title block now reads **3:2**. Also notice the scale of the isometric view has changed due to the Use sheet scale option.

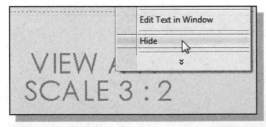

5. Since the sheet scale is labeled as **3:2** in the title block, the view scale note is not necessary. Right click on the note and select **Hide** from the pop-up option menu.

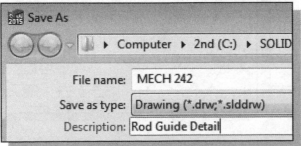

6. On your own, save the drawing file with **MECH 242** as the *File name* and **Rod Guide Detail** as the *Description*.

7. If the SOLIDWORKS pop-up window appears, click **Save All**.

## Completing the Drawing Sheet

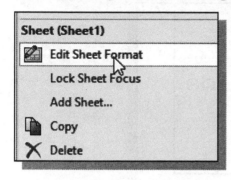

1. Move the cursor into the graphics area and **right-click** to open the pop-up option menu.

2. Select **Edit Sheet Format** from the pop-up option menu to change to *Edit Sheet Format* mode.

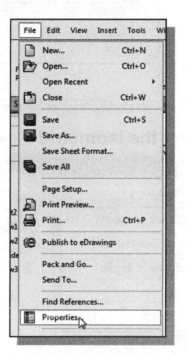

3. On your own, use the **Zoom** and **Pan** commands to adjust the display to work on the title block area. Most of the information is accurate due to **Property Links**.

4. To update some of the information in the title block, select **Properties** from the *File* pull-down menu on the *Menu Bar*.

5. In the *Summary Information* window select the **Custom** tab. Notice the *Custom Properties* appearing in the list. We defined these properties and saved them in the template file. (Note: Enter the custom properties in the table as shown.)

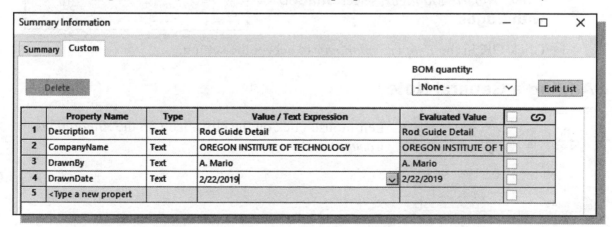

6. Change the DrawnDate Value to the current date.

7. Click **OK** in the *Summary Information* window to accept the change.

8.  On your own, correct any other entries to the title block if necessary.

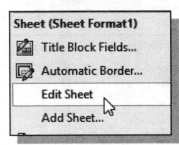

## Editing the Isometric view

1.  Move the cursor into the graphics area and **right-click** to open the pop-up option menu.

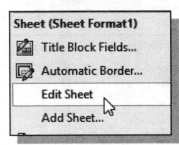

2.  Select **Edit Sheet** from the pop-up option menu to change to *Edit Sheet* mode.

3.  Left-click inside the *isometric* view to bring up the *Drawing View Property Manager*.

4.  In the *Display Style* panel of the *Drawing View Property Manager*, select **Shaded with Edges**.

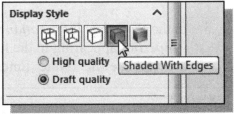

5.  Click **OK** in the *Property Manager* to accept the setting.

## Adding a General Note

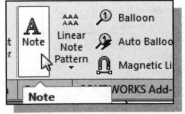

1.  Left-mouse-click on the **Note** icon on the *Annotations Command Manager*.

2.  On your own, create a general note at the lower left corner of the border as shown.

- On your own, save the final version of the drawing file.

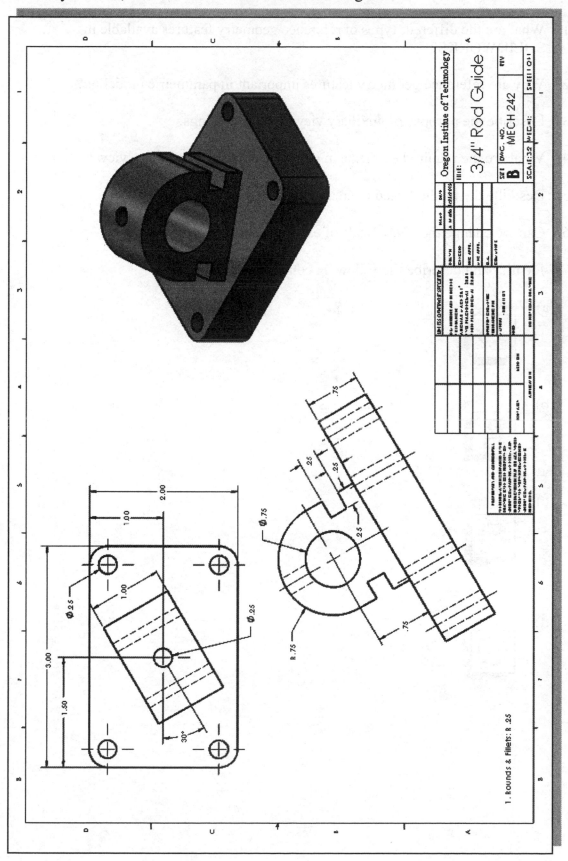

## Questions:

1. What are the different types of reference geometry features available in SOLIDWORKS?

2. Why are reference geometry features important in parametric modeling?

3. Describe the purpose of auxiliary views in 2D drawings.

4. What are the required elements in order to generate an auxiliary view?

5. Describe the method used to create centerlines in the chapter.

6. Can we change the View Scale of existing views? How?

7. Identify and describe the following commands:

   (a)

   (b)

   (c)

   (d)

**Exercises:** (Create the solid models and the associated 2D drawings.)

1.  **Rod Slide** (Dimensions are in inches.)

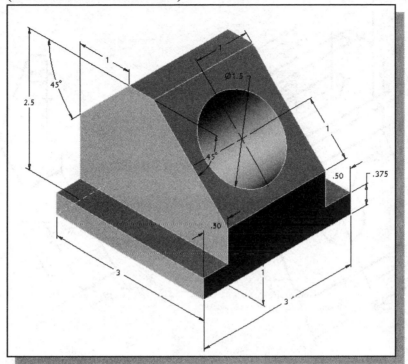

2.  **Angle Support** (Dimensions are in millimeters.)

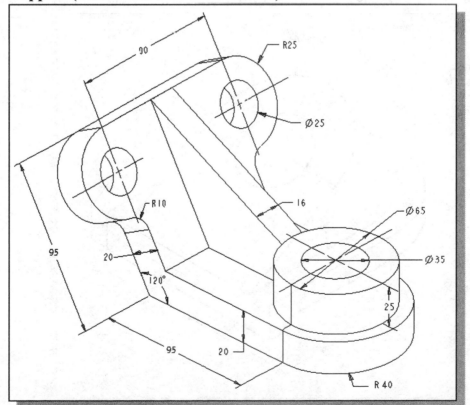

3. **Anchor Base** (Dimensions are in inches.)

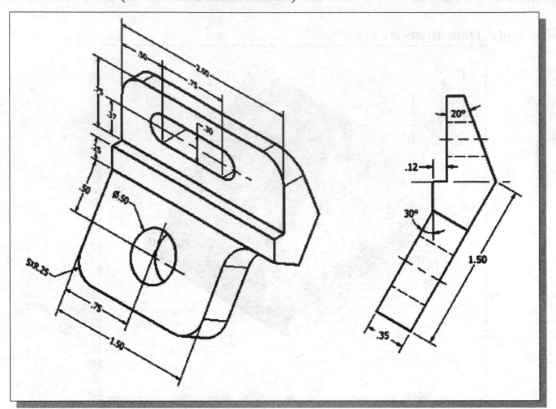

4. **Bevel Washer** (Dimensions are in inches.)

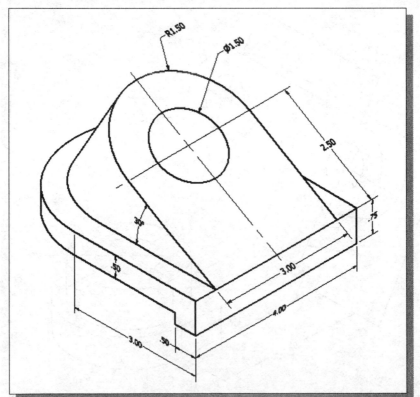

5. **Angle Bracket** (Dimensions arc in inches.)

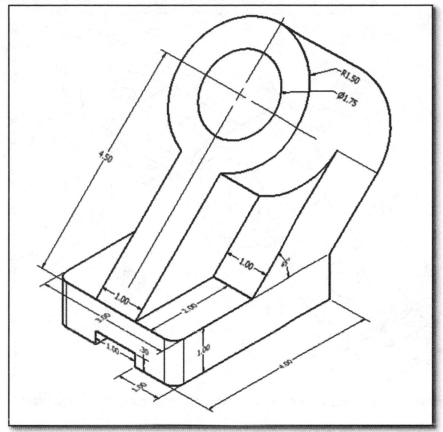

6. **Slider** (Dimensions are in inches.)

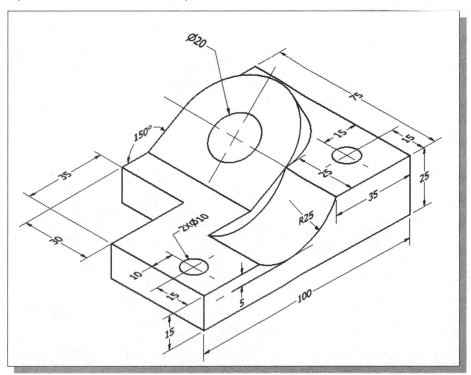

7. **Shaft Support** (Dimensions are in inches.)

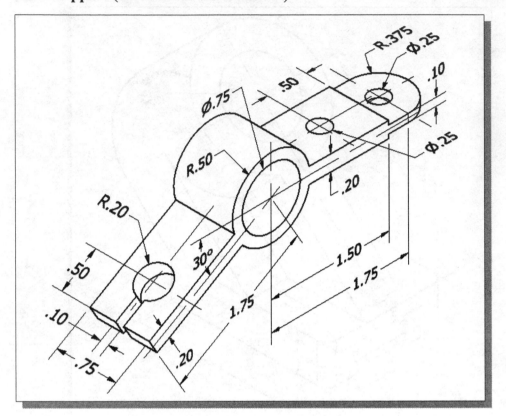

# Chapter 13
# Introduction to 3D Printing

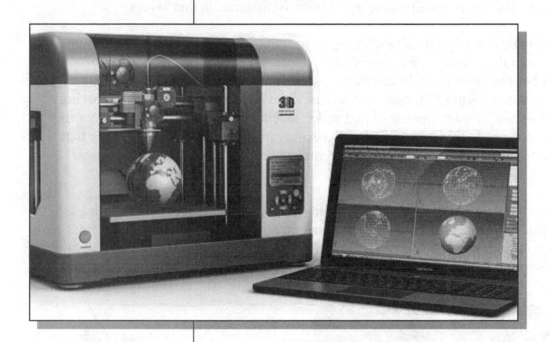

## Learning Objectives

- ♦ Understand the History and Development of 3D Printing
- ♦ Be aware of the Primary types of 3D Printing Technologies
- ♦ Be able to identify the commonly used Filament types for Fused Filament Fabrication
- ♦ Understand the general procedure for 3D Printing

## What is 3D Printing?

**3D Printing** is a type of *Rapid Prototyping* (RP) method.  Rapid prototyping refers to the techniques used to quickly fabricate a design to confirm/validate/improve conceptual design ideas. 3D printing is also known as **"Additive Manufacturing"** and construction of parts or assemblies is usually done by addition of material in thin layers.

Prior to the 1980s, nearly all metalworking was produced by machining, fabrication, forming, and mold casting; the majority of these processes require the removal of material rather than adding it. In contrast to *Additive Manufacturing* technology, the traditional manufacturing processes can be described as **Subtractive Manufacturing**. The term *Additive Manufacturing* gained wider acceptance in the 2010s. As the various additive processes continue to advance and become more mature, it is quite clear that material removal will no longer be the main manufacturing process in the very near future.

The basic principle behind *3D printing* is that it is an additive process. 3D printing is a radically different manufacturing method based on advanced technology that creates parts directly, by adding material layer by layer at the sub millimeter scale. One way to think about 3D Printing is the additive process is really performing **"2D printing over and over again."**

A number of limitations exist to the traditional manufacturing processes, which has been based on human labor and made by hand ideology, including expensive tooling, designing of fixtures, and the assembly of parts. The *3D printing* technology provides a way to create parts with complex geometric shapes quite easily using thin layers. The traditional *subtractive manufacturing* can also be quite wasteful as excess materials are cut and removed from large stock blocks, while the *3D printing* process only uses the material needed for the parts. *3D printing* is an enabling technology that encourages and drives innovation with unprecedented design freedom while being a tooling-less process that reduces costs and lead times. The relatively fast turnaround time also makes *3D printing* ideal for prototyping. Components with intricate geometry and complex features can also be designed specifically for *3D printing* to avoid complicated assembly requirements. *3D printing* is also an energy efficient technology that can provide better environmental friendliness in terms of the manufacturing process itself and the type of materials used for the product. There are quite a few different techniques to 3D print an object. *3D Printing* brings together two fundamental innovations: the manipulation of objects in the digital format and the manufacturing of objects by addition of material in thin layers.

The term **3D-printing** originally refers only to the smaller 3D printers with moveable print heads similar to an inkjet printer. Today, the term **3D-printing** is used interchangeably with **Additive Manufacturing**, as both refer to the technology of creating parts through the process of adding/forming thin layers of materials.

## Development of 3D Printing Technologies

The earliest 3D printing technology was first invented in the 1980s; at that time it was generally called **Rapid Prototyping** (**RP**) technology. This is because the process was originally conceived as a fast and time-effective method for creating prototypes for product development in industry. In 1981, Dr. Hideo Kodama of *Nagoya Municipal Industrial Research Institute* invented two methods of creating three-dimensional plastic models with photo-hardening polymer through the use of a UV Laser. In 1986, the first US patent for a **stereolithography** apparatus (**SLA**) was issued to Charles Hull, who first invented his SLA machine in 1983. Chuck Hull went on to co-found *3D Systems Corporation*, which is one of the largest companies in the 3D printing sector today. Chuck Hull also designed the **STL** (**ST**ereo**L**ithography) file format, which is widely used by 3D printing software performing the digital slicing and infill strategies common to the additive manufacturing processes. The first available commercial RP system, the **SLA-1** by *3D Systems* (as shown in the figure below), was made available in 1987.

The 1980s also mark the birth of many RP technologies worldwide. In 1989, Carl Deckard of *University of Texas* developed the **Selective Laser Sintering (SLS)** process. In 1989, Scott Crump, one of the founders of *Stratasys Inc.*, also created the **Fused Deposition Modeling (FDM)**. In *Europe*, Hans Langer started *EOS GmbH* in Germany; the company focuses on the **Laser Sintering (LS)** process. The *EOS systems Corp.* also developed the **Direct Metal Laser Sintering (DMLS)** process. Today, *3D Systems*, *EOS* and *Stratasys* are still the main leaders in the *Additive Manufacturing* industry.

During the 1990s, the *3D printing* sector started to show signs of distinct diversification with two specific areas of emphasis which are much more clearly defined today. First, there was the high end of 3D printing, still very expensive systems, which were geared towards part production for relatively complex designs. For example, in 1995, *Sciaky Inc.* developed an additive welding process based on its proprietary **Electron Beam Additive Manufacturing (EBAM)** technology. Many *RP* system companies, such as *Solidscape*, *ZCorporation*, *Arcam* and *Objet Geometries* were all launched in 1990s. At the other end of the spectrum, some of the 3D printing system manufacturers started to develop smaller desktop systems in the 1990s.

The idea of creating low-cost desktop 3D printers also intrigued many technology professionals and hobby enthusiasts during the late 1990s. In 2004, a retired professor, Dr. Adrian Bowyer (person on the left in the below photo), started the **RepRap** (*Replication Rapid-Prototyper*) project of an open source, self-replicating 3D printer (**RepRap 1.0 - Darwin**). This set the stage for what was to come in the following years. It was around 2007 that the open source *3D printing* movement started gaining visibility and momentum. In January of 2009, the first commercially available open source 3D printer, the **BFB RapMan** 3D printer, became available. *Makerbot Industries* also came out with their **Makerbot** 3D printer in April of 2009. Since then, a host of low-cost desktop 3D printers have emerged each year.

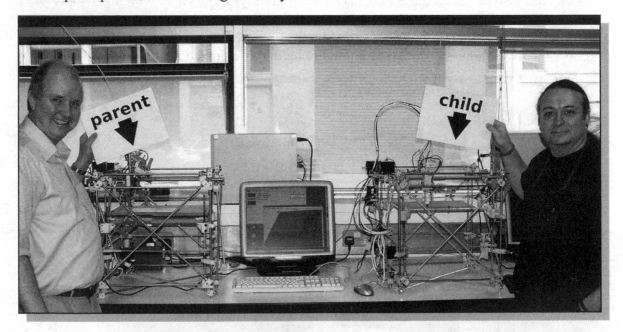

In the beginning of the 2010s, alternative 3D printing processes, such as using the **Polymer Resins** material, became available at the desktop level of the market. The **B9 Creator** by *B9Creations*, using **Digital Light Processing (DLP)** technology, came first in June of 2012, followed by the **Form 1** desktop printer by *Formlabs Inc.* Both 3D printers were launched via KickStarter's crowd-funding website and both enjoyed huge success. 2012 was also the year that many different mainstream media noticed the exciting 3D printing technology, which dramatically increased awareness and uptake to the general public. 2013 was also a year of significant growth and consolidation. One of the most notable moves was the acquisition of Makerbot by Stratasys. In 2016, the new developments in 3D printing concentrate more on multi-color, multi-material using single or multiple extruders and new technologies to shorten the 3D printing time.

As a result of the market divergence, the price of desktop 3D printers continues to go down each year. Today, very capable fully assembled desktop 3D printers, such as *Robo3D R1+, Prusa I3 MK2*, can be acquired for under $1000. Fully assembled smaller desktop 3D printers, such as Xyz-printing's *DA Vinci mini 3D printer* and M3D's *Micro 3D* can be acquired for less than $350. Unassembled desktop 3D printer kits can even be acquired for under $200.

Another trend that happened in the 2010s is the availability of **3D printing Services**. 3D printing services are growing quite rapidly in the US. For example, many public libraries, especially in California, are now providing 3D printing services to the general public and **UPS** started its worldwide 3D printing services in May of 2016. This trend is spreading throughout the US, with many more companies planning to provide 3D printing services in the very near future. It is now quite feasible, and perhaps more economical, to 3D print designs without owning or ever touching a 3D printer, but understanding of the technology is still needed to increase productivity.

As the exponential adoption rate continues on all fronts, more and more technologies, materials, applications, and online services will continue to emerge. It is predicted that the development of 3D printing will continue in the years to come and 3D printing will eventually become the mainstream manufacturing method in industries and in homes.

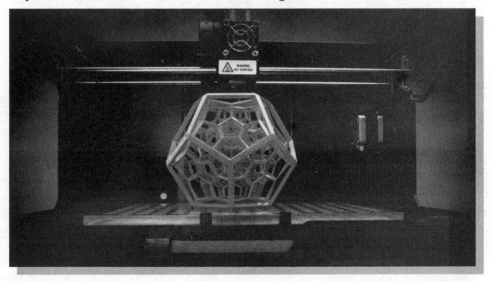

# Primary types of 3D Printing processes

There are quite a few different techniques to 3D print an object. The different types of 3D printers each employ a different technology that processes different materials in different ways. For example, some 3D printers process powdered materials (nylon, plastic, ceramic and metal) which utilize a light/heat source to sinter/melt/fuse layers of the powder together in the defined shape. Others process polymer resin materials and again utilize a light/laser to solidify the resin in thin layers. **Stereolithography (SLA** or **SL), Fused Deposition Modeling (FDM** or **FFF)** and **Laser Sintering (LS** or **SLS)** represent the three primary types of 3D printing processes; the majority of the other 3D printing technologies are variations of the three main types.

## Stereolithography

Stereolithography (**SLA** or **SL**) is widely recognized as the first 3D printing process; it was certainly the first to be commercialized. *SLA* is a laser-based process that works with photopolymer resins. The photopolymer resins react with the laser and cure to form a solid in a very precise way to produce very accurate parts. It is a complex process, but simply put, the photopolymer resin is held in a container with a movable platform inside. A laser beam is directed in the X-Y axes across the surface of the resin according to the 3D data supplied to the machine. The resin hardens precisely as the laser hits the designated area. Once the current layer is completed, the platform within the container drops down by a fraction (in the Z axis) and the subsequent layer is traced out by the laser. This 2D layer tracing continues until the entire object is completed.

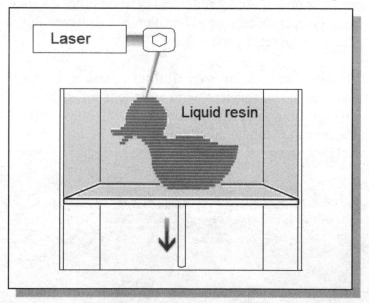

Because of the nature of the SLA process, support structures are needed for some parts, specifically those with overhangs or undercuts. These support structures need to be removed once the part is created. Many 3D printed objects using SLA need to be further cleaned and/or cured. Curing involves subjecting the part to intense light in an oven-like machine to fully harden the resin. SLA is generally accepted as being one of the most accurate 3D printing processes with excellent surface finish.

**Fused Deposition Modeling (FDM) / Fused Filament Fabrication (FFF)**

*3D printing* utilizing the extrusion of thermoplastic material is probably the most popular 3D printing process. The original name for the process is **Fused Deposition Modeling (FDM)**, which was developed in the early 1990s and is a trade name registered by *Stratasys*. However, a similar process, **Fused Filament Fabrication (FFF)**, has emerged since 2009. The majority of desktop 3D printers, both open source and proprietary, utilize the FFF process that is in a more basic extrusion form of FDM.

The FDM and FFF processes work by melting plastic filament that is deposited, via a heated extruder, one layer at a time, onto a build platform according to the 3D data supplied to the 3D printer. Each layer hardens as it cools down and bonds to the previous layer.

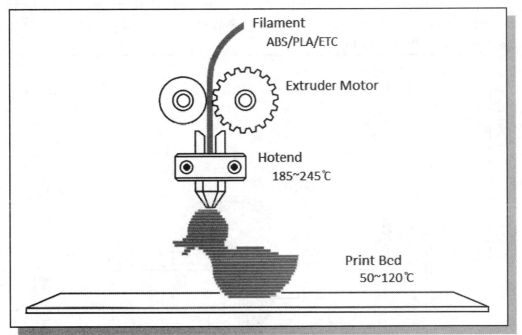

*Stratasys* has developed a range of proprietary industrial grade materials for its FDM process that are suitable for production applications. However, the most common materials for both FDM and FFF 3D printers are **ABS (Acrylonitrile Butadiene Styrene)** and **PLA (Polylactic Acid)**. The FDM and FFF processes require support structures for any applications with overhanging geometries. This generally entails a second, typically water-soluble or breakaway material, which allows support structures to be easily removed once the print is complete.

The FDM and FFF printing processes can be slow for large parts or parts with complex geometries. The layer to layer adhesion can also be a problem, resulting in parts that warp or separate easily. The surface finish of FDM and FFF printed parts might appear a bit rough as the thin layers are generally visible. To improve the appearance, several options are feasible, such as using Acetone, Sanding and/or Spray paint.

## Laser Sintering / Laser Melting

**Laser Sintering (LS)** or **Selective Laser Sintering (SLS)** creates tough and geometrically intricate parts using a high-powered $CO_2$ laser to fuse/sinter/melt powdered thermoplastics. The main advantage of SLS *3D printing* is that as a part is made, it remains encased in powder; this eliminates the need for support structures and allows for very complex 3D geometries to be 3D printed. SLS can be used to produce very strong parts as exceptional materials such as Nylon and metal powders are commonly used.

Laser sintering refers to a laser-based 3D printing process that works with powdered materials. The laser is traced across a powder bed of tightly compacted powdered material, according to the 3D data provided to the machine, in the X-Y axes. As the laser interacts with the powdered material it sinters and fuses the particles to each other forming a solid. As each layer is completed the powder bed drops incrementally and a roller is used to compact the powder over the top surface of the bed prior to the next pass of the laser for the subsequent layer.

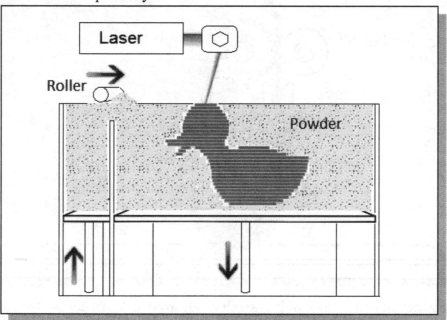

The build chamber is completely sealed as it is necessary to maintain a precise temperature during the process specific to the melting point of the powdered material of choice. One of the key advantages of this process is that the powder bed serves as an in-process support structure for overhangs and undercuts, and therefore complex shapes that could not be manufactured in any other way become possible with this process. Because of the high temperatures required for laser sintering, cooling can take a long time. Porosity is also a common issue with this process; an additional metal infiltration process may be required to improve mechanical characteristics.

Laser sintering can process plastic and metal materials, although metal sintering does require a much higher powered laser and higher in-process temperatures. Parts produced with this process are much stronger than parts made with SLA or FDM, although generally the surface finish and accuracy is not as good.

## Primary 3D Printing Materials for FDM and FFF

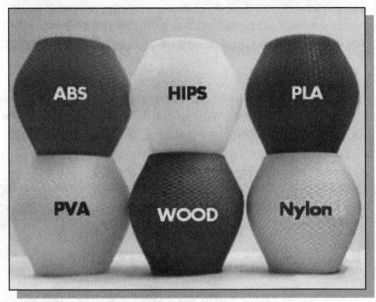

### ABS (Acrylonitrile Butadiene Styrene)

*ABS* is a popular choice for 3D printing. It is a strong thermoplastic that is among one of the most widely used plastics. It is tough with mild flexibility, making it more durable to stress and has a higher heat resistance of up to 200 degree Fahrenheit. However, this material has a tendency to shrink, which can affect the accuracy of designs. ABS has a pretty high melting point, and can experience warping if cooled while printing. Because of this, ABS objects are printed typically on a heated surface. ABS also requires ventilation when in use, as the fumes can be unpleasant. The aforementioned factors make ABS printing difficult for hobbyist printers, though it's the preferred material for professional applications.

### PLA (Polylactic Acid)

*PLA* is a staple and it is becoming one of the most popular choices for 3D printing with good reason. Aside from the fact that it is a biodegradable thermoplastic derived from renewable resources such as corn starch, tapioca roots, chips or starch, or sugarcane, *PLA* is a very rigid material that is easy to use for 3D printing and it is able to withstand a good amount of impact and weight. It also has a glossier finish than ABS and in most scenarios PLA is the preferred material for 3D printing large objects. The main disadvantage of PLA is it's not as heat resistant as ABS, so it should not be placed in environments that exceed 140 degrees Fahrenheit.

### Flexible (Thermoplastic Elastomer)

*Flexible material* is for applications that require incredible rubbery flex in their applications. Flexible filament goes beyond bending; it is more like rubber. When it comes to Flexible filament, it's all about finding a balance between flexibility (softness) and printability. This softness is sometimes indicated with a *Shore* value (like 85A or 60D). Higher Shore value means less flexibility. Harder filaments (less flexible) are easier to 3D print when compared to softer, more flexible filaments.

### PETG (Polyethylene Terephthalate)

*PETG* is a material that is similar to *PLA*, with more attractive characteristics such as being generally a tougher and denser material, and good heat resistance of up to 190 degrees Fahrenheit. It claims to have the strength of *ABS*, while printing as easily as *PLA*.

### HIPS (High Impact Polystyrene) and PVA (Polyvinyl Alcohol)

*HIPS* and *PVA* are relatively new materials that are growing in popularity for their dissolvable properties. They are used for creating support material. Their ability to dissolve under certain liquids means that they can be easily removed. These materials can be hard to print with because they don't stick well to the build plates. Be sure to not print *PVA* too hot either, as it can turn into tar and jam the extruder.

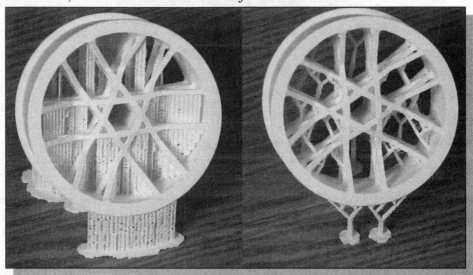

### Wood Fiber

*Wood Fiber* filament contains a mixture of recycled wood and binding polymer. Thus, a 3D printed object can look and smell like real wood. Due to its wooden nature, it's difficult to tell that the object is 3D printed. Using Wood filament is similar to using a thermoplastic filament like ABS or PLA. However, a 3D object having a wooden-like appearance can be created with this material.

# From 3D model to 3D printed Part

To create a 3D printed part, it all starts with making a virtual design of the object. This virtual design may be created with a computer-aided design (CAD) package, via a 3D scanner, or by a digital camera and photogrammetry software. 3D scanning and photogrammetry software can process the collected digital data on the shape and appearance of a real object and create a digital 3D model. The 3D virtual design can generally be modified with 3D CAD packages, allowing verification of the virtual design before it is 3D printed.

Once the virtual design is verified, the 3D data will then be transferred to the 3D printing software. There is a multitude of file formats that 3D printing software supports. However, the most popular are the STL file format and the OBJ file format. The STL file format is the most commonly used file format for 3D printing. Most CAD software has the capability of exporting models in the STL format. The STL file contains only the surface geometry of the modeled object. The OBJ file format is considered to be more complex than the STL file format as it is capable of displaying texture, color and other attributes of the three-dimensional object. However, the STL file format holds the top spot for 3D printing, as this file format is simpler to use, and most CAD packages work better with STL files than OBJ files.

Once the 3D data of the virtual design is transferred into the 3D printing software, further examination and/or repair can be performed if necessary. The 3D printing software will also process the imported 3D data by the special software known as a **Slicer**, which converts the model into a series of thin layers and produces a G-code file containing instructions tailored to a specific type of 3D printer. G-code is the common name for the most widely used numerical control (NC) programming language. It is used mainly in computer-aided manufacturing to control automated machine tools. The generated G-code file can be sent to the 3D printer and create the 3D printed part.

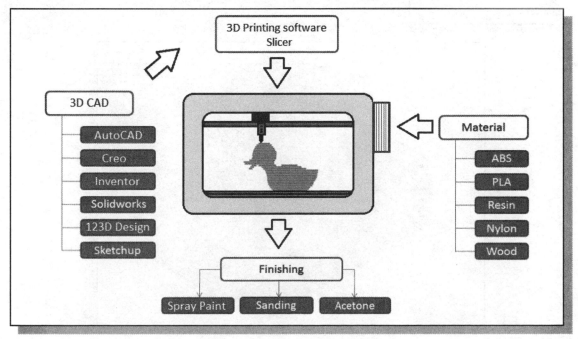

## Starting SOLIDWORKS

1. Select the **SOLIDWORKS** option on the *Start* menu or select the **SOLIDWORKS** icon on the desktop to start SOLIDWORKS. The SOLIDWORKS main window will appear on the screen.

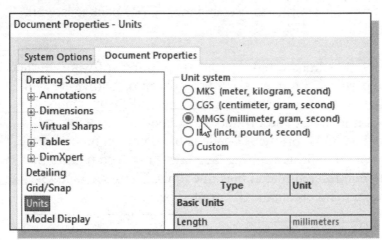

2. In the SOLIDWORKS *Startup* dialog box, select the **Rod-Guide.SLDPRT** file and select **Open a Drawing** with a single click of the left-mouse-button. Use the *browser* to locate the file if it is not displayed in the *File name* list box.

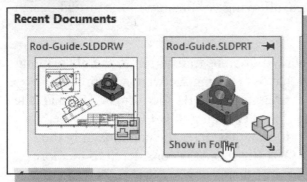

3. On your own, change the *Units* option to millimeter as shown. Note that the majority of the 3D printer settings are measured in millimeters, such as filament diameter, layer height and extruder size.

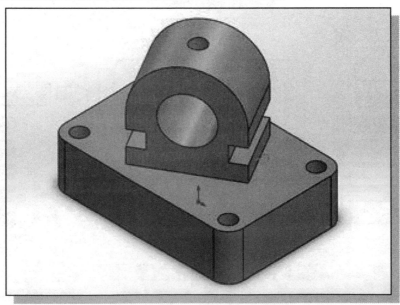

## SOLIDWORKS Print3D command

3D printers will generally accept a 3D model with the STL or OBJ file formats. SOLIDWORKS offers two options to saving the 3D model in STL format: 1. Using the SOLIDWORKS' **Print3D** command or 2. Using the **Save As** option. The Save As option can be used to very quickly export the 3D model, while the *Print3D* command provides more control review options.

1. In the *File Toolbar,* select the **Print3D** command as shown.

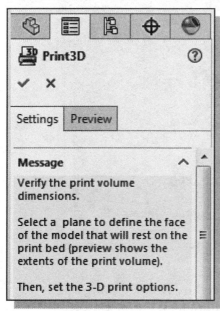

2. In the *Print3D* dialog box, SOLIDWORKS expects us to select a plane on the model that will be used to define the orientation of the 3D model to be aligned to the print bcd of the 3D printer.

3. Select the bottom surface of the Rod-Guide design as shown.

4.  In the graphics window, the orientation of the 3D model is set based on the selected surface. The larger grey box indicates the max print volume of the 3D printer being used; these dimensions can be adjusted in the dialog box as shown.

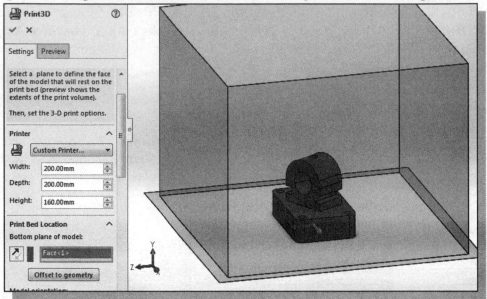

5.  Additional orientation and scale settings are available in the *Print3D* dialog box. On your own, adjust some of the settings to see the effects of the adjustments.

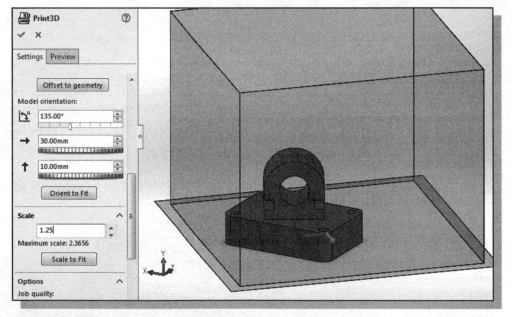

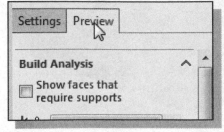

6.  Click **Preview** in the *Print3D* dialog box to switch to the additional 3D print validation section.

- In SOLIDWORKS 2019, the print validation section is now available under the Print3D **Preview** tab. The new print validation section provides additional tools to examine the 3D print model before it is actually printed, thus a **Preview**.  Three options are available: 1. **Show faces that require supports**: this option will show us if any surface needs additional supports. This is typically needed for overhang features. Note that *supports* can be added in the 3D printing/Slicer software.  2. **Show striation lines**: this option can be used to determine whether the print resolution is sufficiently fine to produce the desired output. Adjusting the layer height will change the appearance of the 3D print. However, the layer height is mainly determined by the specific 3D printer in use; check your 3D printer's specs before lowering the layer height. 3. **Thickness/Gap Analysis**: one of the most common causes of a failed 3D print is because there are features in the model that are too small to print, or gaps too small to be recognized. To help prevent these failed builds, there is the *Thickness/Gap analysis check*. This check is particularly useful when scaling down a model to fit on the 3D printer. Small features and gaps can easily be overlooked when a model is scaled down. An additional benefit comes if you are not sure what value of thickness or gap to check for. For FDM/FFF 3D printers, SOLIDWORKS provides a list of materials with ideal wall thicknesses allowable based on the layer height. If the material to be used is not in the list, the Custom Thickness and Gap check box can be used to indicate specific values. This option can be used to quickly check the geometry and SOLIDWORKS will highlight any unprintable features, upfront and before sending the job to the machine. This could save hours of build time and also a lot of material.

7.  In the *Print3D* dialog box, switch **on** the **Show faces that require supports** option. Note that SOLIDWORKS indicates the 3 overhang surfaces of the Rod Guide design will require additional *supports* for the 3D print.

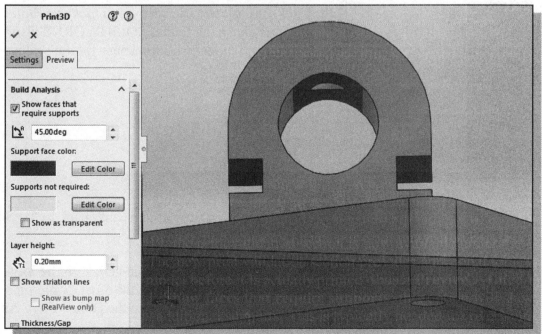

8. Turn on the **Show striation lines** option to view the smoothness of the model using the current layer height. (Hint: Zoom in to see the striation lines.)

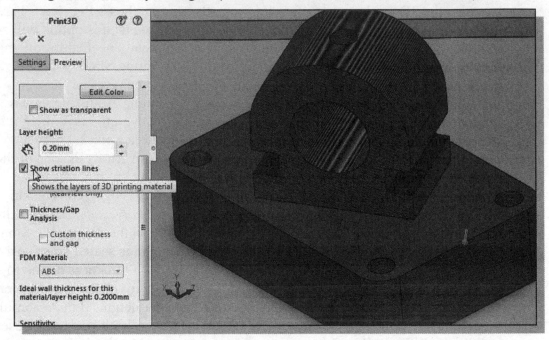

9. On your own, adjust the layer height to 0.1mm and examine the difference in smoothness of the previewed 3D print model by reducing the layer height.

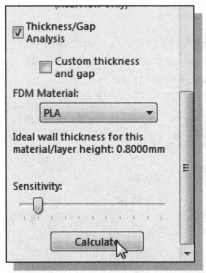

10. Turn on the **Thickness/Gap analysis** option and set the FDM material to PLA.

11. Click **Calculate** to perform the *Thickness/Gap analysis check*. Note that SOLIDWORKS also indicates the ideal wall thickness for the selected material is 0.8mm.

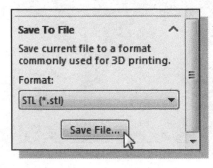

12. In the *Print3D* dialog area, switch back to the **Settings** tab.

13. Set the file format to STL and click on the **Save File** button as shown. Note that the *Save As dialog box* appears; this is redirected to the [**File→Save As**] command.

14. In the **Save As** dialog box, click on the **Options** button as shown.

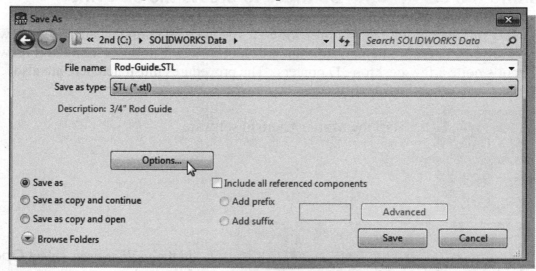

15. Note that SOLIDWORKS automatically sets the file format to STL as shown.

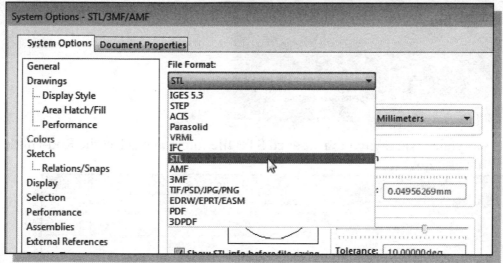

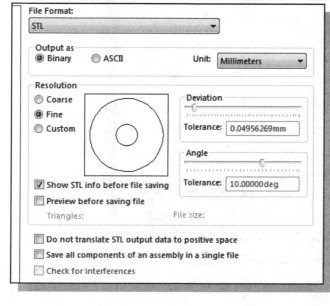

16. On your own, confirm the *output format* is set to **Binary** and *Resolution* to **Fine** as shown.

17. Click **OK** and then **Save** to proceed with saving the file.

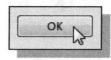

18. Click **Yes** and **OK** to proceed with saving the Rod-Guide.STL file.

## Using the 3D Printing software to create the 3D Print

To 3D print the model, we will open the STL file in the 3D printing software. We will use **Matter Control** to demonstrate the procedure. Note that *Matter Control* (Freeware) supports quite a few desktop 3D printers. The procedures illustrated here are also applicable to other similar software.

1.  Start the **Matter Control** software.

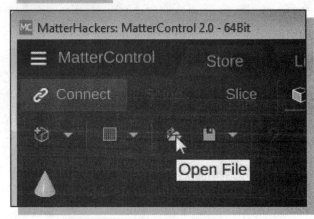

2.  In the *toolbar area*, select **Open File** as shown.

3.  On your own, switch to the saved STL file folder and select the Rod-Guide.STL file as shown.

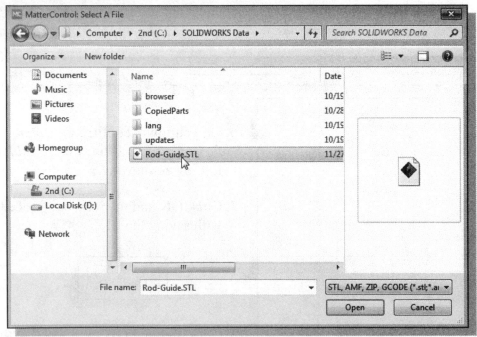

4.  Click **Open** to import the STL file in to Matter Control.

5. Once the STL file is imported into the program, the STL model is displayed in the *View* window. Note that the model is imported with the incorrect orientation of the model; the bottom side of the model is not aligned to the print bed.

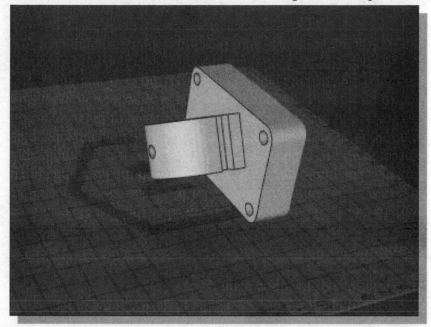

6. To adjust the display of the model, use the three mouse button to perform the **Pan,** Zoom and Rotate functions.

7. Note that the View Cube, located on the right side of the graphics window, is also available to control the viewing direction of the print bed.

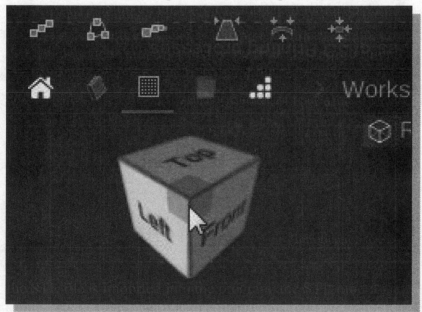

8. Click on the model, with the left mouse button, to enter the *Edit mode* and click & drag the model to reposition the model on the print-bed.

9. To rotate the mode, click on the **Rotate icon** to enter the *Rotate control*; note the associated dial allows more precise rotation.

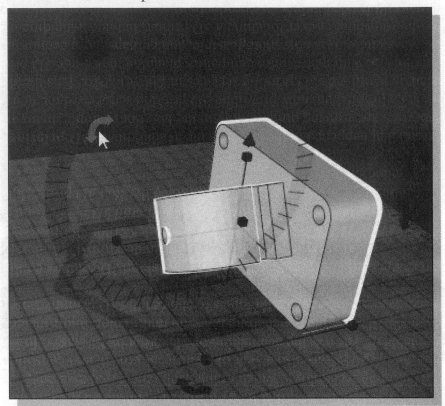

10. On your own, rotate the model so that it is setting vertically as shown.

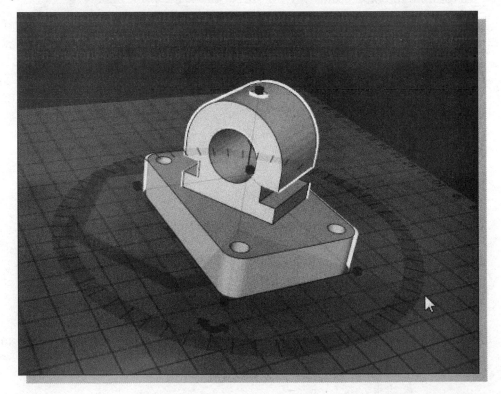

11. Note that additional *Editing tools*, such as **Scale** and **Mirror**, arc also available through the right-mouse click on the model as shown.

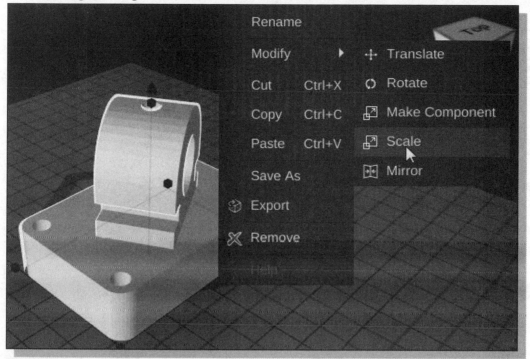

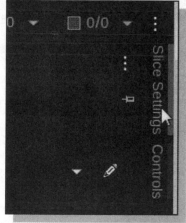

12. In the toolbar area, click **Lay Flat** to align the bottom surface of the model to the print bed.

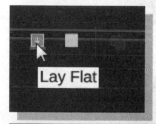

13. Click **Slice Settings** to review the 3D Printing settings. Note that the **Control tab** contain commands to directly control the movements of the 3D printer.

14. Under the **Slice Settings tab,** different settings are available to adjust the 3D printing settings.

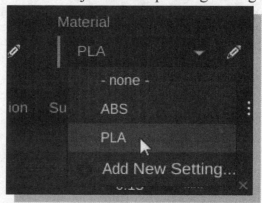

15. Under the **General** tab, a list of layer settings are available, such as the layer thickness, Top and Bottom Solid Layers thickness and the Infill type.

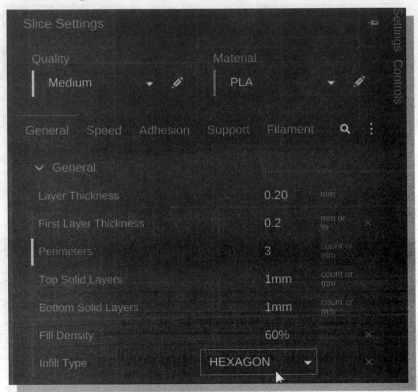

16. Under the **Speed** tab, the list of different speed settings on 3d printing are displayed and can be adjusted.

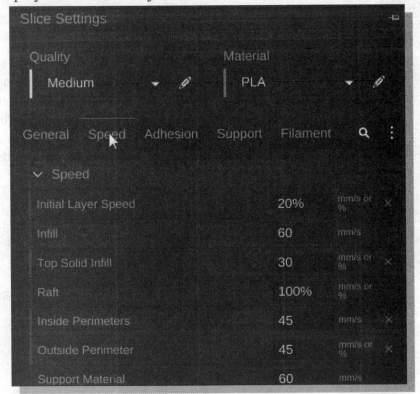

17. Under the **Adhesion** tab, the list of settings on improving adhesion on the first layer, such as Skirt, Raft and Brim, are available.

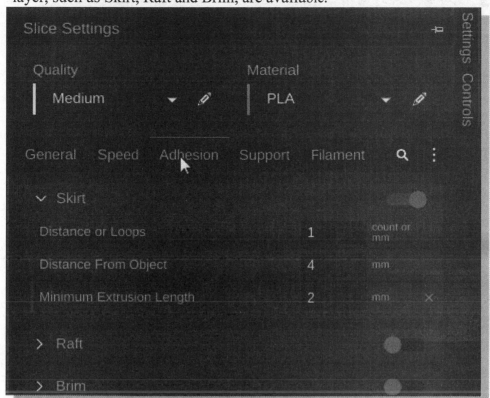

18. Under the **Support tab,** the adding support option can be turned on to *generate Support Material* as shown.

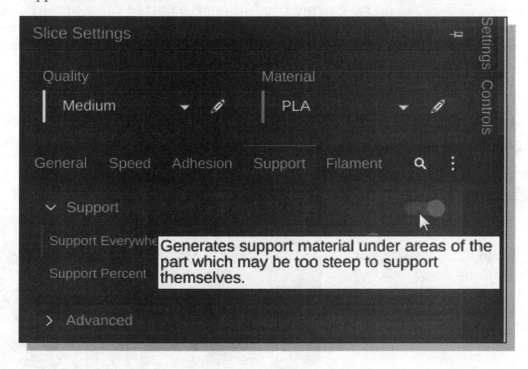

19. Switch to the **Filament tab** and in the **Material list** confirm/modify the filament properties, such as the diameter, to match the actual filament being used.

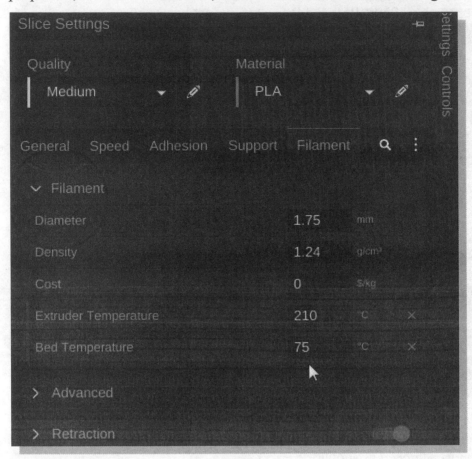

20. The temperature for the **Extruder** and the print bed can be adjusted on this page as well. Note that the temperatures are different based on the types of filaments, as well as the different brands and the type of printer bed. It is necessary to do some testing and/or experimenting when a new roll of filament is used.

21. In the toolbar area, click **Slice** to process the 3D model, which includes slicing and generating the associated G-code for the specific 3D printer.

22. Depending on the size and complexity of the design, it might take several minutes to complete the process.

23. On your own, drag the vertical slider to review the thin layers generated by the slicer.

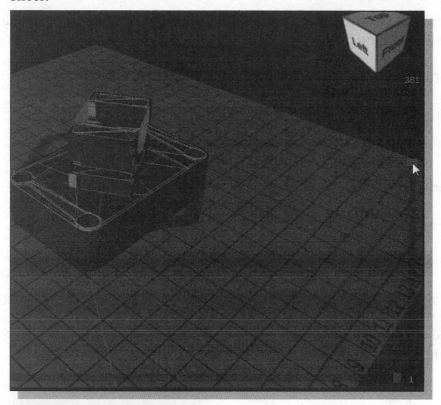

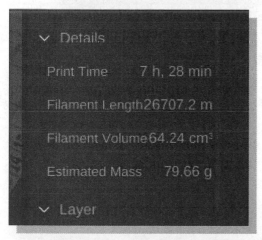

24. Note that with the current settings, it will take 7 hours and 28 minutes to complete the print using 26707.2 mm of filament and the volume of the printed part is 79.66gram.

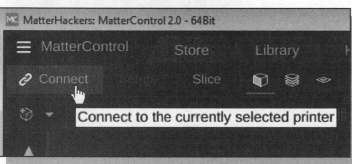

25. To start the 3D print, switch on the 3D printer and click Connect → Print to start the printing of the 3D model.

26. *Matter Control* will now begin printing.

## Questions:

1. What is the main difference between the **Additive Manufacturing** and the traditional **Subtractive Manufacturing** technologies?

2. Which 3D printing process is recognized as the first 3D printing process?

3. Describe the general procedure to create a 3D printed part.

4. What are the three primary types of 3D printing processes?

5. Which 3D printing process is the most popular 3D printing process?

6. What is the main advantage of using PLA over ABS for FFF process?

7. Which are the most popular file formats for 3D printing?

8. What is the main function of a **Slicer** program??

9. List and describe the print validation options available with the SOLIDWORKS **Print3D** command.

# Chapter 14
# Threads and Fasteners

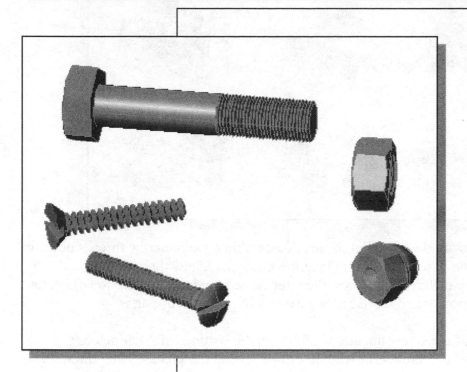

## Learning Objectives

♦ **Understand Screw-Threads Terminology**
♦ **Define and Label the Different Parts of a Screw Thread**
♦ **Draw Detailed, Schematic and Simplified Threads in Section and Standard Views**
♦ **Use the Fastener Clearance Fits**
♦ **Identify Various Fasteners and Describe Their Use**
♦ **Use the Design Library to create Fasteners**

## Introduction

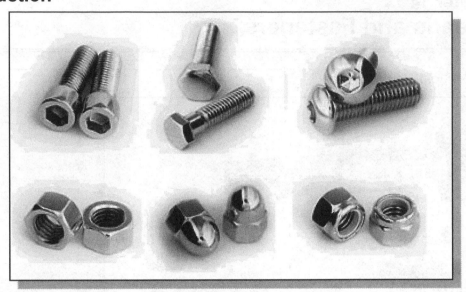

Threads and fasteners are the principal means of assembling parts. Screw threads occur in one form or another on practically all engineering products. Screw threads are designed for many different purposes; the three basic functionalities are to hold parts together, to transmit power and for use as adjustment of locations of parts.

The earliest records of the screw threads are found in the writings of Archimedes (278 to 212 B.C.), the mathematician who described several designs applying the screw principle. Screw thread was a commonly used element in the first century B.C. but was crudely made during that period of time. Machine production of screws started in England during the Industrial Revolution and the standardization of the screw threads was first proposed by Sir Joseph Whitworth in 1841. His system was generally adopted in England but not in the United States during the 1800s. The initial attempt to standardize screw threads in the United States came in 1864 with the adoption of the thread system designed by William Sellers. The "Sellers thread" fulfilled the need for a general-purpose thread, but it became inadequate with the coming of modern devices, such as automobiles and airplanes. Through the efforts of various engineering societies, the National Screw Thread Commission was authorized in 1918. The Unified Screw Thread, a compromise between the American and British systems, was established on November 18, 1948.

The Metric fastener standard was established in 1946, through the cooperative efforts of several organizations: The International Organization for Standardization (ISO), the Industrial Fasteners Institute (IFI) and the American National Standards Institute.

The following sections describe the general thread definitions; for a more complete description refer to the ANSI/ASME standards B1.1, B1.7M, B1.13M, Y14.6 and Y14.6aM. (The letter M is for metric system.)

# Screw-Thread Terminology

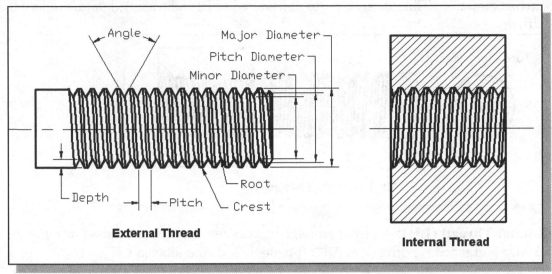

**Screw Thread (Thread)**: A ridge of a uniform section in the form of a helix on the external or internal surface of a cylinder or cone.

**External Thread**: A thread on the external surface of a cylinder or cone.

**Internal Thread**: A thread on the internal surface of a cylinder or cone.

**Major Diameter**: The largest diameter of a screw thread.

**Minor Diameter**: The smallest diameter of a screw thread.

**Pitch Diameter**: The diameter of an imaginary cylinder, the surface of which cuts the thread where the width of the thread and groove are equal.

**Crest**: The outer edge or surface that joins the two sides of a thread.

**Root**: The bottom edge or surface that joins the sides of two adjacent threads.

**Depth of Thread**: The distance between crest and root measured normal to the axis.

**Angle of Thread**: The angle included between the two adjacent sides of the threads.

**Pitch**: The distance between corresponding points on adjacent thread forms measured parallel to the axis. This distance is a measure of the size of the thread form used, which is equal to 1 divided by the number of threads per inch.

**Threads per Inch**: The reciprocal of the pitch and the value specified to govern the size of the thread form.

**Form of Thread**: The profile (cross section) of a thread. The next section shows various forms.

**Right-hand Thread (RH)**: A thread which advances into a nut when turned in a clockwise direction. Threads are always considered to be right-handed unless otherwise specified.

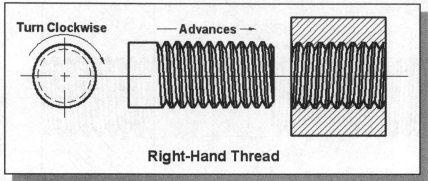

**Left-hand Thread (LH)**: A thread which advances into a nut when turned in a counter-clockwise and receding direction. All left-hand threads are labeled **LH**.

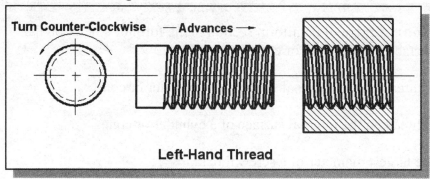

**Lead**: The distance a threaded part moves axially, with respect to a fixed mating part, in one complete revolution. See the definition and examples of multiple thread in the below figures.

**Multiple Threads**: A thread having two or more helical curves of the cylinder running side by side. For a single thread, lead and pitch are identical values; for a double thread, lead is twice the pitch; and for a triple thread, lead is three times the pitch. A multiple thread permits a more rapid advance without a larger thread form. The *slope line* can be used to aid the construction of multiple threads; it is the hypotenuse of a right triangle where the short side equals .5P for single threads, P for double threads and so on.

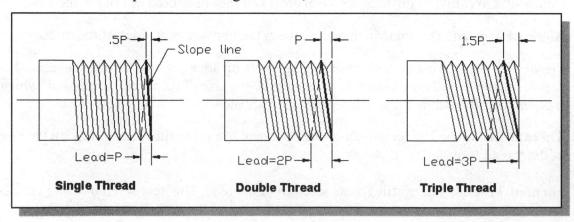

# Thread Forms

Screw threads are used on fasteners to fasten parts together, on devices for making adjustments, and for the transmission of power and motion. For these different purposes, a number of thread forms are in use. In practical usage, clearance must be provided between the external and internal threads.

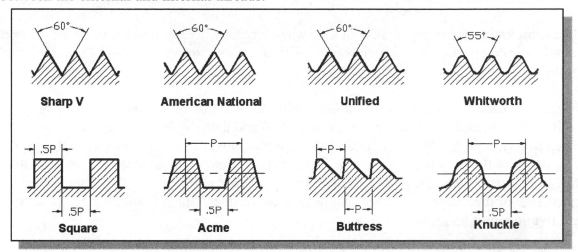

The most common screw thread form is the one with a symmetrical V-Profile. The **Sharp V** is rarely used now, because it is difficult to maintain the sharp roots in production. The form is of interest, however, as the basis of more practical V-type threads; also, because of its simplicity, it is used on drawings as a conventional representation for other (V-profile) threads. The modified V-profile standard thread form in the United States is the **American National**, which is commonly used in fasteners. The V-profile form is prevalent in the **Unified** Screw Thread (UN, UNC, UNF, UNEF) form as well as the ISO/Metric thread. The advantage of symmetrical threads is that they are easier to manufacture and inspect compared to non-symmetrical threads.

The **Unified Screw Thread** is the standard of the United States, Canada, and Great Britain and as such is known as the **Unified** thread. Note that while the crest may be flat or rounded, the root is rounded by design. These are typically used in general purpose fasteners.

The former British standard was the **Whitworth**, which has an angle of 55° with crests and roots rounded. The British Association Standard that uses an angle of 47°, measured in the metric system, is generally used for small threads. The French and the International Metric Standards have a form similar to the American National but are in the metric system.

The V shapes are not desirable for transmitting power since part of the thrust tends to be transmitted to the side direction. **Square thread** is used for this purpose as it transmits all the forces nearly parallel to the axis. The square thread form, while strong, is harder to manufacture. It also cannot be compensated for wear unlike an **Acme** thread. Because of manufacturing difficulties, the square thread form is generally modified by providing a slight taper (5°) to the sides.

The **Acme** is generally used in place of the square thread. It is stronger, more easily produced, and permits the use of a disengaging or split nut that cannot be used on a square thread.

The **buttress**, for transmitting power in one direction, has the efficiency of the square and the strength of the V thread.

The **knuckle** thread is especially suitable when threads are to be molded or rolled in sheet metal. It is commonly used on glass jars and in a shallow form on bases of ordinary light bulbs.

Internal threads are produced by cutting, while external threads are made by cutting or rolling. For internal threads, a hole is first drilled and then the threads are cut using a tap. The tap drill hole is a little bigger than the minor diameter of the mating external thread. The depth of the tap drill is generally deeper than the length of the threads. There are a few useless threads at the end of a normal tap. For external threads, a shaft that is the same size of the major diameter is cut using a die or on a lathe. Chamfers are generally cut to allow easy assembly.

## Thread Representations

The true representation of a screw thread is almost never used in making working drawings. In true representation, the crest and root lines appear as the projections of helical curves, which are extremely tedious to draw.

On practical working drawings, three representation methods are generally used: **detailed**, **schematic** and **simplified**.

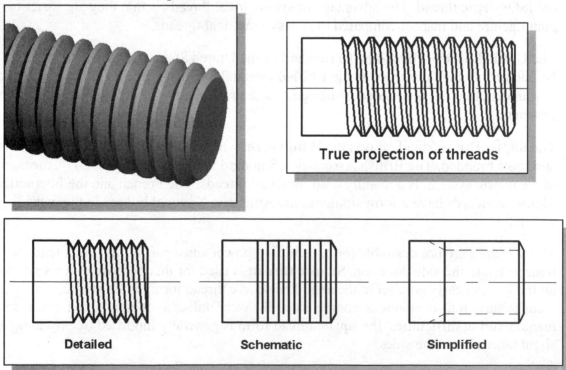

True projection of threads

Detailed                Schematic                Simplified

## Detailed Representation

The detailed representation simplifies the drawing of the thread principally by constructing the projections of the helical curves into straight lines. And where applicable, further simplifications are made. For example, the 29° angle of the Acme is generally drawn as 30°, and the *American National* and *Unified* threads are represented by the sharp V. In general, true pitch should be shown, although a small increase or decrease in pitch is permissible so as to have even measurements in making the drawing. For example, seven threads per inch may be decreased to six. Remember this is only to simplify the drawing—the actual threads per inch must be specified in the drawing.

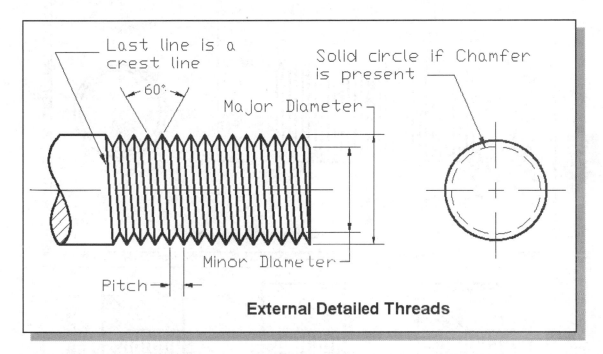

**External Detailed Threads**

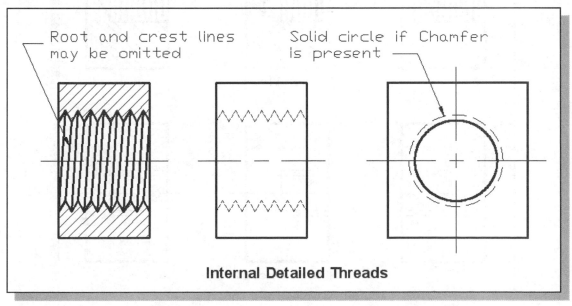

**Internal Detailed Threads**

## Schematic Representation

The schematic representation simplifies the drawing of the thread even further. This method omits the true forms and indicates the crests and roots by lines perpendicular to the main axis. And the schematic representation is nearly as effective as the detailed representation but is much easier to construct. This representation should not be used for section views of external threads or when internal threads are shown as hidden lines; the simplified representation is used instead. Also, to avoid too dense of line pattern, do not use this method when the pitch is less than 1/8 inch or 3 mm.

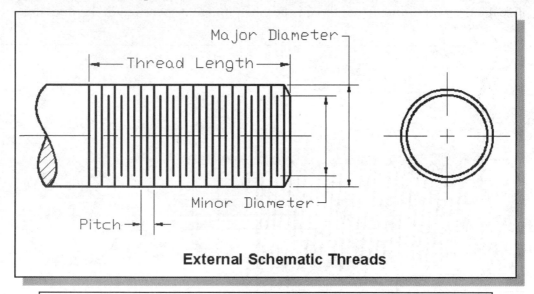

**External Schematic Threads**

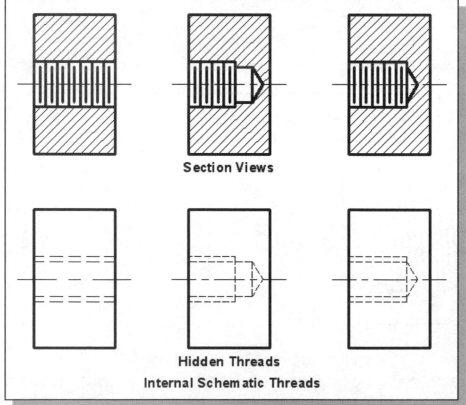

**Section Views**

**Hidden Threads**

**Internal Schematic Threads**

## Simplified Representation

The simplified representation omits both form and crest-lines and indicates the threaded portion by dashed lines parallel to the axis at the approximate depth of thread. The simplified representation is less descriptive than the schematic representation, but they are quicker to draw and for this reason are preferred whenever feasible.

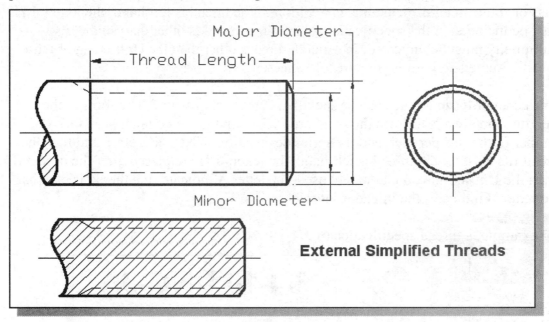

**External Simplified Threads**

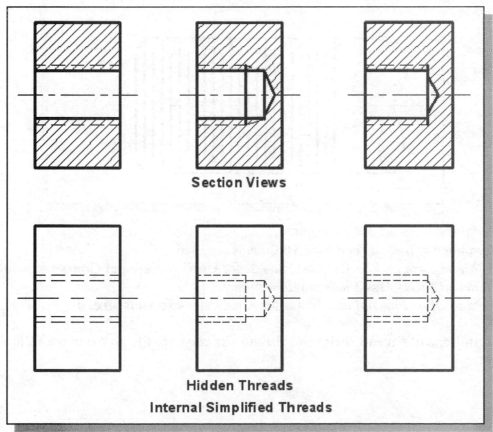

**Section Views**

**Hidden Threads**
**Internal Simplified Threads**

## Thread Specification – English Units

The orthographic views of a thread are necessary in order to locate the position of the thread on the part. In addition, the complete thread specification is normally conveyed by means of a local note or dimensions and a local note. The essential information needed for manufacturing is **form**, **nominal (major) diameter**, **threads per inch**, and **thread class** or **toleranced dimensions**. In addition, if the thread is left-hand, the letters **LH** must be included in the specification; also, if the thread is other than single, its multiplicity must be indicated. In general, threads other than the Unified and Metric threads, Acme and buttress require tolerances.

Unified and Metric threads can be specified completely by note. The form of the specification always follows the same order: The first is the nominal size, then the number of threads per inch and the series designation (UNC, NC, etc.), and then the thread fits class. If the thread is left-hand, the letters **LH** are placed after the class. Also, when the Unified-thread classes are used, the letter **A** indicates the thread is external and letter **B** indicates the thread is internal.

For example, a thread specification of **3/4-10UNC-3A**

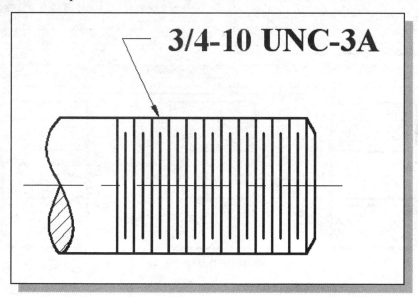

*Major diameter*: **3/4** in. diameter
*Number of threads per inch*: **10** threads per inch
*Thread form and Series*: **UNC** stands for **Unified National Coarse** series thread
*Thread Fits*: Class **3** is for high accuracy
*External or Internal thread*: Letter **A** indicates **external thread**

The descriptions of **thread series** and **thread fits class** are illustrated in the following sections.

## Unified Thread Series

Threads are classified in "series" according to the number of threads per inch used with a specific diameter. The above Unified thread example having 20 threads per inch applied to a 1/4-in. diameter results in a thread belonging to the coarse-thread series; note that one with 28 threads per inch on the same diameter describes a thread belonging to the fine-thread series. In the United States, the Unified and Metric threads, the Acme, pipe threads, buttress, and the knuckle thread all have standardized series designations. Note that only the Unified system has more than one series of standardized forms. The Unified and Metric threads standard covers eleven series of screw threads. In the descriptions of the series which follow, the letters "U" and "N" used in the series designations stand for the words "Unified" and "National," respectively. The three standard series include the **coarse-thread series**, the **fine-thread series** and the **extra fine-thread series**.

- The coarse-thread series, designated "**UNC**" or "**NC**," is recommended for general use where conditions do not require a fine thread.

- The fine-thread series, designated "**UNF**" or "**NF**," is recommended for general use in automotive and aircraft work and where special conditions require a fine thread.

- The extra fine-thread series, designated "**UNEF**" or "**NEF**," is used particularly in aircraft and aeronautical equipment, where an extremely shallow thread or a maximum number of threads within a given length is required.

There are also the eight **constant pitch series**: **4, 6, 8, 12, 16, 20, 28** and **32 threads** with constant pitch. The **8, 12** and **16** series are the more commonly used series.

- The 8-thread series, designated **8UN** or **8N**, is a uniform-pitch series using eight threads per inch for any diameter. This series is commonly used for high-pressure conditions. The constant pitch allows excessive torque to be applied and to maintain a proper initial tension during assembly. Accordingly, the 8-thread series has become the general series used in many types of engineering work and as a substitute for the coarse-thread series.

- The 12-thread series, designated **12UN** or **12N**, is a uniform-pitch series using 12 threads per inch for any diameter. Sizes of 12-pitch threads range from 1/2 to 13/4 in. in diameter. This series is commonly used for machine construction, where fine threads are needed for strength. It also provides continuation of the extra fine-thread series for diameters larger than 11/2 in.

- The 16-thread series, designated **16UN** or **16N**, is a uniform-pitch series using 16 threads per inch for any diameter. This series is intended for applications requiring a very fine thread, such as threaded adjusting collars and bearing retaining nuts. It also provides continuation of the extra-fine-thread series for diameters larger than 2 in.

In addition, there are three special thread series, designated **UNS**, **NS** and **UN** as covered in the standards, that include special combinations of diameter, pitch, and length of engagement.

## Thread Fits

The classes provided by the American Standard (ANSI) are classes 1, 2 and 3. These classes are achieved through toleranced thread dimensions given in the standards.

- **Classes 1 fit**: This class of fit is intended for rapid assembly and easy production. Tolerances and allowance are largest with this class.

- **Classes 2 fit**: This class of fit is a quality standard for the bulk of screws, bolts, and nuts produced and is suitable for a wide variety of applications. A moderate allowance provides a minimum clearance between mating threads to minimize galling and seizure. A thread fit class of **2** is used as the normal production fit; this fit is assumed if none is specified.

- **Classes 3 fit**: This class of fit provides a class where accuracy and closeness of fit are important. No allowance is provided. This class of fit is only recommended where precision tools are used.

## Thread Specification – Metric

Metric threads, similar to the Unified threads, can also be specified completely by note. The form of the specification always follows the same order: The first is the letter M for metric thread, and then the nominal size, then the pitch and then the tolerance class. If the thread is left-hand, the letters **LH** are placed after the class.

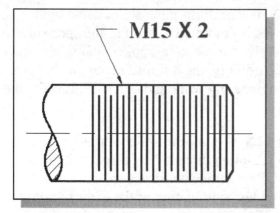

For example, the basic thread note **M15 x 2** is an adequate description for general commercial products.

*Letter M*: **Metric** thread
*Major Diameter*: **15** mm.
*Pitch*: **2** mm.

- **Tolerance Class**: This is used to indicate the tightness or looseness fit between the internal and external threads. In a thread note, the minor or pitch diameter tolerance is stated first followed by the major diameter tolerance if it is different. For general purpose threads, the fit 6H/6g should be used; this fit is assumed if none is specified. For a closer fit, use 6H/5g6g. Note that the number 6 in metric threads is equivalent to the Unified threads class 2 fit. Also, the letter g and G are used for small allowances, where h and H are used for no allowance.

## Thread Notes Examples

### (a) Metric

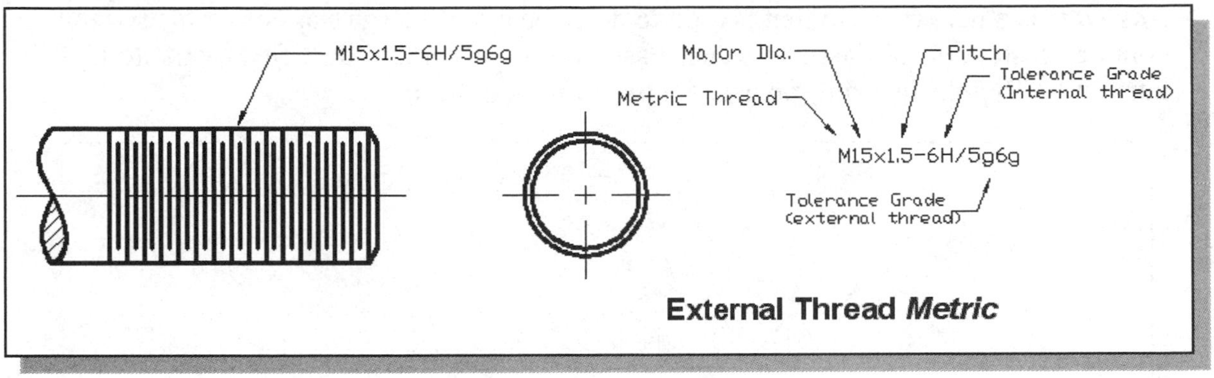
M15x1.5-6H/5g6g
Major Dia.
Pitch
Tolerance Grade (Internal thread)
Metric Thread
M15x1.5-6H/5g6g
Tolerance Grade (external thread)

**External Thread *Metric***

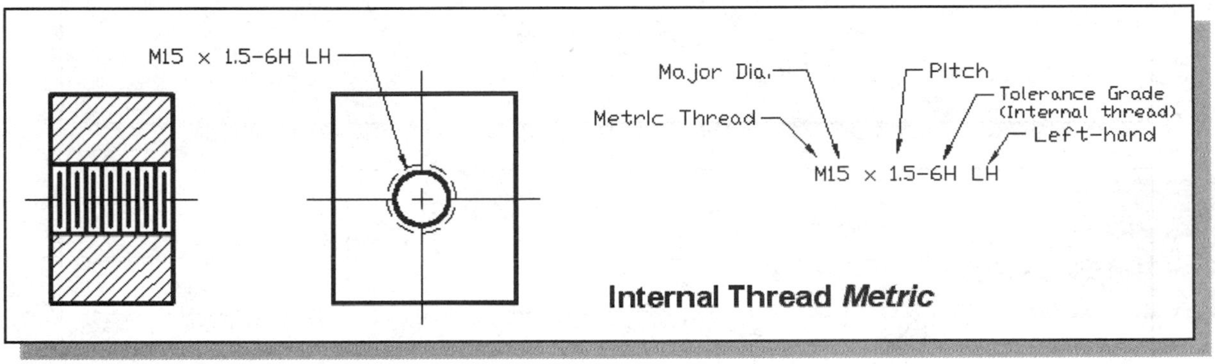
M15 x 1.5-6H LH
Major Dia.
Pitch
Tolerance Grade (Internal thread)
Metric Thread
Left-hand
M15 x 1.5-6H LH

**Internal Thread *Metric***

### (b) Unified

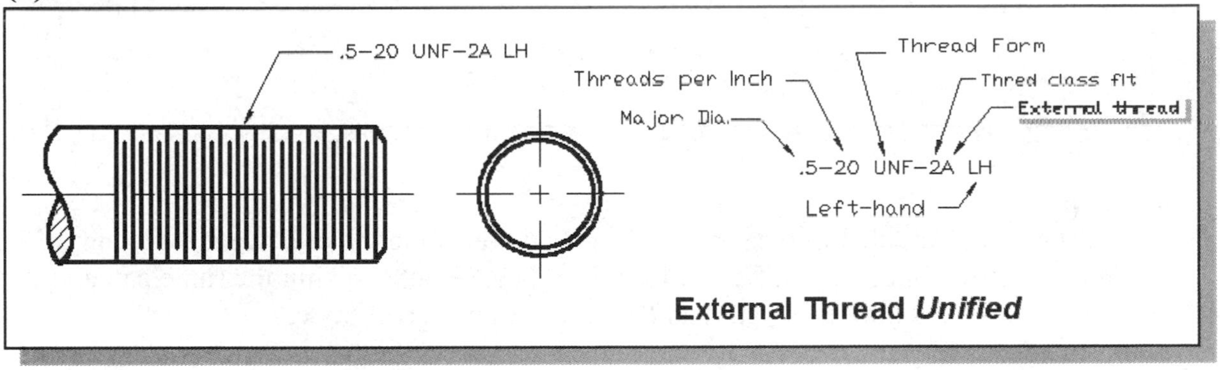
.5-20 UNF-2A LH
Thread Form
Threads per Inch
Thred class fit
External thread
Major Dia.
.5-20 UNF-2A LH
Left-hand

**External Thread *Unified***

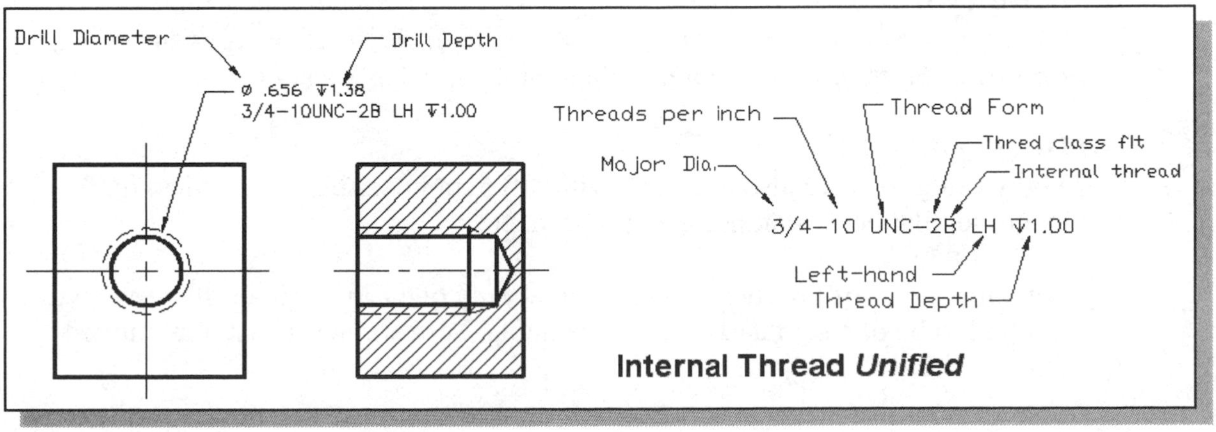
Drill Diameter
Drill Depth
Ø .656 ↧1.38
3/4-10UNC-2B LH ↧1.00
Threads per inch
Thread Form
Thred class fit
Major Dia.
Internal thread
3/4-10 UNC-2B LH ↧1.00
Left-hand
Thread Depth

**Internal Thread *Unified***

## Specifying Fasteners

*Fastener* is a generic term that is used to describe a fairly large class of parts used to connect, fasten, or join parts together. Fasteners are generally identified by the following attributes: **Type**, **Material**, **Size**, and **Thread information**.

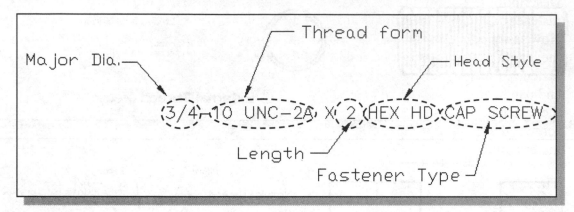

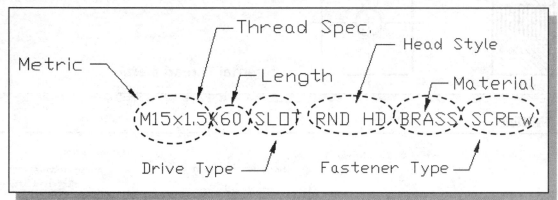

- **Type**
  Fasteners are divided into categories based on their function or design; for example, Wood Screw, Sheet Metal Screw, Hex Bolt, Washer, etc. Within the same category, some variations may exist, such as **Drive Type** and **Head Style**.

  **(a) Drive Type**
  Fasteners in some categories are available with different drive types such as *Philips* or *Slotted*; for example, Philips, Slotted, Allen/Socket etc.

  **(b) Head Style**
  Many categories are also available with different head shapes or styles; for example, Flat head, Pan head, Truss head, etc.

  For some types of fasteners, there is either only one drive type or the head style is implied to be of a standard type; for example, Socket screws have the implied Allen drive.

- **Material**
  Fastener material describes the material from which the fastener was made as well as any material grade; for example, Stainless Steel, Bronze, etc.

- **Size**
  Descriptions of a fastener's size typically include its **diameter** and **length**. The fastener diameter is measured either as a size number or as a direct measurement. How fastener length is measured varies based on the type of head. As a general rule, the length of fasteners is measured from the surface of the material, to the end of the fastener. For fasteners where the head usually sits above the surface such as hex bolts and pan head screws, the measurement is from directly under the head to the end of the fastener. For fasteners that are designed to be counter sunk such as flat head screws, the fastener is measured from the point on the head where the surface of the material will be, to the end of the fastener.

- **Thread Information**
  Thread pitch or thread count is used only on machine thread fasteners. The thread pitch or count describes how fine the threads are.

## Commonly used Fasteners

### Hex bolts

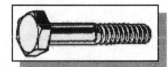

A bolt is a fastener having a head on one end and a thread on the other end. A bolt is used to hold two parts together, by means of passing through aligned clearance holes with a nut screwed on the threaded end. (See Appendix E for specs.)

### Studs

A stud is a rod with threaded ends.

### Cap Screws

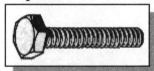

A hexagon cap screw is similar to a bolt except it is used without a nut, and generally has a longer thread. Cap screws are available in a variety of head styles and materials. (See Appendix E for specs.)

### Machine screws

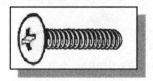

A machine screw is similar to the slot-head cap screw but smaller, available in many styles and materials. A machine screw is also commonly referred to as a stove bolt. (See Appendix E for specs.)

### Wood screws

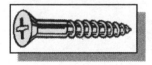

A tapered shank screw is for use exclusively in wood. Wood screws are available in a variety of head styles and materials.

**Sheet metal screws**

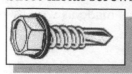

Highly versatile fasteners designed for thin materials. Sheet metal screws can be used in wood, fiberglass and metal, also called self-tapping screws, available in steel and stainless steel.

**Carriage bolts**

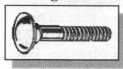

A carriage bolt is mostly used in wood with a domed shape top and a square under the head, which is pulled into the wood as the nut is tightened.

**Socket screws**

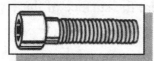

Socket screws, also known as **Allen head**, are fastened with a hexagon Allen wrench, available in several head styles and materials. (See Appendix E for specs.)

**Set Screws**

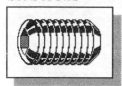

Set screws are used to prevent relative motion between two parts. A set screw is screwed into one part so that its point is pushed firmly against the other part, available in a variety of point styles and materials.

**Nuts**

Nuts are used to attach machine thread fasteners. (See Appendix E for specs.)

**Washers**

Washers provide a greater contact surface under the fastener. This helps prevent a nut, bolt or screw from breaking through the material. (See Appendix E for specs.)

**Keys**

Keys are used to prevent relative motion between shafts and wheels, couplings and similar parts attached to shafts.

**Rivets**

Rivets are generally used to hold sheet metal parts together. Rivets are generally considered as permanent fasteners and are available in a variety of head styles and materials.

## Drawing Standard Bolts

As a general rule, standard bolts and nuts are not drawn in the detail drawings unless they are non-standard sizes. The standard bolts and nuts do appear frequently in assembly drawings, and they may be drawn from the exact dimensions from the ANSI/ASME B18.2.1- 1996 standard (see Appendix E) if accuracy is important. Note that for small or unspecified chamfers, fillets, or rounds, the dimension of 1/16 in. (2mm) is generally used in creating the drawing. A simplified version of a bolt is illustrated in the following figure. Note that many companies now offer thousands of standard fasteners and part drawings in AutoCAD DWG or DXF formats for free. You are encouraged to do a search on the internet and compare the downloaded drawings to the specs as listed in the appendix.

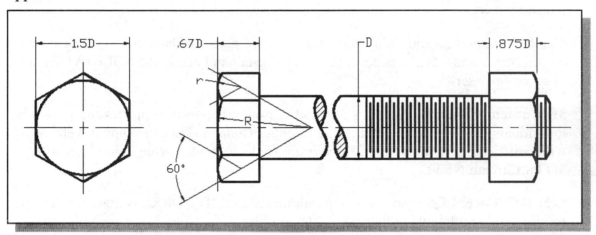

## Bolt and Screw Clearances

Bolts and screws are generally used to hold parts together, and it is frequently necessary to have a clearance hole for the bolt or screw to pass through. The size of the clearance hole depends on the major diameter of the fastener and the type of fit that is desired. For bolts and screws, three standard fits are available: **normal fit, close fit** or **loose fit**. See Appendix for the counter-bore and countersink clearances. (See Appendix G for more details.)

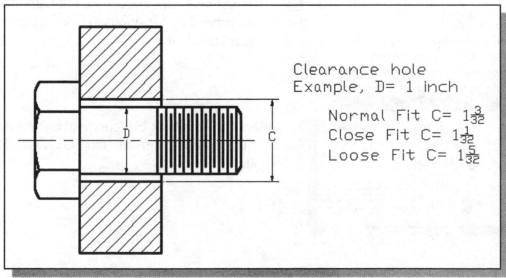

Clearance hole
Example, D= 1 inch

Normal Fit C= $1\frac{3}{32}$
Close Fit C= $1\frac{1}{32}$
Loose Fit C= $1\frac{5}{32}$

## Fasteners using SOLIDWORKS' Design Library

In SOLIDWORKS, we also have the option of using the standard parts library through what is known as the *Design Library*. The *Design Library* consists of multiple libraries of reusable elements such as parts, assemblies and sketches. It is accessed through the *Task Pane* at the right of the SOLIDWORKS work area. Significant amounts of time can be saved by using these parts. Note that we can also create and publish our libraries so that others can reuse our parts. The Design Library Task Pane divides the library into four major sections:

- **Design Library** – Contains subfolders populated by SOLIDWORKS with reusable items such as parts, blocks, and annotations. Additional folders and content can be added by the user.

- **Toolbox** – Contains multiple libraries of standard parts that have been created based on industry standards. To access the content, install and add in the SOLIDWORKS *Toolbox* browser.

- **3D Content Central** – Accesses 3D models from component suppliers and individuals via the internet. It includes *Supplier Content* (links to supplier web sites with certified 3D models) and *User Library* (links to models from individuals using 3D PartStream.NET).

- **SOLIDWORKS Content** – Contains additional SOLIDWORKS content for blocks, routing, and weldments in downloadable .zip files.

## ANSI Inch - Machine Screw

1. Select the **SOLIDWORKS** option on the *Start* menu or select the **SOLIDWORKS** icon on the desktop to start SOLIDWORKS. The SOLIDWORKS main window will appear on the screen.

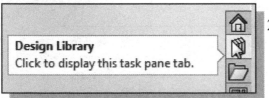

2. In the *task pane*, select the **Design Library** command by left-mouse-clicking the associated tab.

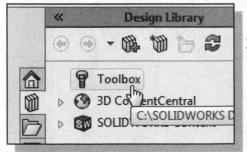

3. Select the **Toolbox** category and click **Add in now** to install and activate the SOLIDWORKS Toolbox.

4. Select the **ANSI Inch** folder by double-clicking on the folder as shown.

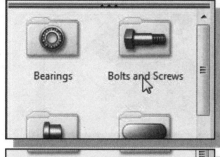

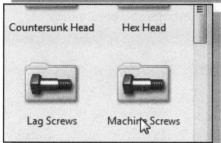

5. Open the **Bolts and Screws** folder by double-clicking on the folder as shown.

6. Open the **Machine Screws** folder by double-clicking on the folder as shown.

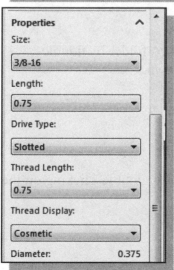

7. *Right-click* once on **Fillister Head Screw** to open the **option menu**.

8. Select **Create Part** in the *option list* as shown.

9. In the *Property Manager* area, set the size to **3/8-16** (Major diameter 3/8 and 16 threads per inch), *Length* to **0.75**, *Drive Type* to **Slotted** and *Thread Display* to **Cosmetic**.

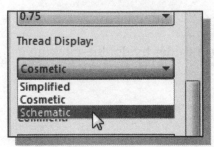

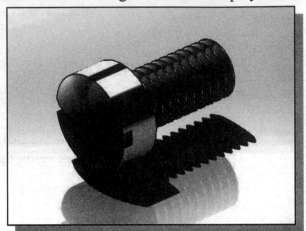

10. Click on the *Thread Display* list and choose **Schematic** to change the threads display as shown.

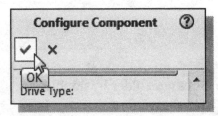

11. Click **OK** to accept the settings and create the machine screw.

## ANSI Metric - Machine Screw

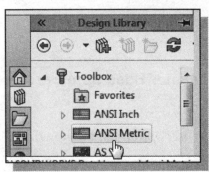

1.  In the *Design Library* area, select **ANSI Metric** as shown.

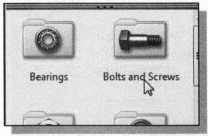

2.  Open the **Bolts and Screws** folder by double-clicking on the folder as shown.

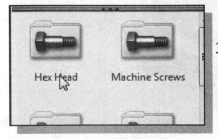

3.  Open the **Hex Head** folder by double-clicking on the folder as shown.

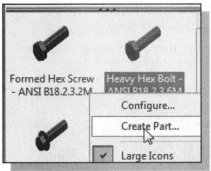

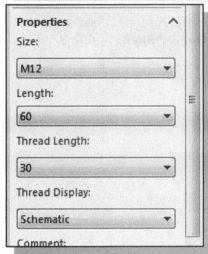

4. *Right-click* once on **Heavy Hex Bolt** to open the **option menu.**

5. Select **Create Part** in the *option list* as shown.

6. In the *Property Manager* area, set the size to **M12** (Metric Major diameter 12), *Length* to **60**, *Thread Length* to **30** and *Thread Display* to **Schematic.**

7. Click **OK** to accept the configuration and create *Hex Bolt*.

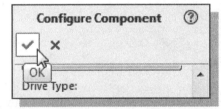

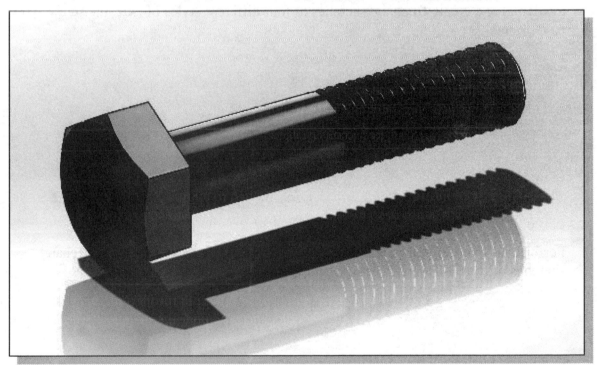

## Questions:

1.  Describe the thread specification of **3/4-16 UNF -3A**.

2.  Describe the thread specification of **M25X3**.

3.  List and describe four different types of commonly used fasteners.

4.  Determine the sizes of a clearance hole for a **0.75** bolt using the *Close fit* option.

5.  Perform an internet search and find information on the different type of **set-screws**, and create freehand sketches of four point styles of set screws you have found.

6.  Perform an internet search and examine the different types of **machine screws** available, and create freehand sketches of four different styles of machine screws you have found.

7.  Perform an Internet search and find information on **locknuts**, and create freehand sketches of four different styles of locknuts you have found.

# Chapter 15
# Assembly Modeling and Working Drawings

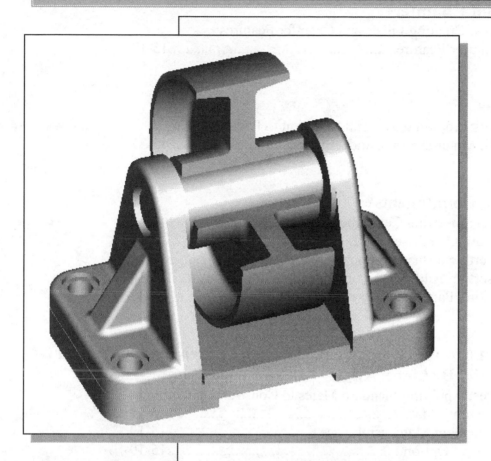

## Learning Objectives

♦ **Understand the Terminology Related to Working Drawings**
♦ **Understand the Assembly Modeling Methodology**
♦ **Understand and Control Degrees of Freedom for Assembly Components**
♦ **Understand and Utilize SOLIDWORKS Assembly Mates**
♦ **Place parts using SOLIDWORKS SmartMates**
♦ **Create Exploded Assemblies**

## Certified SOLIDWORKS Associate Exam Objectives Coverage

### Fillets and Chamfers
Objectives: Creating Fillets and Chamfer Features.

Chamfer Feature...........................................................15-10

### Materials
Objectives: Applying Material Selection to Parts.

Edit Material Command...............................................15-37

### Inserting Components
Objectives: Inserting Components into an Assembly.

Creating an Assembly File.........................................15-13
Inserting a Base Component ......................................15-14
Inserting Additional Components.............................15-16
Editing Parts in an Assembly....................................15-37

### Standard Mates – Coincident, Parallel, Perpendicular, Tangent, Concentric, Distance, Angle
Objectives: Applying Standard Mates to Constrain Assemblies.

Assembly Mates...........................................................15-16
Coincident Mate, using Faces ...................................15-19
Aligned Option............................................................15-19
Anti-Aligned Option ..................................................15-19
Coincident Mate, using Temporary Axes .................15-19
Coincident Mate, using Planes..................................15-22
Concentric Mate..........................................................15-24
SmartMates .................................................................15-26

### Drawing Sheets and Views
Objectives: Creating and Setting Properties for Drawing Sheets; Inserting and Editing Standard Views.

Creating a Drawing from an Assembly ....................15-34

### Annotations
Objectives: Creating Annotations.

AutoBalloon Command ...........................................15-41
Bill of Materials .......................................................15-35

## General Engineering Design Process

Engineering design is the ability to create and transform ideas and concepts into a product definition that meets the desired objective.

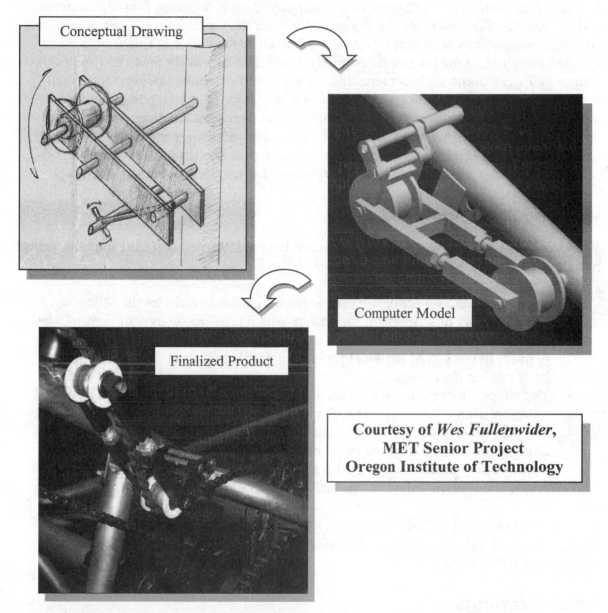

Conceptual Drawing

Computer Model

Finalized Product

**Courtesy of *Wes Fullenwider*,
MET Senior Project
Oregon Institute of Technology**

The general procedure for the design of a new product or improving an existing product involves the following six stages:

1.  **Develop and identify the desired objectives.**
2.  **Conceptual design stage – concepts and ideas of possible solutions.**
3.  **Engineering analysis of components.**
4.  **Computer Modeling and/or prototypes.**
5.  **Refine and finalize the design.**
6.  **Working drawings of the finalized design.**

It is during the **conceptual design** stage that the first drawings, known as **conceptual drawings**, are usually created. The conceptual drawings are typically done in the form of freehand sketches showing the original ideas and concepts of possible solutions to the set objectives. From these conceptual drawings, engineering analyses are performed to improve and confirm the suitability of the proposed design. Working from the sketches and the results of the analyses, the design department then creates prototypes or performs computer simulations to further refine the design. Once the design is finalized, a set of detailed drawings of the proposed design is created. It is accurately made and shows the shapes and sizes of the various parts; this is known as the **detail drawing**. On a detail drawing, all the views necessary for complete shape description of a part are provided, and all the necessary dimensions and manufacturing directions are given. The set of drawings is completed with the addition of an **assembly drawing** and a parts list or bill of material. The assembly drawing is necessary as it provides the location and relationship of the parts. The completed drawing set is known as **working drawings**.

## Working Drawings

Working drawings are the set of drawings used to give information for the manufacturing of a design.

The description given by a set of working drawings generally includes the following:
1.  The **assembly** description of the design, which provides an overall view of what the design is and also shows the location and relationship of the parts required.
2.  A **parts list** or **bill of material** provides a detailed material description of the parts used in the design.
3.  The **shape description** of the individual parts, which provides the full graphical representation of the shape of each part.
4.  The **size description** of the individual parts, which provides the necessary dimensions of the parts used in the design.
5.  **Explanatory notes** are the general and local notes on the individual drawings, giving the specific information, such as material, heat-treatment, finish, etc.

A set of drawings will include, in general, two classes of drawings: **detail drawings** giving details of individual parts and an **assembly drawing** giving the location and relationship of the parts.

## Detail Drawings

A detail drawing is the drawing of the individual parts, giving a complete and exact description of its form, dimensions, and related construction information. A successful detail drawing will provide the manufacturing department all the necessary information to produce the parts. This is done by providing adequate orthographic views together with dimensions, notes, and a descriptive title. Information on a detail drawing usually includes the shape, size, material and finish of a part; specific information of shop operations, such as the limits of accuracy and the number of parts needed are also provided. The detail drawing should be complete and also exact in description, so that a satisfactory part can be produced. The drawings created in the previous chapters are all detail drawings.

## Assembly Drawings

An assembly drawing is, as its name implies, a drawing of the design put together, showing the relative positions of the different parts. The assembly drawing of a finalized design is generally done after the detail drawings are completed. The assembly drawing can be made by tracing from the detail drawings. The assembly drawing can also be drawn from the dimensions of the detail drawings; this provides a valuable check on the correctness of the detail drawings.

The assembly drawing sometimes gives the overall dimensions and dimensions that can be used to aid the assembly of the design. However, many assembly drawings need no dimensions. An assembly drawing should not be overloaded with detail, particularly hidden detail. Unnecessary dashed lines (hidden lines) should not be used on any drawing; this is even more critical on assembly drawings.

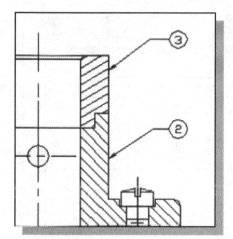

Assembly drawings usually have reference letters or numbers designating the different parts. These "numbers" are typically enclosed in circles ("balloons") with a leader pointing to the part; these numbers are used in connection with the parts list and bill of material.

For complicated designs, besides the assembly drawing, subassembly drawings are generally used. A subassembly is a drawing showing the details of a related group of parts, as it would not be practical to include all the features on a single assembly drawing. Thus, a subassembly is used to aid in clarifying the relations of a subset of an assembly.

## Bill of Materials (BOM) and Parts List

| 4 | Cap screw 3/8 | 2 | STOCK |
|---|---|---|---|
| 3 | Collar | 1 | STEEL |
| 2 | Bearing | 1 | STEEL |
| 1 | Base Plate | 1 | C.I. |
| No | DESCRIPTION | REQ | MATL |

A bill of materials (BOM) is a table that contains information about the parts within an assembly. The BOM can include information such as part names, quantities, costs, vendors, and all of the other information related to building the part. The *parts list*, which is used in an assembly drawing, is usually a partial list of the associated BOM.

## Drawing Sizes

The standard drawing paper sizes are as shown in the below tables.

| American National Standard | International Standard |
|---|---|
| A – 8.5" X 11.0" | A4 – 210 mm X 297 mm |
| B – 11.0" X 17.0" | A3 – 297 mm X 420 mm |
| C – 17.0" X 22.0" | A2 – 420 mm X 594 mm |
| D – 22.0" X 34.0" | A1 – 594 mm X 841 mm |
| E – 34.0" X 44.0" | A0 – 841 mm X 1189 mm |

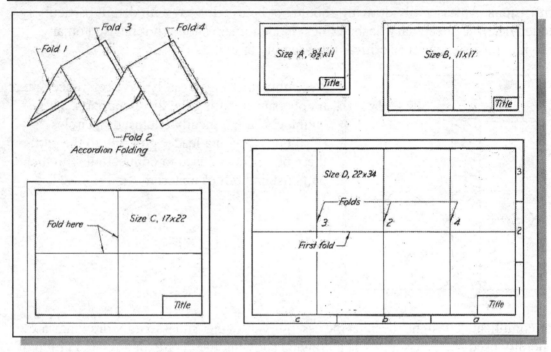

## Drawing Sheet Borders and Revisions Block

The drawing sheet borders are generally drawn at a distance parallel to the edges of the sheet, typically with distance varying from 0.25" to 0.5" or 5 mm to 10 mm.

Once a drawing has been released to the shop, any alterations or changes should be recorded on the drawing and new prints be issued to the production facilities. In general, the upper right corner of a working drawing sheet is the designated area for such records. This is known as the **revisions** block.

| | | REVISIONS | | | |
|---|---|---|---|---|---|
| ZONE | REV | DESCRIPTION | | DATE | APPROVED |
| | | | | | |

## Title Blocks

The title of a working drawing is usually placed in the lower right corner of the sheet. The spacing and arrangement of the space depend on the information to be given. In general, the title of a working drawing should contain the following information:

1. **Name of the company and its location.**
2. **Name of the part represented.**
3. **Signature of the person who made the drawing and the date of completion.**
4. **Signature of the checker and the date of completion.**
5. **Signature of the approving personnel and the date of approval.**
6. **Scale of the drawing.**
7. **Drawing number.**

Other information may also be given in the title blocks area, such as material, heat treatment, finish, hardness, general tolerances; it depends on the company and the peculiarities of the design.

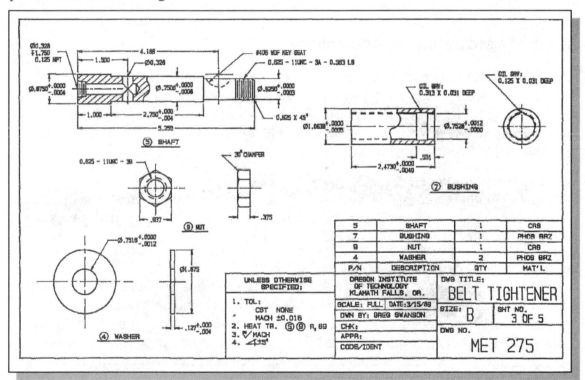

## Working Drawings with SOLIDWORKS

In the previous lessons, we have gone over the fundamentals of creating basic parts and drawings. In this lesson, we will examine the assembly modeling functionality of SOLIDWORKS. We will start with a demonstration on how to create and modify assembly models. The main task in creating an assembly is establishing the assembly relationships between parts. It is a good practice to assemble parts based on the way they would be assembled in the actual manufacturing process. We should also consider breaking down the assembly into smaller subassemblies, which helps the management of parts. In SOLIDWORKS, a subassembly is treated the same way as a single part during assembling. Many parallels exist between assembly modeling and part modeling in parametric modeling software such as SOLIDWORKS.

SOLIDWORKS provides full associative functionality in all design modules, including assemblies. When we change a part model, SOLIDWORKS will automatically reflect the changes in all assemblies that use the part. We can also modify a part in an assembly. Bi-directional full associative functionality is the main feature of parametric solid modeling software that allows us to increase productivity by reducing design cycle time.

## Assembly Modeling Methodology

The SOLIDWORKS assembly modeler provides tools and functions that allow us to create 3D parametric assembly models. An assembly model is a 3D model with any combination of multiple part models. SOLIDWORKS **Mate Relations** can be used to control relationships between parts in an assembly model. SOLIDWORKS can work with any of the assembly modeling methodologies:

**The Bottom Up approach**

The first step in the *bottom up* assembly modeling approach is to create the individual parts. The parts are then pulled together into an assembly. This approach is typically used for smaller projects with very few team members.

**The Top Down approach**

The first step in the *top down* assembly modeling approach is to create the assembly model of the project. Initially, individual parts are represented by names or symbolically. The details of the individual parts are added as the project gets further along. This approach is typically used for larger projects or during the conceptual design stage. Members of the project team can then concentrate on the particular section of the project to which he/she is assigned.

**The Middle Out approach**

The *middle out* assembly modeling approach is a mixture of the bottom-up and top-down methods. This type of assembly model is usually constructed with most of the parts already created and additional parts are designed and created using the assembly for construction information. Some requirements are known and some standard components are used, but new designs must also be produced to meet specific objectives. This combined strategy is a very flexible approach for creating assembly models.

- The different assembly modeling approaches described above can be used as guidelines to manage design projects. Keep in mind that we can start modeling our assembly using one approach and then switch to a different approach without any problems.

## The Shaft Support Assembly

In this lesson, the *bottom up* assembly modeling approach is illustrated through the creation of the shaft support assembly shown in the figure. All of the parts (components) required to form the assembly are created first. SOLIDWORKS has assembly modeling tools that allow us to create complex assemblies by using components that are created in part files or are created in assembly files. A component can be a subassembly or a single part, where features and parts can be modified at any time.

## Parts

- Four parts are required for the assembly: (1) **Collar**, (2) **Bearing**, (3) **Base-Plate** and (4) **Cap-Screw**. Create the four parts as shown below, then save the models as separate part files: *Collar*, *Bearing*, *Base-Plate*, and *Cap-Screw*. (Close all part files or exit SOLIDWORKS after you have created the parts.)

## Creating the Collar with the Chamfer Command

(1) *Collar*

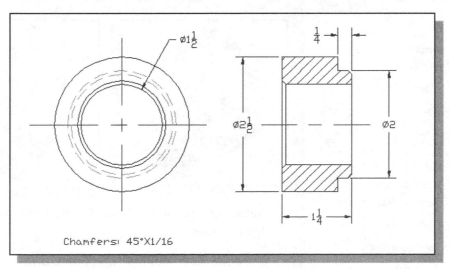

1. On your own, start a new part file using your **Part_IPS_ANSI** template and create the basic *Collar* without the chamfers, as shown below.

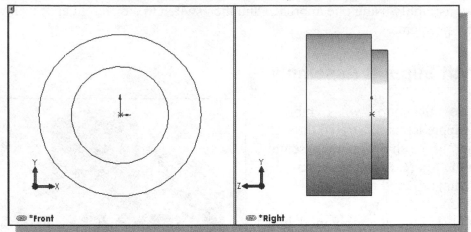

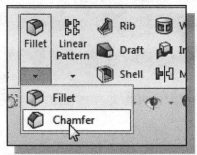

2. In the *Features* toolbar, left-click on the **arrow** next to the **Fillet** icon and select the **Chamfer** command from the pop-up menu.

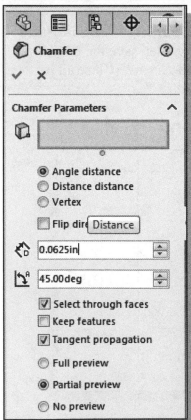

3. In the *Chamfer Parameters* panel of the *Chamfer Property Manager,* select the **Angle distance** option.

4. Set the *Distance* to **1/16 in** (0.0625in).

5. Set the *Angle* to **45°**.

6. Select the two circular edges shown in the figures below.

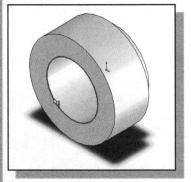

7. Select **OK** in the *Chamfer Property Manager* to create the chamfers.

8. Save the part with the filename ***Collar***.

## Creating the Bearing and Base-Plate

➢ On your own, create and save the *Bearing* and *Base-Plate* part files based on the drawings below. Align the base of each part with the **Top Plane**.

(2) ***Bearing*** (Construct the part with the datum origin aligned to the bottom center.)

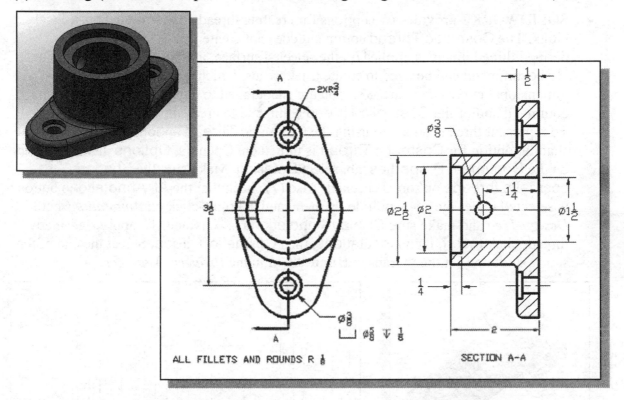

(3) ***Base-Plate*** (Construct the part with the datum origin aligned to the bottom center.)

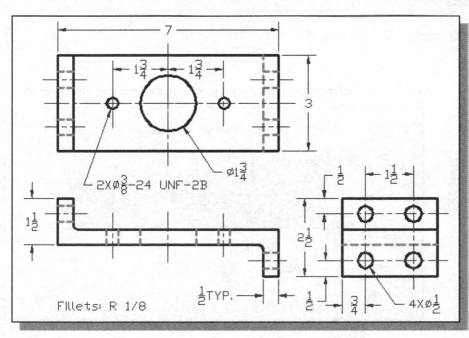

**NOTE:** we will ignore the threads and simply create 3/8″ diameter holes.

## Creating the Cap-Screw

➤ On your own, create and save the *Cap-Screw* part file based on the drawings below. **NOTE:** Threads are not necessary.

### (4) *Cap-Screw*

• SOLIDWORKS provides two options for creating threads: **Cosmetic Thread** and **Helix**. The **Cosmetic Thread** command does not create true 3D threads; a pre-defined thread image is applied on the selected surface, as shown in the figure. The **Helix** command can be used to create true threads, which contain complex three-dimensional curves and surfaces. You are encouraged to experiment with the **Helix** command and/or the **Cosmetic Thread** command to create threads.
To represent threads on a part using the **Cosmetic Thread** option, first make sure the shading option for **Cosmetic Thread** is turned on. Open the **Options** dialog box and on the **Document Properties** tab, select **Detailing**. Make sure the **Shaded cosmetic threads** option is checked. Also (1) check that the **All Annotations** button is selected under the *View* pull-down menu and (2) right-click on *Annotations* in the *Design Tree* and make sure **Display Annotations** is activated. To apply a cosmetic thread, click **Insert**, then **Annotations**, then **Cosmetic Thread**. Select the end of the screw shaft and set the threads to **Blind-0.5 in** in the *Property Manager*.

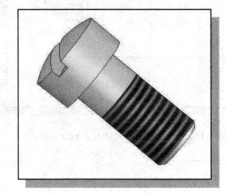

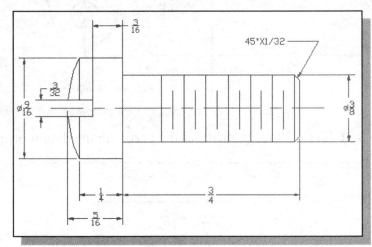

• **HINT:** First create the profile shown here for a revolved feature. Note the centerline is also aligned to the origin.

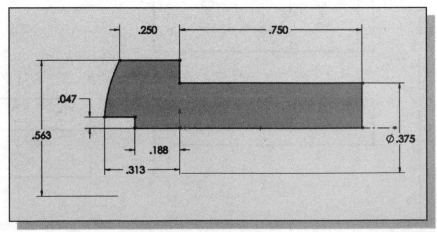

## Starting SOLIDWORKS

1.  Select the **SOLIDWORKS** option on the *Start* menu or select the **SOLIDWORKS** icon on the desktop to start SOLIDWORKS.

2.  Select the **Assembly** icon with a single click of the left-mouse-button on the *Menu Bar* to start the new **Assembly** file.

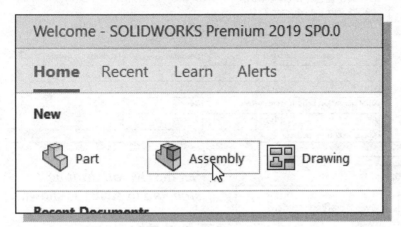

❖  SOLIDWORKS opens an assembly file, and automatically opens the *Begin Assembly Property Manager*. SOLIDWORKS expects us to insert the first component. We will first adjust the document properties to match those of the previously created parts.

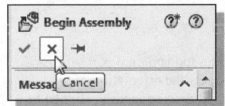

3.  On your own, cancel the open file option and cancel the **Begin Assembly** command by clicking once with the left-mouse-button on the **Cancel** icon as shown.

## Document Properties

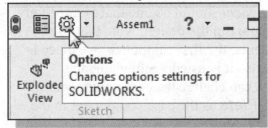

1.  Select the **Options** icon from the *Menu Bar* to open the *Options* dialog box.

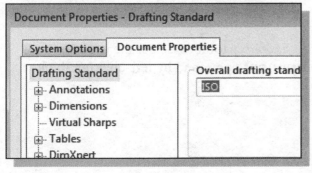

2.  Select the **Document Properties** tab.

3.  Select **ISO** in the pull-down selection window under the *Overall drafting standard* panel to reset the default settings.

4.  Click **Units**, select the **IPS (inch, pound, second)** unit system, and select **.12** in the *Decimals* spin box for the *Length units* as shown.

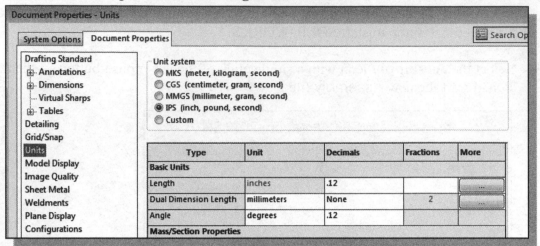

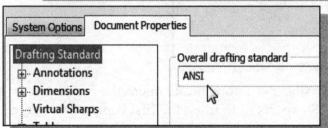

5.  Set the *Overall drafting standard* to **ANSI** as shown.

6.  Click **OK** in the *Document Properties* window to accept the settings.

## Insert the First Component

*   The first component inserted in an assembly should be a fundamental part or subassembly. The first component in an assembly file sets the orientation of all subsequent parts and subassemblies. The origin of the first component is aligned to the origin of the assembly coordinates and the part is grounded (all degrees of freedom are removed). The rest of the assembly is built on the first component, the **base component**. In most cases, this *base component* should be one that is **not likely to be removed** and **preferably a non-moving part** in the design. Note that there is no distinction in an assembly between components; the first component we place is usually considered as the *base component* because it is usually a fundamental component to which others are constrained. We can change the base component if desired. For our project, we will use the ***Base-Plate*** as the base component in the assembly.

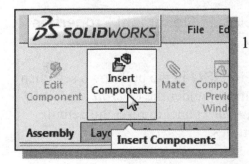

1.  In the *Assembly* toolbar, select the **Insert Component** command by left-clicking the icon.

2.  Use the **browse** option to locate and select the ***Base-Plate*** part file: **Base-Plate.sldprt**.

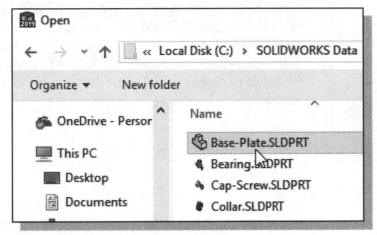

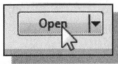

3.  Click on the **Open** button to retrieve the model.

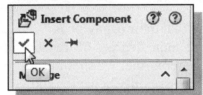

4.  Click **OK** in the *Property Manager* to place the *Base-Plate* aligned to the **origin** of the default datum planes.

## Insert the Second Component

➢   We will retrieve the *Bearing* part as the second component of the assembly model.

1.  In the *Assembly* toolbar, select the **Insert Component** command by left-clicking the icon.

2.  Select the ***Bearing*** design (part file: **Bearing.sldprt**) in the browser. And click on the **Open** button to retrieve the model.

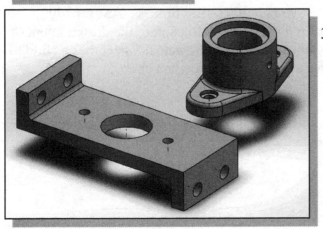

3.  Place the *Bearing* toward the upper right corner of the graphics window, as shown in the figure. Click once with the **left-mouse-button** to place the component. (**NOTE:** Your *Bearing* may initially be oriented differently. We will apply assembly mates to orient the *Bearing*.)

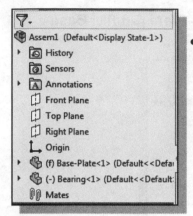

- Inside the *Feature Manager Design Tree*, the retrieved parts are listed in their corresponding order. The **(f)** in front of the *Base-Plate* filename signifies the part is **fixed** and all *six degrees of freedom* are restricted. The number behind the filename is used to identify the number of copies of the same component in the assembly model.

## Degrees of Freedom

- Each component in an assembly has six **degrees of freedom (DOF),** or ways in which rigid 3D bodies can move: movement along the X, Y, and Z axes (translational freedom), plus rotation around the X, Y, and Z axes (rotational freedom). *Translational DOF*s allow the part to move in the direction of the specified vector. *Rotational DOF*s allow the part to turn about the specified axis.

➢ It is usually a good idea to fully constrain components so that their behavior is predictable as changes are made to the assembly. Leaving some degrees of freedom open can sometimes help retain design flexibility. As a general rule, we should use only enough assembly mate relations to ensure predictable assembly behavior and avoid unnecessary complexity.

## Assembly Mates

- We are now ready to assemble the components together. We will start by placing assembly constraints on the *Bearing* and the *Base-Plate* using SOLIDWORKS assembly **Mates**. Mates create geometric relationships between assembly components. Mates define the allowable directions of linear or rotational motion of the components, limiting the degrees of freedom. A component can be moved within its degrees of freedom, visualizing the assembly's behavior.

  To assemble components into an assembly, we need to establish the assembly relationships between components. It is a good practice to assemble components the way they would be assembled in the actual manufacturing process. **Assembly Mates** create a parent/child relationship that allows us to capture the design intent of the assembly. Because the component that we are placing actually becomes a child to the already assembled components, we must use caution when choosing mate types and references to make sure they reflect the intent.

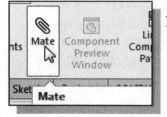

1. In the *Assembly* toolbar, select the **Mate** command by left-clicking once on the icon.

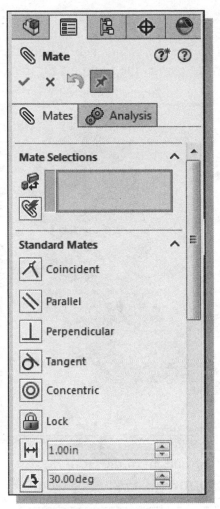

- The *Mate Property Manager* appears on the screen.

- Assembly models are created by applying proper *assembly relations* to the individual components. The Mate relations are used to restrict the movement between parts. Mate relations eliminate rigid body degrees of freedom (**DOF**). A 3D part has *six degrees of freedom* since the part can rotate and translate relative to the three coordinate axes. Each time we add a Mate relation between two parts, one or more DOF is eliminated. The movement of a fully constrained part is restricted in all directions.

- Seven basic types of assembly mates, called Standard Mates, are available in SOLIDWORKS: Coincident, Parallel, Perpendicular, Tangent, Concentric, Distance, and Angle. Each type of mate removes different combinations of rigid body degrees of freedom. Note that it is possible to apply different constraints and achieve the same results.

- Advanced Mates including Gear and Cam Mates are also available.

The Standard Mates are described below:

➢ **Coincident** – positions selected faces, edges, and planes (in combination with each other or combined with a single vertex) so they share the same infinite line. Positions two vertices so they touch.

➢ **Parallel** – places the selected items so they lie in the same direction and remain a constant distance apart from each other.

➢ **Perpendicular** – places the selected items at a 90 degree angle to each other.

➢ **Tangent** – places the selected items in a tangent mate (at least one selection must be a cylindrical, conical, or spherical face).

➢ **Concentric** – places the selections so that they share the same center point.

➢ **Distance** – places the selected items with the specified distance between them.

➢ **Angle** – places the selected items at the specified angle to each other.

## Apply the First Assembly Mate

1.  Note that the *Mate Selections* panel in the *Mate Property Manager* is highlighted. Select the **top horizontal surface** of the *Base-Plate* part as the first part for the Mate command.

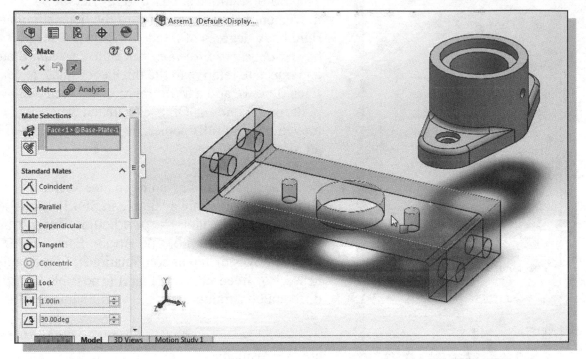

2.  On your own, rotate the displayed model to view the bottom of the *Bearing* part, as shown in the figure below. (**HINT:** Use the up arrow [↑] on the keyboard.)

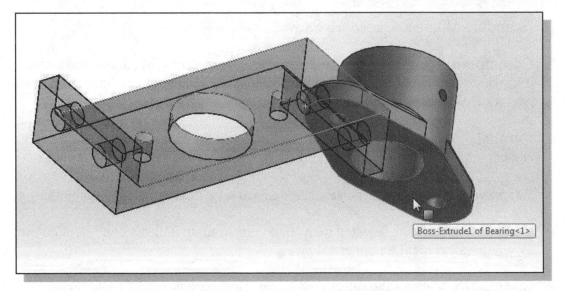

3.  Click on the bottom face of the *Bearing* part as the second part selection to apply the relation. The *Bearing* moves to automatically apply the default Coincident mate.

➤ Notice the two faces selected appear in the *Mate Selections* panel of the *Property Manager*.

➤ Notice the **Tangent** and **Concentric** options are not available. All the mate types are always shown in the *Property Manager*, but only the mates that are applicable to the current selections are available.

➤ Notice the **Coincident** mate is selected. This is the relation we want to apply.

➤ For the **Coincident** mate there are two options for *Mate alignment* – **Aligned** and **Anti-Aligned**. These are displayed at the bottom of the *Standard Mates* panel.

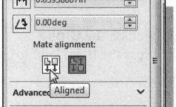

4. On your own, select the **Aligned** option at the bottom of the *Standard Mates* panel and observe the effect on the assembly.

5. Select the **Anti-Aligned** option.

6. Select the **Isometric View** to adjust the display of the assembly model.

7. Click the **Accept** button in the *Property Manager* to accept the settings and create the **Anti-Aligned Coincident** mate.

## Apply a Second Mate

❖ The Mate command can also be used to align axes of cylindrical features.

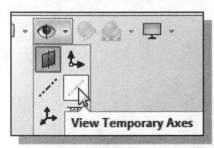

1. On your own, turn *ON* the visibility of the **Planes** and the **Temporary Axes**. (**HINT:** Use the *Hide/Show* pull-down menu on the *Heads-up View* toolbar.)

2. Note that the *Mate Selections* panel in the *Mate Property Manager* is highlighted. Select the **axis** of the right counter bore hole of the *Bearing* part.

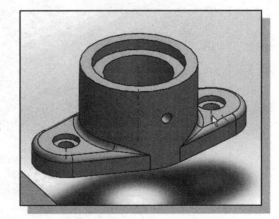

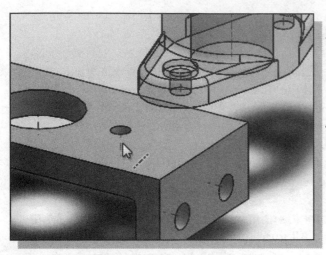

3.  Select the **axis** of the small hole on the *Base-Plate* part, as shown. The *Bearing* moves to automatically apply the default Coincident mate.

4.  Click **Add Mate** in the *Option toolbar* to create the Coincident Mate.

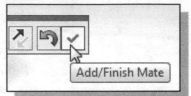

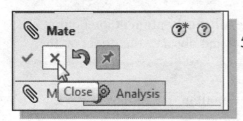

5.  In the *Property Manager*, click on the **Close** button (the red **X** icon), or press the **[Esc]** key, to exit the Mate command.

❖   The *Bearing* part appears to be placed in the correct position. But the DOF for the *Bearing* is not fully restricted; the *Bearing* part can still rotate about the displayed vertical axis.

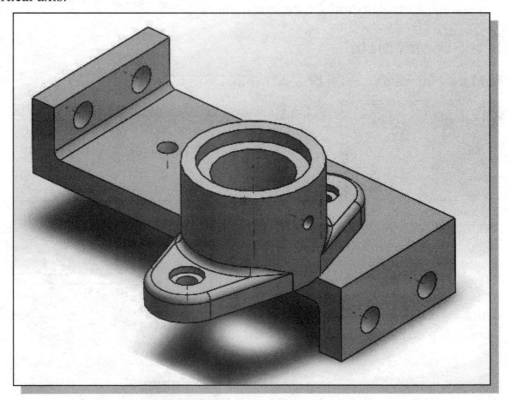

# Constrained Move

❖ To see how well a component is constrained, we can perform a constrained move. A constrained move is done by dragging the component in the graphics window with the left-mouse-button. A constrained move will honor previously applied assembly mates. That is, the selected component and parts mated to the component move together in their constrained positions. A fixed component remains fixed during the move.

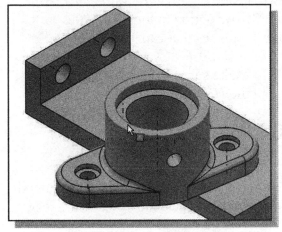

1. Inside the graphics window, move the cursor on top of the top surface of the *Bearing* part as shown in the figure.

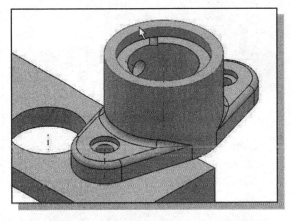

2. Press and hold down the left-mouse-button and drag the *Bearing* part downward.

❖ The *Bearing* part can freely rotate about the mated axis.

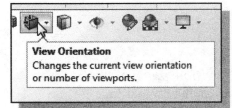

3. On your own, use the rotation commands to view the alignment of the *Bearing* part.

4. Rotate the *Bearing* part and adjust the display roughly as shown in the figure.

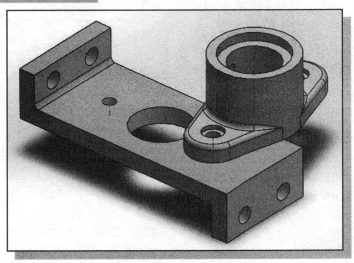

# Apply a Third Mate

- Besides selecting the surfaces of solid models to apply constraints, we can also select the established planes to apply the assembly mates. This is an additional advantage of using the *BORN technique* in creating part models. For the *Bearing* part, we will apply a **Mate** to two of the planes and eliminate the last rotational DOF.

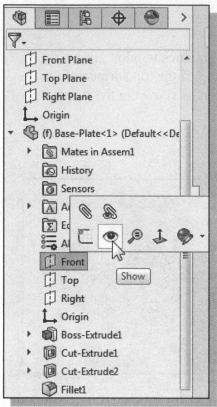

1. On your own, inside the *Feature Manager Design Tree* use the **Show** option to toggle *ON* the visibility for the planes that can be used for alignment of the *Base-Plate* and the *Bearing* parts. (**HINT:** Make sure the visibility of *Planes* is toggled *ON* in the *Hide/Show* pull-down menu.)

2. Press the **[Esc]** key to ensure that no object is selected.

3. In the *Assembly* toolbar, select the **Mate** command by left-clicking once on the icon.

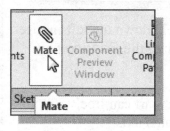

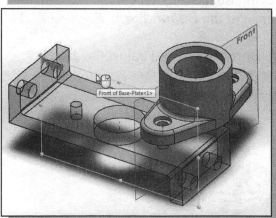

4. Select the *visible plane* of the *Base-Plate* part as the first surface for the **Mate** command.

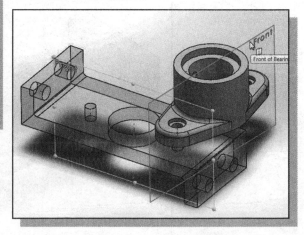

5. Select the *visible plane* of the *Bearing* part as the second surface for the **Mate** command.

❖ The Coincident mate is executed by default and makes two planes coplanar with the **Aligned** option automatically selected. (**NOTE:** If you did not create the base plate with the front plane aligned with the middle of the plate, you will need to select the corresponding plane.)

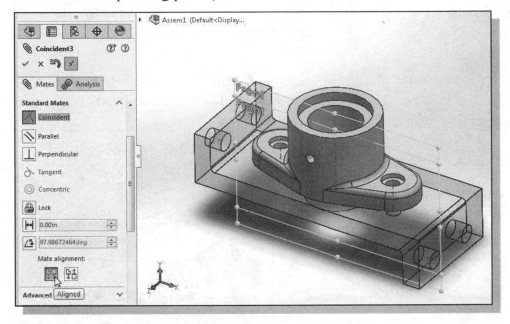

6. On your own, select the **Anti-aligned** option at the bottom of the *Standard Mates* panel and observe the effect on the assembly.

7. Select the **Aligned** option again so that the planes are aligned as shown.

8. Click the **OK** button in the *Property Manager* to accept the settings and create the **Aligned Coincident** mate.

9. In the *Property Manager*, click on the **Close** button (or press the **[Esc]** key) to exit the Mate command.

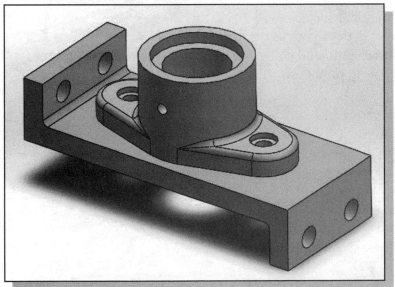

❖ Note the assembly is *fully defined.*

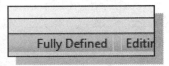

## Inserting the Third Component

➢ We will retrieve the *Collar* part as the third component of the assembly model.

1. In the *Assembly* toolbar, select the **Insert Component** command by left-mouse-clicking the icon.

2. Select the *Collar* design (part file: **Collar.sldprt**) in the browser and click on the **Open** button to retrieve the model.

3. Insert the *Collar* part toward the lower left corner of the graphics window, as shown in the figure.

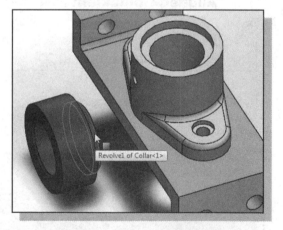

## Applying Concentric and Coincident Mates

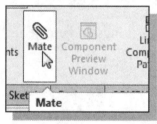

1. In the *Assembly* toolbar, select the **Mate** command by left-clicking once on the icon.

2. Select the external cylindrical surface of the *Collar* part as shown in the figure as the first part for the Mate command. (**HINT:** Use the **arrow** keys on the keyboard to rotate the view.)

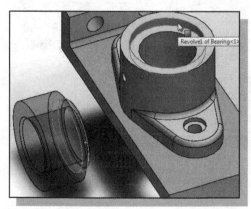

3. Select the internal cylindrical surface of the *Bearing* part as shown in the figure as the second part for the Mate command.

4. Notice in the *Mate Property Manager* that the Concentric and Anti-Aligned options are automatically selected. On your own, select the correct alignment (Aligned or Anti-Aligned) for your assembly. Click the **Accept** button in the *Property Manager* to accept the settings and create the **Concentric** mate.

5. On your own, move the ***Collar*** part to a position like that in the figure. Notice that the move is constrained to maintain the concentric relationship.

6. Activate the **Mate** command and select the top circular surface of the *Bearing* part as shown in the figure as the first part for a new **Mate** command.

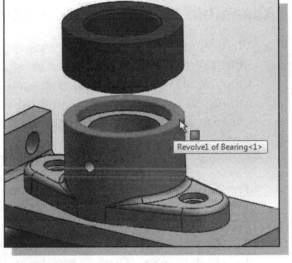

7. Select the circular surface of the *Collar* part as shown in the figure as the second part for a new **Mate** command.

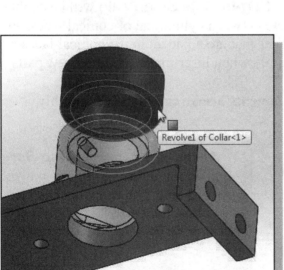

8. Notice in the *Mate Property Manager* that the **Coincident** and **Anti-Aligned** options are automatically selected. Click the **OK** button in the *Property Manager* to accept the settings and create the **Anti-Aligned Coincident** mate.

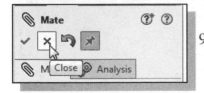

9. In the *Property Manager*, click on the **Close** button (or press the **[Esc]** key) to exit the Mate command.

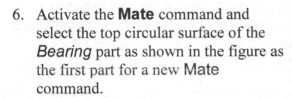

10. On your own, change the view to **Isometric**.

11. Note that the *Collar* part is not fully constrained. One rotational degree of freedom remains open; the *Collar* part can still freely rotate about the vertical axis. Click on the *Collar* and drag with the **left-mouse-button** to observe this constrained motion.

## Assemble the Cap-Screws using SmartMates

❖ We will place two *Cap-Screw* parts to complete the assembly model. We will use SOLIDWORKS **SmartMates** to apply the geometric relations. The **SmartMates** function allows us to create commonly-used mates without using the *Mate Property Manager*. Application of a **SmartMate** involves dragging one part to a mating position with another. In most cases, one mate is created. The type of **SmartMate** created depends on the geometry used to drag the component, and the type of geometry onto which you drop the component.

The **SmartMates** functionality creates multiple mates under certain conditions. If the application finds circular edges to mate, a **Peg-In-Hole** SmartMate will be applied. Two mates are applied: a **Concentric** mate between cylindrical or conical faces, and a **Coincident** mate between the planar faces that are adjacent to the conical faces. This is the type of **SmartMate** we will apply when inserting the *Cap-Screw* parts.

• **Applying a SmartMate when two components are already in the assembly**

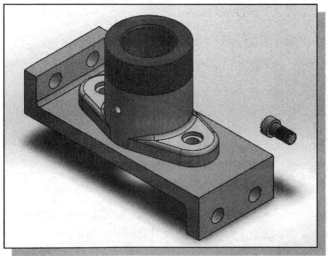

1. On your own, insert a ***Cap-Screw*** part into the assembly as shown.

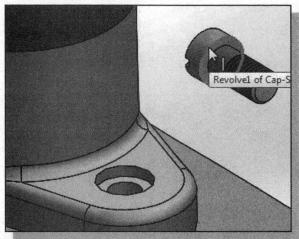

2. Select the circular edge of the *Cap-Screw* head as shown. Press and hold down the left-mouse-button.

3. Press down and hold the **[Alt]** key. While holding the **[Alt]** key, drag the circular edge of the *Cap-Screw* onto the circular edge of the counter sink in the *Bearing* part as shown. Notice the **Peg-In-Hole** mate icon appears.

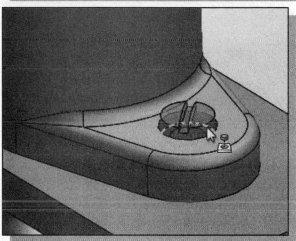

4. Release **[Alt]** while still holding the mouse key down.

5. If the alignment condition is incorrect (for example: screw is upside down), press the **[Tab]** key while still holding the left mouse button down.

6. When the screw is aligned correctly, release the left-mouse-button to perform the SmartMate.

- **Applying a SmartMate to a new component while adding it to the assembly**

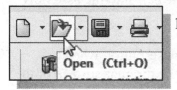

1. Click once with the **left-mouse-button** on the **Open** icon on the *Menu Bar*.

2. Select the *Cap-Screw* design (part file: **Cap-Screw.sldprt**) in the browser, and click on the **Open** button to retrieve the model.

3. Select the **Tile Horizontally** option from the *Window* pull-down menu.

4. In the *Cap-Screw* window, Rotate and Zoom so that the circular edge of the screw head is visible as shown in the figure.

5. In the *Assembly* window, Pan, Rotate, and Zoom so that the counter bore of the *Bearing* appears as shown in the figure.

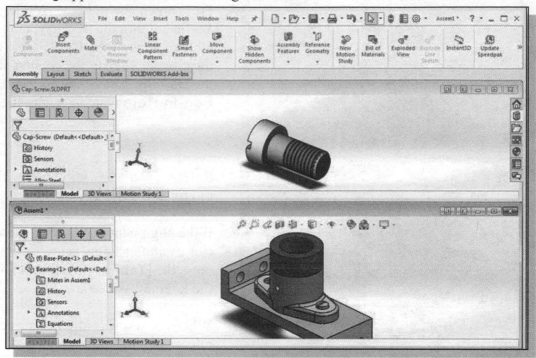

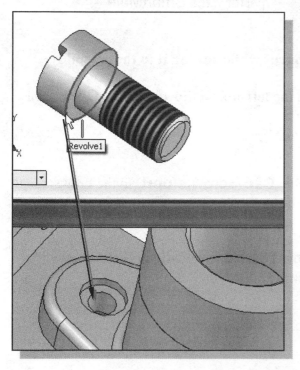

6. Using the left-mouse-button, **click and drag** the **circular edge** of the head of the *Cap-Screw* part into the assembly window.

7. Release the mouse-button when the cursor is over the circular surface of the counter bore, applying the Peg-In-Hole SmartMate.

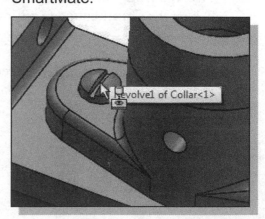

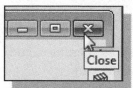

8.  Close the *Cap-Screw* window.

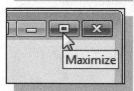

9.  Maximize the *Assem1* window.

10. On your own, change the view to **Isometric**.

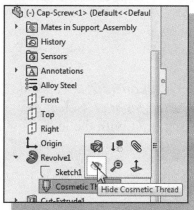

11. Notice the iso-circles representing the bottom of the **Cosmetic Threads** on the screws are visible. In the *Feature Manager Design Tree*, **right-click** on the **Cosmetic Thread1** icon under **Cap-Screw(1)** and select **Hide**.

❖ Notice, this setting affects both *Cap-Screws*.

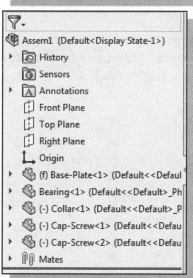

•   Inside the *Feature Manager Design Tree*, the retrieved parts are listed in the order they are placed. The number behind the part name is used to identify the number of copies of the same part in the assembly model. Move the cursor to the last part name and notice the corresponding part is highlighted in the graphics window.

•   The **(f)** symbol indicates the part is fixed. The **(-)** symbol indicates the part has at least one degree of freedom. The *Collar* and the two *Cap Screws* each have one rotational degree of freedom. Click and drag on each of them to observe the rotational freedom.

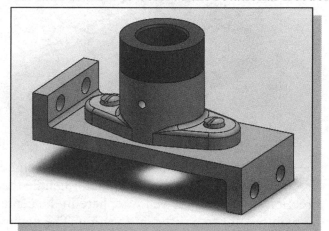

## Exploded View of the Assembly

Exploded assemblies are often used in design presentations, catalogs, sales literature, and in the shop to show all of the parts of an assembly and how they fit together. In SOLIDWORKS, exploded views are created by selecting and dragging parts in the graphics area.

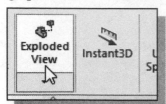

1. In the *Assembly* toolbar, select the **Exploded View** command by left-clicking once on the icon.

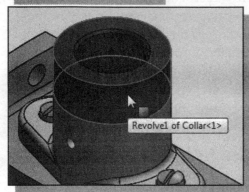

2. The message "*Select component(s)...*" appears in the *Explode Property Manager*. Select the *Collar* part by clicking on it with the left-mouse-button.

3. Notice a manipulator appears in the graphics area. Move the pointer over the manipulator handle that points in the upward direction, and **drag the manipulator** handle to the location shown in the figure. **Right-click** once to place the part at the new location.

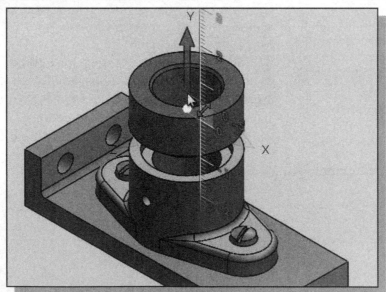

❖ The exploded view component movements can be made along the major axis directions by dragging the manipulator handle that points in the corresponding direction. Dragging the yellow sphere in the center of the manipulator will move the manipulator to a different location. The manipulator can be dropped on an edge or face and an axis of the manipulator will align with the edge or face.

4. On your own, repeat the above steps and create an exploded assembly by repositioning the components as shown in the figure.

5. Click **OK** in the *Property Manager* to close the Exploded View command.

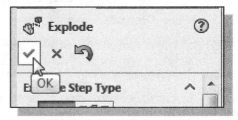

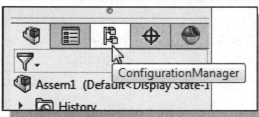

6. Select the **Configuration Manager** tab.

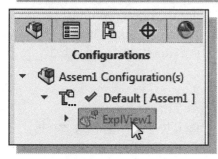

7. In the *Configuration Manager*, expand the tree to reveal the Exploded View icon labeled ExpView1. **Double-click** on the **ExpView1** icon to collapse the exploded view.

❖ Note that the components are reset back to their assembled positions, based on the applied assembly mates.

8. **Double-click** on the **ExpView1** icon again to return to the exploded view.

9. On your own, return to the collapsed view.

10. Select the **Feature Manager Design Tree** tab to return to the *Design Tree*.

## Save the Assembly Model

1.  Select the **Save As** option from the *File* pull-down menu.

2.  In the *Save As* pop-up window, enter **Support_Assembly** for the *File name* and **Shaft-Support Assembly** for the *Description*.

3.  Click **Save** to save the file.

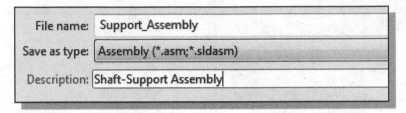

4.  If the SOLIDWORKS window appears, click **Save All**.

## Editing the Components

*   The associative functionality of SOLIDWORKS allows us to change the design at any level, and the system reflects the changes at all levels automatically.

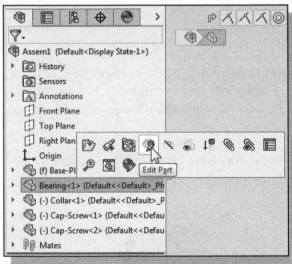

1.  Inside the *Feature Manager Design Tree*, move the cursor on top of the **Bearing** part. Left-click once to bring up the option menu and select **Edit Part** in the option list.

❖  Note that we are automatically switched back to *Part Editing* mode. The Bearing component portion of the *Feature Manager Design Tree* appears blue to denote that the part model is active.

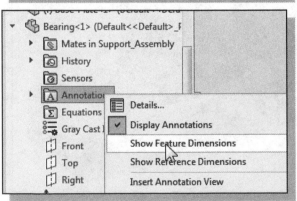

2.  Expand the Bearing portion of the *Design Tree*. Move the cursor over the **Annotations** icon **under the Bearing**. Click once with the right-mouse-button to open the pop-up option menu. Select the **Show Feature Dimensions** option.

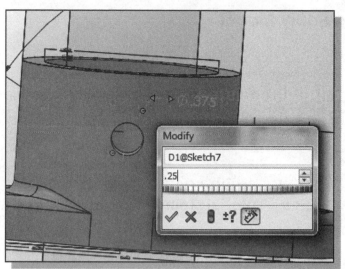

3. On your own, adjust the diameter of the small *Drill Hole* to **0.25** as shown.

4. Click on the **Rebuild** button in the *Menu Bar* to proceed with updating the model.

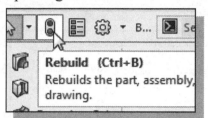

5. On your own, turn *OFF* the visibility of the feature dimensions.

6. Inside the graphics window, click once with the **right-mouse-button** to display the *option menu*.

7. Select **Edit Assembly:Support_Assembly** in the pop-up menu to exit *Part Editing* mode and return to *Assembly* mode.

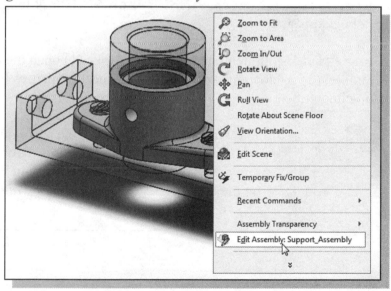

➤ SOLIDWORKS has updated the part in all levels and in the current *Assembly* mode. On your own, open the *Bearing* part to confirm the modification is performed.

8. On your own, save the Support_Assembly file. If the *Save Modified Documents* window appears, select **Save All**.

9. If the SOLIDWORKS *Rebuild* window appears, select **Rebuild and save the document**.

## Set up a Drawing of the Assembly Model

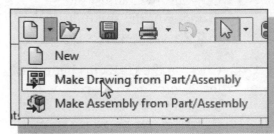

1. Click on the **arrow** next to the **New** icon on the *Menu Bar* and select **Make Drawing from Part/Assembly** in the pull-down menu.

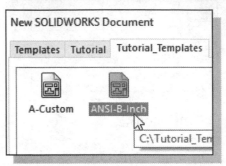

2. In the *New SOLIDWORKS Document* window, select the **Tutorial_Templates** tab.

3. Select the **ANSI-B-Inch** template file icon. This is the custom drawing template you created in Chapter 8. (If your ANSI-B-Inch template is not available, see note below.)

4. Select **OK** in the *New SOLIDWORKS Document* window to open the new drawing file.

➢ The new drawing file opens using the **ANSI-B-Inch.SLDDRW** template. The B-size drawing sheet appears.

❖ **NOTE:** If your ANSI-B-Inch template is not available: (1) open the default **Drawing** template; (2) set document properties and sheet properties following instructions in chapter 8.

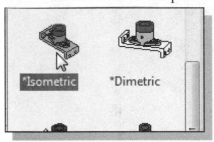

5. The *View Palette* appears in the *task pane*. Views can be dragged from the *View Palette* onto the drawing sheet to create a base drawing view. Click on the **Isometric** view in the *View Palette* as shown.

6. Hold the **left-mouse-button** down and **drag** to place the **isometric view** as shown.

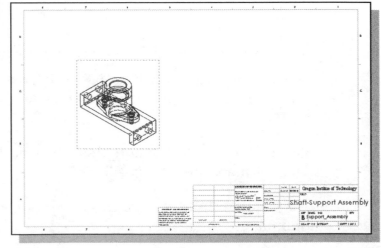

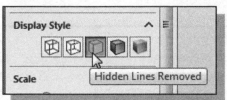

7. In the *Projected View Property Manager*, select the **Hidden Lines Removed** option for the *Display Style*.

8. Left-click on the **OK** icon in the *Drawing View Property Manager* to accept the view.

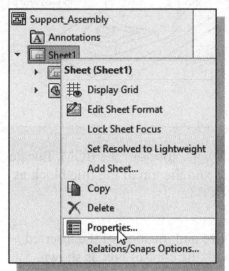

9. In the *Feature Manager Design Tree*, **right-click** on the **Sheet** icon to open the pop-up option menu and select **Properties**.

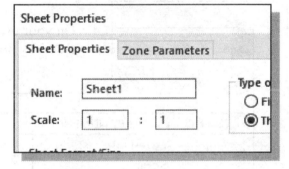

10. In the Sheet Properties window, set the scale to **1:1** and click **Apply Changes**.

## Creating a Bill of Materials

A *Bill of Materials (BOM)*, or parts list, can be automatically inserted into a drawing using SOLIDWORKS. A Bill of Materials is an example of a SOLIDWORKS *Annotation Table*. A SOLIDWORKS Annotation Table can be located in the drawing by clicking in the graphics area or by using a **Table** Anchor.

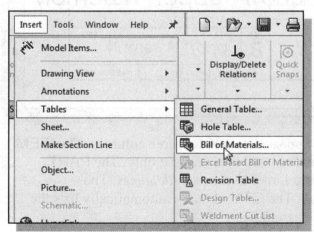

1. Under the **Insert** pull-down menu, select **Tables**, then **Bill of Materials**.

2. The message *"Select a drawing view to specify the model for creating the Bill of Materials"* appears in the *Bill of Materials Property Manager*. Select on the **shaft-support assembly view** in the graphics area by clicking on it once with the **left-mouse-button**.

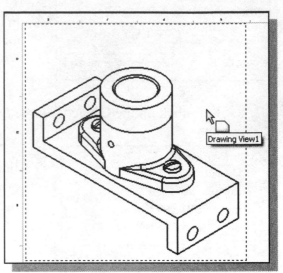

3. Click on **OK** in the *Bill of Materials Property Manager* to insert the BOM. Locate the BOM so that it is aligned with the right border and the top of the title block as shown. Click the left-mouse-button to insert the BOM.

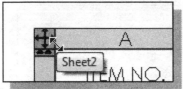

4. On your own, adjust the size and location of the inserted BOM by drag and drop with the control box as shown.

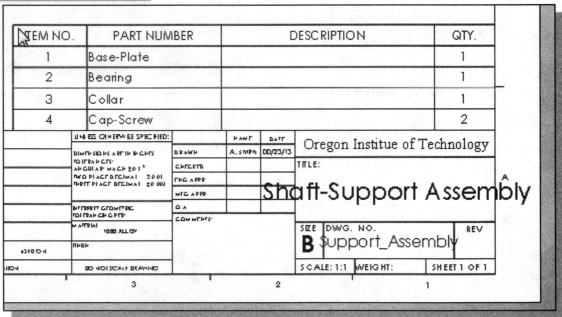

| ITEM NO. | PART NUMBER | DESCRIPTION | QTY. |
|---|---|---|---|
| 1 | Base-Plate | | 1 |
| 2 | Bearing | | 1 |
| 3 | Collar | | 1 |
| 4 | Cap-Screw | | 2 |

Oregon Institue of Technology

TITLE:

**Shaft-Support Assembly**

SIZE: **B**  DWG. NO. Support_Assembly  REV

SCALE: 1:1  WEIGHT:  SHEET 1 OF 1

❖ The default SOLIDWORKS BOM appears. It includes three columns. The **ITEM NO.** column lists the individual parts inserted into the assembly. The **PART NUMBER** column includes **Property Links** to the part *filenames*. The **DESCRIPTION** column is left blank. The **QTY.** column automatically lists the quantity of each part included in the assembly.

## Editing the Bill of Materials

There are many ways in which the BOM table can be modified. We will add an additional Material column. This could be done simply by entering text. However, we will use SOLIDWORKS Property Links. We will first set the material for each part in the part files. Then we will place the appropriate Property Links in the BOM.

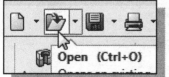

1. Click once with the **left-mouse-button** on the **Open** icon on the *Menu Bar*.

2. Select the ***Base-Plate*** design (part file: **Base-Plate.sldprt**) in the browser, and click on the **Open** button to retrieve the model.

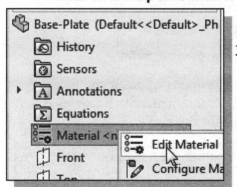

3. In the *Feature Manager Design Tree* for the part, *right click* on the *Material* icon, then select **Edit Material**.

4. In the *Material* pop-up window, expand the **Steel** folder of the selection tree, and select **Alloy Steel**.

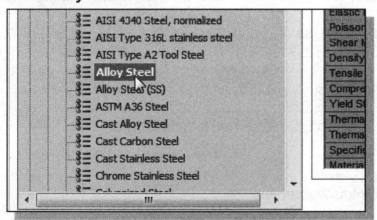
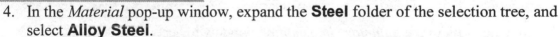

5. Click **Apply** in the *Material* pop-up window to accept the setting.

6. Select **Close** to close the window.

❖ The material property has been selected for the *Base-Plate* and is defined as a SOLIDWORKS user-defined custom property.

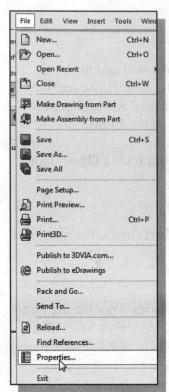

7. Select **Properties** from the *File* pull-down menu.

● The *Summary Information* window appears with the table of *Custom Properties* blank.

8. Click in the box under *Property Name* and enter **Material**.

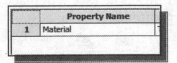

9. Click in the box under *Value/Text Expression* and use the pull-down option list to select **Material**.

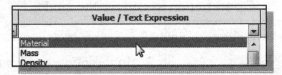

❖ Notice the **Material** custom property appears and the **Evaluated Value** in the table automatically becomes Alloy Steel, the material selected for the part.

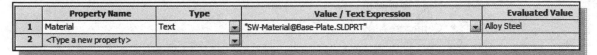

| | Property Name | Type | Value / Text Expression | Evaluated Value |
|---|---|---|---|---|
| 1 | Material | Text | "SW-Material@Base-Plate.SLDPRT" | Alloy Steel |
| 2 | <Type a new property> | | | |

10. Click **OK** in the *Summary Information* window to accept the settings.

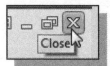

11. On your own, **save** and **close** the *Base-Plate* part file. If the SOLIDWORKS warning window appears, select **Yes**.

12. On your own, repeat the steps above to set the **Material** and the **Material Custom Property** for the *Bearing*, *Collar*, and *Cap-Screw* parts. Use the following: *Bearing* – Grey Cast Iron; *Collar* – Tin Bearing Bronze; *Cap-Screw* – Alloy Steel.

❖ The SOLIDWORKS custom property **Material** for each part is now available for use in **Property Links**. We will now use the **Property Links** in the BOM.

13. Move the cursor over the top of the third column in the BOM. When the **Column** icon appears, **right-click** to open the *option menu*.

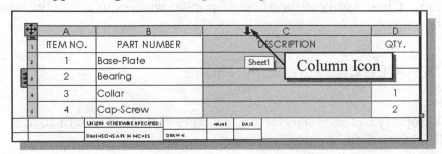

14. Select Insert → Column Right to add a new column to the right side.

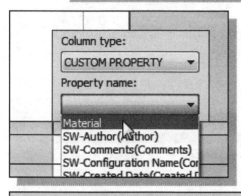

15. In the Column type dialog box, choose Material as the Custom Property Name as shown.

16. Click on the sheet, away from the *Column Property* panel to accept the settings.

➤ Notice the corresponding material appears for each part due to the property link.

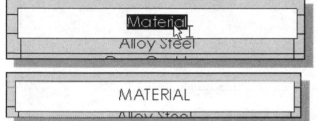

17. **Double-click** on the note entry Material in the new BOM.

18. Type over the word in all caps.

19. Press [**Enter**] to accept the change.

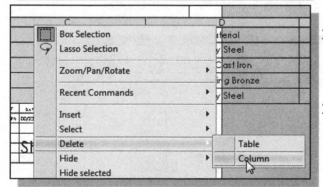

20. Click on the **Description** cell and **delete** the entire Column as shown.

21. On your own, adjust, reposition and resize the BOM so that it is as shown in the below figure.

| ITEM NO. | PART NUMBER | MATERIAL | QTY. |
|---|---|---|---|
| 1 | Base-Plate | Alloy Steel | 1 |
| 2 | Bearing | Gray Cast Iron | 1 |
| 3 | Collar | Tin Bearing Bronze | 1 |
| 4 | Cap-Screw | Alloy Steel | 2 |

| UNLESS OTHERWISE SPECIFIED: | | NAME | DATE | Oregon Institue of Technology |
|---|---|---|---|---|
| DIMENSIONS ARE IN INCHES TOLERANCES: | DRAWN | A. smith | DD/23/13 | |

22. Select the BOM table by clicking on the upper left corner as shown in the figure.

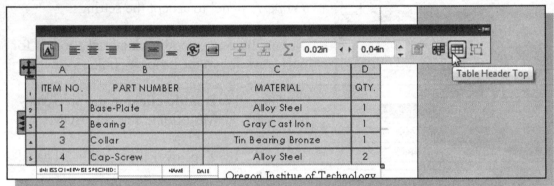

23. In the *pop-up toolbar*, click on the **Table Header Top** icon to set the location of the *Table Header*.

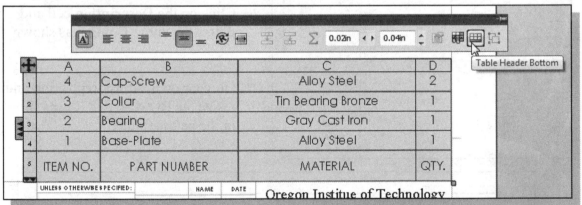

- The Table Header has been adjusted to the bottom and the Item Numbers are also adjusted accordingly as shown.

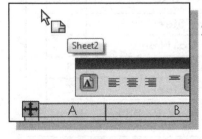

24. Click on the drawing sheet to exit the edit BOM table mode as shown.

## Completing the Assembly Drawing

1.  Press the **[F]** key to fit the drawing to the graphics area.

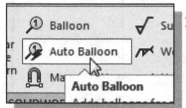

2.  In the *Annotation* toolbar, click on the **Auto Balloon** button. (Note the Balloon command above can also be used to add individual balloons.)

3.  In the *Auto Balloon Property-Manager*, the message "*Select drawing sheet/view(s) to insert Auto Balloon*" is displayed. Click on the assembly view in the graphics area.

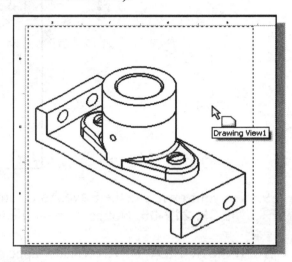

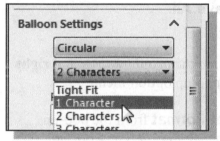

4.  Select **1 Character** for the *Size* option on the *Balloon Setting* panel in the *Property Manager*.

5.  Click **OK** in the *Property Manager* to create the balloons.

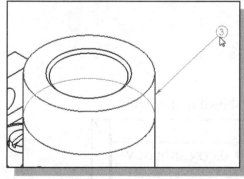

6.  Move the cursor over the balloon labeling the *Collar* part. Click and drag with the **left-mouse-button** to move the balloon to the new location shown in the figure.

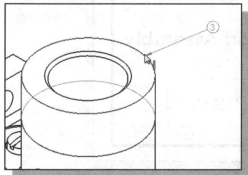

7.  Click and drag to move the arrowhead to the new location shown in the figure.

8. On your own, repeat the above steps to reposition the balloons as shown.

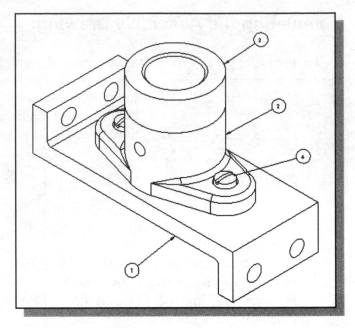

9. On your own, use the **Save As** command and save the drawing with the filename **MECH 212-05**. Notice the new filename appears in the title block.

10. If the SOLIDWORKS window appears, select **Save All**.

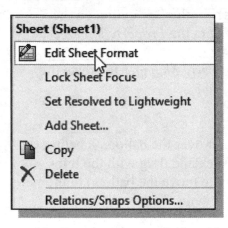

11. Move the cursor into the *graphics area* and **right-click** to open the pop-up option menu.

12. Select **Edit Sheet Format** from the pop-up option menu to change to *Edit Sheet Format* mode.

13. On your own, edit the title block to appear as shown in the figure.

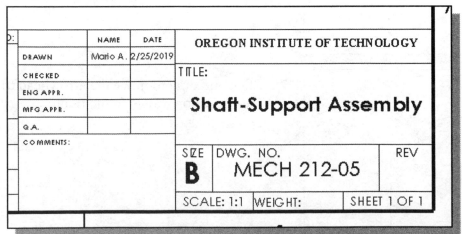

14. On your own, exit the *Edit Sheet Format* mode.

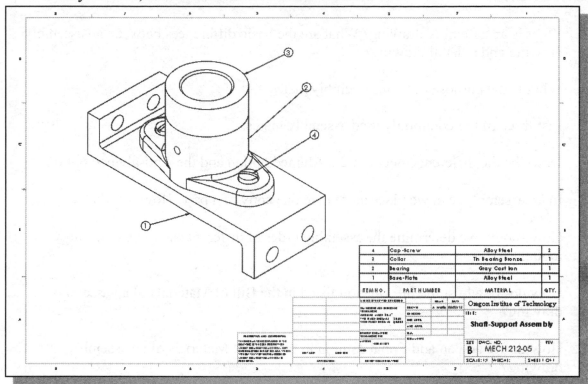

## Exporting the Bill of Materials

❖  The bill of materials can be exported as an *Excel* file.

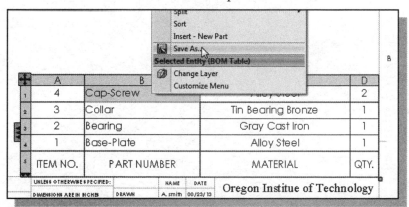

1. Right click the bill of materials on the drawing sheet.

2. Select **Save As** from the *option* menu.

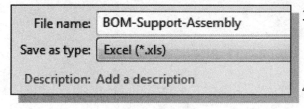

3. Select **Excel** or another appropriate Excel file type for the *Save as type:* option.

4. Enter a *File name* and click **Save**.

5. On your own, open the file using *Microsoft Excel* and examine the BOM.

## Questions:

1. What is an assembly drawing? What are the basic differences between an assembly drawing and a detail drawing?

2. What is the purpose of using assembly mates?

3. List three of the commonly used assembly mates.

4. Describe the difference between the Aligned option and the Anti-aligned option.

5. In an assembly, can we place more than one copy of a part? How is it done?

6. How should we determine the assembly order of different parts in an assembly model?

7. How do we adjust the information listed in the Bill of Materials of an assembly drawing?

8. How do we add an additional column in the Bill of Materials of an assembly drawing?

## Exercises:

1.  **Wheel Assembly** (Create a set of detail and assembly drawings. All dimensions are in mm.)

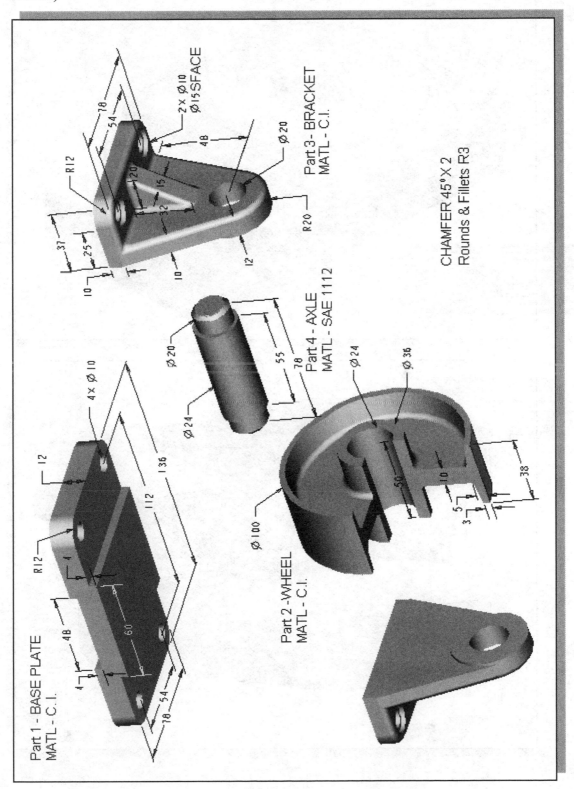

2. **Leveling Assembly** (Create a set of detail and assembly drawings. All dimensions are in mm.)

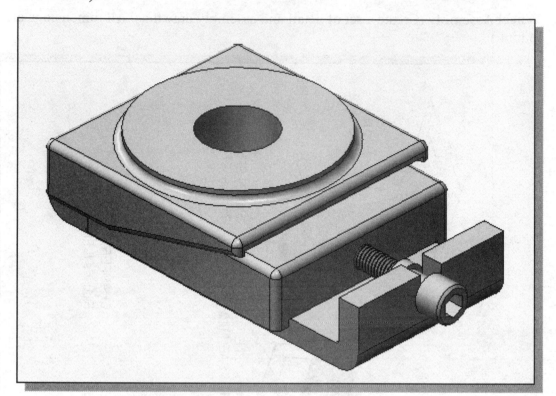

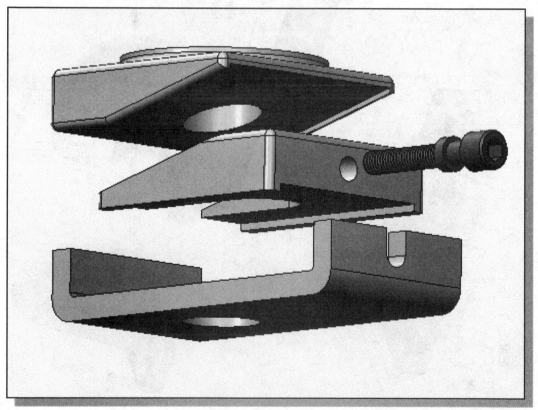

**(a) Base Plate**

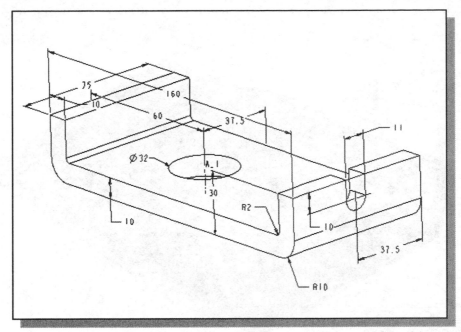

**(b) Sliding Block** (Rounds & Fillets: R3)

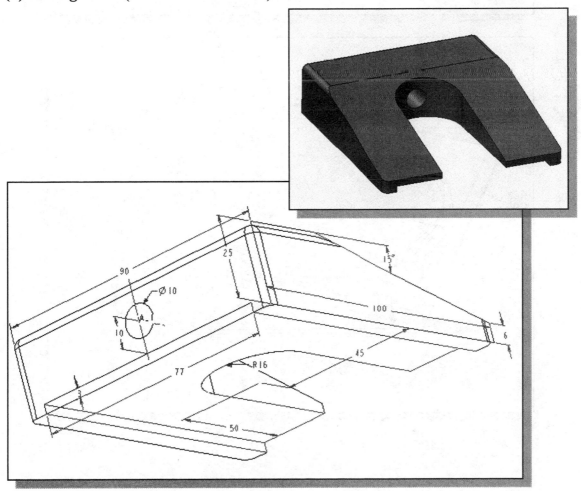

**(c) Lifting Block** (Rounds & Fillets: R3)

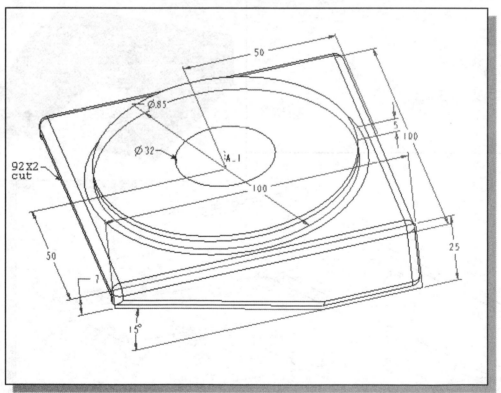

**(d) Adjusting Screw** (M10 × 1.5) (Use the **Threads** or **Coil** command to create the threads.)

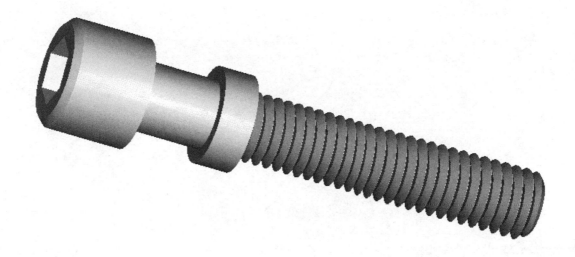

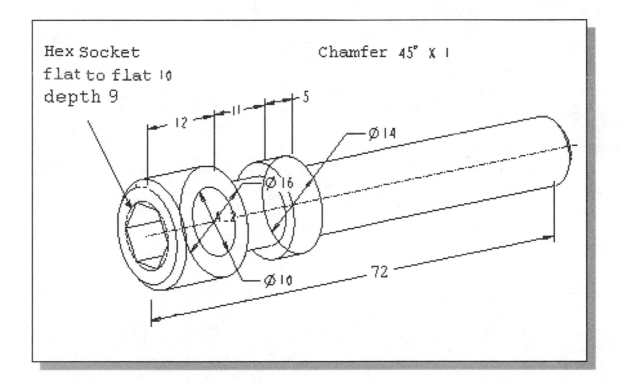

Hex Socket
flat to flat 10
depth 9

Chamfer 45° X 1

12  11  5  Ø14  Ø16  Ø10  72

**Notes:**

# Chapter 16
# CSWA Exam Preparation

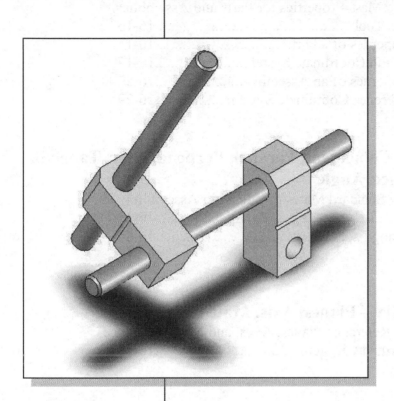

## Learning Objectives

♦ **Control Part Orientation**
♦ **Create a New View Orientation**
♦ **Use the SOLIDWORKS Mass Properties Tool**
♦ **Obtain the Mass and Center of Mass of a Part**
♦ **Obtain the Mass and Center of Mass of an Assembly**
♦ **Use Distance and Angle Mates**
♦ **Create and Use Reference Axes**

Certified Associate Reference Guide

## Certified SOLIDWORKS Associate Exam Objectives Coverage

### Mass Properties

Objectives: Obtaining Mass Properties for Parts and Assemblies.

Mass Properties Tool ..................................................16-16
View Mass Properties of a Part.................................16-17
Relative to Default Coordinate System.....................16-17
View Mass Properties of an Assembly ......................16-29
Relative to Reference Coordinate System .................16-29

### Standard Mates – Coincident, Parallel, Perpendicular, Tangent, Concentric, Distance, Angle

Objectives: Applying Standard Mates to Constrain Assemblies.

Distance Mate .............................................................16-23
Angle Mate, using Faces.............................................16-25

### Reference Geometry – Planes, Axis, Mate References

Objectives: Creating Reference Planes, Axes, and Mate References.

Reference Coordinate System....................................16-28

## Tips about Taking the Certified SOLIDWORKS Associate (CSWA) Examination

1. **Study**: The first step to maximize your potential on an exam is to sufficiently prepare for it. You need to be familiar with the SOLIDWORKS package, and this can only be achieved by doing designs and exploring the different commands available. The Certified SOLIDWORKS Associate (CSWA) exam is designed to measure your familiarity with the SOLIDWORKS software. You must be able to perform the given task and answer the exam questions correctly and quickly.

2. **Make Notes**: Take notes on what you learn either while attending classroom sessions or going through study material. Use these notes as a review guide before taking the actual test.

3. **Time Management**: The examination has a time limit. Manage the time you spend on each question. Always remember you do not need to score 100% to pass the exam. Also keep in mind that some questions are weighed more heavily and may take more time to answer. You can flip back and forth to view different problems during the test time by using the arrow buttons. If you encounter a question you cannot answer in a reasonable amount of time, use the *Save As* feature in SOLIDWORKS to save a copy of the file, and move on to the next question. You can return to any question and enter or change the answer as long as you do not hit the [**End Examination**] button.

4. **Use the SOLIDWORKS Help System**: If you get confused and can't think of the answer, remember the SOLIDWORKS Help System is a great tool to confirm your considerations. In preparing for the exam, familiarize yourself with the help utility organization (e.g., Contents, Index, Search options).

5. **Use Internet Search**: Use of an internet search utility is allowed during the test. If a test question requires general knowledge, for example definitions of engineering or drafting concepts (stress, yield strength, auxiliary view, etc.), remember the internet is available as a tool to assist in your considerations.

6. **Use Common Sense**: If you are unable to get the correct answer and unable to eliminate all distracters, then you need to select the best answer from the remaining selections. This may be a task of selecting the best answer from amongst several correct answers, or it may be selecting the least incorrect answer from amongst several poor answers.

7. **Be Cautious and Don't Act in Haste**: Devote some time to ponder and think of the correct answer. Ensure that you interpret all the options correctly before selecting from available choices. Don't go into panic mode while taking a test. Use the *Arrow Buttons* to review each question. When you are confident that you have answered all questions, end the examination using the [**End Examination**] button to submit your answers for scoring. You will receive a score report once you have submitted your answers.

8. **Relax before exam**: In order to avoid last minute stress, make sure that you arrive 10 to 15 minutes early and relax before taking the exam.

Certified Associate Reference Guide

## Introduction

In this lesson, we will create and examine part and assembly files in a manner consistent with the format of the CSWA exam. This is a multiple-choice exam. You are asked to build complex models and assemblies and your work is assessed in a multiple-choice format by asking the value of parameters such as the mass or center of mass. These can be retrieved using the SOLIDWORKS **Mass Properties** tool. In answering questions related to the location of the center of mass, it is important to ensure that the origin and coordinate system axes are consistent between your model and the exam problem as posed.

We will first create a part based on a dimensioned drawing with the origin and coordinate axes annotated. We will create the part, apply the correct material properties, and retrieve the mass and center of mass data. In building the model, we will ensure that it is oriented relative to the default coordinate system in the manner indicated in the problem. While the part is relatively simple, the procedure – taking care to locate the part, applying material properties, and retrieving mass properties – is the same as should be followed for many CSWA exam questions. The part requires the creation of an inclined **Reference Plane** to use in creating a sketch.

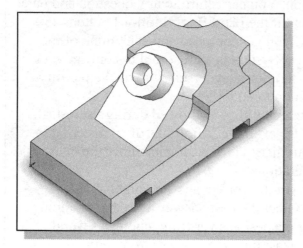

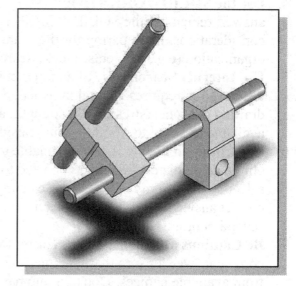

We will then create an assembly using multiple copies of two mating parts. The assembly will require the use of **Distance** and **Angle** mates, based on the dimensions given in the problem. In this case, we will not orient the parts or assembly based on the problem definition. Instead, after the assembly has been completed, we will create **Reference Axes** to match those in the problem statement and use these reference axes when retrieving the center of mass.

## The Part Problem

We will model the part shown below and use the **Mass Properties** tool to determine the mass and center of mass of the part. The problem is presented below in a manner similar to that encountered on the CSWA exam.

Build the part shown below.
Unit system: MMGS (millimeter, gram, second)
Part origin: As shown.
Part material: Ductile iron     Density: 0.0071 g/mm$^3$

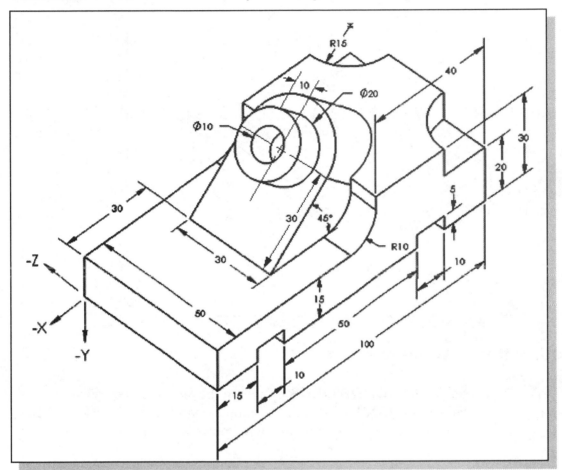

    (1) What is the overall mass of the part in grams?
    (2) Where is the center of mass of the part?

Before starting to build the model, it is important to select a strategy for controlling the origin and orientation of the X, Y, and Z axes since the center of mass must be defined relative to a known reference system. In this case the axis directions are defined in the drawing. We can either (1) build the part aligned to the default or global coordinate system as specified in the figure, or (2) build the part independent of this constraint, and after the part is complete, create a reference coordinate system with axes aligned with the part as shown in the figure. We will use the first strategy for the part problem. The second strategy will be used in the second problem to be done later in the chapter.

## Strategy for Aligning the Part to the Default Axis System

The first feature to be created will be the base feature of the part. We must determine which plane to use to define the first sketch in order to align the part correctly. In the dimensioned drawing defining the part, we can see that the base rests on the X-Z plane (i.e., the Top Plane) with the positive X-axis aligned with the long dimension and the positive Z-axis aligned with the short dimension. We therefore want to create the base feature as shown below, with the origin located at the * and the X, Y and Z axes aligned as shown.

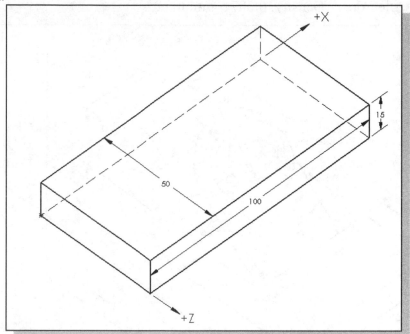

## Creating the Base Feature

1. Select the **SOLIDWORKS** option on the *Start* menu or select the **SOLIDWORKS** icon on the desktop to start SOLIDWORKS. The SOLIDWORKS main window will appear on the screen.

2. Select the **Part** icon with a single click of the left-mouse-button to start a new part document.

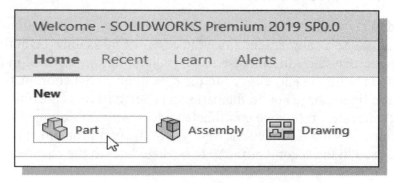

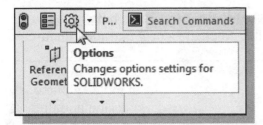

3. Select the **Options** icon from the *Menu Bar* to open the *Options* dialog box.

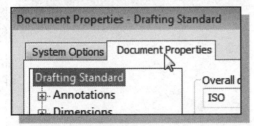

4. Select the **Document Properties** tab.

5. Set the *Overall drafting standard* to **ISO** to reset the default setting.

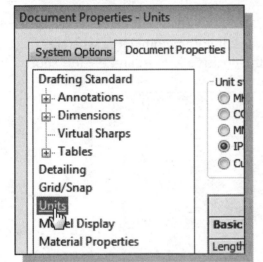

6. Click **Units** as shown in the figure.

7. Select **MMGS (millimeter, gram, second)** under the *Unit system* options.

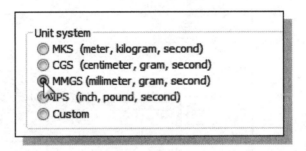

8. Select **None** in the *Decimals* spin box for the *Length units* as shown to define the display of digits after the decimal point.

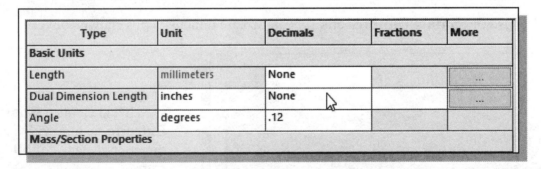

| Type | Unit | Decimals | Fractions | More |
|------|------|----------|-----------|------|
| **Basic Units** | | | | |
| Length | millimeters | None | | ... |
| Dual Dimension Length | inches | None | | ... |
| Angle | degrees | .12 | | |
| **Mass/Section Properties** | | | | |

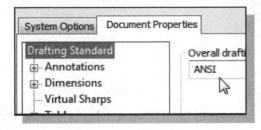

9. Select **ANSI** to set the *Overall drafting standard*. Click **OK** to accept the settings.

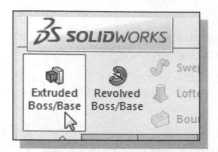

10. Select the **Extruded Boss/Base** button on the *Features* toolbar to create a new extruded feature.

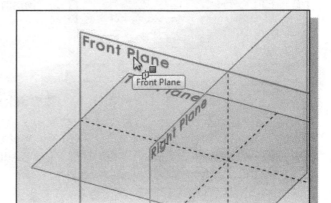

11. Select the **Front** (XY) **Plane** as the sketch plane for the new sketch.

12. On your own, create the sketch with the lower left corner coincident with the **origin**, the bottom long edge aligned with the positive X-axis, and the left short edge aligned with the positive Y-axis as shown below.

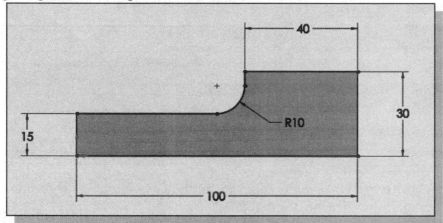

13. Use the **Smart Dimensions** tool and create dimensions as shown above.

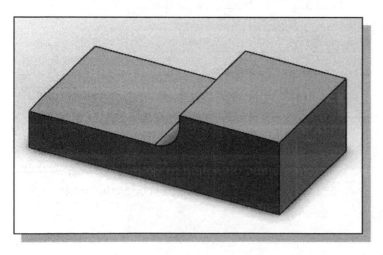

14. On your own, exit the sketch and create a **Boss Extrude** feature with a Z-direction distance of **50 mm**.

# Creating a New View Orientation

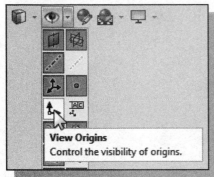

1.  Use the *Hide/Show* pull-down menu on the *Heads-up View* toolbar to turn *ON* the **origin's** visibility.

2.  On your own, set the *View Orientation* to **Isometric** as shown.

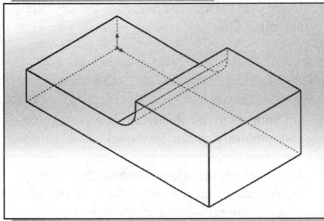

> ➤ Notice the axes are aligned correctly relative to the part, but the default isometric view does not correspond to the view orientation as defined in the problem statement. We will create a new view orientation which better matches the view described in the problem statement (page 16-5).

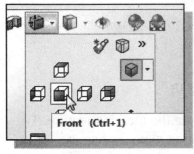

3.  Left-click on the **View Orientation** icon on the *Heads-up View* toolbar to reveal the *View Orientation* pull-down menu. Select the **Front** command.

4.  Hit the right arrow button [→] **three times** to rotate the view 45°.

5.  Hit the down arrow button [↓] **three times** to rotate the view 45°.

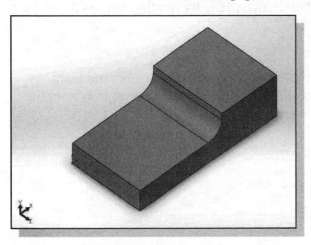

> ➤ The view should appear as shown here. This view orientation corresponds to the one in the problem definition. We will want to return to this view as we continue working on the part. SOLIDWORKS allows the user to define a new view orientation for this purpose.

6.  Select **New View** in the *Heads-up View* display toolbar as shown.

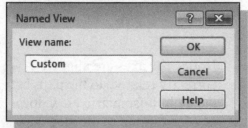

7.  Enter **Custom** as the name for the new view and click **OK**.

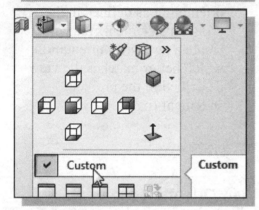

8.  On your own, use the dynamic rotate option and randomly rotate the model.

9.  Left-click on the **View Orientation** icon on the *Heads-up View* toolbar to reveal the *View Orientation* pull-down menu. Select the **Custom** view we created to reset the view orientation.

## Create Reference Planes and a Reference Axis

➢ We will now create the reference geometry needed to create the inclined feature. To create an inclined reference plane, we must select an existing plane and an axis of rotation. The strategy we will use for creating the inclined reference plane involves three steps: (1) create a vertical reference plane offset from the left face by 30 mm; (2) create a reference axis at the intersection of the offset plane and the top face of the base feature; and (3) create the inclined reference plane by rotating the top face of the base feature about the reference axis.

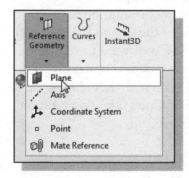

1.  Select the **Reference Geometry** command from the *Features* toolbar, and select the **Plane** option from the pull-down menu.

2.  Move the cursor over the left face of the base feature as shown below. Click once with the **left-mouse-button** to select the face.

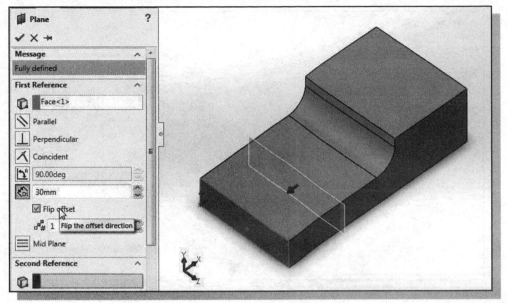

3.  Notice the face, called **Face<1>**, appears in the *First Reference* panel of the *Plane Property Manager*, and the **Offset Distance** option is automatically selected.

4.  Check the **Flip** option box and enter **30 mm** as the offset distance.

5.  Click **OK** in the *Property Manager* to accept the settings and create the new reference plane.

6.  Press the **[Esc]** key once to unselect the newly created reference plane.

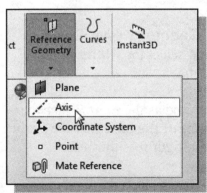

7.  In the *Features* toolbar, select the **Reference Geometry** command by left-clicking the icon.

8.  In the pull-down option menu of the Reference Geometry command, select the **Axis** option.

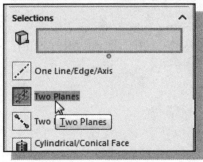

9.  In the *Selections* panel of the *Axis Property Manager*, select the **Two Planes** option. (We will define the reference axis as the intersection of the offset reference plane and top face of the base feature.)

10. Select the offset reference plane, **Plane1**, in the graphics area or the *Design Tree* and the **horizontal face of the base feature** in the graphics area.

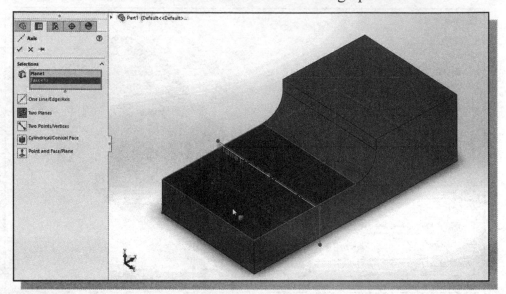

11. Click **OK** in the *Property Manager* to create the new reference axis.

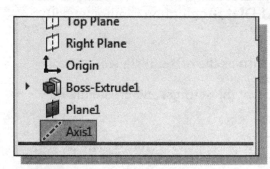

• Notice the newly created reference axis is pre-selected.

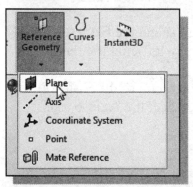

12. Select the **Reference Geometry** command from the *Features* toolbar, and select the **Plane** option from the pull-down menu.

13. The pre-selected axis is used as the *First Reference*. Select the **horizontal face of the base feature** (same plane as shown in the top figure) as the *Second Reference*.

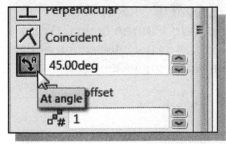

14. In the *Second Reference* panel of the *Plane Property Manager* select the **At Angle** option button.

15. Enter **45 deg** as the angle in the *Property Manager*. The preview of the new plane should appear as in the figure below.

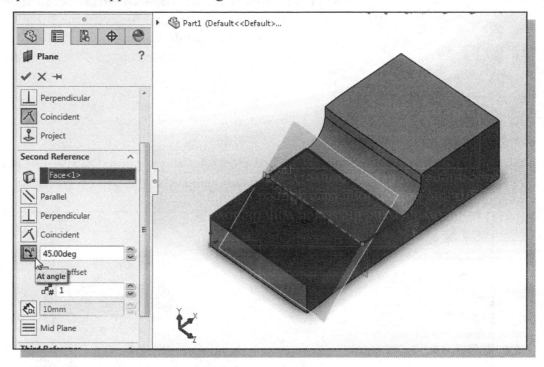

16. Click **OK** in the *Property Manager* to create the new reference plane.

➤ We now have the correct inclined reference plane on which to sketch the inclined feature.

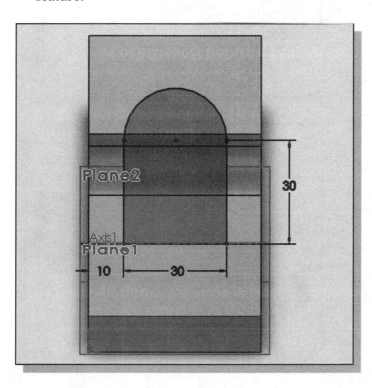

17. On your own, create an **Extruded Boss/Base** feature, with the sketch shown, using the inclined reference plane, **Plane2**, as the sketch plane. (**NOTE:** Use the **Normal to** view orientation to view the sketch plane as shown.)

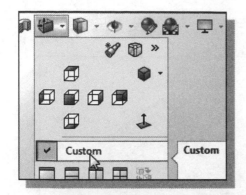

18. Left-click on the **View Orientation** icon on the *Heads-up View* toolbar to reveal the *View Orientation* pull-down menu. Select the **Custom** view that we created earlier.

19. On your own, complete an **Extruded Boss** feature as shown below. (**NOTE:** Use the **Reverse Direction** option if necessary. Use the **Up To Next** option for the *end condition*.)

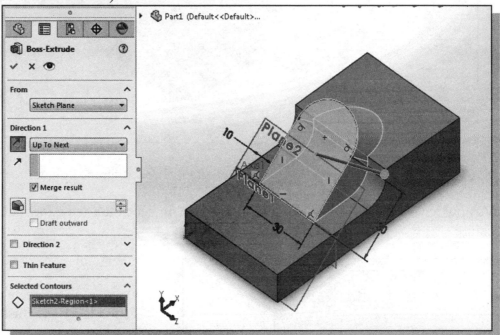

20. On your own, create another **Extruded Boss** feature with the sketch as shown using the inclined face as the sketch plane.

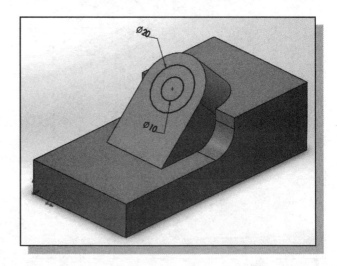

21. Complete the *10 mm* **Extruded Boss** feature as shown below.

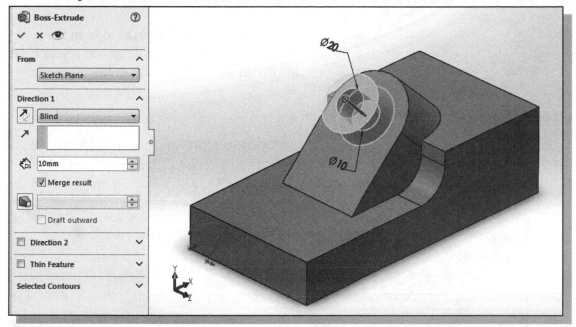

22. On your own, create the extruded cut features necessary to complete the part. (See page 16-5 for dimensions.)

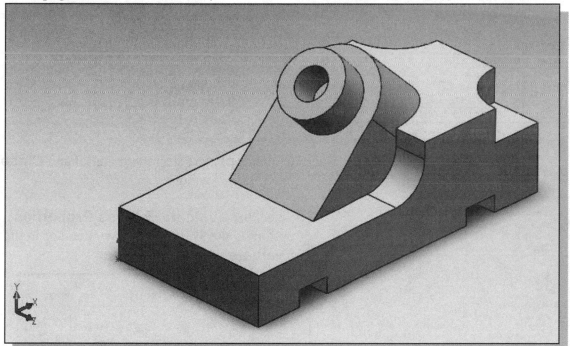

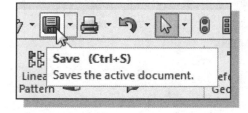

23. On your own, save the part with the filename ***Review-Prob1.SLDPRT***.

## Selecting the Material and Viewing the Mass Properties

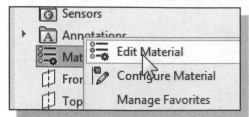

1. Right-click on the **Material** icon in the *Feature Manger Design Tree* and select **Edit Material** from the pop-up menu.

2. In the *Material* pop-up window, select **Ductile Iron** (in the Iron folder) as the material type. Notice the density listed in the *Material Property* table is 0.0071 g/mm$^3$, as required in the problem statement.

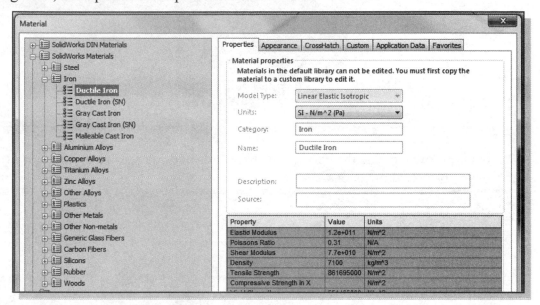

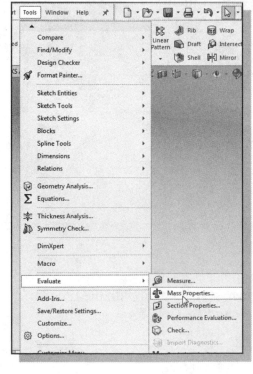

3. Click **Apply** to select the material and **Close** the *Material* window.

4. Select the **Evaluate → Mass Properties** option on the *Tools* pull-down menu or in the **Evaluate** tab as shown.

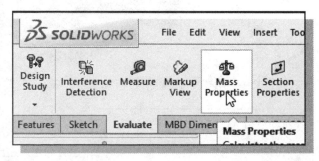

➢ The *Mass Properties* window is displayed. Notice the *Output coordinate system* is the default system.

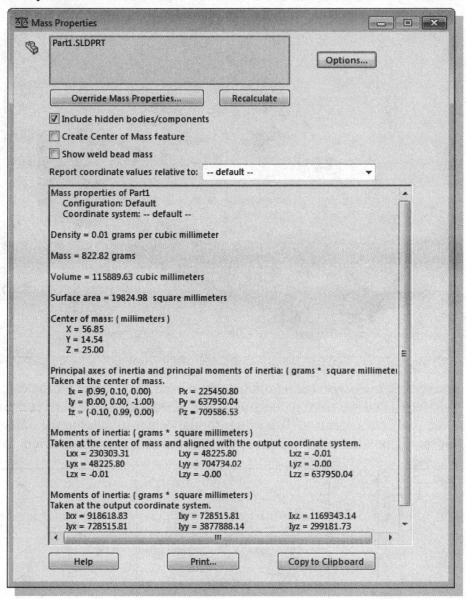

➢ Notice the *Mass* is **822.82 grams**.

➢ Notice the *Center of mass* is:
   **X = 56.85 mm**
   **Y = 14.54 mm**
   **Z = 25.00 mm**

➢ Look again at the problem on page 16-5. Note that the *Center of mass* coordinates above are based on the correct origin and along the correct X, Y, Z directions.

➢ Notice that *Moments of Inertia* are also reported.

➢ Notice the center of mass and the principal axes are displayed. On your own, view the part from different directions and observe the location of the center of mass. Minimize the *Mass Properties* window if necessary.

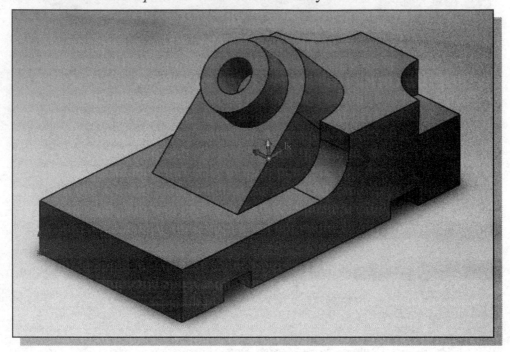

➢ Also displayed are the origin and coordinate axes used as the basis for the calculated properties. Verify that we have successfully created our part in the correct orientation based on the problem statement. The values for the center of mass in the *Mass Properties* dialog box therefore represent the correct values for the problem as stated.

  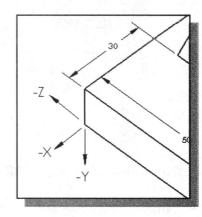

5. **Close** the *Mass Properties* window.

6. On your own, save the part with the filename ***Review-Prob1.SLDPRT***.

➢ We have successfully completed the problem. The part was created matching the coordinate axes defined in the problem statement and the mass and center of mass were obtained. The SOLIDWORKS Mass Properties tool provides an efficient method to obtain properties of parts and/or assemblies with complex shapes.

# The Assembly Problem

We will model the assembly shown below and use the **Mass Properties** tool to determine the mass and center of mass of the assembly. The problem is presented below in a manner similar to that encountered on the CSWA exam.

Build the assembly shown.

It contains two *Brackets* and two *Rods*.

Unit system: MMGS (millimeter, gram, second)

Assembly origin: as shown.

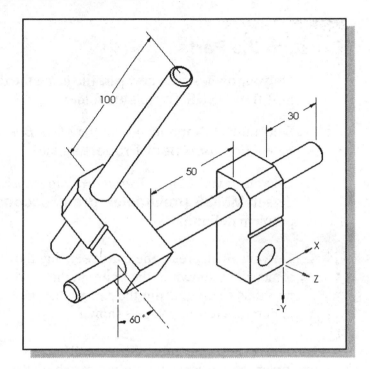

*Brackets:* A dimensioned figure appears to the right (holes through all). The 1 mm radius notch is semicircular and located with its center 25 mm from the bottom edge of the bracket. Material: 1060 Aluminum Alloy; Density = 0.0027 g/mm³.

*Rods:* 10 mm in diameter and 150 mm long, with 1mm x 45° chamfers at each end. Material: 201 Annealed Stainless Steel (SS); Density = 0.00786 g/mm³.

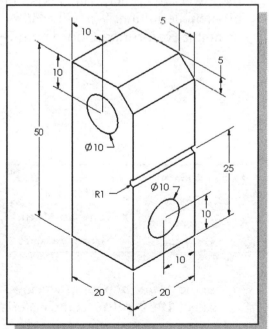

## Questions:

(1) What is the overall mass of the assembly in grams?

(2) What is the center of mass of the assembly?

➢ Before starting to build the model, it is important to select a strategy for controlling the origin and orientation of the X, Y, and Z axes since the center of mass must be defined relative to a known reference system. The axis directions are defined in the drawing. We will build the part independent of this constraint and, after the part is complete, create a **reference coordinate system** aligned with the part as shown in the figure.

## Creating the Parts

1. On your own, start a new part file using the default **Part** template in the SOLIDWORKS Templates folder.

2. Select the **Options** icon from the *Menu Bar* to open the *Options* dialog box. Select the **Document Properties** tab.

3. On your own, set the *Overall drafting standard* to **ANSI** and the *Units* to the default **MMGS (millimeter, gram, second)** unit system as instructed in the problem definition.

4. On your own, create the *Bracket* using the dimensions shown in the figure in the previous section. Build the part so that the isometric view appears as shown.

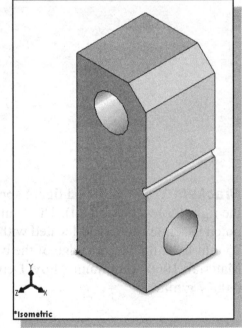

➢ Note the axis alignment is intentionally set to the lower left corner, which is not matching the problem definition. We will address this later using the **Reference Coordinate System** tool.

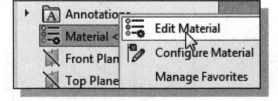

6. In the *Material* pop-up window, select **1060 Alloy** as the material type.

7. Save the part with the filename **Bracket.SLDPRT**.

5. Right-click on the **Material** icon in the *Feature Manger Design Tree* and select **Edit Material** from the pop-up menu.

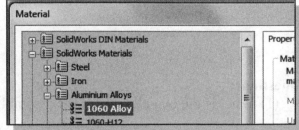

8.  On your own, start a new part file using the default **Part** template in the SOLIDWORKS **Templates** folder, and set the *Overall drafting standard* to **ANSI** and the *Units* to the default **MMGS (millimeter, gram, second)** unit system.

9.  On your own, create the *Rod* using the description in the previous section. Build the part so that the isometric view appears as shown. (**HINT:** Draw a circle on the Front Plane and extrude.)

10. On your own, set the material to **201 Annealed Stainless Steel (SS)** (in the Steel folder).

11. Save the part with the filename *Rod.SLDPRT*.

## Creating the Assembly

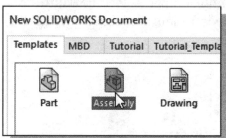

1.  Select the **New** icon with a single click of the left-mouse-button on the *Menu Bar*. The *New SOLIDWORKS Document* dialog box appears.

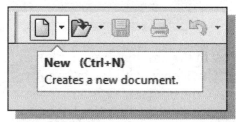

2.  Select the **Templates** tab and select the **Assembly** icon as shown.

3.  Click **OK** to open the new Assembly file.

➢   SOLIDWORKS opens an assembly file, and automatically opens the *Begin Assembly Property Manager*. The message *"Select a part or assembly to insert ..."* appears in the *Property Manager*. SOLIDWORKS expects you to insert the first component.

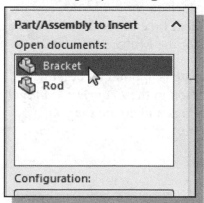

➢   The *Part/Assembly to Insert* panel displays *Open documents*, including the *Bracket* and *Rod* parts.

4.  Select the *Bracket* part as the base part for the assembly by clicking on **Bracket** in the *Part/Assembly to Insert* panel. (**NOTE:** If you have closed the *Bracket* part, it will not appear. In that case, click the **Browse** button and use the browser to open the *Bracket* part file.)

5. By default, the component is automatically aligned to the origin of the assembly coordinates. Click **OK** in the *Property Manager* to place the *Bracket* at the **origin**.

6. On your own, set the *Overall drafting standard* to **ANSI** and the *Units* to the default **MMGS (millimeter, gram, second)** unit system. (**HINT**: Select Tools → Option → Document Properties.)

7. In the *Assembly* toolbar, select the **Insert Component** command by left-mouse-clicking the icon.

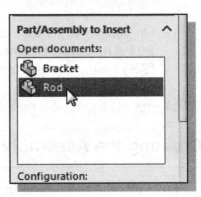

8. Select the **Rod** part in the *Part/Assembly to Insert* panel. (**NOTE:** If you have closed the *Rod* part, it will not appear. In that case, click the **Browse** button and use the browser to open the *Rod* part file.)

9. Place the *Rod* in the graphics area as shown in the figure. Click once with the **left-mouse-button** to place the component.

10. In the *Assembly* toolbar, select the **Mate** command by left-mouse-clicking once on the icon.

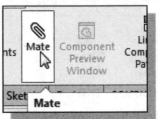

11. Select the **external cylindrical face** of the **Rod** as the first surface for the Mate, and the **interior cylindrical face** of the **Bracket** as the second face for the Mate, as shown.

12. A Concentric mate is applied by default. Click the **OK** button in the *Property Manager* to accept the settings and create the **Concentric** mate.

❖ We will now add a **Distance** mate to define the position of the *Rod* relative to the *Bracket* as defined in the problem statement.

13. Rotate the view (e.g., using the arrow keys) to display the reverse side of the *Rod* and *Bracket* as shown below.

14. Select the **edge face** of the ***Rod*** and the **side face** of the ***Bracket*** as shown below.

➤ SOLIDWORKS will apply a **Coincident** mate by default. We want to apply a mate which is offset by a specific distance (30 mm as defined by the problem statement).

15. Select the *Distance* option by clicking the **Distance** button in the *Standard Mates* panel of the *Property Manager*.

16. Enter **30 mm** as the distance.

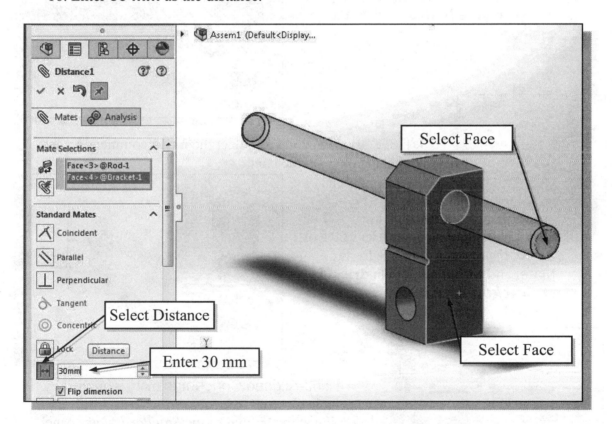

17. Click the **OK** button in the *Property Manager* to accept the settings and create the **Distance** mate.

18. Click **OK** again (or hit the [**Esc**] key) to exit the **Mate** command.

19. On your own, return to the **Isometric** view.

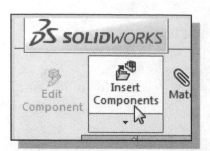

20. In the *Assembly* toolbar, select the **Insert Component** command by left-clicking the icon.

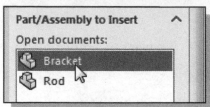

21. Select the *Bracket* part as the base part for the assembly by clicking on **Bracket** in the *Part/Assembly to Insert* panel.

22. Place the *Bracket* in the graphics area as shown in the figure. Click once with the **left-mouse-button** to place the component.

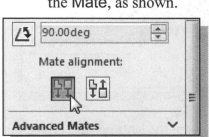

23. In the *Assembly* toolbar, select the **Mate** command by left-clicking once on the icon.

24. Select the **external cylindrical face** of the **Rod** as the first surface for the Mate, and the **interior cylindrical face** of **Bracket<2>** as the second face for the Mate, as shown.

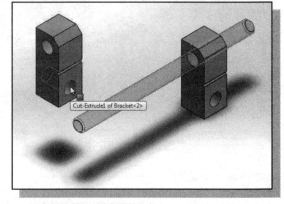

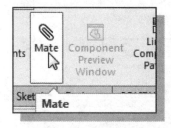

25. An Anti-Aligned Concentric mate is applied by default. On your own, select the **Anti-Aligned** option at the bottom of the *Standard Mates* panel and observe the effect on the assembly.

26. Select the **Aligned** option. Look at the drawing in the problem statement to verify that it is the Aligned option that is required.

27. Click the **OK** button in the *Property Manager* to accept the settings and create the **Aligned Concentric** mate.

❖ We will now add a **Distance** mate to define the position of *Bracket<2>* relative to *Bracket<1>* as defined in the problem statement.

28. Select the **face** of ***Bracket<1>*** as shown.

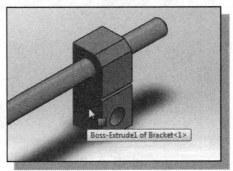

29. Rotate the view and select the **face** of ***Bracket<2>*** as shown.

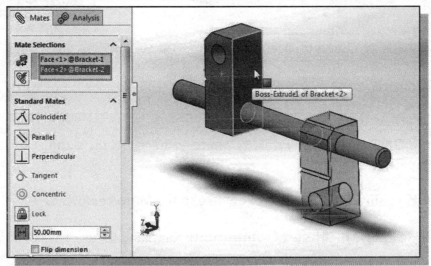

30. Select the *Distance* option by clicking the **Distance** button in the *Standard Mates* panel and enter **50 mm** as the distance.

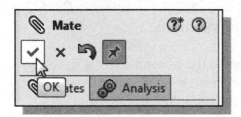

31. Click the **OK** button in the *Property Manager* to accept the settings and create the **Distance** mate.

❖ Next, we will add an **Angle** mate to define the angular orientation of *Bracket<2>* relative to *Bracket<1>* as defined in the problem statement.

32. Select the **face** of ***Bracket<1>*** as shown.

33. Select the *Angle* option by clicking the **Angle** button in the *Standard Mates* panel and enter **60°** as the angle value.

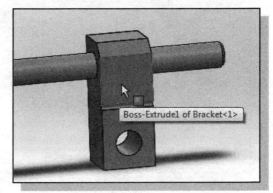

34. Select the **face** of **Bracket<1>** as shown, and check the **Flip dimension** box **if necessary** to set the angle as shown in the figure below.

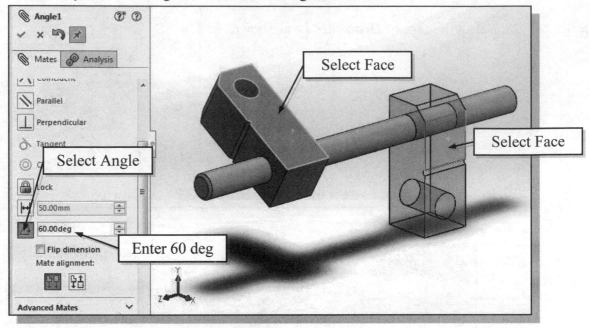

35. Click the **OK** button in the *Property Manager* to accept the settings and create the **Angle** mate as shown.

36. Click **OK** again (or hit the [**Esc**] key) to exit the **Mate** command.

37. On your own, use the **Insert Component** command to insert a second **Rod** part.

38. On your own, create a **concentric** matc to align the **Rod** to **Bracket<2>**.

39. On your own, create a **Distance-100mm** mate to define the position of **Rod<2>** relative to **Bracket<2>** as defined in the problem statement.

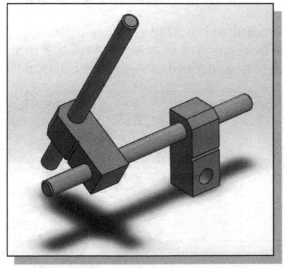

40. Compare the assembly to the drawing in the problem statement on page 16-19. Note that the locations and orientations of the brackets (particularly the locations of the chamfers and notches) and rods are correct.

41. On your own, save the assembly with the filename **Review-Prob2.SLDASM**.

# Creating a Reference Coordinate System

We have created the assembly without aligning it with the origin and coordinate axes defined in the problem statement. In order to calculate the center of mass relative to the correct origin and coordinate axes we will create a Reference Coordinate System.

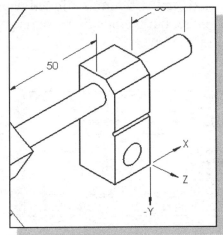

➤ The Reference Coordinate System needs to be defined in alignment with the edges of the base bracket (*Bracket<1>* in our assembly) as shown in the figure defining the problem.

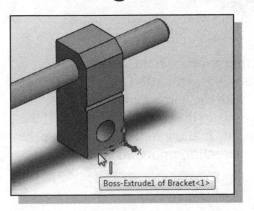

1. In the *Assembly* toolbar, select the **Reference Geometry** command by left-clicking the icon.

2. In the pull-down option menu of the Reference Geometry command, select the **Coordinate System** option.

3. The *Coordinate System Property Manager* opens with the *Origin* selection window active. Select the **lower right corner** of *Bracket<1>* as the origin, as shown.

➤ The origin selection appears in the *Property-Manager* as **Vertex<1>@Bracket-1**.

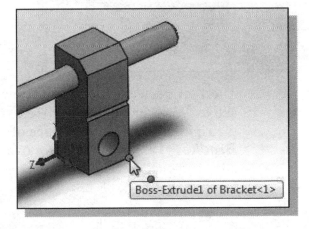

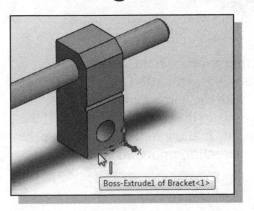

4. The *X-axis* selection window is now active. We want to align the X-axis with the **bottom edge on the right face** of *Bracket<1>*. Move the cursor over the edge and click once with the left-mouse-button to **select** it.

➤ The X-axis selection appears in the *Property Manager* as Edge<1>@Bracket-1. Look at the coordinate icon which is now attached to the newly defined reference origin. Notice the X-axis in the figure is aligned with the correct edge, but the positive X-axis points in the wrong direction.

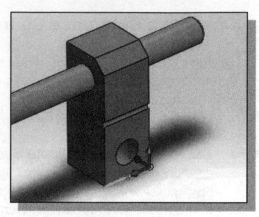

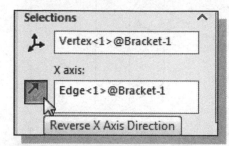

5. In the *Property Manager*, select the **Reverse X Axis Direction** option by clicking once with the left-mouse-button on the toggle button located to the left of the *X axis* selection window (if necessary).

6. Verify that the X-axis is now aligned with the correct edge and points in the correct direction.

7. The *Y-axis* selection window is now active. We want to align the Y-axis with the **right edge of the right face** of **Bracket<1>**. Move the cursor over the edge and click once with the left-mouse-button to **select** it.

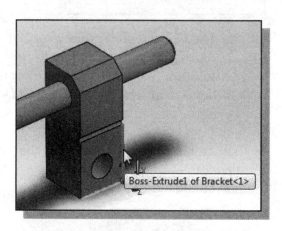

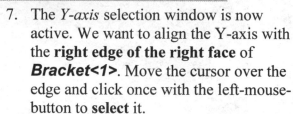
Boss-Extrude1 of Bracket<1>

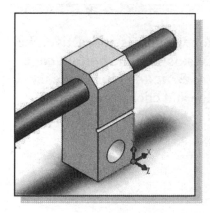

8. The Y-axis selection appears in the *Property Manager* as Edge<2>@Bracket-1. Verify that the Y-axis is aligned with the correct edge, and the positive Y-axis points in the correct direction. (If the positive Y-axis points in the wrong direction, use the Reverse Y Axis Direction option to correct it.)

9. The origin and coordinate axes are now aligned as defined in the problem statement. Click **OK** in the *Property Manager* to accept the setting and create the **Reference Coordinate System**.

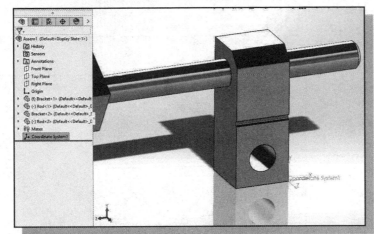

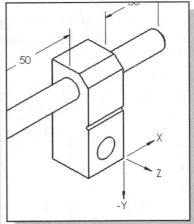

➤ Notice the **Reference Coordinate System** appears in the *Feature Manager Design Tree* as **Coordinate System1**, and that the reference axes correspond to the axes defined in the problem statement. We are now ready to calculate the *mass properties*.

## View the Mass Properties

1. Select the **Mass Properties** option through the *Evaluate* tab.

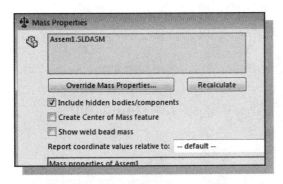

2. The *Mass Properties* window is displayed. Notice the *Report coordinate system* is the **default** system.

➤ In the graphics area, the **default** coordinate system is shown. The values appearing in the *Mass Properties* window were calculated relative to this coordinate system. These are **not** the results we need for the solution to the problem.

➤ We must change the *Report coordinate system* to the **Reference Coordinate System** we created.

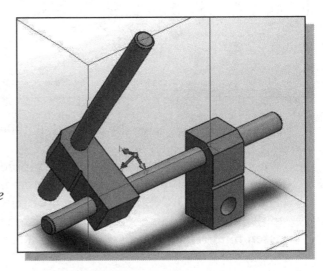

3. Inside the *Mass Property* window, **left-click** on the arrow at the right end of the *Report coordinate system:* selection window, and select **Coordinate System1** from the selection list.

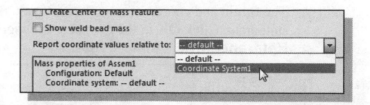

➤ **Coordinate System1** is now the basis for the mass property calculations. Notice the calculated values for the center of mass (and other properties) change to those for the new coordinate system.

➤ Notice the coordinate system displayed in the graphics area is the correct system for the problem.

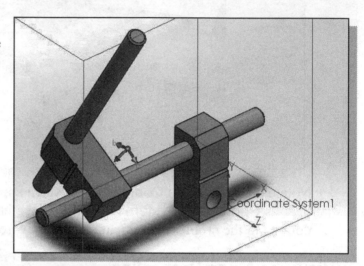

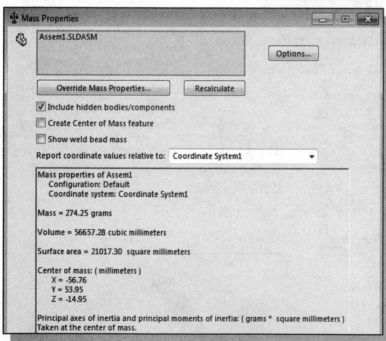

➤ Notice the *Mass* is **274.25 grams**.

➤ Notice the *Center of mass* is:

**X = -56.76 mm**
**Y = 53.95 mm**
**Z = -14.95 mm**

4. **Close** the *Mass Properties* window.

5. On your own, save the part with the filename ***Review-Prob2.SLDASM***.

➤ We have successfully completed the problem. The correct alignment of the coordinate system was accomplished through the creation of a Reference Coordinate System. The SOLIDWORKS **Reference Coordinate System** and **Mass Properties** tools can be used in combination to obtain mass properties of complex parts and assemblies relative to a specific coordinate system.

## Questions:

1. How do we save a new View Orientation for future use?

2. Describe the effect of using the Up to Next end condition for an extruded boss feature.

3. How can we view the mass of a part or assembly?

4. Describe the steps in creating a Reference Coordinate System.

5. What mass properties are dependent on the selection of the output coordinate system?

6. What mass properties are independent of the selection of the output coordinate system?

7. How do we control the output coordinate system used in calculating mass properties?

8. How can we apply an assembly mate establishing an offset distance between two planes or faces?

## Exercises:

1.  Build the part shown below.
    Unit system: IPS (inch, pound, second)
    Part origin: As shown.
    Part material: AISI 1020
    Density: 0.285406 lb/in$^3$

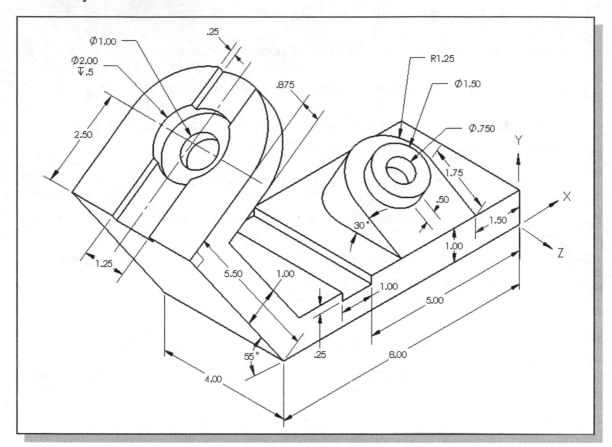

What is the overall mass of the part in pounds?

What is the center of mass of the part?


Answer:

Mass:
    18.58 pounds

Center of mass:
    X = -6.07 inches
    Y = 1.91 inches
    Z = -1.98 inches

2. Build the assembly in SOLIDWORKS. The assembly includes 8 components – 1 base, 1 jaw, 2 keys, 1 screw, 1 handle rod, 2 handle knobs.
   Use the IPS (inch, pound, second) unit system.

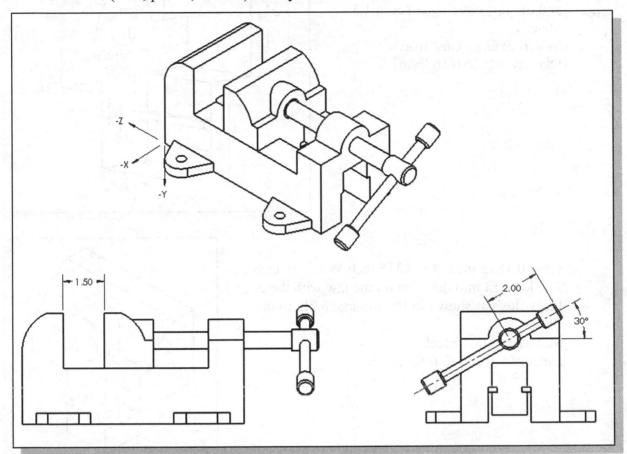

**Base:** A dimensioned figure appears to the right. The 1.5 inch wide and 1.25 inch wide slots are cut through the entire base.
Material: Gray Cast Iron
Density = 0.260116 lb/in³

**Jaw:** A dimensioned figure appears to the right. The shoulder of the jaw rests on the flat surface of the base and the jaw opening is set to 1.5 inches.
Material: Gray Cast Iron
Density = 0.260116 lb/in$^3$

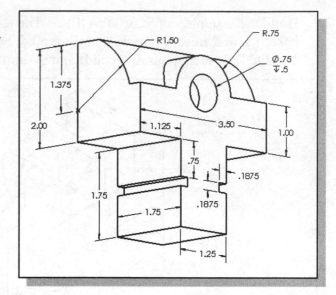

**Key:** 0.1875 inch H x 0.375 inch W x 1.75 inch L. The keys fit into the slots on the jaw with the edge faces flush as shown in the sub-assembly to the right.
Material: Alloy Steel
Density = 0.27818 lb/in$^3$.

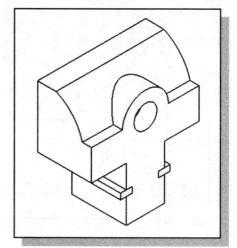

**Screw:** A dimensioned figure appears to the right. There is one chamfered edge (0.0625 inch x 45°). The flat ∅ 0.75″ edge of the screw is flush with the corresponding recessed ∅ 0.75 face on the jaw.
Material: Alloy Steel
Density = 0.27818 lb/in$^3$

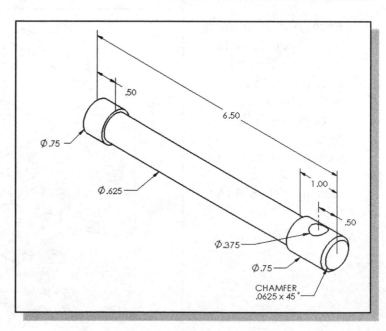

**Handle Rod:** ∅ 0.375″ x 5.0″ L. The handle rod passes through the hole in the screw and is rotated to an angle of 30° with the horizontal as shown in the assembly view. The flat ∅ 0.375″ edges of the handle rod are flush with the corresponding recessed ∅ 0.735 faces on the handle knobs.
Material: Alloy Steel
Density = 0.27818 lb/in$^3$.

**Handle Knob:** A dimensioned figure appears to the right. There are two chamfered edges (0.0625 inch x 45°). The handle knobs are attached to each end of the handle rod. The resulting overall length of the handle with knobs is 5.50″. The handle is aligned with the screw so that the outer edge of the upper knob is 2.0″ from the central axis of the screw.
Material: Alloy Steel
Density = 0.27818 lb/in$^3$

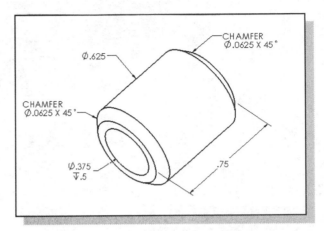

What is the mass of the assembly?

What is the center of mass of the assembly with respect to the coordinate system illustrated in the assembly view?

Answer:

Mass:
    17.52 pounds

Center of mass:
    X = 1.74 inches
    Y = 1.75 inches
    Z = 3.98 inches

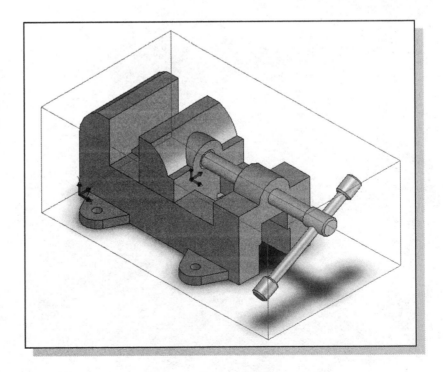

**Notes:**

# APPENDIX A

## Running and Sliding Fits – American National Standard (Inches)

Basic Hole System, Limits are in thousandths of an inch.
Limits for hole and shaft are applied to the basic size to obtain the limits of sizes for the parts.

| Nominal Size Range, Inches Over To | Class RC1 | | | Class RC2 | | | Class RC3 | | | Class RC4 | | |
|---|---|---|---|---|---|---|---|---|---|---|---|---|
| | Limits of Clearance | Standard Limits | | Limits of Clearance | Standard Limits | | Limits of Clearance | Standard Limits | | Limits of Clearance | Standard Limits | |
| | | Hole H5 | Shaft g4 | | Hole H6 | Shaft g5 | | Hole H7 | Shaft f6 | | Hole H8 | Shaft f7 |
| 0 – 0.12 | 0.1 | +0.2 | -0.1 | 0.1 | +0.25 | -0.1 | 0.3 | +0.4 | -0.3 | 0.3 | +0.6 | -0.3 |
| | 0.45 | +0 | -0.25 | 0.55 | 0 | -0.3 | 0.95 | 0 | -0.55 | 1.3 | 0 | -0.7 |
| 0.12 – 0.24 | 0.15 | +0.2 | -0.15 | 0.15 | +0.3 | -0.15 | 0.4 | +0.5 | -0.4 | 0.4 | +0.7 | -0.4 |
| | 0.5 | 0 | -0.3 | 0.65 | 0 | -0.35 | 1.12 | 0 | -0.7 | 1.6 | 0 | -0.9 |
| 0.24 – 0.40 | 0.2 | +0.25 | -0.2 | 0.2 | +0.4 | -0.2 | 0.5 | +0.6 | -0.5 | 0.5 | +0.9 | -0.5 |
| | 0.6 | 0 | -0.35 | 0.85 | 0 | -0.45 | 1.5 | 0 | -0.9 | 2.0 | 0 | -1.1 |
| 0.40 – 0.71 | 0.25 | +0.3 | -0.25 | 0.25 | +0.4 | -0.25 | 0.6 | +0.7 | -0.6 | 0.6 | +1.0 | -0.6 |
| | 0.75 | +0 | -0.45 | 0.95 | 0 | -0.55 | 1.7 | 0 | -1.0 | 2.3 | 0 | -1.3 |
| 0.71 – 1.19 | 0.3 | +0.4 | -0.3 | 0.3 | +0.5 | -0.3 | 0.8 | +0.8 | -0.8 | 0.8 | +1.2 | -0.8 |
| | 0.95 | 0 | -0.55 | 1.2 | 0 | -0.7 | 2.1 | 0 | -1.3 | 2.8 | 0 | -1.6 |
| 1.19 – 1.97 | 0.4 | +0.4 | -0.4 | 0.4 | +0.6 | -0.4 | 1.0 | +1.0 | -1.0 | 1.0 | +1.6 | -1.0 |
| | 1.1 | 0 | -0.7 | 1.4 | 0 | -0.8 | 2.6 | 0 | -1.6 | 3.6 | 0 | -2.0 |
| 1.97 – 3.15 | 0.4 | +0.5 | -0.4 | 0.4 | +0.7 | -0.4 | 1.2 | +1.2 | -1.2 | 1.2 | +1.8 | -1.2 |
| | 1.2 | 0 | -0.7 | 1.6 | 0 | -0.9 | 3.1 | 0 | -1.9 | 4.2 | 0 | -2.4 |
| 3.15 – 4.73 | 0.5 | +0.6 | -0.5 | 0.5 | +0.9 | -0.5 | 1.4 | +1.4 | -1.4 | 1.4 | +2.2 | -1.4 |
| | 1.5 | 0 | -0.9 | 2.0 | 0 | -1.1 | 3.7 | 0 | -2.3 | 5.0 | 0 | -2.8 |
| 4.73 – 7.09 | 0.6 | +0.7 | -0.6 | 0.6 | +1.0 | -0.6 | 1.6 | +1.6 | -1.6 | 1.6 | +2.5 | -1.6 |
| | 1.8 | 0 | -1.1 | 2.3 | 0 | -1.3 | 4.2 | 0 | -2.6 | 5.7 | 0 | -3.2 |
| 7.09 – 9.85 | 0.6 | +0.8 | -0.6 | 0.6 | +1.2 | -0.6 | 2.0 | +1.8 | -2.0 | 2.0 | +2.8 | -2.0 |
| | 2.0 | 0 | -1.2 | 2.6 | 0 | -1.4 | 5.0 | 0 | -3.2 | 6.6 | 0 | -3.8 |
| 9.85 – 12.41 | 0.8 | +0.9 | -0.8 | 0.8 | +1.2 | -0.7 | 2.5 | +2.0 | -2.5 | 2.5 | +3.0 | -2.2 |
| | 2.3 | 0 | -1.4 | 2.9 | 0 | -1.6 | 5.7 | 0 | -3.7 | 7.5 | 0 | -4.2 |
| 12.41 – 15.75 | 1.0 | +1.0 | -1.0 | 1.0 | +1.4 | -0.7 | 3.0 | +2.2 | -3.0 | 3.0 | +3.5 | -2.5 |
| | 2.7 | 0 | -1.7 | 3.4 | 0 | -1.7 | 6.6 | 0 | -4.4 | 8.7 | 0 | -4.7 |
| 15.75 – 19.69 | 1.2 | +1.0 | -1.2 | 0.8 | +1.6 | -0.8 | 4.0 | +2.5 | -4.0 | 2.8 | +4.0 | -2.8 |
| | 3.0 | 0 | -2.0 | 3.4 | 0 | -1.8 | 8.1 | 0 | -5.6 | 9.3 | 0 | -5.3 |

## APPENDIX A  (Running and Sliding Fits Continued)

## Running and Sliding Fits – American National Standard (Inches)

Basic Hole System, Limits are in thousandths of an inch.
Limits for hole and shaft are applied to the basic size to obtain the limits of sizes for the parts.

| Nominal size Range, Inches Over To | Class RC5 Limits of Clearance | Class RC5 Hole H8 | Class RC5 Shaft e7 | Class RC6 Limits of Clearance | Class RC6 Hole H9 | Class RC6 Shaft e8 | Class RC7 Limits of Clearance | Class RC7 Hole H9 | Class RC7 Shaft d8 | Class RC8 Limits of Clearance | Class RC8 Hole H10 | Class RC8 Shaft c9 | Class RC9 Limits of Clearance | Class RC9 Hole H11 | Class RC9 Shaft |
|---|---|---|---|---|---|---|---|---|---|---|---|---|---|---|---|
| 0 – 0.12 | 0.6 | +0.6 | -0.6 | 0.6 | +1.0 | +0.6 | 1.0 | +1.0 | -1.0 | 2.5 | +1.6 | -2.5 | 4.0 | +2.5 | -4.0 |
|  | 1.6 | 0 | -1.0 | 2.2 | 0 | -1.2 | 2.6 | 0 | -1.6 | 5.1 | 0 | -3.5 | 8.1 | 0 | -5.6 |
| 0.12 – 0.24 | 0.8 | +0.7 | -0.8 | 0.8 | +1.2 | -0.8 | 1.2 | +1.2 | -1.2 | 2.8 | +1.8 | -2.8 | 4.5 | +3.0 | -4.5 |
|  | 2.0 | 0 | -1.3 | 2.7 | 0 | -1.5 | 3.1 | 0 | -1.9 | 5.8 | 0 | -4.0 | 9.0 | 0 | -6.0 |
| 0.24 – 0.40 | 1.0 | +0.9 | -1.0 | 1.0 | +1.4 | -1.0 | 1.6 | +1.4 | -1.6 | 3.0 | +2.2 | -3.0 | 5.0 | +3.5 | -5.0 |
|  | 2.5 | 0 | -1.6 | 3.3 | 0 | -1.9 | 3.9 | 0 | -2.5 | 6.6 | 0 | -4.4 | 10.7 | 0 | -7.2 |
| 0.40 – 0.71 | 1.2 | +1.0 | -1.2 | 1.2 | +1.6 | -1.2 | 2.0 | +1.6 | -2.0 | 3.5 | +2.8 | -3.5 | 6.0 | +4.0 | -6.0 |
|  | 2.9 | 0 | -1.9 | 3.8 | 0 | -2.2 | 4.6 | 0 | -3.0 | 7.9 | 0 | -5.1 | 12.8 | 0 | -8.8 |
| 0.71 – 1.19 | 1.6 | +1.2 | -1.6 | 1.6 | +2.0 | -1.6 | 2.5 | +2.0 | -2.5 | 4.5 | +3.5 | -4.5 | 7.0 | +5.0 | -7.0 |
|  | 3.6 | 0 | -2.4 | 4.8 | 0 | -2.8 | 5.7 | 0 | -3.7 | 10.0 | 0 | -6.5 | 15.5 | 0 | -10.5 |
| 1.19 – 1.97 | 2.0 | +1.6 | -2.0 | 2.0 | +2.5 | -2.0 | 3.0 | +2.5 | -3.0 | 5.0 | +4.0 | -5.0 | 8.0 | +6.0 | -8.0 |
|  | 4.6 | 0 | -3.0 | 6.1 | 0 | -3.6 | 7.1 | 0 | -4.6 | 11.5 | 0 | -7.5 | 18.0 | 0 | 12.0 |
| 1.97 – 3.15 | 2.5 | +1.8 | -2.5 | 2.5 | +3.0 | -2.5 | 4.0 | +3.0 | -4.0 | 6.0 | +4.5 | -6.0 | 9.0 | +7.0 | -9.0 |
|  | 5.5 | 0 | -3.7 | 7.3 | 0 | -4.3 | 8.8 | 0 | -5.8 | 13.5 | 0 | -9.0 | 20.5 | 0 | -13.5 |
| 3.15 – 4.73 | 3.0 | +2.2 | -3.0 | 3.0 | +3.5 | -3.0 | 5.0 | +3.5 | -5.0 | 7.0 | +5.0 | -7.0 | 10.0 | +9.0 | -10.0 |
|  | 6.6 | 0 | -4.4 | 8.7 | 0 | -5.2 | 10.7 | 0 | -7.2 | 15.5 | 0 | -10.5 | 24.0 | 0 | -15.0 |
| 4.73 – 7.09 | 3.5 | +2.5 | -3.5 | 3.5 | +4.0 | -3.5 | 6.0 | +4.0 | -6.0 | 8.0 | +6.0 | -8.0 | 12.0 | +10.0 | -12.0 |
|  | 7.6 | 0 | -5.1 | 10.0 | 0 | -6.0 | 12.5 | 0 | -8.5 | 18.0 | 0 | -12.0 | 28.0 | 0 | -18.0 |
| 7.09 – 9.85 | 4.0 | +2.8 | -4.0 | 4.0 | +4.5 | -4.0 | 7.0 | +4.5 | -7.0 | 10.0 | +7.0 | -10.0 | 15.0 | +12.0 | -15.0 |
|  | 8.6 | 0 | -5.8 | 11.3 | 0 | -6.8 | 14.3 | 0 | -9.8 | 21.5 | 0 | -14.5 | 34.0 | 0 | -22.0 |
| 9.85–12.41 | 5.0 | +3.0 | -5.0 | 5.0 | +5.0 | -5.0 | 8.0 | +5.0 | -8.0 | 12.0 | +8.0 | -12.0 | 18.0 | +12.0 | -18.0 |
|  | 10.0 | 0 | -7.0 | 13.0 | 0 | -8.0 | 16.0 | 0 | -11 | 25.0 | 0 | -17.0 | 38.0 | 0 | -26.0 |
| 12.41– 15.75 | 6.0 | +3.5 | -6.0 | 6.0 | +6.0 | -6.0 | 10.0 | +6.0 | -10 | 14.0 | +9.0 | -14.0 | 22.0 | +14.0 | -22.0 |
|  | 11.7 | 0 | -8.2 | 15.5 | 0 | -9.5 | 19.5 | 0 | -13.5 | 29.0 | 0 | -20.0 | 45.0 | 0 | -31.0 |
| 15.75– 19.69 | 8.0 | +4.0 | -8.0 | 8.0 | +6.0 | -8.0 | 12.0 | +6.0 | -12.0 | 16.0 | +10.0 | -16.0 | 25.0 | +16.0 | -25.0 |
|  | 14.5 | 0 | -10.5 | 18.0 | 0 | -12.0 | 22.0 | 0 | -16.0 | 32.0 | 0 | -22.0 | 51.0 | 0 | -35.0 |

USAS/ASME B4.1 – 1967 (R2004) Standard. For larger diameters, see the standard. ASME/ANSI B18.3.5M – 1986 (R2002) Standard. Reprinted from the standard listed by permission of the American Society of Mechanical Engineers. All rights reserved.

# Locational clearance fits – American National Standard (Inches)

Basic Hole System, Limits are in thousandths of an inch.
Limits for hole and shaft are applied to the basic size to obtain the limits of sizes for the parts.

| Nominal Size Range Inches | Class LC1 Standard Limits | | Class LC2 Standard Limits | | Class LC3 Standard Limits | | Class LC4 Standard Limits | |
|---|---|---|---|---|---|---|---|---|
| Over    To | Hole | Shall | Hole | Shalt | Hole | Shaft | Hole | Shalt |
| 0   -  0.12 | +0.25 | 0 | +0.4 | 0 | +0.6 | 0 | +1.6 | 0 |
|             | 0 | -0.2 | 0 | -0.25 | 0 | -0.4 | 0 | -1.0 |
| 0.12 - 0.24 | +0.3 | 0 | +0.5 | 0 | +0.7 | 0 | +1.8 | 0 |
|             | 0 | -0.2 | 0 | -0.3 | 0 | -0.5 | 0 | -1.2 |
| 0.24 - 0.40 | +0.4 | 0 | +0.6 | 0 | +0.9 | 0 | +2.2 | 0 |
|             | 0 | -0.25 | 0 | -0.4 | 0 | -0.6 | 0 | -1,4 |
| 0,40 - 0.71 | +0.4 | 0 | +0.7 | 0 | +1,0 | 0 | +2.6 | 0 |
|             | 0 | -0.3 | 0 | -0.4 | 0 | -0.7 | 0 | -1.6 |
| 0.71 - 1.10 | +0.5 | 0 | +0.8 | 0 | +1.2 | 0 | +3.5 | 0 |
|             | 0 | -0.4 | 0 | -0.5 | 0 | -0.8 | 0 | -2.0 |
| 1.19 - 1.97 | +0.6 | 0 | +1.0 | 0 | +1.6 | 0 | +4.0 | 0 |
|             | 0 | -0.4 | 0 | -0.6 | 0 | -1.0 | 0 | -2.5 |
| 1.97 - 3.15 | +0.7 | 0 | +1.2 | 0 | +1.8 | 0 | +4.5 | 0 |
|             | 0 | -0.5 | o | -0.7 | 0 | -1.2 | 0 | -3.0 |
| 3.15 - 4.73 | +0.9 | 0 | +1.4 | 0 | +2.2 | 0 | +5.0 | 0 |
|             | 0 | -0.6 | 0 | -0.9 | 0 | -1.4 | 0 | -3.5 |
| 4.73 - 7.09 | +1.0 | 0 | +1.6 | 0 | +2.5 | 0 | +6.0 | 0 |
|             | 0 | -0.7 | 0 | -1.0 | 0 | -1.6 | 0 | -4.0 |
| 7.09 - 9.85 | +1.2 | 0 | +1.8 | 0 | +2.8 | 0 | +7.0 | 0 |
|             | 0 | -0.8 | 0 | -1.2 | 0 | -1.8 | 0 | -4.5 |
| 9.85 - 12.41 | +1.2 | 0 | +2.0 | 0 | +3.0 | 0 | +8.0 | 0 |
|             | 0 | -0.9 | 0 | -1.2 | 0 | -2.0 | 0 | -5.0 |
| 12.41 - 15.75 | +1.4 | 0 | +2.2 | 0 | +3.5 | 0 | +9.0 | 0 |
|             | 0 | -1.0 | 0 | -1.4 | 0 | -2.2 | 0 | -6.0 |
| 15.75 - 19.69 | +1.6 | 0 | +2.5 | 0 | +4.0 | 0 | +10.0 | 0 |
|             | 0 | -1.0 | 0 | -1.6 | 0 | -2.5 | 0 | -6.0 |

USAS/ASME B4.1 – 1967 (R2004) Standard. For larger diameters, see the standard.  ASME/ANSI
B18.3.5M – 1986 (R2002) Standard.  Reprinted from the standard listed by permission of the American
Society of Mechanical Engineers. All rights reserved.

**APPENDIX A  (Locational clearance fits Continued)**

## Locational clearance fits – American National Standard (Inches)

Basic Hole System, Limits are in thousandths of an inch.
Limits for hole and shaft are applied to the basic size to obtain
the limits of sizes for the parts.

| Nominal Size Range Inches | Class LC5 | | Class LC6 | | Class LC7 | | Class LC8 | |
|---|---|---|---|---|---|---|---|---|
| | Standard Limits | | Standard Limits | | Standard Limits | | Standard Limits | |
| Over    To | Hole | Shaft | Hole | Shaft | Hole | Shaft | Hole | Shaft |
| 0   -  0.12 | +0.4 | -0.1 | +1.0 | -0.3 | +1.6 | -0.8 | +1.6 | -1.0 |
| | 0 | -0.35 | 0 | -0.9 | 0 | -1.6 | 0 | -2.0 |
| 0.12 - 0.24 | +0.5 | -0.15 | +1.2 | -0.4 | +1.8 | -0.8 | +1.8 | -1.2 |
| | 0 | -0.45 | 0 | -1.1 | 0 | -2.0 | 0 | -2.4 |
| 0.24 - 0.40 | +0.6 | -0.2 | +1.4 | -0.5 | +2.2 | -1.0 | +2.2 | -1.6 |
| | 0 | -0.6 | 0 | -1.4 | 0 | -2.4 | 0 | -3.0 |
| 0.40- 0.71 | +0.7 | -0.25 | +1.6 | -0.6 | +2.8 | -1.2 | +2.8 | -2.0 |
| | 0 | -0.65 | 0 | -1.6 | 0 | -2.8 | 0 | -3.6 |
| 0.71 - 1.19 | +0.8 | -0.3 | +2.0 | -0.8 | +3.5 | -1.6 | -3.5 | -2.5 |
| | 0 | -0.8 | 0 | -2.0 | 0 | -3.6 | 0 | -4.5 |
| 1.19- 1.97 | +1.0 | -0.4 | +2.5 | -1.0 | +4.0 | -2.0 | +4.0 | -3.0 |
| | 0 | -1.0 | 0 | -2.6 | 0 | -4.5 | 0 | -5.5 |
| 1.97- 3.15 | +1.2 | -0.4 | +3.0 | -1.2 | +4.5 | -2.5 | +4.5 | -4.0 |
| | 0 | -1 1 | 0 | -3.0 | 0 | -5.5 | 0 | -7.0 |
| 3.15- 4.73 | +1.4 | -0.5 | +3.5 | -1.4 | +5.0 | -3.0 | +5.0 | -5.0 |
| | 0 | -1.4 | 0 | -3.6 | 0 | -6.5 | 0 | -8.5 |
| 4.73- 7.09 | +1.6 | -0.6 | +4.0 | -1 .6 | +6.0 | -3.5 | +6.0 | -6.0 |
| | 0 | -1.6 | 0 | -4.1 | 0 | -7.5 | 0 | -10.0 |
| 7.09- 9.85 | +1.6 | -0.6 | +4.5 | -2.0 | +7.0 | -4.0 | +7.0 | -7.0 |
| | 0 | -1.8 | 0 | -4.8 | 0 | -8.5 | 0 | -11.5 |
| 9.85 - 12.41 | +2.0 | -0.7 | +5.0 | -2.2 | +8.0 | -4.5 | +8.0 | -7.0 |
| | 0 | -1.9 | 0 | -5.2 | 0 | -9.5 | 0 | -12.0 |
| 12 41 - 15.75 | +2.2 | -0.7 | +6.0 | -2.5 | +9.0 | -5.0 | +9.0 | -8.0 |
| | 0 | -2.1 | 0 | -6.0 | 0 | -11.0 | 0 | -14.0 |
| 15.75 - 19.69 | +2.5 | -0.8 | +6.0 | -2.8 | +10.0 | -5.0 | +10.0 | -9.0 |
| | 0 | -2.4 | 0 | -6.8 | 0 | -11.0 | 0 | -15.0 |

USAS/ASME B4.1 – 1967 (R2004) Standard. For larger diameters, see the standard.  ASME/ANSI B18.3.5M – 1986 (R2002) Standard.  Reprinted from the standard listed by permission of the American Society of Mechanical Engineers. All rights reserved.

**APPENDIX A** (Locational clearance fits Continued)

## Locational clearance fits – American National Standard (Inches)

Basic Hole System, Limits are in thousandths of an inch.
Limits for hole and shaft are applied to the basic size to obtain
the limits of sizes for the parts.

| Nominal Size Range Inches | | Class LC9 | | Class LC10 | | Class LC11 | |
|---|---|---|---|---|---|---|---|
| | | Standard Limits | | Standard Limits | | Standard Limits | |
| Over | To | Hole | Shaft | Hole | Shaft | Hole | Shaft |
| 0 | - 0.12 | +2.5 | -2.5 | +4.0 | -4.0 | +6.0 | -5.0 |
| | | 0 | -4.1 | 0 | -8.0 | 0 | -11.0 |
| 0.12- | 0.24 | +3.0 | -3.0 | +5.0 | -4.5 | +7.0 | -6.0 |
| | | 0 | -5.2 | 0 | -9.5 | 0 | -13.0 |
| 0.25- | 0.40 | +3.5 | -3.5 | +6.0 | -5.0 | +9.0 | -7.0 |
| | | 0 | -6.3 | 0 | -11.0 | 0 | -16.0 |
| 0.40- | 0.71 | +4.0 | -4.5 | +7.0 | -6.0 | +10.0 | -8.0 |
| | | 0 | -8.0 | 0 | -13.0 | 0 | -18.0 |
| 0.71 - | 1.19 | +5.0 | -5.0 | +8.0 | -7.0 | +12.0 | -10.0 |
| | | 0 | -9.0 | 0 | -15.0 | 0 | -22.0 |
| 1.19- | 1.97 | +6.0 | -6.0 | +10.0 | -8.0 | +16.0 | -12.0 |
| | | 0 | -10.5 | 0 | -18.0 | 0 | -28.0 |
| 1.97- | 3.15 | +7.0 | -7.0 | +12.0 | -10.0 | +18.0 | -14.0 |
| | | 0 | -12.0 | 0 | -22.0 | 0 | -32.0 |
| 3.15- | 4.73 | +9.0 | -8.0 | +14.0 | -11.0 | +22.0 | -16.0 |
| | | 0 | -14.0 | 0 | -25.0 | 0 | -38.0 |
| 4.73- | 7.09 | +10.0 | -10.0 | +16.0 | -12.0 | +25.0 | -18.0 |
| | | 0 | -17.0 | 0 | -28.0 | 0 | -43.0 |
| 7.09- | 9.85 | +12.0 | -12.0 | +18.0 | -16.0 | +25.0 | -22.0 |
| | | 0 | -20.0 | 0 | -34.0 | 0 | -50.0 |
| 9.85 - | 12.41 | +12.0 | -0.7 | +20.0 | -20.0 | +30.0 | -28.0 |
| | | 0 | -1.9 | 0 | -40.0 | 0 | -58.0 |
| 12 41 - | 15.75 | +14.0 | -14.0 | +22.0 | -22.0 | +35.0 | -30.0 |
| | | 0 | -23.0 | 0 | -44.0 | 0 | -65.0 |
| 15.75 - | 19.69 | +16.0 | --16.0 | +25.0 | -25.0 | +40.0 | -35.0 |
| | | 0 | -26.0 | 0 | -50.0 | 0 | -75.0 |

USAS/ASME B4.1 – 1967 (R2004) Standard. For larger diameters, see the standard.  ASME/ANSI
B18.3.5M – 1986 (R2002) Standard.  Reprinted from the standard listed by permission of the American
Society of Mechanical Engineers. All rights reserved.

**APPENDIX A**

## Force and Shrink fits – American National Standard (Inches) - Partial List

Basic Hole System, Limits are in thousandths of an inch.
Limits for hole and shaft are applied to the basic size to obtain the limits of sizes for the parts.

❖  Values shown below are in thousandths of an inch

| Nominal Size Range, Inches | | Class FN 1 | | | Class FN 2 | | | Class FN 3 | | | Class FN 4 | | | Class FN 5 | | |
|---|---|---|---|---|---|---|---|---|---|---|---|---|---|---|---|---|
| | | Interference Limits | Standard Limits | | Interference Limits | Standard Limits | | Interference Limits | Standard Limits | | Interference Limits | Standard Limits | | Interference Limits | Standard Limits | |
| Over | To | | Hole | Shaft | | Hole | Shaft | | Hole | Shaft | | Hole | Shaft | | Hole | Shaft |
| | | | H6 | | | H7 | s6 | | H7 | t6 | | H7 | u6 | | H8 | x7 |
| 0 | 0.12 | +0.05 | +0.25 | +0.5 | +0.2 | +0.4 | +0.85 | | | | +0.3 | +0.4 | +0.95 | +0.3 | +0.6 | +1.3 |
| | | +0.5 | 0 | +0.3 | +0.85 | 0 | +0.6 | | | | +0.95 | 0 | +0.7 | +1.3 | 0 | +0.9 |
| 0.12 | 0.24 | +0.1 | +0.3 | +0.6 | +0.2 | +0.5 | +1.0 | | | | +0.4 | +0.5 | +1.2 | +0.5 | +0.7 | +1.7 |
| | | +0.6 | 0 | +0.4 | +1.0 | 0 | +0.7 | | | | +1.2 | 0 | +0.9 | +1.7 | 0 | +1.2 |
| 0.24 | 0.4 | +0.1 | +0.4 | +0.75 | +0.4 | +0.6 | +1.4 | | | | +0.6 | +0.6 | +1.6 | +0.5 | +0.9 | +2.0 |
| | | +0.75 | 0 | +0.5 | +1.4 | 0 | +1.0 | | | | +1.6 | 0 | +1.2 | +2.0 | 0 | +1.4 |
| 0.40 | 0.56 | +0.1 | +0.4 | +0.8 | +0.5 | +0.7 | +1.6 | | | | +0.7 | +0.7 | +1.8 | +0.6 | +1.0 | +2.3 |
| | | +0.8 | 0 | +0.5 | +1.6 | 0 | +1.2 | | | | +1.8 | 0 | +1.4 | +2.3 | 0 | +1.6 |
| 0.56 | 0.71 | +0.2 | +0.4 | +0.9 | +0.5 | +0.7 | +1.6 | | | | +0.7 | +0.7 | +1.8 | +0.8 | +1.0 | +2.5 |
| | | +0.9 | 0 | +0.6 | +1.6 | 0 | +1.2 | | | | +1.8 | 0 | +1.4 | +2.5 | 0 | +1.8 |
| 0.71 | 0.95 | +0.2 | +0.5 | +1.1 | +0.6 | +0.8 | +1.9 | | | | +0.8 | +0.8 | +2.1 | +1.0 | +1.2 | +3.0 |
| | | +1.1 | 0 | +0.7 | +1.9 | 0 | +1.4 | | | | +2.1 | 0 | +1.6 | +3.0 | 0 | +2.2 |
| 0.95 | 1.19 | +0.3 | +0.5 | +1.2 | +0.6 | +0.8 | +1.9 | +0.8 | +0.8 | +2.1 | +1.0 | +0.8 | +2.3 | +1.3 | +1.2 | +3.3 |
| | | +1.2 | 0 | +0.8 | +1.9 | 0 | +1.4 | +2.1 | 0 | +1.6 | +2.3 | 0 | +1.8 | +3.3 | 0 | +2.5 |
| 1.19 | 1.58 | +0.3 | +0.6 | +1.3 | +0.8 | +1.0 | +2.4 | +1.0 | +1.0 | +2.6 | +1.5 | +1.0 | +3.1 | +1.4 | +1.6 | +4.0 |
| | | +1.3 | 0 | +0.9 | +2.4 | 0 | +1.8 | +2.6 | 0 | +2.0 | +3.1 | 0 | +2.5 | +4.0 | 0 | +3.0 |
| 1.58 | 1.97 | +0.4 | +0.6 | +1.4 | +0.8 | +1.0 | +2.4 | +1.2 | +1.0 | +2.8 | +1.8 | +1.0 | +3.4 | +2.4 | +1.6 | +5.0 |
| | | +1.4 | 0 | +1.0 | +2.4 | 0 | +1.8 | +2.8 | 0 | +2.2 | +3.4 | 0 | +2.8 | +5.0 | 0 | +4.0 |
| 1.97 | 2.56 | +0.6 | +0.7 | +1.8 | +0.8 | +1.2 | +2.7 | +1.3 | +1.2 | +3.2 | +2.3 | +1.2 | +4.2 | +3.2 | +1.8 | +6.2 |
| | | +1.8 | 0 | +1.3 | +2.7 | 0 | +2.0 | +3.2 | 0 | +2.5 | +4.2 | 0 | +3.5 | +6.2 | 0 | +5.0 |
| 2.56 | 3.15 | +0.7 | +0.7 | +1.9 | +1.0 | +1.2 | +2.9 | +1.8 | +1.2 | +3.7 | +2.8 | +1.2 | +4.7 | +4.2 | +1.8 | +7.2 |
| | | +1.9 | 0 | +1.4 | +2.9 | 0 | +2.2 | +3.7 | 0 | +3.0 | +4.7 | 0 | +4.0 | +7.2 | 0 | +6.0 |
| 3.15 | 3.94 | +0.9 | +0.9 | +2.4 | +1.4 | +1.4 | +3.7 | +2.1 | +1.4 | +4.4 | +3.6 | +1.4 | +5.9 | +4.8 | +2.2 | +8.4 |
| | | +2.4 | 0 | +1.8 | +3.7 | 0 | +2.8 | +4.4 | 0 | +3.5 | +5.9 | 0 | +5.0 | +8.4 | 0 | +7.0 |
| 3.94 | 4.73 | +1.1 | +0.9 | +2.6 | +1.6 | +1.4 | +3.9 | +2.6 | +1.4 | +4.9 | +4.6 | +1.4 | +6.9 | +5.8 | +2.2 | +9.4 |
| | | +2.6 | 0 | +2.0 | +3.9 | 0 | +3.0 | +4.9 | 0 | +4.0 | +6.9 | 0 | +6.0 | +9.4 | 0 | +8.0 |

USAS/ASME B4.1 – 1967 (R2004) Standard. For larger diameters, see the standard.  ASME/ANSI B18.3.5M – 1986 (R2002) Standard.  Reprinted from the standard listed by permission of the American Society of Mechanical Engineers. All rights reserved.

# APPENDIX B – METRIC LIMITS AND FITS

## Hole Basis Clearance Fits

Preferred Hole Basis Clearance Fits.  Dimensions in mm.

| Basic Size | Loose Running | | Free Running | | Close Running | | Sliding | | Locational Clearance | |
|---|---|---|---|---|---|---|---|---|---|---|
| | Hole H11 | Shaft c11 | Hole H9 | Shaft d9 | Hole H8 | Shaft f7 | Hole H7 | Shaft g6 | Hole H7 | Shaft h6 |
| 1 max | 1.060 | 0.940 | 1.025 | 0.980 | 1.014 | 0.994 | 1.010 | 0.998 | 1.010 | 1.000 |
| min | 1.000 | 0.880 | 1.000 | 0.955 | 1.000 | 0.984 | 1.000 | 0.992 | 1.000 | 0.994 |
| 1.2 max | 1.260 | 0.940 | 1.225 | 1.180 | 1.214 | 1.194 | 1.210 | 1.198 | 1.210 | 1.200 |
| min | 1.200 | 0.880 | 1.200 | 1.155 | 1.200 | 1.184 | 1.200 | 1.192 | 1.200 | 1.194 |
| 1.6 max | 1.660 | 1.540 | 1.625 | 1.580 | 1.614 | 1.594 | 1.610 | 1.598 | 1.610 | 1.600 |
| min | 1.600 | 1.480 | 1.600 | 1.555 | 1.600 | 1.584 | 1.600 | 1.592 | 1.600 | 1.594 |
| 2 max | 2.060 | 1.940 | 2.025 | 1.980 | 2.014 | 1.994 | 2.010 | 1.998 | 2.010 | 2.000 |
| min | 2.000 | 1.880 | 2.000 | 1.955 | 2.000 | 1.984 | 2.000 | 1.992 | 2.000 | 1.994 |
| 2.5 max | 2.560 | 2.440 | 2.525 | 2.480 | 2.514 | 2.494 | 2.510 | 2.498 | 2.510 | 2.500 |
| min | 2.500 | 2.380 | 2.500 | 2.455 | 2.500 | 2.484 | 2.500 | 2.492 | 2.500 | 2.494 |
| 3 max | 3.060 | 2.940 | 3.025 | 2.980 | 3.014 | 2.994 | 3.010 | 2.998 | 3.010 | 3.000 |
| min | 3.000 | 2.880 | 3.000 | 2.955 | 3.000 | 2.984 | 3.000 | 2.992 | 3.000 | 2.994 |
| 4 max | 4.075 | 3.930 | 4.030 | 3.970 | 4.018 | 3.990 | 4.012 | 3.996 | 4.012 | 4.000 |
| min | 4.000 | 3.855 | 4.000 | 3.940 | 4.000 | 3.987 | 4.000 | 3.988 | 4.000 | 3.992 |
| 5 max | 5.075 | 4.930 | 5.030 | 4.970 | 5.018 | 4.990 | 5.012 | 4.998 | 5.012 | 5.000 |
| min | 5.000 | 4.855 | 5.000 | 4.940 | 5.000 | 4.978 | 5.000 | 4.988 | 5.000 | 4.992 |
| 6 max | 6.075 | 5.930 | 6.030 | 5.970 | 6.018 | 5.990 | 6.012 | 5.996 | 6.012 | 6.000 |
| min | 6.000 | 5.855 | 6.000 | 5.940 | 6.000 | 5.978 | 6.000 | 5.988 | 6.000 | 5.992 |
| 8 max | 8.090 | 7.920 | 8.036 | 7.960 | 8.022 | 7.987 | 8.015 | 7.995 | 8.015 | 8.000 |
| min | 8.000 | 7.830 | 8.000 | 7.924 | 8.000 | 7.972 | 8.000 | 7.986 | 8.000 | 7.991 |
| 10 max | 10.090 | 9.920 | 10.036 | 9.960 | 10.022 | 9.987 | 10.015 | 9.995 | 10.015 | 10.000 |
| min | 10.000 | 9.830 | 10.000 | 9.924 | 10.000 | 9.972 | 10.000 | 9.986 | 10.000 | 9.991 |
| 12 max | 12.110 | 11.905 | 12.043 | 11.950 | 12.027 | 11.984 | 12.018 | 11.994 | 12.018 | 12.000 |
| min | 12.000 | 11.795 | 12.000 | 11.907 | 12.000 | 11.966 | 12.000 | 11.983 | 12.000 | 11.989 |
| 16 max | 16.110 | 15.905 | 16.043 | 15.950 | 16.027 | 15.984 | 16.018 | 15.994 | 16.018 | 16.000 |
| min | 16.000 | 15.795 | 16.000 | 15.907 | 16.000 | 15.966 | 16.000 | 15.983 | 16.000 | 15.989 |
| 20 max | 20.130 | 19.890 | 20.052 | 19.935 | 20.033 | 19.980 | 20.021 | 19.993 | 20.021 | 20.000 |
| min | 20.000 | 19.760 | 20.000 | 19.883 | 20.000 | 19.959 | 20.000 | 19.980 | 20.000 | 19.987 |
| 25 max | 25.130 | 24.890 | 25.052 | 24.935 | 25.033 | 24.980 | 24.993 | 25.021 | 25.021 | 25.000 |
| min | 25.000 | 24.760 | 25.000 | 24.883 | 25.000 | 24.959 | 24.980 | 25.000 | 25.000 | 24.987 |
| 30 max | 30.130 | 29.890 | 30.052 | 29.935 | 30.033 | 29.980 | 30.021 | 29.993 | 30.021 | 30.000 |
| min | 30.000 | 29.760 | 30.000 | 29.883 | 30.000 | 29.959 | 30.000 | 29.980 | 30.000 | 29.987 |
| 40 max | 40.160 | 39.880 | 40.062 | 39.920 | 40.039 | 39.975 | 40.025 | 39.991 | 40.025 | 40.000 |
| min | 40.000 | 39.720 | 40.000 | 39.858 | 40.000 | 39.950 | 40.000 | 39.975 | 40.000 | 39.984 |

**APPENDIX B – METRIC LIMITS AND FITS (Continued)**

# Hole Basis Transition and Interference Fits

Preferred Hole Basis Clearance Fits. Dimensions in mm.

| Basic Size | | Locational Transition | | Locational Transition | | Locational Transition | | Medium Drive | | Force | |
|---|---|---|---|---|---|---|---|---|---|---|---|
| | | Hole H7 | Shaft k6 | Hole H7 | Shaft n6 | Hole H7 | Shaft p6 | Hole H7 | Shaft s6 | Hole H7 | Shaft u6 |
| 1 | max | 1.010 | 1.006 | 1.010 | 1.010 | 1.010 | 1.012 | 1.010 | 1.020 | 1.010 | 1.024 |
| | min | 1.000 | 1.000 | 1.000 | 1.004 | 1.000 | 1.006 | 1.000 | 1.014 | 1.000 | 1.018 |
| 1.2 | max | 1.210 | 1.206 | 1.210 | 1.210 | 1.210 | 1.212 | 1.210 | 1.220 | 1.210 | 1.224 |
| | min | 1.200 | 1.200 | 1.200 | 1.204 | 1.200 | 1.206 | 1.200 | 1.214 | 1.200 | 1.218 |
| 1.6 | max | 1.610 | 1.606 | 1.610 | 1.610 | 1.610 | 1.612 | 1.610 | 1.620 | 1.610 | 1.624 |
| | min | 1.600 | 1.600 | 1.600 | 1.604 | 1.600 | 1.606 | 1.600 | 1.614 | 1.600 | 1.618 |
| 2 | max | 2.010 | 2.006 | 2.010 | 2.020 | 2.010 | 2.012 | 2.010 | 2.020 | 2.010 | 2.024 |
| | min | 2.000 | 2.000 | 2.000 | 2.004 | 2.000 | 2.006 | 2.000 | 1.014 | 2.000 | 2.018 |
| 2.5 | max | 2.510 | 2.510 | 2.510 | 2.510 | 2.510 | 2.512 | 2.510 | 2.520 | 2.510 | 2.524 |
| | min | 2.500 | 2.500 | 2.500 | 2.504 | 2.500 | 2.506 | 2.500 | 2.514 | 2.500 | 2.518 |
| 3 | max | 3.010 | 3.010 | 3.010 | 3.010 | 3.010 | 3.012 | 3.010 | 3.020 | 3.010 | 3.024 |
| | min | 3.000 | 3.000 | 3.000 | 3.004 | 3.000 | 3.006 | 3.000 | 3.014 | 3.000 | 3.018 |
| 4 | max | 4.012 | 4.012 | 4.012 | 4.016 | 4.012 | 4.020 | 4.012 | 4.027 | 4.012 | 4.031 |
| | min | 4.000 | 4.000 | 4.000 | 4.008 | 4.000 | 4.012 | 4.000 | 4.019 | 4.000 | 4.023 |
| 5 | max | 5.012 | 5.009 | 5.012 | 5.016 | 5.012 | 5.020 | 5.012 | 5.027 | 5.012 | 5.031 |
| | min | 5.000 | 5.001 | 5.000 | 5.008 | 5.000 | 5.012 | 5.000 | 5.019 | 5.000 | 5.023 |
| 6 | max | 6.012 | 6.009 | 6.012 | 6.016 | 6.012 | 6.020 | 6.012 | 6.027 | 6.012 | 6.031 |
| | min | 6.000 | 6.001 | 6.000 | 6.008 | 6.000 | 6.012 | 6.000 | 6.019 | 6.000 | 6.023 |
| 8 | max | 8.015 | 8.010 | 8.015 | 8.019 | 8.015 | 8.024 | 8.015 | 8.032 | 8.015 | 8.037 |
| | min | 8.000 | 8.001 | 8.000 | 8.010 | 8.000 | 8.015 | 8.000 | 8.023 | 8.000 | 8.028 |
| 10 | max | 10.015 | 10.010 | 10.015 | 10.019 | 10.015 | 10.024 | 10.015 | 10.032 | 10.015 | 10.037 |
| | min | 10.000 | 10.001 | 10.000 | 10.010 | 10.000 | 10.015 | 10.000 | 10.023 | 10.000 | 10.028 |
| 12 | max | 12.018 | 12.012 | 12.018 | 12.023 | 12.018 | 12.029 | 12.018 | 12.039 | 12.018 | 12.044 |
| | min | 12.000 | 12.001 | 12.000 | 12.012 | 12.000 | 12.018 | 12.000 | 12.028 | 12.000 | 12.033 |
| 16 | max | 16.018 | 16.012 | 16.018 | 16.023 | 16.018 | 16.029 | 16.018 | 16.039 | 16.018 | 16.044 |
| | min | 16.000 | 16.001 | 16.000 | 16.012 | 16.000 | 16.018 | 16.000 | 16.028 | 16.000 | 16.033 |
| 20 | max | 20.021 | 20.015 | 20.021 | 20.028 | 20.021 | 20.035 | 20.021 | 20.048 | 20.021 | 20.054 |
| | min | 20.000 | 20.002 | 20.000 | 20.015 | 20.000 | 20.022 | 20.000 | 20.035 | 20.000 | 20.041 |
| 25 | max | 25.021 | 25.015 | 25.021 | 25.028 | 25.021 | 25.035 | 25.021 | 25.048 | 25.021 | 25.061 |
| | min | 25.000 | 25.002 | 25.000 | 25.015 | 25.000 | 25.022 | 25.000 | 25.035 | 25.000 | 25.048 |
| 30 | max | 30.021 | 30.015 | 30.021 | 30.028 | 30.021 | 30.035 | 30.021 | 30.048 | 30.021 | 30.061 |
| | min | 30.000 | 30.002 | 30.000 | 30.015 | 30.000 | 30.022 | 30.000 | 30.035 | 30.000 | 30.048 |
| 40 | max | 40.025 | 40.018 | 40.025 | 40.033 | 40.025 | 40.042 | 40.025 | 40.059 | 40.025 | 40.076 |
| | min | 40.000 | 40.002 | 40.000 | 40.017 | 40.000 | 40.026 | 40.000 | 40.043 | 40.000 | 40.060 |

**APPENDIX B – METRIC LIMITS AND FITS (Continued)**

# Shaft Basis Clearance Fits

Preferred Shaft Basis Clearance Fits.  Dimensions in mm.

| Basic Size | Loose Running | | Free Running | | Close Running | | Sliding | | Locational Clearance | |
|---|---|---|---|---|---|---|---|---|---|---|
| | Hole C11 | Shaft h11 | Hole D9 | Shaft h9 | Hole F8 | Shaft h7 | Hole G7 | Shaft h6 | Hole H7 | Shaft h6 |
| 1 max | 1.120 | 1.000 | 1.045 | 1.000 | 1.020 | 1.000 | 1.012 | 1.000 | 1.010 | 1.000 |
| min | 1.060 | 0.940 | 1.020 | 0.975 | 1.006 | 0.990 | 1.002 | 0.994 | 1.000 | 0.994 |
| 1.2 max | 1.320 | 1.200 | 1.245 | 1.200 | 1.220 | 1.200 | 1.212 | 1.200 | 1.210 | 1.200 |
| min | 1.260 | 1.140 | 1.220 | 1.175 | 1.206 | 1.190 | 1.202 | 1.194 | 1.200 | 1.194 |
| 1.6 max | 1.720 | 1.600 | 1.645 | 1.600 | 1.620 | 1.600 | 1.612 | 1.600 | 1.610 | 1.600 |
| min | 1.660 | 1.540 | 1.620 | 1.575 | 1.606 | 1.590 | 1.602 | 1.594 | 1.600 | 1.594 |
| 2 max | 2.120 | 2.000 | 2.045 | 2.000 | 2.020 | 2.000 | 2.012 | 2.000 | 2.010 | 2.000 |
| min | 2.060 | 1.940 | 2.020 | 1.975 | 2.006 | 1.990 | 2.002 | 1.994 | 2.000 | 1.994 |
| 2.5 max | 2.620 | 2.500 | 2.545 | 2.500 | 2.520 | 2.500 | 2.512 | 2.500 | 2.510 | 2.500 |
| min | 2.560 | 2.440 | 2.520 | 2.475 | 2.506 | 2.490 | 2.502 | 2.494 | 2.500 | 2.494 |
| 3 max | 3.120 | 3.000 | 3.045 | 3.000 | 3.020 | 3.000 | 3.012 | 3.000 | 3.010 | 3.000 |
| min | 3.060 | 2.940 | 3.020 | 2.975 | 3.006 | 2.990 | 3.002 | 2.994 | 3.000 | 2.994 |
| 4 max | 4.145 | 4.000 | 4.060 | 4.000 | 4.028 | 4.000 | 4.016 | 4.000 | 4.012 | 4.000 |
| min | 4.070 | 3.925 | 4.030 | 3.970 | 4.010 | 3.988 | 4.004 | 3.992 | 4.000 | 3.992 |
| 5 max | 5.145 | 5.000 | 5.060 | 5.000 | 5.028 | 5.000 | 5.016 | 5.000 | 5.012 | 5.000 |
| min | 5.070 | 4.925 | 5.030 | 4.970 | 5.010 | 4.988 | 5.004 | 4.992 | 5.000 | 4.992 |
| 6 max | 6.145 | 6.000 | 6.060 | 6.000 | 6.028 | 6.000 | 6.016 | 6.000 | 6.012 | 6.000 |
| min | 6.070 | 5.925 | 6.030 | 5.970 | 6.010 | 5.988 | 6.004 | 5.992 | 6.000 | 5.992 |
| 8 max | 8.170 | 8.000 | 8.076 | 8.000 | 8.035 | 8.000 | 8.020 | 8.000 | 8.015 | 8.000 |
| min | 8.080 | 7.910 | 8.040 | 7.964 | 8.013 | 7.985 | 8.005 | 7.991 | 8.000 | 7.991 |
| 10 max | 10.170 | 10.000 | 10.076 | 10.000 | 10.035 | 10.000 | 10.020 | 10.000 | 10.015 | 10.000 |
| min | 10.080 | 9.910 | 10.040 | 9.964 | 10.013 | 9.985 | 10.005 | 9.991 | 10.000 | 9.991 |
| 12 max | 12.205 | 12.000 | 12.093 | 12.000 | 12.043 | 12.000 | 12.024 | 12.000 | 12.018 | 12.000 |
| min | 12.095 | 11.890 | 12.050 | 11.957 | 12.016 | 11.982 | 12.006 | 11.989 | 12.000 | 11.989 |
| 16 max | 16.205 | 16.000 | 16.093 | 16.000 | 16.043 | 16.000 | 16.024 | 16.000 | 16.018 | 16.000 |
| min | 16.095 | 15.890 | 16.050 | 15.957 | 16.016 | 15.982 | 16.006 | 15.989 | 16.000 | 15.989 |
| 20 max | 20.240 | 20.000 | 20.117 | 20.000 | 20.053 | 20.000 | 20.028 | 20.000 | 20.021 | 20.000 |
| min | 20.110 | 19.870 | 20.065 | 19.948 | 20.020 | 19.979 | 20.007 | 19.987 | 20.000 | 19.987 |
| 25 max | 25.240 | 25.000 | 25.117 | 25.000 | 25.053 | 25.000 | 25.028 | 25.000 | 25.021 | 25.000 |
| min | 25.110 | 24.870 | 25.065 | 24.948 | 25.020 | 24.979 | 25.007 | 24.987 | 25.000 | 24.987 |
| 30 max | 30.240 | 30.000 | 30.117 | 30.000 | 30.053 | 30.000 | 30.028 | 30.000 | 30.021 | 30.000 |
| min | 30.110 | 29.870 | 30.065 | 29.948 | 30.020 | 29.979 | 30.007 | 29.987 | 30.000 | 29.987 |
| 40 max | 40.280 | 40.000 | 40.142 | 40.000 | 40.064 | 40.000 | 40.034 | 40.000 | 40.025 | 40.000 |
| min | 40.120 | 39.840 | 40.080 | 39.938 | 40.025 | 39.975 | 40.009 | 39.984 | 40.000 | 39.984 |

**APPENDIX B - METRIC LIMITS AND FITS (Continued)**

## Shaft Basis Transition and Interference Fits

Preferred Shaft Basis Transition and Interference Fits.  Dimensions in mm.

| Basic Size | Locational Transition | | Locational Transition | | Locational Interference | | Medium Drive | | Force | |
|---|---|---|---|---|---|---|---|---|---|---|
| | Hole K7 | Shaft h6 | Hole N7 | Shaft h6 | Hole P7 | Shaft h6 | Hole S7 | Shaft h6 | Hole U7 | Shaft h6 |
| **1 max** | 1.000 | 1.000 | 0.996 | 1.000 | 0.994 | 1.000 | 0.986 | 1.000 | 0.982 | 1.000 |
| min | 0.990 | 0.994 | 0.986 | 0.994 | 0.984 | 0.994 | 0.976 | 0.994 | 0.972 | 0.994 |
| **1.2 max** | 1.200 | 1.200 | 1.196 | 1.200 | 1.194 | 1.200 | 1.186 | 1.200 | 1.182 | 1.200 |
| min | 1.190 | 1.194 | 1.186 | 1.194 | 1.184 | 1.194 | 1.176 | 1.194 | 1.172 | 1.194 |
| **1.6 max** | 1.600 | 1.600 | 1.596 | 1.600 | 1.594 | 1.600 | 1.586 | 1.600 | 1.582 | 1.600 |
| min | 1.590 | 1.594 | 1.586 | 1.594 | 1.584 | 1.594 | 1.576 | 1.594 | 1.572 | 1.594 |
| **2 max** | 2.000 | 2.000 | 1.996 | 2.000 | 1.994 | 2.000 | 1.986 | 2.000 | 1.982 | 2.000 |
| min | 1.990 | 1.994 | 1.986 | 1.994 | 1.984 | 1.994 | 1.976 | 1.994 | 1.972 | 1.994 |
| **2.5 max** | 2.500 | 2.500 | 2.496 | 2.500 | 2.494 | 2.500 | 2.486 | 2.500 | 2.482 | 2.500 |
| min | 2.490 | 2.494 | 2.486 | 2.494 | 2.484 | 2.494 | 2.476 | 2.494 | 2.472 | 2.494 |
| **3 max** | 3.000 | 3.000 | 2.996 | 3.000 | 2.994 | 3.000 | 2.986 | 3.000 | 2.982 | 3.000 |
| min | 2.990 | 2.994 | 2.986 | 2.994 | 2.984 | 2.994 | 2.976 | 2.994 | 2.972 | 2.994 |
| **4 max** | 4.003 | 4.000 | 3.996 | 4.000 | 3.992 | 4.000 | 3.985 | 4.000 | 3.981 | 4.000 |
| min | 3.991 | 5.992 | 3.984 | 5.992 | 3.980 | 5.992 | 3.973 | 5.992 | 3.969 | 5.992 |
| **5 max** | 5.003 | 5.000 | 4.996 | 5.000 | 4.992 | 5.000 | 4.985 | 5.000 | 4.981 | 5.000 |
| min | 4.991 | 4.992 | 4.984 | 4.992 | 4.980 | 4.992 | 4.973 | 4.992 | 4.969 | 4.992 |
| **6 max** | 6.003 | 6.000 | 5.996 | 6.000 | 5.992 | 6.000 | 5.985 | 6.000 | 5.981 | 6.000 |
| min | 5.991 | 5.992 | 5.984 | 5.992 | 5.980 | 5.992 | 5.973 | 5.992 | 5.969 | 5.992 |
| **8 max** | 8.005 | 8.000 | 7.996 | 8.000 | 7.991 | 8.000 | 7.983 | 8.000 | 7.978 | 8.000 |
| min | 7.990 | 7.991 | 7.981 | 7.991 | 7.976 | 7.991 | 7.968 | 7.991 | 7.963 | 7.991 |
| **10 max** | 10.005 | 10.000 | 9.996 | 10.000 | 9.991 | 10.000 | 9.983 | 10.000 | 9.978 | 10.000 |
| min | 9.990 | 9.991 | 9.981 | 9.991 | 9.976 | 9.991 | 9.968 | 9.991 | 9.963 | 9.991 |
| **12 max** | 12.006 | 12.000 | 11.995 | 12.000 | 11.989 | 12.000 | 11.979 | 12.000 | 11.974 | 12.000 |
| min | 11.988 | 11.989 | 11.977 | 11.989 | 11.971 | 11.989 | 11.961 | 11.989 | 11.956 | 11.989 |
| **16 max** | 16.006 | 16.000 | 15.995 | 16.000 | 15.989 | 16.000 | 15.979 | 16.000 | 15.974 | 16.000 |
| min | 15.988 | 15.989 | 15.977 | 15.989 | 15.971 | 15.989 | 15.961 | 15.989 | 15.956 | 15.989 |
| **20 max** | 20.006 | 20.000 | 19.993 | 20.000 | 19.986 | 20.000 | 19.973 | 20.000 | 19.967 | 20.000 |
| min | 19.985 | 19.987 | 19.972 | 19.987 | 19.965 | 19.987 | 19.952 | 19.987 | 19.946 | 19.987 |
| **25 max** | 25.006 | 25.000 | 24.993 | 25.000 | 24.986 | 25.000 | 24.973 | 25.000 | 24.960 | 25.000 |
| min | 24.985 | 24.987 | 24.972 | 24.987 | 24.965 | 24.987 | 24.952 | 24.987 | 24.939 | 24.987 |
| **30 max** | 30.006 | 30.000 | 29.993 | 30.000 | 29.986 | 30.000 | 29.973 | 30.000 | 29.960 | 30.000 |
| min | 29.985 | 29.987 | 29.972 | 29.987 | 29.965 | 29.987 | 29.952 | 29.987 | 29.939 | 29.987 |
| **40 max** | 40.007 | 40.000 | 39.992 | 40.000 | 39.983 | 40.000 | 39.966 | 40.000 | 39.949 | 40.000 |
| min | 39.982 | 39.984 | 39.967 | 39.984 | 39.958 | 39.984 | 39.941 | 39.984 | 39.924 | 39.984 |

ANSI B4.2 – 1978 (R2004) Standard.  ASME/ANSI B18.3.5M – 1986 (R2002) Standard.  Reprinted from the standard listed by permission of the American Society of Mechanical Engineers.  All rights reserved.

# APPENDIX C – UNIFIED NATIONAL THREAD FORM

(External Threads) Approximate Minor diameter = D – 1.0825P      P = Pitch

| Nominal Size, in. | Basic Major Diameter (D) | Coarse UNC | | Fine UNF | | Extra Fine UNEF | |
|---|---|---|---|---|---|---|---|
| | | Thds. Per in. | Tap Drill Dia. | Thds Per in. | Tap Drill. Dia. | Thds. Per in. | Tap Drill Dia. |
| #0 | 0.060 | ... | ... | 80 | 3/64 | ... | ... |
| #1 | 0.0730 | 64 | 0.0595 | 72 | 0.0595 | ... | ... |
| #2 | 0.0860 | 56 | 0.0700 | 64 | 0.0700 | ... | ... |
| #3 | 0.0990 | 48 | 0.0785 | 56 | 0.0820 | ... | ... |
| #4 | 0.1120 | 40 | 0.0890 | 48 | 0.0935 | ... | ... |
| #5 | 0.1250 | 40 | 0.1015 | 44 | 0.1040 | ... | ... |
| #6 | 0.1380 | 32 | 0.1065 | 40 | 0.1130 | ... | ... |
| #8 | 0.1640 | 32 | 0.1360 | 36 | 0.1360 | ... | ... |
| #10 | 0.1900 | 24 | 0.1495 | 32 | 0.1590 | ... | ... |
| #12 | 0.2160 | 24 | 0.1770 | 28 | 0.1820 | 32 | 0.1850 |
| 1/4 | 0.2500 | 20 | 0.2010 | 28 | 0.2130 | 32 | 7/32 |
| 5/16 | 0.3125 | 18 | 0.257 | 24 | 0.272 | 32 | 9/32 |
| 3/8 | 0.3750 | 16 | 5/16 | 24 | 0.332 | 32 | 11/32 |
| 7/16 | 0.4375 | 14 | 0.368 | 20 | 25/64 | 28 | 13/32 |
| 1/2 | 0.5000 | 13 | 27/64 | 20 | 29/64 | 28 | 15/32 |
| 9/16 | 0.5625 | 12 | 31/64 | 18 | 33/64 | 24 | 33/64 |
| 5/8 | 0.6250 | 11 | 17/32 | 18 | 37/64 | 24 | 37/64 |
| 11/16 | 0.675 | ... | ... | ... | ... | 24 | 41/64 |
| 3/4 | 0.7500 | 10 | 21/32 | 16 | 11/16 | 20 | 45/64 |
| 13/16 | 0.8125 | ... | ... | ... | ... | 20 | 49/64 |
| 7/8 | 0.8750 | 9 | 49/64 | 14 | 13/16 | 20 | 53/64 |
| 15/16 | 0.9375 | ... | ... | ... | ... | 20 | 57/64 |
| 1 | 1.0000 | 8 | 7/8 | 12 | 59/64 | 20 | 61/64 |
| 1 1/8 | 1.1250 | 7 | 63/64 | 12 | 1 3/64 | 18 | 1 5/64 |
| 1 1/4 | 1.2500 | 7 | 1 7/64 | 12 | 1 11/64 | 18 | 1 3/16 |
| 1 3/8 | 1.3750 | 6 | 1 7/32 | 12 | 1 19/64 | 18 | 1 5/16 |
| 1 1/2 | 1.5000 | 6 | 1 11/32 | 12 | 1 27/64 | 18 | 1 7/16 |
| 1 5/8 | 1.6250 | ... | ... | ... | ... | 18 | 1 9/16 |
| 1 3/4 | 1.7500 | 5 | 1 9/16 | ... | ... | ... | ... |
| 1 7/8 | 1.8750 | ... | ... | ... | ... | ... | ... |
| 2 | 2.0000 | 4 1/2 | 1 25/32 | ... | ... | ... | ... |
| 2 1/4 | 2.2500 | 4 1/2 | 2 1/32 | ... | ... | ... | ... |
| 2 1/2 | 2.5000 | 4 | 2 1/4 | ... | ... | ... | ... |
| 2 3/4 | 2.7500 | 4 | 2 1/2 | ... | ... | ... | ... |

# APPENDIX D – METRIC THREAD FORM

(External Threads) Approximate Minor diameter = D – 1.2075P    P = Pitch
Preferred sizes for commercial threads and fasteners are shown in boldface type.

| Coarse (general purpose) | | Fine | |
|---|---|---|---|
| Nominal Size & Thread Pitch | Tap Drill Diameter, mm | Nominal Size & Thread Pitch | Tap Drill Diameter, mm |
| **M1.6 x 0.35** | 1.25 | --- | --- |
| M1.8 x 0.35 | 1.45 | --- | --- |
| **M2 x 0.4** | 1.6 | --- | --- |
| M2.2 x 0.45 | 1.75 | --- | --- |
| **M2.5 x 0.45** | 2.05 | --- | --- |
| **M3 x 0.5** | 2.5 | --- | --- |
| M3.5 x 0.6 | 2.9 | --- | --- |
| **M4 x 0.7** | 3.3 | --- | --- |
| M4.5 x 0.75 | 3.75 | --- | --- |
| **M5 x 0.8** | 4.2 | --- | --- |
| **M6 x 1** | 5.0 | --- | --- |
| M7 x 1 | 6.0 | --- | --- |
| **M8 x 1.25** | 6.8 | **M8 x 1** | 7.0 |
| M9 x 1.25 | 7.75 | --- | --- |
| **M10 x 1.5** | 8.5 | **M10 x 1.25** | 8.75 |
| M11 x 1.5 | 9.50 | --- | --- |
| **M12 x 1.75** | 10.30 | **M12 x 1.25** | 10.5 |
| M14 x 2 | 12.00 | **M14 x 1.5** | 12.5 |
| **M16 x 2** | 14.00 | **M16 x 1.5** | 14.5 |
| M18 x 2.5 | 15.50 | **M18 x 1.5** | 16.5 |
| **M20 x 2.5** | 17.5 | **M20 x 1.5** | 18.5 |
| M22 x 2.5* | 19.5 | **M22 x 1.5** | 20.5 |
| **M24 x 3** | 21.0 | **M24 x 2** | 22.0 |
| M27 x 3* | 24.0 | **M27 x 2** | 25.0 |
| **M30 x 3.5** | 26.5 | **M30 x 2** | 28.0 |
| M33 x 3.5 | 29.5 | M33 x 2 | 31.0 |
| **M36 x 4** | 32.0 | **M36 x 2** | 33.0 |
| M39 x 4 | 35.0 | M39 x 2 | 36.0 |
| **M42 x 4.5** | 37.5 | **M42 x 2** | 39.0 |
| M45 x 4.5 | 40.5 | M45 x 1.5 | 42.0 |
| **M48 x 5** | 43.0 | **M48 x 2** | 45.0 |
| M52 x 5 | 47.0 | M52 x 2 | 49.0 |
| **M56 x 5.5** | 50.5 | **M56 x 2** | 52.0 |
| M60 x 5.5 | 54.5 | M60 x 1.5 | 56.0 |
| **M64 x 6** | 58.0 | **M64 x 2** | 60.0 |
| M68 x 6 | 62.0 | M68 x 2 | 64.0 |
| **M72 x 6** | 66.0 | **M72 x 2** | 68.0 |
| **M80 x 6** | 74.0 | **M80 x 2** | 76.0 |
| **M90 x 6** | 84.0 | **M90 x 2** | 86.0 |
| **M100 x 6** | 94.0 | **M100 x 2** | 96.0 |

*Only for high strength structural steel fasteners
ASME B1.13M – 2001 Standard. Reprinted from the standard listed by permission of the American Society of Mechanical Engineers. All rights reserved.

## APPENDIX E – FASTENERS (INCH SERIES)

**Important!** All fastener dimensions have a tolerance; each dimension has a maximum and minimum value. Only one size for each dimension is given in this appendix. Refer to the standards noted for the complete listing of values.

## Regular Hex Head Bolts

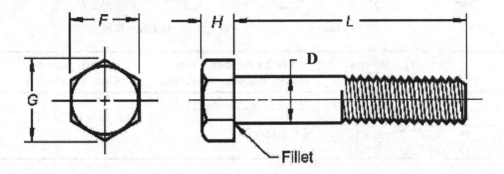

| Size (D) | Head Height Basic | Width Across Flats Basic Adjust to sixteenths | Width Across Corners Max. | Thread Length 6 in. or shorter | Thread Length Over 6 in. |
|---|---|---|---|---|---|
| 1/4 | H = 0.625 D + 0.016 | F = 1.500 D + 0.062 | | 0.75 | 1.00 |
| 5/16 – 7/16 | H = 0.625 D + 0.016 | | | 1.00 | 1.25 |
| 1/2 – 7/8 | H = 0.625 D + 0.031 | | G = 1.1547 F | 1.25 | 1.50 |
| 1 – 1 7/8 | H = 0.625 D + 0.062 | F = 1.500 D | | 2.50 | 2.75 |
| 2 – 3 3/4 | H = 0.625 D + 0.125 | | | 5.25 | 5.50 |
| 4 | H = 0.25 D + 0.188 | | | 6.25 | 6.50 |

- Radius of Fillet: D less than ½ in.: R 0.01~0.03,
  D larger than ½ in. but less than 1 in.: R 0.02~0.06
  D larger than 1 in.: R 0.03~0.09

## Heavy Hex Head Bolts

| Size (D) | Head Height Basic* | Width Across Flats Basic Adjust to sixteenth | Width Across Corners Max. |
|---|---|---|---|
| 1/2 – 3 | Same as for regular hex head bolts. | F = 1.500 D + 0.125 | Max. G =1.1547 F |

*Size to 1 in. adjusted to sixty-fourths. 1 1/8 through 2 1/2 in. sizes adjusted upward to thirty-seconds. 2 3/4 thru 4 in. sizes adjusted upward to sixteenths.

ASME B18.2.1 - 1996 Standard. Reprinted from the standard listed by permission of the American Society of Mechanical Engineers. All rights reserved.

## APPENDIX E – FASTENERS (INCH SERIES) Continued

## Hex Nuts and Hex Jam Nuts

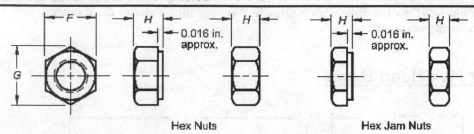

Hex Nuts            Hex Jam Nuts

| Nut Size (D) | Nut Thickness Basic | Width Across Flats Basic (Adjust to Sixteenths) | Width Across Corners (Max.) |
|---|---|---|---|
| 1/4 | H = 0.875D | F=1.500 D + 0.062 | |
| 5/6 – 5/8 | H = 0.875D | | Max. G = 1.1547 F |
| 3/4 – 1 1/8 | H = 0.875D – 0.016 | F = 1.500 D | |
| 1 1/4 – 1 1/2 | H = 0.875D – 0.031 | | |

## Hex Thick Nuts

| Nut Size (D) | Width Across Flats Basic (Adjust to Sixteenths) | Width Across Corners (Max.) | Nut Thickness Basic |
|---|---|---|---|
| ¼ | F = 1.500 D + 0.062 | | |
| 5/6 – 5/8 | F = 1.500 D | Max. G = 1.1547 F | See Table |
| 3/4 – 1 ½ | F = 1.500 D | | |

| Nut Size (D) | 1/4 | 5/16 | 3/8 | 7/16 | 1/2 | 9/16 | 5/8 |
|---|---|---|---|---|---|---|---|
| Nut Thickness Basic | 9/32 | 21/64 | 13/32 | 29/64 | 9/16 | 39/64 | 23/32 |

| Nut Size (D) | 3/4 | 7/8 | 1 | 1 1/8 | 1 1/4 | 1 3/8 | 1 1/2 |
|---|---|---|---|---|---|---|---|
| Nut Thickness Basic | 13/16 | 29/32 | 1 | 1 5/32 | 1 1/4 | 1 3/8 | 1 1/2 |

## Hex Jam Nut

| Nut Size (D) | Nut Thickness Basic | Width Across Flats Basic (Adjust to Sixteenths) | Width Across Corners (Max.) |
|---|---|---|---|
| 1/4 | See Table | F=1.500 D + 0.062 | |
| 5/6 – 5/8 | See Table | | Max. G = 1.1547 F |
| 3/4 – 1 1/8 | H = 0.500D – 0.047 | F = 1.500 D | |
| 1 1/4 – 1 1/2 | H = 0.500D – 0.094 | | |

| Nut Size (D) | 1/4 | 5/16 | 3/8 | 7/16 | 1/2 | 9/16 | 5/8 |
|---|---|---|---|---|---|---|---|
| Nut Thickness Basic | 5/32 | 3/16 | 7/32 | 1/4 | 5/16 | 5/16 | 3/8 |

ASME/ANSI B18.2.2 -1987 (R1999) Standard.  Reprinted from the standard listed by permission of the American Society of Mechanical Engineers.  All rights reserved.

**APPENDIX E – FASTENERS (INCH SERIES) Continued**

## Hexagon and Spline Socket Head Cap Screws

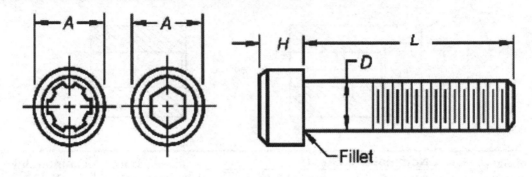

| Screw Size (D) | Head Diameter | Head Height |
|---|---|---|
| # 0 — # 10 | See Table | Max. H = D |
| 1/4 – 1 1/2 | Max. A = 1.50 D | |

| Screw Size (D) | # 0 | # 1 | # 2 | # 3 | # 4 | #5 | # 6 | # 8 | # 10 |
|---|---|---|---|---|---|---|---|---|---|
| Max. Head Diameter | 0.096 | 0.018 | 0.140 | 0.161 | 0.183 | 0.205 | 0.226 | 0.270 | 0.312 |

## Hexagon and Spline Socket Flat Countersunk Head Cap Screws

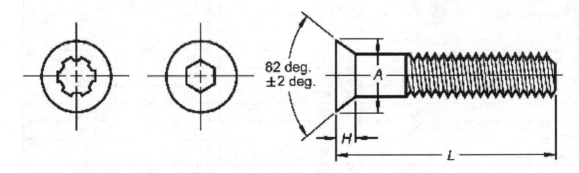

| Screw Size (D) | Head Diameter (A) Theor. Sharp | Head Height (H) |
|---|---|---|
| # 0 — # 10 | See Table | |
| # 4 — 3/8 | Max. A = 2 D + 0.031 | Max.H = 0.5 (Max. A – D) x cot(41° ) |
| 7/16 | Max. A = 2 D – 0.031 | |
| 1/2 – 1 1/2 | Max. A = 2 D – 0.062 | |

**APPENDIX E – FASTENERS (INCH SERIES) Continued**

# Drill and Counterbore Sizes for Socket Head Cap Screws

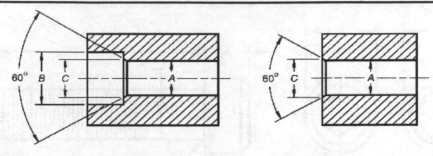

| Nominal Size of Screw (D) | Nominal Drill Size (A) | | Counterbore Diameter (B) | Countersink (C) |
|---|---|---|---|---|
| | Close Fit | Normal Fit | | |
| #0 (0.0600) | (#51) 0.067 | (#49) 0.073 | 1/8 | 0.074 |
| #1 (0.0730) | (#46) 0.081 | (#43) 0.089 | 5/32 | 0.087 |
| #2 (0.0860) | 3/32 | (#36) 0.106 | 3/16 | 0.102 |
| #3 (0.0990) | (#36) 0.106 | (#31) 0.120 | 7/32 | 0.115 |
| #4 (0.1120) | 1/8 | (#29) 0.136 | 7/32 | 0.130 |
| #5 (0.1250) | 9/64 | (#23) 0.154 | 1/4 | 0.145 |
| #6 (0.1380) | (#23) 0.154 | (#18) 0.170 | 9/32 | 0.158 |
| #8 (0.1640) | (#15) 0.180 | (#10) 0.194 | 5/16 | 0.188 |
| #10 (0.1900) | (#5) 0.206 | (#2) 0.221 | 3/8 | 0.218 |
| 1/4 | 17/64 | 9/32 | 7/16 | 0.278 |
| 5/16 | 21/64 | 11/32 | 17/32 | 0.346 |
| 3/8 | 25/64 | 13/32 | 5/8 | 0.415 |
| 7/16 | 29/64 | 15/32 | 23/32 | 0.483 |
| 1/2 | 33/64 | 17/32 | 13/16 | 0.552 |
| 5/8 | 41/64 | 21/32 | 1 | 0.689 |
| 3/4 | 49/64 | 25/32 | 1 3/16 | 0.828 |
| 7/8 | 57/64 | 29/32 | 1 3/8 | 0.963 |
| 1 | 1 1/64 | 1 1/32 | 1 5/8 | 1.100 |
| 1 1/4 | 1 9/32 | 1 5/16 | 2 | 1.370 |
| 1 1/2 | 1 17/32 | 1 9/16 | 2 3/8 | 1.640 |
| 1 3/4 | 1 25/32 | 1 13/16 | 2 3/4 | 1.910 |
| 2 | 2 1/32 | 2 1/16 | 3 1/8 | 2.180 |

(1)  Countersink.  It is considered good practice to countersink or break the edges of holes that are smaller than F (max.) in parts having a hardness which approaches, equals, or exceeds the screw hardness. The countersink or corner relief, however, should not be larger than is necessary to ensure that the fillet on the screw is cleared. Normally, the diameter of countersink does not have to exceed F (max.).  Countersinks or corner reliefs in excess of this diameter reduce the effective bearing area and introduce the possibility of embedment or brinelling or flaring of the heads of the screws.

(2)  Close Fit.  The close fit is normally limited to holes for those lengths of screws that are threaded to the head in assemblies where only one screw is to be used or where two or more screws are to be used and the mating holes are to be produced either at assembly or by matched and coordinated tooling.

(3)  Normal Fit.  The normal fit is intended for screws of relatively long length or for assemblies involving two or more screws where the mating holes are to be produced by conventional tolerancing methods.  It provides for the maximum allowable eccentricity of the longest standard screws and for certain variations in the parts to be fastened, such as deviations in hole straightness, angularity between the axis of the tapped hole and that of the hole for the shank, differences in center distances of the mating holes, etc.

**APPENDIX E – FASTENERS (INCH SERIES) Continued**

## Slotted Flat Countersunk Head Cap Screws

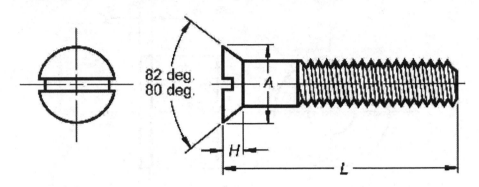

| Screw Size (D) | Head Diameter (A) Theor. Sharp | Head Height (H) |
|---|---|---|
| 1/4 through 3/8 | Max. A = 2.000 D | Max. H = 0.596 D |
| 7/16 | Max. A = 2.000 D – 0.063 | Max. H = 0.596 D – 0.0375 |
| 1/4 through 3/8 | Max. A = 2.000 D – 0.125 | Max. H = 0.596 D – 0.075 |
| 1 1/8 through 1 1/2 | Max. A = 2.000 D – 0.188 | Max. H = 0.596 D – 0.112 |

## Slotted Round Head Cap Screws

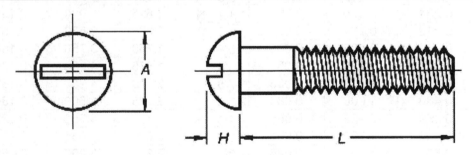

| Screw Size (D) | Head Diameter (A) Theor. Sharp | Head Height (H) |
|---|---|---|
| 1/4 through 5/16 | Max. A = 2.000 D – 0.063 | Max. H = 0.875 D – 0.028 |
| 3/8 through 7/16 | Max. A = 2.000 D – 0.125 | Max. H = 0.875 D – 0.055 |
| 1/2 through 9/16 | Max. A = 2.000 D – 0.1875 | Max. H = 0.875 D – 0.083 |
| 5/8 through 3/4 | Max. A = 2.000 D – 0.250 | Max. H = 0.875 D – 0.110 |

**APPENDIX E – FASTENERS (INCH SERIES) Continued**

# Preferred Sizes of Type A Plain Washers

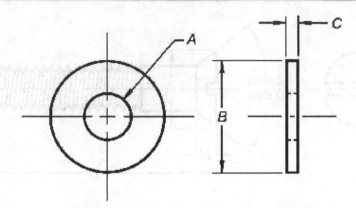

| Nominal Washer Size * | Inside Diameter (A) Basic | Outside Diameter (B) Basic | Thickness (C) | Nominal Washer Size* | Inside Diameter (A) Basic | Outside Diameter (B) Basic | Thickness (C) |
|---|---|---|---|---|---|---|---|
| | 0.078 | 0.188 | 0.020 | 1 N | 1.062 | 2.000 | 0.134 |
| | 0.094 | 0.250 | 0.020 | 1 W | 1.062 | 2.500 | 0.165 |
| | 0.125 | 0.312 | 0.032 | 1 1/8 N | 1.250 | 2.250 | 0.134 |
| #6 (0.138) | 0.156 | 0.375 | 0.049 | 1 1/8 W | 1.250 | 2.750 | 0.165 |
| #8 (0.164) | 0.188 | 0.438 | 0.049 | 1 1/4 N | 1.375 | 2.500 | 0.165 |
| #10 (0.190) | 0.219 | 0.500 | 0.049 | 1 1/4 W | 1.375 | 3.000 | 0.165 |
| 3/16 | 0.250 | 0.562 | 0.049 | 1 3/8 N | 1.500 | 2.750 | 0.165 |
| #12 (0.216) | 0.250 | 0.562 | 0.065 | 1 3/8 W | 1.500 | 3.250 | 0.180 |
| 1/4 N | 0.281 | 0.625 | 0.065 | 1 1/2 N | 1.625 | 3.000 | 0.165 |
| 1/4 W | 0.312 | 0.734 | 0.065 | 1 1/2 W | 1.625 | 3.500 | 0.180 |
| 5/16 N | 0.344 | 0.688 | 0.065 | 1 5/8 | 1.750 | 3.750 | 0.180 |
| 5/16 W | 0.375 | 0.875 | 0.083 | 1 3/4 | 1.875 | 4.000 | 0.180 |
| 3/8 N | 0.406 | 0.812 | 0.065 | 1 7/8 | 2.000 | 4.250 | 0.180 |
| 3/8 W | 0.438 | 1.000 | 0.083 | 2 | 2.125 | 4.500 | 0.180 |
| 7/16 N | 0.469 | 0.922 | 0.065 | 2 1/4 | 2.375 | 4.750 | 0.220 |
| 7/16 W | 0.500 | 1.250 | 0.083 | 2 1/2 | 2.625 | 5.000 | 0.238 |
| 1/2 N | 0.531 | 1.062 | 0.095 | 2 3/4 | 2.875 | 5.250 | 0.259 |
| 1/2 W | 0.562 | 1.375 | 0.109 | 3 | 3.125 | 5.500 | 0.284 |
| 9/16 N | 0.594 | 1.156 | 0.095 | | | | |
| 9/16 W | 0.625 | 1.469 | 0.109 | | | | |
| 5/8 N | 0.656 | 1.312 | 0.095 | | | | |
| 5/8 W | 0.688 | 1.750 | 0.134 | | | | |
| 3/4 N | 0.812 | 1.469 | 0.134 | | | | |
| 3/4 W | 0.812 | 2.000 | 0.148 | | | | |
| 7/8 N | 0.938 | 1.750 | 0.134 | | | | |
| 7/8 W | 0.938 | 2.250 | 0.165 | | | | |

*Nominal washer sizes are intended for use with comparable nominal screw or bolt sizes.
ANSI B18.22.1 - 1965 (R2003) Standard.  Reprinted from the standard listed by permission of the American Society of Mechanical Engineers.  All rights reserved.

**APPENDIX E – FASTENERS (INCH SERIES) Continued**

# Regular Helical Spring-Lock Washers

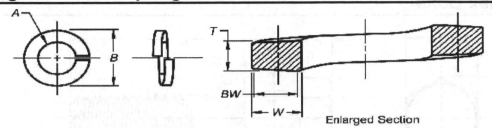

Enlarged Section

| Nominal Washer Size | Min. Inside Diameter (A) | Max. Outside Diameter (B) | Mean Section Thickness (T) | Min. Section Width (W) | Min. Bearing Width (BW) |
|---|---|---|---|---|---|
| #2 (0.086) | 0.088 | 0.172 | 0.020 | 0.035 | 0.024 |
| #3 (0.099) | 0.101 | 0.195 | 0.025 | 0.040 | 0.028 |
| #4 (0.112) | 0.114 | 0.209 | 0.025 | 0.040 | 0.028 |
| #5 (0.125) | 0.127 | 0.236 | 0.031 | 0.047 | 0.033 |
| #6 (.0138) | 0.141 | 0.250 | 0.031 | 0.047 | 0.033 |
| #8 (0.164) | 0.167 | 0.293 | 0.040 | 0.055 | 0.038 |
| #10 (0.190) | 0.193 | 0.334 | 0.047 | 0.062 | 0.043 |
| #12 (0.216) | 0.220 | 0.377 | 0.056 | 0.070 | 0.049 |
| 1/4 | 0.252 | 0.487 | 0.062 | 0.109 | 0.076 |
| 5/16 | 0.314 | 0.583 | 0.078 | 0.125 | 0.087 |
| 3/8 | 0.377 | 0.680 | 0.094 | 0.141 | 0.099 |
| 7/16 | 0.440 | 0.776 | 0.109 | 0.156 | 0.109 |
| 1/2 | 0.502 | 0.869 | 0.125 | 0.171 | 0.120 |
| 9/16 | 0.564 | 0.965 | 0.141 | 0.188 | 0.132 |
| 5/8 | 0.628 | 1.073 | 0.156 | 0.203 | 0.142 |
| 11/16 | 0.691 | 1.170 | 0.172 | 0.219 | 0.153 |
| 3/4 | 0.753 | 1.265 | 0.188 | 0.234 | 0.164 |
| 13/16 | 0.816 | 1.363 | 0.203 | 0.250 | 0.175 |
| 7/8 | 0.787 | 1.459 | 0.219 | 0.266 | 0.186 |
| 15/16 | 0.941 | 1.556 | 0.234 | 0.281 | 0.197 |
| 1 | 1.003 | 1.656 | 0.250 | 0.297 | 0.208 |
| 1 1/16 | 1.066 | 1.751 | 0.266 | 0.312 | 0.218 |
| 1 1/8 | 1.129 | 1.847 | 0.281 | 0.328 | 0.230 |
| 1 3/16 | 1.192 | 1.943 | 0.297 | 0.344 | 0.241 |
| 1 1/4 | 1.254 | 2.036 | 0.312 | 0.359 | 0.251 |
| 1 5/16 | 1.317 | 2.133 | 0.328 | 0.375 | 0.262 |
| 1 3/8 | 1.379 | 2.219 | 0.344 | 0.391 | 0.274 |
| 1 7/16 | 1.442 | 2.324 | 0.359 | 0.406 | 0.284 |
| 1 1/2 | 1.504 | 2.419 | 0.375 | 0.422 | 0.295 |
| 1 5/8 | 1.633 | 2.553 | 0.389 | 0.424 | 0.297 |
| 1 3/4 | 1.758 | 2.679 | 0.389 | 0.424 | 0.297 |
| 1 7/8 | 1.883 | 2.811 | 0.422 | 0.427 | 0.299 |
| 2 | 2.008 | 2.936 | 0.422 | 0.427 | 0.299 |
| 2 1/4 | 2.262 | 3.221 | 0.440 | 0.442 | 0.309 |
| 2 1/2 | 2.512 | 3.471 | 0.440 | 0.422 | 0.309 |
| 2 3/4 | 2.762 | 3.824 | 0.458 | 0.491 | 0.344 |
| 3 | 3.012 | 4.074 | 0.458 | 0.491 | 0.344 |

## APPENDIX F – METRIC FASTENERS

## Metric Hex Bolts

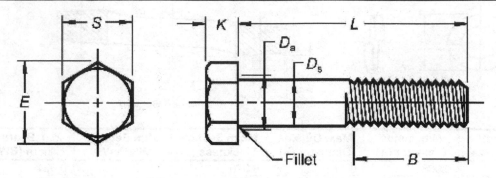

| D | Ds | S | E | K | Da | Thread Length (B) | | |
|---|---|---|---|---|---|---|---|---|
| Nominal Bolt Diameter And Thread Pitch | Max. Body Dia. | Max. Width Across Flats | Max. Width Across Corners | Max. Head Height | Fillet Transition Diameter | Bolt Lengths ≤ 125 | Bolt Lengths > 125 and ≤ 200 | Bolt Lengths >200 |
| M5 x 0.8 | 5.48 | 8.00 | 9.24 | 3.88 | 5.7 | 16 | 22 | 35 |
| M6 x 1 | 6.19 | 10.00 | 11.55 | 4.38 | 6.8 | 18 | 24 | 37 |
| M8 x 1.25 | 8.58 | 13.00 | 15.01 | 5.68 | 9.2 | 22 | 28 | 41 |
| M10 x 1.5 | 10.58 | 16.00 | 18.48 | 6.85 | 11.2 | 26 | 32 | 45 |
| M12 x 1.75 | 12.70 | 18.00 | 20.78 | 7.95 | 13.7 | 30 | 36 | 49 |
| M14 x 2 | 14.70 | 21.00 | 24.25 | 9.25 | 15.7 | 34 | 40 | 53 |
| M16 x 2 | 16.70 | 24.00 | 27.71 | 10.75 | 17.7 | 38 | 44 | 57 |
| M20 x 2.5 | 20.84 | 30.00 | 34.64 | 13.40 | 22.4 | 46 | 52 | 65 |
| M24 x 3 | 24.84 | 36.00 | 41.57 | 15.90 | 26.4 | 54 | 60 | 73 |
| M30 x 3.5 | 30.84 | 46.00 | 53.12 | 19.75 | 33.4 | 66 | 72 | 85 |
| M36 x 4 | 37.00 | 55.00 | 63.51 | 23.55 | 39.4 | 78 | 84 | 97 |
| M42 x 4.5 | 43.00 | 65.00 | 75.06 | 27.05 | 45.4 | 90 | 96 | 109 |
| M48 x 5 | 49.00 | 75.00 | 86.60 | 31.07 | 52.0 | 102 | 108 | 121 |
| M56 x 5.5 | 57.00 | 85.00 | 98.15 | 36.20 | 62.0 | | 124 | 137 |
| M64 x 6 | 65.52 | 95.00 | 109.70 | 41.32 | 70.0 | | 140 | 153 |
| M72 x 6 | 73.84 | 105.00 | 121.24 | 46.45 | 78.0 | | 156 | 169 |
| M80 x 6 | 82.16 | 115.00 | 132.79 | 51.58 | 86.0 | | 172 | 185 |
| M90 x 6 | 92.48 | 130.00 | 150.11 | 57.74 | 96.0 | | 192 | 205 |
| M100 x 6 | 102.80 | 145.00 | 167.43 | 63.90 | 107.0 | | 212 | 225 |

ASME B18.3 - 2003 Standard. Reprinted from the standard listed by permission of the American Society of Mechanical Engineers. All rights reserved.

**APPENDIX F – METRIC FASTENERS (Continued)**

## Metric Hex Nuts

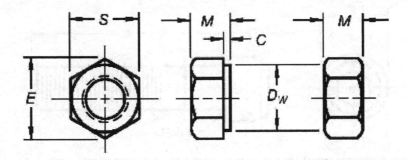

| D | S | E | M | DW | C |
|---|---|---|---|---|---|
| Nominal Bolt Diameter and Thread Pitch | Max. Width Across Flats | Max. Width Across Corners | Max. Thickness | Min. Bearing Face Diameter | Max. Washer Face Thickness |
| M1.6 x 0.35 | 3.20 | 3.70 | 1.30 | 2.3 | |
| M2 x 0.4 | 4.00 | 4.62 | 1.60 | 3.1 | |
| M2.5 x 0.45 | 5.00 | 5.77 | 2.00 | 4.1 | |
| M3 x 0.5 | 5.50 | 6.35 | 2.40 | 4.6 | |
| M3.5 x 0.6 | 6.00 | 6.93 | 2.80 | 5.1 | |
| M4 x 0.7 | 7.00 | 8.08 | 3.20 | 6.0 | |
| M5 x 0.8 | 8.00 | 9.24 | 4.70 | 7.0 | |
| M6 x 1 | 10.00 | 11.55 | 5.20 | 8.9 | |
| M8 x 1.25 | 13.00 | 15.01 | 6.80 | 11.6 | |
| M10 x 1.5 | 15.00 | 17.32 | 9.10 | 13.6 | |
| M10 x 1.5 | 16.00 | 18.45 | 8.40 | 14.6 | |
| M12 x 1.75 | 18.00 | 20.78 | 10.80 | 16.6 | |
| M14 x 2 | 21.00 | 24.25 | 12.80 | 19.4 | |
| M16 x 2 | 24.00 | 27.71 | 14.80 | 22.4 | |
| M20 x 2.5 | 30.00 | 34.64 | 18.00 | 27.9 | 0.8 |
| M24 x 3 | 36.00 | 41.57 | 21.50 | 32.5 | 0.8 |
| M30 x 3.5 | 46.00 | 53.12 | 25.60 | 42.5 | 0.8 |
| M36 x 4 | 55.00 | 63.51 | 31.00 | 50.8 | 0.8 |

**APPENDIX F – METRIC FASTENERS (Continued)**

# Metric Socket Head Cap Screws

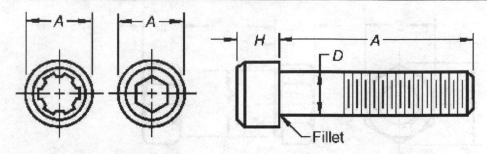

| Screw Size (D) | Head Diameter (A) | Head Height (H) |
|---|---|---|
| 1.6 through 2.5 | See Table | |
| 3 through 8 | Max. A = 1.5 D + 1 | Max. H = D |
| > 10 | Max. A = 1.5 D | |

| Screw Size (D) | 1.6 | 2 | 2.5 |
|---|---|---|---|
| Max. Head Diameter (A) | 3.00 | 3.80 | 4.50 |

ASME/ANSI B18.3.1M - 1986 (R2002) Standard.  Reprinted from the standard listed by permission of the American Society of Mechanical Engineers.  All rights reserved.

# Metric Countersunk Socket Head Cap Screws

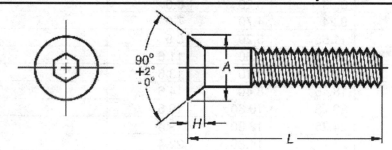

| Basic Screw Diameter and Thread Pitch | Head Diameter (A) Theor. Sharp | Head Height (H) |
|---|---|---|
| M3 x 0.5 | 6.72 | 1.86 |
| M4 x 0.7 | 8.96 | 2.48 |
| M5 x 0.8 | 11.20 | 3.10 |
| M6 x 1 | 13.44 | 3.72 |
| M8 x 1.25 | 17.92 | 4.96 |
| M10 x 1.5 | 22.40 | 6.20 |
| M12 x 1.75 | 26.88 | 7.44 |
| M14 x 2 | 30.24 | 8.12 |
| M16 x 2 | 33.60 | 8.80 |
| M20 x 2.5 | 40.32 | 10.16 |

ASME/ANSI B18.3.5M - 1986 (R2002) Standard.  Reprinted from the standard listed by permission of the American Society of Mechanical Engineers.  All rights reserved.

**APPENDIX F – METRIC FASTENERS (Continued)**

# Drill and Counterbore Sizes for Socket Head Cap Screws

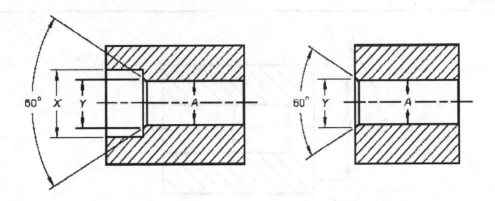

| Nominal Size or Basic Screw Diameter | A | | X | Y |
|---|---|---|---|---|
| | Nominal Drill Size | | Counterbore Diameter | Countersink Diameter |
| | Close Fit | Normal Fit | | |
| M1.6 | 1.80 | 1.95 | 3.50 | 2.0 |
| M2 | 2.20 | 2.40 | 4.40 | 2.6 |
| M2.5 | 2.70 | 3.00 | 5.40 | 3.1 |
| M3 | 3.40 | 3.70 | 6.50 | 3.6 |
| M4 | 4.40 | 4.80 | 8.25 | 4.7 |
| M5 | 5.40 | 5.80 | 9.75 | 5.7 |
| M6 | 6.40 | 6.80 | 11.25 | 6.8 |
| M8 | 8.40 | 8.80 | 14.25 | 9.2 |
| M10 | 10.50 | 10.80 | 17.25 | 11.2 |
| M12 | 12.50 | 12.80 | 19.25 | 14.2 |
| M14 | 14.50 | 14.75 | 22.25 | 16.2 |
| M16 | 16.50 | 16.75 | 25.50 | 18.2 |
| M20 | 20.50 | 20.75 | 31.50 | 22.4 |
| M24 | 24.50 | 24.75 | 37.50 | 26.4 |
| M30 | 30.75 | 31.75 | 47.50 | 33.4 |
| M36 | 37.00 | 37.50 | 56.50 | 39.4 |
| M42 | 43.00 | 44.00 | 66.00 | 45.6 |
| M48 | 49.00 | 50.00 | 75.00 | 52.6 |

**APPENDIX F – METRIC FASTENERS (Continued)**

# Drill and Countersink Sizes for Flat Countersunk Head Cap Screws

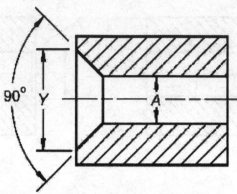

| D | A | Y |
|---|---|---|
| Nominal Screw Size | Nominal Hole Diameter | Min. Countersink Diameter |
| M3 | 3.5 | 6.72 |
| M4 | 4.6 | 8.96 |
| M5 | 6.0 | 11.20 |
| M6 | 7.0 | 13.44 |
| M8 | 9.0 | 17.92 |
| M10 | 11.5 | 22.40 |
| M12 | 13.5 | 26.88 |
| M14 | 16.0 | 30.24 |
| M16 | 18.0 | 33.60 |
| M20 | 22.4 | 40.32 |

ASME/ANSI B18.3.5M - 1986 (R2002) Standard.  Reprinted from the standard listed by permission of the American Society of Mechanical Engineers.  All rights reserved.

# APPENDIX G – FASTENERS

## BOLT AND SCREW CLEARANCE HOLES

### (1)  Inch Clearance Holes

| Nominal Screw Size | Fit Classes | | |
|---|---|---|---|
| | Normal | Close | Loose |
| | Nominal Drill Size | | |
| #0 (0.06) | #48 (0.0760) | #51 (0.0670) | 3/32 |
| #1 (0.073) | #43 (0.0890) | #46 (0.0810) | #37 (0.1040) |
| #2 (0.086) | #38 (0.1015) | 3/32 | #32 (0.1160) |
| #3 (0.099) | #32 (0.1160) | #36 (0.1065) | #30 (0.1285) |
| #4 (0.112) | #30 (0.1285) | #31 (0.1200) | #27 (0.1440) |
| #5 (0.125) | 5/32 | 9/64 | 11/64 |
| #6 (0.138) | #18 (0.1695) | #23 (0.1540) | #13 (0.1850) |
| #8 (0.164) | #9 (0.1960) | #15 (0.1800) | #3 (0.2130) |
| #10 (0.190) | #2 (0.2210) | #5 (0.2055) | B (0.238) |
| 1/4 | 9/32 | 17/64 | 19/64 |
| 5/16 | 11/32 | 21/64 | 23/64 |
| 3/8 | 13/32 | 25/64 | 27/64 |
| 7/16 | 15/32 | 29/64 | 31/64 |
| 1/2 | 9/16 | 17/32 | 39/64 |
| 5/8 | 11/16 | 21/32 | 47/64 |
| 3/4 | 13/16 | 25/32 | 29/32 |
| 7/8 | 15/16 | 29/32 | 1 1/32 |
| 1 | 1 3/32 | 1 1/32 | 1 5/32 |
| 1 1/8 | 1 7/32 | 1 5/32 | 1 5/16 |
| 1 1/4 | 1 11/32 | 1 9/32 | 1 7/16 |
| 1 3/8 | 1 1/2 | 1 7/16 | 1 39/64 |
| 1 1/2 | 1 5/8 | 1 9/16 | 1 47/64 |

ASME B18.2.8 - 1999 Standard.  Reprinted from the standard listed by permission of the American Society of Mechanical Engineers.  All rights reserved.

**APPENDIX G – FASTENERS**

**BOLT AND SCREW CLEARANCE HOLES (Continued)**

**(2)  Metric Clearance Holes**

| Nominal Screw Size | Fit Classes | | |
|---|---|---|---|
| | Normal | Close | Loose |
| | Nominal Drill Size | | |
| M1.6 | 1.8 | 1.7 | 2 |
| M2 | 2.4 | 2.2 | 2.6 |
| M2.5 | 2.9 | 2.7 | 3.1 |
| M3 | 3.4 | 3.2 | 3.6 |
| M4 | 4.5 | 4.3 | 4.8 |
| M5 | 5.5 | 5.3 | 5.8 |
| M6 | 6.6 | 6.4 | 7 |
| M8 | 9 | 8.4 | 10 |
| M10 | 11 | 10.5 | 12 |
| M12 | 13.5 | 13 | 14.5 |
| M14 | 15.5 | 15 | 16.5 |
| M16 | 17.5 | 17 | 18.5 |
| M20 | 22 | 21 | 24 |
| M24 | 26 | 25 | 28 |
| M30 | 33 | 31 | 35 |
| M36 | 39 | 37 | 42 |
| M42 | 45 | 43 | 48 |
| M48 | 52 | 50 | 56 |
| M56 | 62 | 58 | 66 |
| M64 | 70 | 66 | 74 |
| M72 | 78 | 74 | 82 |
| M80 | 86 | 82 | 91 |
| M90 | 96 | 93 | 101 |
| M100 | 107 | 104 | 112 |

## APPENDIX H – REFERENCES

- ASME B1.1 - 2003:  Unified Inch Screw Threads (UN and UNR Thread Form)
- ASME B1.13M - 2001:  Metric Screw Threads: M Profile
- USAS/ASME B4.1 - 1967 (R2004):  Preferred Limits and Fits for Cylindrical Parts
- ANSI B4.2 - 1978 (R2004):  Preferred Metric Limits and Fits
- ASME B18.2.1 - 1996:  Square and Hex Bolts and Screws (Inch Series)
- ASME/ANSI B18.2.2 - 1987 (R1999):  Square and Hex Nuts (Inch Series)
- ANSI B18.2.3.5M - 1979 (R2001):  Metric Hex Bolts
- ASME B18.2.4.1M - 2002:  Metric Hex Nuts, Style 1
- ASME B18.2.8 - 1999:  Clearance Holes for Bolts, Screws, and Studs
- ASME B18.3 - 2003:  Socket Cap, Shoulder, and Set Screws, Hex and Spline Keys (Inch Series)
- ASME/ANSI B18.3.1M - 1986 (R2002):  Socket Head Cap Screws (Metric Series)
- ASME/ANSI B18.3.5M - 1986 (R2002):  Hexagon Socket Flat Countersunk Head Cap Screws (Metric Series)
- ASME 18.6.2 - 1998:  Slotted Head Cap Screws, Square Head Set Screws, and Slotted Headless Set Screws (Inch Series)
- ASME B18.21.1 - 1999:  Lock Washers (Inch Series)
- ANSI B18.22.1 - 1965 (R2003):  Plain Washers
- ASME Y14.2M - 1992 (R2003):  Line Conventions and Lettering
- ASME Y14.3 - 2003:  Multiview and Sectional view Drawings
- ASME Y14.4M - 1989 (R1999): Pictorial Drawings
- ASME Y14.5M - 1994:  Dimensioning and Tolerancing
- ASME Y14.6 - 2001:  Screw Thread Representation
- ASME Y14.100 - 2000:  Engineering Drawing Practices

**Notes:**

# INDEX

## A

A Line Tangent to a Circle, 4-9
ABS, 13-9
Acme, 14-6
Add Sheet, 7-78
Add Symbol, 8-33
Aligned Option, 15-23
Aligned Section, 11-8
Allowance, 9-4
Alphabet of Lines, 7-43
American National Standard Limits and
    Fits – Inches, 9-8
Angle option, 4-22
Annotations, 7-63
ANSI, 2-6, 8-4
Anti-aligned, 15-23
Arc, 4-20
Assembly drawing, 15-4
Assembly Mates, 15-16
Assembly Modeling Methodology, 15-8
Auto Balloon, 15-41
Auxiliary view, 12-4, 12-30
Axonometric projection, 10-4

## B

Balloons, 15-41
Baseline dimensioning, 8-15
Base Feature, 3-10
Basic hole system, 9-7
Basic shaft system, 9-7
Basic size, 9-4
Bill of Materials, 15-5, 15-35
Binary Tree, 3-4
Bisection of an Angle, 4-5
Bisection of a Line or Arc, 4-4
Boolean operations, 3-3
BORN technique, 7-47
Bottom Up approach, 15-8
Boundary representation, 1-7
Break Line, 7-44
Broken-Out Section, 11-7

BSI, 8-4
Buttons, Mouse, 1-17
Buttress, 14-6

## C

Cavalier oblique, 10-6
Cabinet oblique, 10-6
CAD files folder, 1-19
Canceling commands, 1-17
Cap Screws, 14-15
Carriage bolts, 14-16
Cartesian coordinate system, 2-7
Center line, 7-44
Center Mark, 8-27
Center of mass, 16-30
Chain dimensioning, 8-15
Chamfer, 15-10
CIRCLE command, 3-14
Circle of a Given Radius Tangent to
    Two Given Lines, 4-11
Circle of a Given Radius Tangent to an
    Arc and a Line, 4-12
Circle of a Given Radius Tangent to
    Two Arcs, 4-13
Circle through Three Points, 4-29
Circular Center Mark, 11-38
Circular Pattern, 11-24
Clearance, 9-4
Clearance fit, 9-6
Close, 5-7
Coincident, 2-12
Command Manager, 1-12, 1-16
Computer Aided Design, 1-3, 1-4
Computer Aided Drafting, 1-4, 1-5
Computer Aided Engineering, 1-4
Computer Aided Manufacturing, 1-4
Computer geometric modeling, 1-4
Concentric, 2-12
Conceptual drawings, 15-3
Concurrent engineering, 1-7
Configuration Manager, 15-31

Constrained Move, 15-21
Constructive solid geometry, 1-7
Contour Rule, 8-6
Conventional Breaks, 11-10
Convert Entities, 6-23
Cosmetic Thread, 15-12
Crest, 14-3
Cut, 3-3
Cutting plane line, 7-44

**D**

Dashed line, 2-12
Degrees of Freedom, 15-16
Delete, 2-16
Design Intent, 2-3
Design Library, 14-18
Design Tree, 1-14
Detail drawing, 15-4
Drafting Standards, 3-8
Drawing Mode, 7-72
Drawing sizes, 15-6
Dimension, Smart, 2-13
Dimension Line, 7-44
Dimensioning, 8-4
Dimetric projections, 10-4
DIN, 2-6, 8-4
Display Style, 2-25
Dividing a Given Line into a Number of
    Equal Parts, 4-7
Document Properties, 2-6
Drafting machine, 1-3
Drag and Drop, 6-8
Driven dimensions, 4-41
Dynamic Mirror, 7-49
Dynamic Viewing, 2-22

**E**

Edit Sheet Format, 8-32
Edit Sketch, 5-24
Engineering Design process, 15-3
Engineering drawing, 1-2
Engineering Graphics, 1-2
Equations, 4-26
ESC, 1-17
EXIT command, 1-18
Exit Sketch, 2-17

Exploded View command, 15-30
EXTEND command, 6-11
Extension line, 7-44
Extruded Boss/Base, 2-7
Extruded Cut, 2-31

**F**

Fastener, 14-14
FDM, 13-7
Feature-based, 1-7, 1-8, 2-3
Feature suppression, 7-65
Features Toolbar, 1-13
FFF, 13-7
FILLET command, 4-33
Fits, 9-6
First angle projection, 7-6
Folding Line Method, 12-6
Force or Shrink Fits, 9-12
Fully Define Sketch, 4-43
Full Section, 11-6

**G**

Geometric Construction, 4-3, 4-16
    CAD Method, 4-16
Geometric Relations, 2-12, 4-18
Glass Box, 7-26
Graphics area, 1-15
Grid Intervals setup, 3-9

**H**

Half Section, 11-6
Half Views, 11-8
Heads-Up View Toolbar, 1-13, 2-24
Help System, 1-18
Hex bolts, 14-15
Hidden line, 7-44
HIPS, 13-10
History-based part modification, 5-23
Hole-Basis system, 9-19
Hole Wizard, 3-20
Horizontal, 2-12

**I**

Insert Component, 15-14
Interference fit, 9-6
International tolerance grade, 9-17

Intersect, 3-3
ISO, 2-6, 8-4
ISO Metric Fits, 9-17
Isometric projections, 10-5
Isometric Sketching, 10-7
Isometric View, 2-19

**J**

JIS, 2-6, 8-4
Join, 3-3

**K**

Keys, 14-16

**L**

Laser Sintering, 13-8
Lead, 14-4
Least Material Condition, 9-4
Left-hand Thread, 14-4
LINE command, 5-12
Line Fit, 9-6
Lines of Sight, 7-4
Line Style, 7-81
Line Tangent To A circle from A Given-
    Point, 4-10
Locational Clearance Fits, 9-9
Locational Interference Fits, 9-11

**M**

Machine screws, 14-15
Major Diameter, 14-3
Make Drawing from Part, 7-72
Mass Properties, 16-16
Mate command, 15-16
Maximum Material Condition, 9-4
Menu Bar, 1-10
Mid Plane, 5-13
Middle Out approach, 15-8
Midpoint, 5-15
Minor Diameter, 14-3
Mirror Entities, 7-53
Model View, 11-31
Mouse Buttons, 1-17
Multiple Threads, 14-4
Multiview drawing, 7-3

**N**

NEW file, 1-11
Normal To, 2-25
Note, 11-27
Nuts, 14-16

**O**

OBJ, 13-11, 13-13
Oblique Projection, 10-5
Oblique Sketching, 10-29
OFFSET command, 6-23
Offset Plane, 12-26
Offset Section, 11-7
One-point perspective, 10-44
Online Help, 1-18
Open a drawing, 9-22
Options, 2-5
Origin, 7-47
Orthographic projection, 7-4
Orthographic views, 7-3

**P**

PAN Realtime, 2-15
Paper sizes, 15-6
Parallel, 2-12
Partial Views, 12-10
Parts list, 15-5
Perimeter Circle, 4-29
Perpendicular, 2-12
Perspective Projection, 10-6
Perspective Sketching, 10-43
PETG, 13-10
Phantom line, 7-44
Pictorials, 10-2
Pitch, 14-3
Pitch Diameter, 14-3
PLA, 13-9
Point, 4-30
Precedence of Lines, 7-45
Precision, 9-2
Primitive Solids, 3-3
Principles of Projection, 7-4
Principal views, 7-5
Profile Sketch, 6-18
Projection, 7-4
Properties, 7-74

Property Links, 8-35
Property Manager, 1-14
Proximity Rule, 8-6
PVA, 13-10

**Q**

Quick Access Toolbar, 1-13

**R**

RECTANGLE command, 3-11
Reference Coordinate System, 16-27
Reference Plane Method, 12-8
Reference Geometry, 12-16
Relations, 4-18
Removed Section, 11-9
Rename, 5-17
Reposition Dimensions, 3-12
Reverse Direction, 2-30
Revisions Block, 15-6
Revolved Boss, 11-13
Revolved Section, 11-9
Ribs and Webs in Sections, 11-10
Right-hand Thread, 14-4
Rivets, 14-16
Root, 14-3
Rotate, 2-19
Rough Sketches, 2-9
Running or Sliding Clearance Fits, 9-8

**S**

SAVE command, 2-35
Screen, SOLIDWORKS, 1-12
Screw Thread, 14-3
Section Drawing Types, 11-6
Section line, 7-44
Section views, 11-3, 11-32
Selected Contours, 6-19, 6-21
Selective Assembly, 9-7
Set Screws, 14-16
Shaft-Basis system, 9-19
Sheet metal screws, 14-16
Show Flat Tree View, 5-26
Show Feature Dimensions, 8-22
Six principal views, 7-26
Sketch Toolbar, 1-13
Sketching, 10-2

Sketching Plane, 2-7
SLA, 13-6
Slicer, 13-11
Smart Dimension, 2-13
SmartMates, 15-26
Snap Intervals setup, 3-9
Socket screws, 14-16
SOLIDWORKS Resources, 1-10
Square thread, 14-5
Standard drawing paper sizes, 15-6
Starting SOLIDWORKS, 1-9
Startup Option, 1-10
Status Bar, 1-15
STL, 13-3, 13-11, 13-16
Studs, 14-15
Suppression, Feature, 7-65
Surface modelers, 1-7

**T**

T-square, 1-2
Tangent, 2-12, 4-45
Tangent Edges Removed, 8-27
Task Pane, 1-12
Technical Drawing, 1-2
Templates, 5-5
Template File, 5-6
Thin Sections, 11-8
Third angle projection, 7-23
Thread Fits, 14-12
Thread Forms, 14-3, 14-5
Threads per Inch, 14-3
Thread Representations, 14-6
Thread, Screw, 14-3
Thread Spec., English Units, 14-10
Thread Spec., Metric Units, 14-12
Three-point perspective, 10-6
Title Block, 15-7
Tolerance, 8-16, 9-2
    Accumulation, 8-17
    Bilateral tolerances, 9-3
    Limits, 8-17, 9-3
    Metric, 9-17
    Nomenclature, 9-4
    Unilateral tolerances, 9-3
Toolbars, 1-13
Tools, 4-27

Top Down approach, 15-8
Transfer of an Angle, 4-6
Transition fit, 9-6
Fits, 9-10
TRIM command, 6-11
Trimetric projections, 10-4
Two-point perspective, 10-45

**U**
Undo command, 2-16
Unified Thread Series, 14-11
Unified Screw Thread, 14-5
Union, 3-3
Units, 2-6
Units Setup, 2-6
Unsuppress, 7-65
Up To Next, 3-24
Up To Surface, 5-16

**V**
Vanishing point, 10-6
Vertical, 2-12
View Origins, 7-47
View Palette, 7-82
View Planes, 12-13
View Selector, 3-25
View Sketch Relations, 4-23
View Temporary Axes, 11-24
Viewing Orientation Toolbar, 2-19, 2-25
Visible Line, 7-44

**W**
Washers, 14-16
Whitworth, 14-5
Wireframe Modeler, 1-6
Wood screws, 14-15
Working Drawings, 15-4
World coordinate system, 7-47

**Z**
Zoom, 2-15

**Notes:**